Statistical Mechanics, Protein Structure, and Protein Substrate Interactions

NATO ASI Series

Advanced Science Institutes Series

A series presenting the results of activities sponsored by the NATO Science Committee, which aims at the dissemination of advanced scientific and technological knowledge, with a view to strengthening links between scientific communities.

The series is published by an international board of publishers in conjunction with the NATO Scientific Affairs Division

A	Life Sciences	Plenum Publishing Corporation
B	Physics	New York and London
C	Mathematical and Physical Sciences	Kluwer Academic Publishers
D	Behavioral and Social Sciences	Dordrecht, Boston, and London
E	Applied Sciences	
F	Computer and Systems Sciences	Springer-Verlag
G	Ecological Sciences	Berlin, Heidelberg, New York, London,
H	Cell Biology	Paris, Tokyo, Hong Kong, and Barcelona
I	Global Environmental Change	

Recent Volumes in this Series

Series B: Physics

Statistical Mechanics, Protein Structure, and Protein Substrate Interactions

Edited by

Sebastian Doniach

Stanford University
Stanford, California

Plenum Press
New York and London
Published in cooperation with NATO Scientific Affairs Division

Proceedings of a NATO Advanced Research Workshop on
Statistical Mechanics, Protein Structure, and Protein Substrate Interactions,
held June 1–5, 1993,
in Cargèse, Corsica, France

NATO-PCO-DATA BASE

The electronic index to the NATO ASI Series provides full bibliographical references (with keywords and/or abstracts) to more than 30,000 contributions from international scientists published in all sections of the NATO ASI Series. Access to the NATO-PCO-DATA BASE is possible in two ways:

—via online FILE 128 (NATO-PCO-DATA BASE) hosted by ESRIN, Via Galileo Galilei, I-00044 Frascati, Italy

—via CD-ROM "NATO Science and Technology Disk" with user-friendly retrieval software in English, French, and German (©WTV GmbH and DATAWARE Technologies, Inc. 1989). The CD-ROM also contains the AGARD Aerospace Database.

The CD-ROM can be ordered through any member of the Board of Publishers or through NATO-PCO, Overijse, Belgium.

Library of Congress Cataloging-in-Publication Data

Statistical mechanics, protein structure, and protein substrate
  interactions / edited by Sebastian Doniach.
      p.   cm. -- (NATO ASI series. Series B, Physics ; v. 325)
   Includes bibliographical references and index.
   ISBN 0-306-44728-2
   1. Proteins--Conformation--Congresses. 2. Proteins--Structure-
-Congresses. 3. Protein folding--Congresses.   I. Doniach, S.
  II. Series.
QP551.S74 1994
574.19'245--dc20   94-30896
                                                            CIP

ISBN 0-306-44728-2

©1994 Plenum Press, New York
A Division of Plenum Publishing Corporation
233 Spring Street, New York, N.Y. 10013

Printed in the United States of America

PREFACE

A number of factors have come together in the last couple of decades to define the emerging interdisciplinary field of structural molecular biology. First, there has been the considerable growth in our ability to obtain atomic-resolution structural data for biological molecules in general, and proteins in particular. This is a result of advances in technique, both in x-ray crystallography, driven by the development of electronic detectors and of synchrotron radiation x-ray sources, and by the development of NMR techniques which allow for inference of a three-dimensional structure of a protein in solution. Second, there has been the enormous development of techniques in DNA engineering which makes it possible to isolate and clone specific molecules of interest in sufficient quantities to enable structural measurements. In addition, the ability to mutate a given amino acid sequence at will has led to a new branch of biochemistry in which quantitative measurements can be made assessing the influence of a given amino acid on the function of a biological molecule.

A third factor, resulting from the exponential increase in computing power available to researchers, has been the emergence of a growing body of people who can take the structural data and use it to build atomic-scale models of biomolecules in order to try and simulate their motions in an aqueous environment, thus helping to provide answers to one of the most basic questions of molecular biology: the relation of structure to function. A fourth factor, resulting from this outpouring of structural data, coupled with an ever growing data base of primary sequence information, has been the development of a second field of computer based research: that of solving the "second genetic code" - the relation of the primary sequence to the tertiary structure of proteins. All these factors and more come together in the "protein folding problem" in which one can ask both about the equilibrium folded structure of a protein in solution and also about the kinetic pathway by which an unfolded protein finds its way to its global free energy minimum.

This book is a collection of articles written by lecturers and some of the participants at a NATO Advanced Research Workshop which was held at the Institut d'Études Scientifiques de Cargèse, Corsica, June 1-5, 1993. The aim of the workshop was to bring together people from various disciplines working on some of the basic aspects of the problems in structural molecular biology. It was organized by a committee made up of Ken Dill (UC San Francisco), Sebastian Doniach (Stanford), Henri Orland (Saclay) and Shoshana Wodak (Bruxelles). We were fortunate to gain a slot in the Cargèse Institute's program and obtain partial support from the NATO Scientific Affairs Division. Additional support from the United States Navy through the Office of Naval Research was much appreciated.

The workshop managed to assemble quite an eclectic group of people including both experimentalists and theorists (see table 1). They came from many of the NATO countries including Belgium, Canada, France, Germany, Greece, Italy, the UK and the US, and in addition, from Austria, Sweden, and Switzerland.

Table 1. Rough distribution of disciplines represented at the workshop.

Lecturers	Physicists	Chemists	Biochemists	Computer Scientists
Experiment	1	1	9	
Theory	5	5	8	1
Participants				
Experiment			8	
Theory	3	2	12	

On the behalf of the organizing committee and all the members of the workshop I would like to record our thanks to Marie-France Hanseler and her capable staff at the Institute. Their effective organization and warm hospitality (including excellent food!) together with the great natural beauty of the Corsican coast provided a wonderful setting for a very intense scientific experience. I would also like to express our thanks to Prof. L. Sertorio and Mrs E. Cowan of the NATO Scientific Affairs Division, and to Mike Marron of ONR for their help in making this meeting possible.

Sebastian Doniach
Stanford, November 1993

CONTENTS

STATISTICAL MECHANICAL MODELS AND PROTEIN STATES

EXPLORING CONFORMATION SPACE

NMR DATA AND PROTEIN STRUCTURE

MEMBRANE PROTEINS

PATHWAYS AND MECHANISM OF PROTEIN FOLDING

Robert L. Baldwin

Biochemistry Department
Beckman Center
Stanford University Medical School
Stanford, CA 94305-5307

ABSTRACT

A short review is presented of models for pathways and mechanism of protein folding. The models are compared with current knowledge of the structures and properties of folding intermediates, with emphasis on molten globule intermediates.

INTRODUCTION

As papers presented at this meeting demonstrate, impressive progress is being made in predicting the folded structures of proteins. Moreover, there is interest in simulating the folding process itself, because a procedure that can reproduce the folding process should be well equipped to predict the final folded structure of a protein. The purpose of this paper is to review older ideas about mechanism and pathway of folding and to consider how new facts have changed these ideas. Much is known now about some molten globule intermediates, notably cytochrome c, apomyoglobin and α-lactalbumin. They should provide suitable benchmarks for theorists interested in predicting the folding process. Since these intermediates are stable in specified conditions, it should be possible to predict their structures by approaches similar to the ones being used now to predict the final folded structures of proteins.

Only noncovalent folding reactions are discussed here, that is, reactions in which no covalent bonds are made or broken. Disulfide bonds appear to be fasteners that stabilize a correctly folded structure but do not determine its folding pathway. On the other hand, once the correct disulfide bonds are formed and are left intact during an unfolding/refolding cycle, they undoubtedly play a major role in determining these pathways.

Levinthal's Two Paradoxes

As is well known, Levinthal (1968) pointed out that it would take an impossibly long time for a small protein to fold to its native conformation by means of a random search of all possible conformations. This observation has become known as "Levinthal's paradox." In 1968 I listened to Cy Levinthal present a seminar at Stanford entitled "How to fold gracefully." He discussed the use of computers to predict protein structures and he suggested that it might be necessary to understand pathways of protein folding in order to predict protein structures. His talk helped to motivate me to study folding pathways, a project I started two years later. At his lecture, I was sitting next to the late Paul Flory.

Statistical Mechanics, Protein Structure, and Protein Substrate Interactions
Edited by S. Doniach, Plenum Press, New York, 1994

1

When Levinthal presented his now famous paradox, Flory nudged me and whispered, "so there must be folding intermediates." Levinthal's paradox stimulated many proposals of folding mechanisms that should allow folding in a reasonable time without requiring observable folding intermediates.

There is a second paradox concerning the mechanism of protein folding that Levinthal was deeply interested in, although his name is not associated with it. The paradox is: how does a protein reach its thermodynamically most stable conformation through folding by the fastest pathway? He suggested (quoted in Baldwin, 1987) that this result is achieved by means of evolution operating on protein folding pathways, although he could not suggest a mechanism on which evolution operates to achieve this result. Flory (1969) once suggested that each protein has only one compact, stable structure: if this is true, there need not be any paradox because the protein has no choice. Today we know, however, that this is not true: the serpin family is characterized by each member having alternate conformations and there are large structural differences between these conformations (Carrel et al., 1991; Creighton, 1992). Also, a stable folding intermediate, that is unable to complete folding to the native form, is produced by α-lytic protease when it attempts to refold without the help of its pro-sequence (Baker et al., 1992).

Nucleation Model of Folding

When Levinthal presented his paradox in 1968, protein chemists were just beginning to take seriously the evidence that small, single-domain proteins unfold and refold without observable intermediates inside the equilibrium transition zone between native and unfolded forms (see Lumry et al., 1966). A natural response was to assume that proteins fold by a nucleation mechanism, akin to the crystallization of ice caused by dropping a seed crystal into supercooled H_2O. It was fashionable in the 1970s to propose nucleation sites for the folding of various proteins (see Matheson & Scheraga, 1978) and Tanford's analysis of the properties of unfolded proteins in 6 M GdmCl (see Tanford, 1968) was motivated in part by the wish to demonstrate that completely unfolded proteins are able to refold: they need not retain any folded nucleation site in order to refold.

Modern studies of refolding intermediates show that protein refolding kinetics are very different from what was envisaged in the nucleation model. Rather stable intermediates are formed rapidly and they have structures that are demonstrably related to those of the fully folded forms because they have protected amide protons at the sites of secondary structural units in the native forms (see review, Baldwin, 1993). The slowest stage or stages of folding come after these early intermediates are formed, in complete contradiction to the nucleation model of folding.

The Framework Model of Folding

The framework model of folding (see Ptitsyn & Rashin, 1975; Kim & Baldwin, 1982) proposed that a stable framework of secondary structure is assembled early in folding, before the tertiary structure is put together by fitting side chains together in a close-packed arrangement. The framework model was considered at first simply as a possible mechanism for folding proteins within a reasonable time. Early experiments on kinetic folding intermediates indicated that secondary structure is formed before the assembly of tertiary structure (Kim & Baldwin, 1982). This remains true today. The original framework model envisaged, however, that the secondary structure would be stabilized by means of close packing interactions, involving interdigitation of side chains between adjacent α-helices or between helices and β-strands. These close packing interactions appear to be absent in the folding intermediates studied thus far. Instead, a different mechanism of stabilizing secondary structures in these folding intermediates appears to be involved (see below).

Microdomain and Subdomain Models of Folding

The diffusion-collision model of folding (Karplus & Weaver, 1976, 1979) resolves Levinthal's paradox by allowing microdomains (contiguous segments of perhaps 10-20 amino acids) to fold independently although only transiently. Two such units that are adjacent in the final folded structure can stabilize each other by diffusing together and coalescing. Because the folding of a microdomain involves searching a relatively small number of conformations, each microdomain can fold rapidly. A basic assumption in this model is that a microdomain knows when it has folded correctly: i.e., that a folded microdomain is stabilized to some extent when its conformation is that of the fully folded protein.

The subdomain model of folding (Kim & Baldwin, 1982; Staley & Kim, 1990) states that units of folding smaller than the domain exist and give rise to important folding intermediates. The principal evidence for the subdomain model comes from NMR data on peptide models of folding intermediates in BPTI (pancreatic trypsin inhibitor). These peptide models are only about one half the size of BPTI; nevertheless, NMR data suggest that their tertiary structures, as well as secondary structures, closely resemble the structure of BPTI (Oas & Kim, 1988; Staley & Kim, 1990).

Molten Globule Model

The molten globule model of folding states that equilibrium molten globule forms are principal intermediates in the kinetics of folding in physiological conditions. Because molten globule forms can exist at equilibrium only when the native form is unstable, they are found in highly non-physiological conditions, especially at acid pH. The first example underlying this model is Kuwajima's study of the folding of bovine α-lactalbumin. He showed that the acid form, also known today as the equilibrium molten globule form, appears to be the same as, or very similar to, the kinetic refolding intermediate that is formed rapidly in folding at neutral pH (Kuwajima, 1977; 1989; Ikeguchi et al., 1986). Recently Jennings & Wright (1993) reported that the rapidly formed intermediate in the kinetics of refolding of apomyoglobin at pH 6 has the same protected amide protons as the equilibrium molten globule form that is found at pH 4.

The realization that molten globules form a class of folding intermediates with common properties derives from the studies of Ptitsyn and coworkers (see Dolgikh et al., 1981). The list of common properties has been modified over time and may be stated today by saying that molten globule forms have: (1) substantial secondary structure indicated by far-UV CD, (2) little fixed tertiary structure, measured by NMR chemical shifts or by near-UV CD spectra of aromatic side chains, (3) compact conformations relative to their unfolded forms, measured by intrinsic viscosity, rotational relaxation, or X-ray scattering in solution, and (4) structures that are stabilized by a loose hydrophobic interaction. Item 4 has been added recently and is more controversial than items 1-3.

An important feature of the molten globule model of folding is that acid molten globules are stable relative to their unfolded forms at neutral pH. This was shown originally for α-lactalbumin by Kuwajima (1977, Ikeguchi et al., 1986) and was demonstrated recently for apomyoglobin by Barrick & Baldwin (1993a). This feature of the molten globule model received strong support from Shortle's studies of truncated mutants of staphylococcal nuclease, which showed that these mutants assume structure in physiological folding conditions although they are unable to fold completely to the native form (Shortle & Meeker, 1989; Shortle et al., 1990).

There is a widely applicable prescription for producing a molten globule form of a protein, provided by Goto et al. (1990). They prescribe a pH low enough to unfold the native protein (e.g. pH 2) followed by raising the anion concentration sufficiently to cause the molten globule to form. At pH 2, the protein has a net positive charge that results in unfavorable charge repulsion when the compact conformation of the molten globule is formed. This may be overcome by anion binding and lower concentrations are needed of strongly bound anions (e.g. trichloroacetate) than of weakly bound anions (e.g. chloride).

Some molten globule forms contain secondary structure characterized by protected amide protons at the sites of helices in the native form: e.g. guinea pig α-lactalbumin (Baum et al., 1989), cytochrome $\underline{c}$ (Jeng et al., 1990) and apomyoglobin (Hughson et al., 1990). Thermally unfolded ribonuclease A has a far-UV CD spectrum suggestive of some helical secondary structure, but it shows no measurable protection of amide protons (Robertson & Baldwin, 1991). Whether it differs from the structural molten globules above in degree or in kind has not yet been established.

Loose Hydrophobic Interaction In Molten Globules

Until experimental data on the subject became available, it was commonly believed that all folding intermediates are partial replicas of the native form that contain some native structure and some unfolded segments. Essentially all predictions of folding pathways are based on this hypothesis. In particular, it was widely assumed that no new interactions would be found in folding intermediates; instead, it was thought that some of the interactions found in the native protein would still be present and others would be missing.

This hypothesis was tested on a folding intermediate of apomyoglobin by using site-directed mutagenesis (Hughson et al., 1991). Earlier, an amide proton exchange study had shown protected amide protons in the folding intermediate at the sites of the A, G and H helices of myoglobin, with marginal protection at a few sites in the B helix (Hughson et al., 1990). Because the A, G and H helices form a compact subdomain of myoglobin, we supposed at first that this subdomain is present in the folding intermediate. Mutations were made at helix-helix pairing sites that should push the two helices apart. For example, Ala 130 in the A•H pairing site was replaced with a larger nonpolar residue (Leu) or a charged residue (Lys^+). Urea unfolding curves measured on native apomyoglobin showed that these mutations destabilized the native form, as expected.

To our surprise, substituting a larger nonpolar residue did not destabilize the folding intermediate (Hughson et al., 1991) and, in several cases, actually gave a small stabilization (Barrick & Baldwin, 1993b). Consequently we proposed that a loose hydrophobic interaction, involving the hydrophobic faces of two helices, is responsible for stabilizing the folding intermediate (Hughson et al., 1991). It had been suggested earlier that coupling between a loose hydrophobic interaction and the formation of secondary structure could explain how the α-helices of the native protein are stabilized in molten globule folding intermediates (Baldwin, 1989). `

Support for the notion that a loose hydrophobic interaction between helices may be important in stabilizing folding intermediates has come from various sources. First, Richmond & Richards (1978) pointed out that the hydrophobic interaction between two helices is potentially a long-range interaction. They analyzed the approach of two helices of myoglobin with fixed side-chain geometry and showed that exclusion of water begins when the two helices are 6Å away from their final position in the native protein. Second, model compounds have been made and tested in which the helical structure was designed to be stabilized by a loose hydrophobic interaction. Goto & Aimoto (1991) designed a 51-residue hairpin helix in which hydrophobic faces of the two helices, that contain only leucine residues, were designed to be in contact. They showed that the helical structure of the model peptide is stabilized by anion binding (the only charged side chains are Lys^+ residues) in a manner that closely resembles stabilization of molten globules at pH 2. Kuroda (1993) used a disulfide bond to crosslink the N and C-terminal helices of cytochrome c and obtained a striking enhancement of helicity. Vuilleumeier and Mutter (1993) attached covalently 4 copies of an 11-residue peptide to the lysine side chains of a carrier or template peptide. The test peptide has the amino acid sequence of a helical segment of hen lysozyme. It shows only modest helix formation as a monomer in solution but, when 4 copies are attached in proximity to each other, helix formation is enhanced and resembles that found in 20% trifluoroethanol. Finally, Raleigh and DeGrado (1992) point out that helical proteins of de novo design often have properties resembling molten globules, suggesting that both may be stabilized by a loose hydrophobic interaction.

CONCLUDING REMARKS

Some important topics have been omitted from this review. Nothing has been said about folding intermediates of all-β-sheet proteins. This subject is only now beginning to open up. It seems clear that new models will be needed to describe the folding of all-β-proteins. In the case of α-helical proteins, it is possible to make a separation between the formation of secondary and tertiary structures. A β-sheet, on the other hand, is both a secondary and a tertiary structure.

Another important topic is the formation of isolated α-helices. Most theoretical treatments of the folding process consider all isolated helices to be unstable and also assume that the helix propensities of all amino acids are equal. Both assumptions are known to be seriously wrong. It seems inevitable that any realistic modeling of the folding process must consider the formation of isolated helices as a preliminary to predicting the structures and stabilities of molten globule intermediates and must assign individual helix propensities to the different amino acids. Corresponding data for β-sheet propensities are only now beginning to be available. For a review of the formation of isolated helices, see Scholtz & Baldwin (1992).

Dill and coworkers have proposed a model ("hydrophobic zippers") for coupling between the hydrophobic interaction and the formation of α-helices and β-strands (Dill et al., 1993). This represents the kind of modeling of the folding process that experimentalists hope to see more of, in which factors that are known to be important experimentally are taken explicitly into account. Dill's proposal that hydrophobic side-chain contacts within a single α-helix can stabilize the helix should be experimentally testable.

A main conclusion from this meeting at Cargèze is that theory and experiment now begin to interact closely in the area of protein folding. It is a very welcome development and it suggests that future meetings like this one will continue to be highly profitable to both groups.

Acknowledgments

This work was supported by NIH Grant GM 19988.

Abbreviations

GdmCl, guanidinium chloride; CD, circular dichroism; amide protons refer to peptide NH protons; NMR, nuclear magnetic resonance.

REFERENCES

Baker, D., Sohl, J.L. and Agard, D.A. (1992) Nature 356, 263-265.
Baldwin, R.L., 1987, "Protein Structure, Folding and Design," D.L. Oxender, ed., p.313-320, A.R. Liss, New York.
Baldwin, R.L. (1989) Trends Biochem. Sci. 7, 291-294.
Baldwin, R.L. (1993) Current Opin. Struct. Biol. 3, 84-91.
Barrick, D. and Baldwin, R.L. (1993a) Biochemistry 32, 3790-3796.
Barrick, D. and Baldwin, R.L. (1993b) Prot. Sci. 2, 869-876.
Baum, J., Dobson, C.M., Evans, P.A. and Hanley, C. (1989) Biochemistry 28, 7-13.
Carrell, R.W., Evans, D.L. and Stein, P.E. (1991) Nature 353, 576-578.
Creighton, T.E. (1992) Nature 356, 194-195.
Dill, K.A., Fiebig, K.M. and Chan, H.S. (1993) Proc. Natl. Acad. Sci. USA 90, 1942-1946.
Dolgikh, D.A., Gilmanshin, R.I., Brazhnikov, E.V., Bychkova, V.E., Semisotnov, G.V., Venyaminov, S.Y. and Ptitsyn, O.B. (1981) FEBS Lett. 136, 311-315.
Flory, P.J. (1969) "Statistical Mechanics of Chain Molecules", Interscience, p. 303.
Goto, Y., Calciano, L.J. and Fink, A.L. (1990) Proc. Natl. Acad. Sci. USA 87, 573-577.
Goto, Y. and Aimoto, S. (1991) J. Mol. Biol. 218, 387-396.
Hughson, F.M., Wright, P.E. and Baldwin, R.L. (1990) Science 249, 1544-1548.
Hughson, F.M., Barrick, D. and Baldwin, R.L. (1991) Biochemistry 30, 4113-4118.
Ikeguchi, M., Kuwajima, K., Mitani, M. and Sugai, S. (1986) Biochemistry 25, 6965-6972.
Jeng, M.-F., Englander, S.W., Elöve, G.A., Wand, A.J. and Roder, H. (1990) Biochemistry 29, 10433-10437.
Jennings, P.A. and Wright, P.E. (1993) Science, in press.
Karplus, M. and Weaver, D.L. (1976) Nature 260, 404-406.
Karplus, M. and Weaver, D.L. (1979) Biopolymers 18, 1421-1437.
Kim, P.S. and Baldwin, R.L. (1982) Ann. Rev. Biochem. 51, 459-489.
Kuroda, Y. (1993) Biochemistry 32, 1219-1224.
Kuwajima, K. (1977) J. Mol. Biol. 114, 241-258.
Kuwajima, K. (1989) Proteins Struct. Funct. Genet. 6, 87-103.

Levinthal, C. (1968) J. Chim. Phys. 65, 44-45.

Lumry, R., Biltonen, R. and Brandts, J.F. (1966) Biopolymers 4, 917-944.

Matheson, R.R.Jr. and Scheraga, H.A. (1978) Macromolecules 11, 819-829.

Oas, T.G. and Kim, P.S. (1988) Nature 336, 42-48.

Ptitsyn, O.B. and Rashin, A.A. (1975) Biophys. Chem. 3, 1-20.

Raleigh, D.P. and DeGrado, W.F. (1992) J. Am. Chem. Soc. 114, 10079-10081.

Richmond, T.J. and Richards, F.M. (1978) J. Mol. Biol. 119, 537-555.

Robertson, A.D. and Baldwin, R.L. (1991) Biochemistry 30, 9907-9914.

Scholtz, J.M. and Baldwin, R.L. (1992) Ann. Rev. Biophys. Biomol. Struct. 21, 95-118.

Shortle, D. and Meeker, A.K. (1989) Biochemistry 28, 936-944.

Shortle, D., Stites, W.E. and Meeker, A.K. (1990) Biochemistry 29, 8033-8041.

Staley, J.S. and Kim, P.S. (1990) Nature, 344, 685-688.

Tanford, C. (1968) Adv. Protein Chem. 23, 121-282.

Vuillemeier, S. and Mutter, M. (1993) Biopolymers 33, 389-400.

THE FOLDING PATHWAY OF APOMYOGLOBIN

Patricia A. Jennings, H. Jane Dyson and Peter E. Wright*

Department of Molecular Biology, The Scripps Research Institute
10666 North Torrey Pines Road, La Jolla, California 92037

INTRODUCTION

The detailed molecular mechanism of protein folding is currently a subject of intense interest. It is generally accepted that folding proceeds via the formation of intermediate species rather than through a random conformational search (1) However, the highly cooperative nature of the folding transition for most small proteins precludes structural characterization of intermediate species under equilibrium conditions. Spectroscopic studies of peptide fragments of proteins at equilibrium have been used to help identify regions of a protein with a high propensity to form structure in solution, and thus possible folding initiation sites (2). Kinetic studies of folding reactions have demonstrated the appearance of transient intermediates, but the lifetimes of these species are usually not compatible with detailed structural studies. Nonetheless, stopped-flow studies have provided valuable insights into the formation of folding intermediates. The development of pulsed amide proton exchange experiments in combination with two-dimensional (2D) NMR methods has facilitated the acquisition of detailed structural information on transiently populated folding intermediates (3-10) and has provided considerable impetus to the elucidation of protein folding pathways and the generalized mechanism of protein folding.

Myoglobin is a good model for analysis of packing interactions and theoretical predictions of protein folding pathways (11-17) because of its simple architecture, eight helices packed to form a hydrophobic core and binding pocket for the heme prosthetic group (Fig. 1). Apomyoglobin (apoMb), myoglobin without the heme, is a typical globular protein and is well suited for studies of protein folding mechanisms. The native apoprotein is populated at neutral pH and although the structure has not been determined, fluorescence spectroscopy (18), calorimetry (19), hydrogen exchange measurements (20), molecular dynamics simulations (22) and 2D NMR experiments (23, 24) indicate that it has a compact hydrophobic core that resembles the tertiary structure of the holoprotein. The differences

---

*To whom correspondence should be addressed

Statistical Mechanics, Protein Structure, and Protein Substrate Interactions
Edited by S. Doniach, Plenum Press, New York, 1994

7

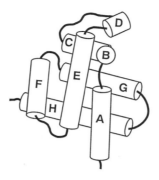

Figure 1. Schematic drawing of Mb with the eight α-helices (A-H) indicated (reproduced with permission from (21))

between the apo and holoproteins appear to be largely confined to the F-helix region which is associated with the heme and provides the proximal histidine ligand for the iron (20, 24) Removal of the heme destabilizes the native protein and leads to the population of a stable intermediate in acid-induced unfolding (19, 20, 25, 26). This intermediate state is compact and contains substantial secondary structure (~ 35% helix) which appears to result from the folding of the A-, G-, and H-helices (20) and has the properties of a molten globule intermediate (27, 28).

Molten globules have been proposed to be key intermediates in the kinetic folding pathways of all proteins (29). These species correspond to a partially folded state which is compact and only slightly expanded in volume relative to the native protein, has a relatively high content of secondary structure, and few, if any, fixed tertiary interactions (27, 29, 30). The equilibrium molten globules of several proteins have been characterized by 2D NMR methods (20, 31-33) establishing the presence of stable elements of hydrogen bonded secondary structure. Although equilibrium molten globules are detected for a number of proteins under appropriate conditions (34-36), whether they actually participate in the kinetic folding pathway has been unclear. To investigate the kinetic folding pathway(s) and the role of the molten globule state in folding, we have undertaken a series of experiments on the folding of apoMb and myoglobin fragments.

THE ApoMb MOLTEN GLOBULE

ApoMb forms a stable, partially folded state during acid induced denaturation (19, 25). The native apoprotein is most stable at 27 °C , and can be destabilized substantially at both higher and lower temperatures. Unlike the cooperative nature of most folding transitions, the acid unfolding transition of apoMb monitored by CD spectroscopy at low and high temperatures occurs in two discrete stages (Fig. 2). The protein first unfolds partially to form a compact intermediate (the I state) with a helix content of ~35%, subsequent additions of acid produces a more fully denatured state with little residual secondary structure (19, 25). The I state is also populated by acid denaturation at high temperature (19) by increasing ionic strength at pH 2 (37) or by addition of denaturants (38, 39).

Additional structural characterization of the apoMb I state was accomplished with amide proton trapping experiments and these studies indicate that the helical structure detected by CD spectroscopy is localized in the A-, G-, and H-helices, with some residual structure possible in the B-helix (20). Dye binding and fluorescence spectroscopic experiments indicate that the I state has a hydrophobic core (37, 40) in which the tryptophan and tyrosine side chains are substantially buried. Mutagenesis experiments (26) suggest that the I state is stabilized by loose and nonspecific hydrophobic interactions.

CHARACTERIZATION OF THE ApoMb FOLDING PATHWAY

The molten globule hypothesis predicts that a kinetic intermediate with the same structure as the equilibrium molten globule should occur at an early stage of folding. In order to test this hypothesis for the refolding of apoMb, we undertook a kinetic study of the refolding pathway (21). The optimal conditions found for kinetic folding studies of apoMb were urea denaturation at pH 6.1 and 5 °C. CD spectra indicate that apoMb contains no detectable helix in 6 M urea under these conditions and the unfolding transition is fully reversible, with complete recovery of the native helical content and prosthetic group binding affinity when the urea concentration is diluted from 6 to 0.8 M. Stopped-flow CD measurements indicate that a substantial fraction of ellipticity at 222 nm (64 % of total change) is regained within the dead time of the instrument in a burst of ellipticity (5 ms; Fig. 3A). This corresponds to a helix content of ~35% for this species. A comparison of the burst phase signal as a function of final denaturant concentration (Fig. 3B) with equilibrium CD unfolding experiments shows clearly that the folding of the burst phase species is cooperative, although less so than the native apoMb. The amplitude of the burst-phase signal may be taken as a measure of the concentration of the species populated in this phase since its formation is kinetically uncoupled from subsequent folding events by at least two orders of magnitude in rate. The urea-dependence of the signal can thus be taken as a measure of the stability of the kinetic intermediate (41). Fitting the data of Fig. 3B to a two-state model gives an estimated stability of 2.5 ± 0.5 kcal/mol relative to the unfolded protein. A CD spectrum of the burst phase species was constructed through a series of stopped-flow experiments in which the wavelength of detection was varied in the range of 230-217 nm. Comparion of the generated spectrum with that obtained at equilibrium in the final solution conditions after refolding (0.8 M urea) indicates that the kinetic intermediate is helical (data not shown).

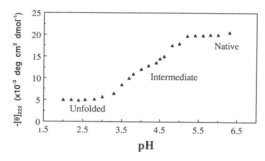

Figure 2. The acid-induced unfolding transition of apoMb indicated by the ellipticity at 222 nm.

HYDROGEN EXCHANGE PULSE LABELING

To further dissect the kinetic folding pathway of apoMb, rapid mixing and amide proton exchange techniques were used to label structured intermediates, with subsequent analysis by 2D NMR methods (3, 4). The strategy for these experiments is shown schematically in Fig. 4. The fully unfolded, fully protonated material is introduced into native conditions via rapid dilution of the denaturant. The protein is allowed to refold for a time period, t_f, during which a subset of amide protons become involved in stable, hydrogen bonded secondary structure.

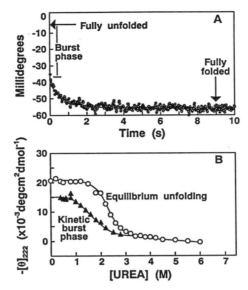

Figure 3. (A) Refolding kinetics of apoMb at pH 6.1 and 5°C monitored by stopped-flow CD spectroscopy at 222 nm. The left arrow indicates the ellipticity of the unfolded protein and was obtained by linear extrapolation of data collected above 4 M urea in equilibrium CD experiments. The arrow to the right indicates the ellipticity of the folded protein in 0.8 M urea as determined from equilibrium experiments. The solid line is a single exponential fit of the data. (B) The equilibrium unfolding of apoMb (O) monitored by the change in ellipticity at 222 nm as a function of urea concentration at pH 6.1 and 5°C. The burst phase amplitudes $[\Theta]_o$ observed in refolding experiments are also indicated (Δ). (Reproduced with permission from (21)).

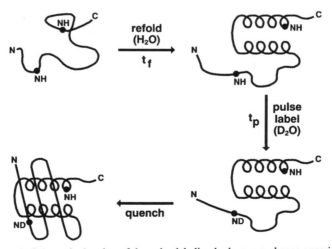

Figure 4. Schematic drawing of the pulse-labeling hydrogen exchange experiment.

The protein is then subjected to a high pH labeling pulse of D_2O buffer for a time, t_p, and any protons not protected from hydrogen exchange by the prior formation of stable hydrogen bonded structure are exchanged for deuterons. Conditions are adjusted such that the labeling pulse is strong enough to label anything that is not involved in stable structure, yet gentle enough not to label amide protons involved in structure formation. Typically, conditions are such that anything not protected more than ~40-50 fold over the intrinsic exchange rate will be completely labeled during the pulse. The exchange reaction is then quenched by rapidly lowering the pH to slow exchange conditions and folding is allowed to proceed to completion. The above techniques effectively result in a "snapshot" of what has happened up until the application of the labeling pulse. The procedure is repeated for various folding times ranging from ms to s, and a folding pathway is mapped out. The samples are analyzed by 2D NMR methods and the amide proton occupancy determined (3, 4, 42). The data are normalized with respect to non-exchangeable protons to account for protein concentration differences between the various samples, and to a native and unfolded control sample to determine the expected occupancy change (3, 4).

One requirement for these experiments is the assignment of the amide proton resonances in the NMR spectrum. In addition, the amide protons in the native sample must be sufficiently stable to allow for characterization. For these reasons, we reconstituted the more stable carbonmonoxy complex of myoglobin, for which we have the amide proton assignments, from the apo protein by addition of hemin and carbon monoxide gas. A portion of the NH-C$^\alpha$H fingerprint region of the double quantum spectrum of reconstituted MbCO is shown in Fig. 5 for the folded control and the 6.1 ms folding experiment. Thirty eight amide protons, well distributed throughout the protein, exchange sufficiently slowly under the experimental conditions to be used as probes of the folding process. A substantial fraction of the available probes become protected at the shortest delay acheivable under the conditions of the pulse labeling experiments (6.1 ms).

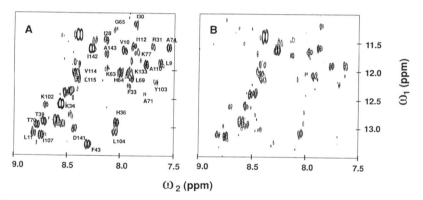

Figure 5. A portion of the fingerprint region of double quantum spectra in D_2O showing the amide proton resonances for the native control (A) and 6.1 ms (B) refolding experiments.

The amide protons of residues in the A-, G- and H-helices are fully protected within 6.1 ms after initiation of the refolding reaction (21).. Several residues in the B-helix, L29, I30 and F33 also exhibit complete protection of their amide protons within 6.1 ms. Additional residues in the B-helix, (I28 and R31), become fully protected within 1 second after exposure to refolding conditions. The amide protons of the residues at the C-terminal ends of helices A (E18) and B (K34) show protection kinetics slightly slower than other amides in these helices, possibly reflecting helix fraying prior to the formation of the native fold (32).

Finally, the amide protons of the C- and E-helices are fully protected from exchange after an estimated ~2-4 seconds and demonstrate protection kinetics very similar to one another. The data indicate that completion of folding of the B-helix slightly precedes that of the C- and E-helices. The observation of protection profiles for the amide protons in the A-, G-, H- and B-helices which are similar to one another and yet are kinetically distinct from those observed for the rest of the protein provides clear indication of the regions of the protein that participate in the cooperatively folded kinetic intermediate observed by CD.

Several residues in the B- and E-helices show partial protection (~20-30%) from exchange at the earliest time point. Partial protection of amide protons within the dead time can arise in several ways: the residues are protected via a loose association with the folded subdomain, a folding phase occurs in the same time frame as the length of the labeling pulse (43), or there is a bifurcation of the folding pathway such that a proportion of protein molecules folds completely and a second subset folds more slowly (42). In principle, changing the pH or the duration of the labeling pulse can discriminate between a parallel folding pathway and a partially associated state (3, 42, 43). Experiments were carried out in which the length of the labeling pulse was varied after a fixed (10.2 ms) folding time. In all cases the proton occupancies were within experimental error when the labeling pulse was lengthened 2-3 fold, suggesting a bifurcation in the folding pathway in the regions of the B- and part of the E-helices. However, since folding continues to different points during the pulse period and the increased pulse may perturb the folding pathway, the results are not definitive. A preferred method involves variation of the pulse pH because the folding time is then identical in each experiment; this method is not feasible for apoMb due to the instability of the apoprotein at pH > 10.2 and the tendency to aggregate at pH < 9.8. Since amide protons in the C-helix and CD- loop and in much of the E-helix show no protection at the earliest time points the most likely explanation involves a dynamic process in which helix B and part of the E-helix become associated with the folded subdomain in a folding phase that is in the same time frame as the labeling pulse (43). Investigations are currently under way to determine the correct interpretation.

PEPTIDE STUDIES IDENTIFY POSSIBLE FOLDING INITIATION SITES

Most folding studies are limited to the ms or slower time scales, owing to technical difficulties in mixing solutions of differing viscosity. Although recent developments using laser photolysis (William Eaton, these proceedings) offer dramatic improvements in time resolution of protein folding reactions, this technique is not widely applicable at present. In order to delineate interactions that are essential prior to the formation of intermediates on the folding pathway, we have used peptide fragments of several proteins as models of key sites that may play a significant role in directing folding (2, 44). Although the majority of peptides show no evidence for conformational preferences for elements of secondary structure, conformational preferences for turns, helices and nascent helices have been observed by CD and NMR (reviewed in (45)). In general, the conformational preferences of peptide fragments reflect the secondary structure of the sequence in the intact protein, that is, peptides derived from helical segments of proteins will form helix or nascent helix, while those from β-sheet are in general unstructured (46, 47). The helix and turn structures that are formed in peptides in aqueous solution are in rapid equilibrium with unstructured forms, so that we can infer that their formation occurs on the submicrosecond time scale, as it is thought to do in the earliest stages in the folding of intact proteins. Thus, understanding the conformational preferences of a particular sequence in the absence of long range tertiary interactions may help pinpoint regions of the protein which are likely to form first, and subsequently guide the formation of the native fold.

The helical hairpin of myoglobin, formed from the G-helix (residues 100-118) and H-helix (residues 124-150) and the intervening 5 residue turn, has been proposed as a potential folding initiation site (48) or as an intermediate on the kinetic folding pathway (15), based on theoretical studies of the myoglobin sequence and structure. Since the helical hairpin is apparently formed in the A-G-H kinetic intermediate, one or more of the secondary structure elements may be involved in directing folding. To identify transient structures which are potentially important in the earliest folding events, we have studied the conformational preferences of peptide fragments corresponding to the G- and H-helices, as well as the G-H turn, a 5 residue intervening loop that forms a turn like structure in native myoglobin (49-53).

The peptide corresponding to the sequence of the G-helix of myoglobin (Mb-G) shows very little evidence for structure formation in aqueous solution as judged from both CD and NMR spectroscopies (51), although helix can be induced by addition of the stabilizing cosolvent trifluoroethanol (52, 53). On the other hand, the H-helix peptide, Mb-H, is significantly helical in aqueous solution in the monomeric state, and undergoes an association reaction at high concentration to form a highly structured tetrameric species with a helical content approaching 100% (51). These results suggest that the H-helix may be considered an autonomous folding unit (54, 55) and can be formed early in the folding process, although it is undoubtedly further stabilized by tertiary interactions. A lower limit on the rate of interconversion of helical and unfolded forms of the H-helix can be inferred from NMR experiments. Since only one set of resonances are observed for the monomeric Mb-H peptide, interconversion must be fast on the NMR time-scale, and is estimated to be $> 10^4 \text{ s}^{-1}$ (51). It thus seems likely that an equilibrium population of secondary structure in the region of the H-helix must be formed long before the formation of the stabilized A-G-H intermediate in the intact protein.

Experimental and theoretical experiments have suggested that the formation of turns can preceed that of helix and thus may be involved in initiating folding (56, 57). The time scale for turn formation is on the order of 1 ns, based on molecular dynamics simulations (58, 59). These rates are 2-3 orders of magnitude faster than measured rates of helix-coil transitions in homopolypeptides (60, 61), confirming the probable importance of turn formation in early folding events. A five residue peptide, HPGDF, corresponding to the sequence of the connecting loop between the G- and H-helices was synthesized and characterized by CD and NMR spectroscopy in aqueous solution. This peptide adopts a highly populated turn conformation in aqueous solution (52). The high propensity of this fragment to adopt ordered structure implicates its formation as an important event in the early steps of folding. The turn propensity is unaltered in longer peptides corresponding to the five residue sequence and segments of the G- and H-helices, suggesting that the turn is likely to form in the intact, unfolded protein under native conditions. Comparitive studies of peptide Mb-GH25 (containing 10 residues, 109-118 and 124-133, on either side of the turn corresponding to 2-3 turns of the G- and H-helices, respectively) and Mb-AA25 (representing the same residues as in Mb-GH25 except that the central P-G sequence of the turn was replaced by A-A) suggest that the turn also serves as a helix stop signal in the intact protein, which allows for the formation of the G-H helical hairpin (52).

A stable G-H helical hairpin does not, however, appear to be formed in isolation in the peptide system. A 51-residue peptide corresponding to the entire G-H sequence, including the G-H turn, showed no evidence either for association between the G and H helix segments or the formation of additional helix that could be imputed to even a slight transient association between the helices (53). The stability of the tetramer formed by Mb-H at high concentration shows that the helical structure is definitely stabilized by association, in this case intermolecular. We therefore conclude that the interaction with the A-helix is likely to be a necessary condition for the formation of the intermediate in myoglobin folding.

STRUCTURE OF THE KINETIC INTERMEDIATE

The first detectable kinetic event (< 5 ms) in the folding of intact apoMb is the formation of a highly helical intermediate (21), with a helix content equivalent to that of the acid-induced equilibrium molten globule (20). Pulse-labeling NMR experiments indicate that probes localized in the A-, G- and H-helices and a region of the B-helix are completely protected from exchange at this time (21). The identical protons are protected in the acid-induced equilibrium molten globule (20). The protection factors are greater in the transient species, however, owing to the difference in stability of the equilibrium and kinetic intermediate under different solution conditions. The equilibrium molten globule state is studied under partially denaturing conditions whereas the kinetic folding intermediate is studied under strongly native conditions. The apoMb equilibrium molten globule state is ~1 kcal/mol more stable under the conditions used for our kinetic studies, than at pH 4.2, where it is maximally populated (39).

The observation that a given amide proton is protected from exchange reports not only on the protected amide itself but also on its hydrogen bonding partner. Contrary to concerns of non-native hydrogen bond formation in folding intermediates (62), currently available evidence suggests protection from exchange as a result of native-like structure formation (28). Our data provide strong evidence that the protection of amide protons in the kinetic folding intermediate results from stable secondary structure formation. Unlike reports of burst phase species detected by CD that do not result in amide proton protection (9, 63, 64), the helical intermediate detected during apoMb refolding is stable to exchange. Presumably, the stability is a result of packing interactions. These conclusions are supported by the peptide studies discussed above, which indicate that the H-helix can form in isolation but is only marginally stable under folding conditions and provides little protection of amide protons from exchange. Further stabilization of helical secondary structure in the G and H sequences appears to require formation of long-range tertiary hydrophobic interactions (26, 51, 53). The cooperative folding of the kinetic intermediate, as judged from the urea dependence of the initial amplitude of the kinetic trace (Fig. 3B), probably results from both helix formation and from interactions between helices (21). Stopped-flow fluorescence studies suggest that the two tryptophans (residues 7 and 14 in the A-helix) are rapidly sequestered from solvent (unpublished data; (65), implying that the A-helix is packed onto the G-H helical hairpin in the kinetic folding intermediate. Thus, the kinetic intermediate has many of the characteristics of a molten globule: secondary structure sufficiently stable to allow complete protection of amide protons from exchange, and formation of a hydrophobic core and loose hydrophobic interactions.

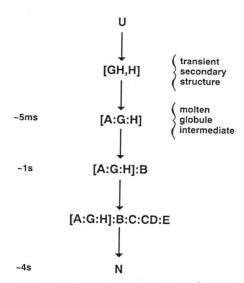

Figure 6. The dominant folding pathway of apoMb at 5°C (Adapted from (21).

KINETIC FOLDING PATHWAY OF ApoMb

The peptide studies suggest that the G-H turn and H-helix probably form spontaneously in a sub-μs time frame, and are likely initiation sites for folding of the apoprotein (51, 52). The G-H helical hairpin appears to be stabilized by association with the A-helix to form the molten globule intermediate, in <5 ms. A portion of the B-helix is also protected on this time scale. With the exception of the C-terminal end (K34), the remainder of the B-helix folds within 1 s. These results suggest that the B-helix assembles onto the already folded A-G-H helical core. The C-helix, CD loop, and E-helix fold even more slowly, the estimated time for complete protection of amide protons from exchange being ~2-4 s. The dominant pathway of folding of apoMb, as deduced from the hydrogen exchange pulse labeling experiments and the NMR and CD studies of the conformational propensities of the Mb peptides is shown in Fig. 6. No information is available on the folding kinetics of the D- and F-helices, since neither of these contains slowly exchanging amide proton probes in native apoMb. Studies of the conformational preferences of peptide fragments corresponding to these helices are currently underway.

IMPLICATIONS FOR PROTEIN FOLDING MECHANISMS

The molten globule model of protein folding predicts the formation of a kinetic folding intermediate with the same structure as the corresponding equilibrium molten globule early in the pathway (27, 30, 34-36) . Optical studies of α-lactalbumin refolding (66) supported this model, but were not conclusive. The folding studies of apoMb reported here, however, provide definitive evidence in support of the molten globule folding model by clearly establishing a close structural similarity between the earliest detectable kinetic folding intermediate and the equilibrium molten globule state.

Acknowledgements

We thank C. Robert Matthews for the use of the stopped-flow CD spectrometer (supported by BRS Shared Instrument Grant RR04953 from the National Institutes of Health) and Linda Tennant for technical assistance. We also thank Doug Barrick, Melanie Cocco, Gulnur Elöve, Fred Hughson, Bryan Jones, Craig Mann, Gene Merutka, Dimitri Morikis, Hang-Cheol Shin, Yves Theriault and Jon Waltho for helpful discussions. Supported by grants DK34909 (P.E.W.), GM38794 (H.J.D.) and postdoctoral award GM14541 (P.A.J.) from the National Institutes of Health.

REFERENCES

1. P.S. Kim and R.L. Baldwin, Specific intermediates in the folding reactions of small proteins and the mechanism of protein folding. *Ann. Rev. Biochem.*, **51**, 459-489 (1982).
2. H.J. Dyson and P.E. Wright, Peptide conformation and protein folding. *Curr. Opin. Struct. Biol.*, **3**, 60-65 (1993).
3. J.B. Udgaonkar and R.L. Baldwin, NMR evidence for an early framework intermediate on the folding pathway of ribonuclease A. *Nature*, **335**, 694-699 (1988).
4. H. Roder, G.A. Elöve and S.W. Englander, Structural characterization of folding intermediates in cytochrome C by H-exchange labelling and proton NMR. *Nature*, **335**, 700-704 (1988).
5. J.B. Udgaonkar and R.L. Baldwin, Early folding intermediate of ribonuclease A. *Proc. Natl. Acad. Sci. USA*, **87**, 8197-8201 (1990).
6. M. Bycroft, A. Matouschek, J.T. Kellis Jr., L. Serrano and A.R. Fersht, Detection and characterization of a folding intermediate in barnase by NMR. *Nature*, **346**, 488-490 (1990).
7. A. Matouschek, J. Kellis J. T., L. Serrano, M. Bycroft and A.R. Fersht, Transient folding intermediates characterized by protein engineering. *Nature*, **346**, 440-445 (1990).
8. M.S. Briggs and H. Roder, Early hydrogen-bonding events in the folding reaction of ubiquitin. *Proc. Natl. Acad. Sci. USA*, **89**, 2017-2021 (1992).
9. S.E. Radford, C.M. Dobson and P.A. Evans, The folding of hen lysozyme involves partially structured intermediates and multiple pathways. *Nature*, **358**, 302-307 (1992).
10. J. Lu and F.W. Dahlquist, Detection and characterization of an early folding intermediate of T4 lysozyme using pulse hydrogen exchange and two-dimensional NMR. *Biochemistry*, **31**, 4749-4756 (1992).
11. O.B. Ptitsyn and A.A. Rashin, A model of myoglobin self-organization. *Biophys. Chem.*, **3**, 1 (1975).
12. T.J. Richmond and F.M. Richards, Packing of α-helices: geometrical constraints and contact areas. *J. Mol. Biol.*, **119**, 537-555 (1978).

13. F.E. Cohen, T.J. Richmond and F.M. Richards, Protein folding: evaluation of some simple rules for the assembly of helices into tertiary structure with myoglobin as an example. *J. Mol. Biol.*, **132**, 275-288 (1979).

14. F.E. Cohen, M.J.E. Sternberg, D.C. Phillips, I.D. Kuntz and P.A. Kollman, A diffusion-collision-adhesion model for the kinetics of myoglobin refolding. *Nature*, **286**, 632-634 (1980).

15. M. Gerritsen, K.-C. Chou, G. Nemethy and H.A. Scheraga, Energetics of multihelix interactions in protein folding: application to myoglobin. *Biopolymers*, **24**, 1271-1291 (1985).

16. D. Bashford, F.E. Cohen, M. Karplus, I.D. Kuntz and D.L. Weaver, Diffusion-collision model for the folding kinetics of myoglobin. *Proteins*, **4**, 211-227 (1988).

17. G. Chelvanayagam, Z. Reich, R. Bringas and P. Argos, Prediction of protein folding pathways. *J. Mol. Biol.*, **227**, 901-916 (1992).

18. S.R. Anderson, M. Brunori and G. Weber, Fluorescence studies of *Aplysia* and sperm whale apomyoglobins. *Biochemistry*, **9**, 4723-4729 (1970).

19. Y.V. Griko, P.L. Privalov, S.Y. Venyaminov and V.P. Kutyshenko, Thermodynamic study of the apomyoglobin structure. *J. Mol. Biol.*, **202**, 127-138 (1988).

20. F.M. Hughson, P.E. Wright and R.L. Baldwin, Structural characterization of a partly folded apomyoglobin intermediate. *Science*, **249**, 1544-1548 (1990).

21. P.A. Jennings and P.E. Wright, Formation of a molten globule intermediate early in the kinetic folding pathway of apomyoglobin. *Science*, **262**, 892-896 (1993).

22. C.L. Brooks, Characterization of "native" apomyoglobin by molecular dynamics simulation. *J. Mol. Biol.*, **227**, 375-380 (1992).

23. M.J. Cocco and J.T.J. Lecomte, Characterization of hydrophobic cores in apomyoglobin: a proton NMR spectroscopy study. *Biochemistry*, **29**, 11067-11072 (1990).

24. M.J. Cocco, Y.-H. Kao, A.T. Phillips and J.T.J. Lecomte, Structural comparison of apomyoglobin and metaquomyoglobin: pH titration of histidines by NMR spectroscopy. *Biochemistry*, **31**, 6481-6491 (1992).

25. F.M. Hughson and R.L. Baldwin, Use of site-directed mutagenesis to destabilize native apomyoglobin relative to folding intermediates. *Biochemistry*, **28**, 4415-4422 (1989).

26. F.M. Hughson, D. Barrick and R.L. Baldwin, Probing the stability of a partly folded apomyoglobin intermediate by site-directed mutagenesis. *Biochemistry*, **30**, 4113-4118 (1991).

27. K. Kuwajima, The molten globule state as a clue for understanding the folding and cooperativity of globular-protein structure. *Proteins*, **6**, 87-103 (1989).

28. R.L. Baldwin, Molten globules:specific or nonspecific intermediates?. *Chemtracts*, **2**, 379-389 (1991).

29. O.B. Ptitsyn, Stage mechanism of self organization of protein molecules. *Dokl. Akad. Nauk. SSSR*, **210**, 1213-1215 (1973).

30. H. Christensen and R.H. Pain, Molten globule intermediates and protein folding. *Eur. Biophys. J.*, **19**, 221-229 (1991).

31. M.-F. Jeng, S.W. Englander, G.A. Elöve, A.J. Wand and H. Roder, Structural description of acid-denatured cytochrome c by hydrogen exchange and 2D NMR. *Biochemistry*, **29**, 10433-10437 (1990).

32. M.-F. Jeng and S.W. Englander, Stable submolecular folding units in a non-compact form of cytochrome c. *J. Mol. Biol.*, **221**, 1045-1061 (1991).

33. J. Baum, C.M. Dobson, P.A. Evans and C. Hanley, Characterization of a partly folded protein by NMR methods: Studies on the molten globule state of guinea pig α-lactalbumin. *Biochemistry*, **28**, 7-13 (1989).

34. O.B. Ptitsyn, How does protein synthesis give rise to the 3-D structure? *FEBS Lett.*, **285**, 176-181 (1991).

35. O.B. Ptitsyn, R.H. Pain, G.V. Semisotnov, E. Zerovnik and O.I. Razgulyaev, Evidence for a molten globule state as a general intermediate in protein folding. *FEBS Lett.*, **262**, 20-24 (1990).

36. R.L. Baldwin. Experimental studies of pathways of protein folding. in *Protein Conformation, Ciba Foundation Symposium*. 1991. Chichester: Wiley. pp 190-205.

37. Y. Goto, L.J. Calciano and A.L. Fink, Acid-induced folding of proteins. *Proc. Natl. Acad. Sci. USA*, **87**, 573-577 (1990).

38. G. Irace, E. Bismuto, F. Savy and G. Colona, Unfolding pathway of myoglobin: molecular properties of intermediate forms. *Arch. Biochem. Biophys.*, **244**, 459-469 (1986).

39. D. Barrick and R.L. Baldwin, Three-state analysis of sperm whale apomyoglobin folding. *Biochemistry*, **32**, 3790-3796 (1993).

40. E. Bismuto, I. Sirangelo and G. Irace, Salt-induced refolding off myoglobin at acidic pH: molecular properties of a partly folded intermediate. *Arch. Biochem. Biophys.*, **298**, 624-629 (1992).

41. K. Kuwajima, Y. Hiraoka, M. Ikeguchi and S. Sugai, Comparison of the transient folding intermediates in lysozyme and α-lactalbumin. *Biochemistry*, **24**, 874-881 (1985).

42. G.A. Elöve and H. Roder, Structure and stability of cytochrome c folding intermediates, in "Protein Refolding", G. Georgiou and E. De Bernardez-Clark, ed., American Chemical Society, Washington D.C. (1991) pp. 50-62.

43. S.W. Englander and L. Mayne, Protein folding studied using hydrogen-exchange labeling and two-dimensional NMR. *Annu. Rev. Biophys. Biomol. Struct.*, **21**, 243-265 (1992).

44. P.E. Wright, H.J. Dyson and R.A. Lerner, Conformation of peptide fragments of proteins in aqueous solution: Implications for initiation of protein folding. *Biochemistry*, **27**, 7167-7175 (1988).

45. H.J. Dyson and P.E. Wright, Defining solution conformations of small linear peptides. *Annu. Rev. Biophys. Biophys. Chem.*, **20**, 519-538 (1991).

46. H.J. Dyson, G. Merutka, J.P. Waltho, R.A. Lerner and P.E. Wright, Folding of peptide fragments comprising the complete sequence of proteins: Models for initation of protein folding. I. Myohemerythrin. *J. Mol. Biol.*, **226**, 795-817 (1992).

47. H.J. Dyson, J.R. Sayre, G. Merutka, H.-C. Shin, R.A. Lerner and P.E. Wright, Folding of peptide fragments comprising the complete sequence of proteins: Models for initiation of protein folding. II. Plastocyanin. *J. Mol. Biol.*, **226**, 819-835 (1992).

48. R.R. Matheson and H.A. Scheraga, A method for predicting nucleation sites for protein folding based on hydrophobic contacts. *Macromolecules*, **11**, 819-829 (1978).

49. J.P. Waltho, V.A. Feher, R.A. Lerner and P.E. Wright, Conformation of a T-cell stimulating peptide in aqueous solution. *FEBS Lett.*, **250**, 400-404 (1989).

50. J.P. Waltho, V.A. Feher and P.E. Wright, The H-helix of myoglobin as a potential independent protein folding domain. *Curr. Res. Prot. Chem.*, 283-293 (1990).

51. J.P. Waltho, V.A. Feher, G. Merutka, H.J. Dyson and P.E. Wright, Peptide models of protein folding initiation sites. 1. Secondary structure formation by peptides corresponding to the G- and H- helices of myoglobin. *Biochemistry*, **32**, 6337-6347 (1993).

52. H.-C. Shin, G. Merutka, J.P. Waltho, P.E. Wright and H.J. Dyson, Peptide models of protein folding initiation sites. 2. The G-H turn region of myoglobin acts as a helix stop signal. *Biochemistry*, **32**, 6348-6355 (1993).

53. H.-C. Shin, G. Merutka, J.P. Waltho, L.L. Tennant, H.J. Dyson and P.E. Wright, Peptide models of protein folding initiation sites. 3. The G-H helical hairpin of myoglobin. *Biochemistry*, **32**, 6356-6364 (1993).

54. K.R. Shoemaker, P.S. Kim, E.J. York, J.M. Stewart and R.L. Baldwin, Tests of the helix dipole model for stabilization of α-helices. *Nature*, **326**, 563-567 (1987).

55. K.R. Shoemaker, R. Fairman, P.S. Kim, E.J. York, J.M. Stewart and R.L. Baldwin, The C-peptide helix from ribonuclease A considered as an autonomous folding unit. *Cold Spring Harbor Symp. Quant. Biol.*, **LII**, 391-398 (1987).

56. J. Skolnick and A. Kolinski, Simulations of the folding of a globular protein. *Science*, **250**, 1121-1125 (1990).

57. J. Skolnick and A. Kolinski, Dynamic Monte Carlo simulation of a new lattice model of globular protein folding, structure and dynamics. *J. Mol. Biol.*, **221**, 499-531 (1991).

58. P.E. Wright, H.J. Dyson, V.A. Feher, L.L. Tennant, J.P. Waltho, R.A. Lerner and D.A. Case, Folding of peptide fragments of proteins in aqueous solution, in "Frontiers of NMR in Molecular Biology", D. Live, I.M. Armitage and D. Patel, ed., Alan R. Liss, Inc., New York. (1990) pp. 1-13.

59. D.J. Tobias and C.L. Brooks, Thermodynamics and mechanism of α-helix initiation in alanine and valine peptides. *Biochemistry*, **30**, 6059-6070 (1991).

60. A.L. Cummings and E.M. Eyring, Helix-coil transition kinetics in aqueous poly(α,L-glutamic acid). *Biopolymers*, **14**, 2107-2114 (1975).

61. R. Lumry, R. Legare and W.G. Miller, The dynamics of the helix-coil transition in poly-α,L-glutamic acid. *Biopolymers*, **2**, 489-500 (1964).

62. T.E. Creighton, Stability of folded conformations. *Curr. Opin. Struct. Biol.*, **1**, 5-16 (1991).

63. G.A. Elöve, A.F. Chaffotte, H. Roder and M.E. Goldberg, Early steps in cytochrome c folding probed by time-resolved circular dichroism and fluorescence spectroscopy. *Biochemistry*, **31**, 6876-6883 (1992).

64. P. Varley, A.M. Gronenborn, H. Christensen, P.T. Wingfield, R.H. Pain and G.M. Clore, Kinetics of folding of the all-β sheet protein interleukin-1β. *Science*, **260**, 1110-1113 (1993).

65. L.L. Shen and J. Hermans Jr, Kinetics of conformation change of sperm-whale myoglobin II Characterization of the rapidly and slowly formed denatured species. *Biochemistry*, 11, 1842-1849 (1972).

66. M. Ikeguchi, K. Kuwajima, M. Mitani and S. Sugai, Evidence for identity between the equilibrium unfolding intermediate and a transient folding intermediate: a comparative study of the folding reaction of α-lactoalbumin and lysozyme. *Biochemistry*, 25, 6965-6972 (1986).

CIRCULAR DICHROISM STOPPED-FLOW STUDIES

OF FOLDING INTERMEDIATES

Michel E. Goldberg and Alain F. Chaffotte

Unité de Biochimie Cellulaire
Institut Pasteur
Paris, France

INTRODUCTION

It is a now widely accepted concept that the renaturation *in vitro* of an unfolded polypeptide chain proceeds through a set of intermediates that "prevent" the molecule from exploring the entire conformational space and "force" it to reach the native conformation in an unexpectedly short time interval. Whether the intermediates appear along a unique folding pathway rather than along different possible pathways is still controversial. Yet, more and more evidence seem to accumulate that support the multiple pathways model (Murry-Brelier & Goldberg, 1989; Elöve & Roder, 1991; Radford et al., 1992).

It is also generally accepted that the slow steps in protein folding correspond to structural changes that require a high activation energy in spite of the fact that they involve only relatively small adjustments of the conformation. Conversely, the folding events that seem most important for rapidly leading the polypeptide chain towards the native conformation correspond to the formation of early folding intermediates. And indeed, several structural features of the native protein have been shown to be already present very early in the game (Udgaonkar & Baldwin, 1988; Roder et al., 1988; Bycroft et al., 1990; Udgaonkar & Baldwin, 1990; Radford et al., 1992). Hence the need to study and characterize early folding intermediates. One such intermediate is currently the subject of much attention: the "molten globule". As early as 1973, it was predicted that a state with the following properties should exist as an early intermediate in the folding of a polypeptide chain: it would share with the native conformation a condensed globular structure with a hydrophobic core, a nearly native secondary structure, and an approximatly correct (as compared to the native state) relative arrangement of the secondary structure elements; it would however differ from the native state in that the hydrophobic inside of the molten globule would not be tightly packed (Ptitsyn, 1973). Equilibrium studies on the unfolding of several proteins at intermediate denaturing agent concentrations have revealed the presence of states that are neither native nor fully unfolded, and that exhibit many of the properties predicted for a molten globule (for a review see Kuwajima, 1989). For instance, their Stokes radius is only slightly larger than that of the native state, indicating a globular state; they have a hydrophobic, yet not tightly packed, internal core as shown by

Statistical Mechanics, Protein Structure, and Protein Substrate Interactions
Edited by S. Doniach, Plenum Press, New York, 1994

19

their ability to bind the bulky hydrophobic fluorescent probe ANS; their far UV circular dichroism spectrum is similar to that of the native protein, indicating that the secondary structure is indeed native-like; they exhibit no near UV circular dichroism signal, and low tryptophan fluorescence polarization, which indicates that the aromatic residues are highly mobile. Furthermore, transient intermediates with some of the properties reported above have also been observed during the folding of several proteins (reviewed by Kuwajima, 1989).

This has lead to the conclusion that a molten globule state may be a general intermediate in protein folding (Ptitsyn et al., 1990). Two conflicting views of the molten globule exist implicitly in the litterature. In one of them, the elements of secondary structure present in the molten globule must be the same as those which ultimately will be found in the native protein (we shall call this a "specific" molten globule). Alternatively, molten globules could exhibit significant amounts of non native secondary structure elements, i.e. α helices or β strands that do not exist in the native state (we shall call such an intermediate a "non specific" molten globule). To distinguish between these two models, a structural characterisation of transient molten globules is needed.

Because of the cooperativity and rapidity of the folding process, studying the structural properties of early folding intermediates is a difficult task. Folding intermediates have been characterized by observing folding kinetics with different conformational probes. Far UV circular dichroism to analyse their content in secondary structure (Pflumm et al., 1986; Kuwajima et al., 1985; Kuwajima et al., 1991); pulsed amide proton exchange to monitor the presence of stable hydrogen-bonded structures (Udgaonkar & Baldwin, 1988; Roder et al., 1988; Bycroft et al., 1990; Radford et al., 1992); fluorescence properties such as spectral shifts or energy transfer to follow the condensation of the polypeptide chain (Tsong, 1976; Blond & Goldberg, 1986; Miranker et al., 1991); immunoreactivity towards specific monoclonal antibodies to detect the formation of native-like local structural patterns on the protein surface (Murry-Brelier & Goldberg, 1988; Blond-Elguindi & Goldberg, 1990). Among these methods, the two first ones provide information that can be directly interpreted in terms of 3D structure. Pulsed amide proton exchange followed by NMR analysis of the protected protons provides a precise description of the kinetics of formation of individual hydrogen bonds. But it can analyze only those amide groups that are protected from exchange in the native state and can detect only intermediates which already have stable hydrogen-bonded structures. Consequently, non native structural elements, or elements not yet stable enough under the labelling conditions, will escape observation by the pulsed amide proton exchange method. Thus, the pulsed amide proton method would fail to detect "non specific" molten globule intermediates if such transient structures were to exist. On the contrary, the peptide far UV circular dichroism provides a measure of the overall content in secondary structure, and can detect the presence of secondary structure elements even if they are not stable or non native. This is the reason why, because of recent developments in the capabilities of commercially available circular dichroism stopped-flow equipments, we undertook investigations on the kinetics of regain of the far UV CD spectra during the refolding *in vitro* of different polypeptide chains: the C-terminal domain of the *Escherichia coli* tryptophan synthase β_2 subunit, hen egg white lysozyme and cytochrome c.

We shall describe and compare here the results we obtained with these three different polypeptide chains of similar size (about 100 residues), but endowed with different structural features. We shall not describe in detail the experiments, which have been reported in previous publications (Chaffotte et al., 1992a and 1992b, Elöve et al. 1992).

THE C-TERMINAL DOMAIN OF E. COLI TRYPTOPHAN SYNTHASE

In a recent report, we suggested that a conformation with non native secondary and supersecondary structures might exist as a folding intermediate, at least for the C-terminal end of the β chain of the tryptophan synthase molecule. Indeed, we showed that a 101 residue polypeptide fragment, F2-V8, corresponding to the C-terminal extremity of the β chain of *E. coli* tryptophan synthase (Friguet et al., 1989; Kaufmann et al., 1991) folds into a stable globular condensed state which, however, contains secondary and supersecondary structures that differ considerably from those of the F2-V8 domain in the native protein (Chaffotte et al.,1991). This clearly ruled out that isolated F2-V8 might share major native-like structural features with the native state. Yet, the large far UV CD signal, the absence of near UV CD signal, the lack of effective protection of amide protons against exchange with deuterium, the absence of induced chemical shift in the NMR spectrum, and the very low enthalpy change and lack of cooperativity of the unfolding transition exhibited by isolated F2-V8 are features commonly observed for molten globules (Kuwajima, 1989). This led us to propose that the conformation of isolated F2-V8 might result from a hydrophobic collapse of the polypeptide chain into a loosely packed, fluctuating, globule that might be a precursor of the specific molten globule on the folding pathway (Chaffotte et al., 1991).

To further test this model, it seemed important to study the kinetics of folding of isolated F2-V8, and to compare them with the kinetics of some of the steps already identified in protein folding. This was done (Chaffotte et al., 1992a) with stopped-flow methods, using the far UV CD to monitor the appearance of secondary structure elements, and with fluorescence to monitor the appearance of the molten globule hydrophobic core able to bind the hydrophobic probe ANS (anilino-naphtalene-sulfonate).

The F2-V8 fragment was unfolded in 6M GuHCl in potassium phosphate buffer at pH 7.8. Under such conditions, F2-V8 is completely unfolded as judged from a far UV CD spectrum characteristic of a random coil. Using a stopped-flow, the unfolded fragment was diluted 80 fold in GuHCl-free buffer. The flow rate was such that the dead time of the stopped-flow apparatus was 4 msec. When the ellipticity at 222 nm (which reflects the presence of secondary structure elements and primarily that of α-helices) was monitored, it was found that the final signal was reached within the 4 msec dead time, and did not change thereafter. Thus, the secondary structure of the non specific F2-V8 molten globule appeared to be formed in less than 4 msec. Similarly, when the binding of the hydrophobic fluorescent probe ANS was monitored in the stopped-flow under the same experimental conditions, all the fluorescence signal was already present at 4 msec, and did not exhibit any later change, indicating that the hydrophobic core of the F2-V8 molten globule was also formed in less than 4 msec.

From these results we concluded that the F2-V8 fragment collapses, in less than 4 msec, into a globular state that exhibits significant amounts of secondary structure, a hydrophobic core and no stable tertiary structure (as judged from the absence of near UV CD). Together with the fact that the far UV CD spectrum of this state suggests the presence of non native secontary structure elements, this led us to propose that this globular state may well represent an early precursor of the specific molten globule on the pathway of folding of the tryptophan synthase β chain C-terminal domain.

HORSE CYTOCHROME C

Cytochrome c is a polypeptide chain of 104 residues that, in the native state, appears to be wrapped around its prosthetic group, the heme. It contains four approximately α-helical regions: the N-terminal, 60's, 70's and C-terminal helices. The refolding of cytochrome c had been thouroughly investigated using a variety of methods, the most informative of which are tryptophan fluorescence (that is strongly quenched by the heme when these two molecules come close to one another - Tsong, 1976; Briggs & Roder, 1992) and pulsed amide proton exchange followed by NMR identification of the protected protons (Roder et al., 1988). In particular, the latter method showed the successive formation of elements of secondary structure (first the N- and C-terminal helices, and later the 60's and 70's helices) that occur in a time range of about 100 milliseconds for the former and of about a second for the latter. Yet, by using CD stopped-flow measurements, Kuwajima et al (1987) showed that, during the refolding of cytochrome c, over 80% of the native far UV ellipticity (and hence, probably, of the native secondary structure) was already regained during the 18 msec dead time of their stopped-flow. This apparent contradiction could be interpreted in two ways: either the CD signal reflected the appearance of secondary structure elements that failed to be detected by pulsed amide proton exchange, or the difference in the kinetics of secondary structure formation as observed by CD and by pulsed amide proton exchange resulted from differences in the experimental conditions chosen for the folding experiments. Indeed, the temperature, the pH and the concentration of residual denaturing agent (GuHCl) were different in the two sets of experiments, and were likely to affect seriously the folding kinetics. To solve this alternative, we studied the kinetics of regain of the UV ellipticity, using the CD stopped-flow to refold cytochrome c under rigorously the same experimental conditions (10°C, pH 6.2 and 0.7 M residual GuHCl) as those used by Roder et al. (1988) for their pulsed proton exchange studies.

The results we obtained can be summarized as follows:

-During the 4 msec dead time of the stopped-flow we used, about 45% of the native ellipticity at 222 nm was already restored, while at this stage only about 15% of the amide protons that are non exchangeable in the native protein were effectively protected. Thus, CD signals detect secondary structure elements that are either not "native like", or not sufficiently stable to protect amide protons against exchange. Such non native-like or non hydrogen bonded helical structures may be of the type predicted by Dill (1985) in his model for describing the initial hydrophobic collapse of an unfolded polypeptide chain in aqueous solvent.

-The intermediate species already formed within the first 4 msec of folding have the following characteristics: they are collapsed structures, as shown by the strong quenching of the tryptophan fluorescence by the heme; they exhibit significant amounts of secondary structure, as evidenced by the far UV CD (see above); they have no stable tertiary structure, as judged from the absence of near UV CD signal (Elöve et al., 1992) and from the very small number of protected tertiary hydrogen bonds (Roder et al., 1988); finally, they have only few non exchangeable amide protons (Roder et al., 1988). Thus, this intermediate shows several of the characteristics of a molten globule. The presence of some native-like N- and C-terminal helices detected by pulsed amide proton exchange suggests that these intermediates may resemble a "specific molten globule".

-After the initial "burst" of folding resulting in the 4 msec intermediates, three other folding phases have been identified. A rapid one, with a rate constant of about 25 s^{-1}, corresponds to the regain of an additional 33% of the native ellipticity at 222 nm, with still no dichroic signal in the near UV. Thus, at the end of that phase, a molten globule with about 80% of the native ellipticity is present. The next phase (rate constants of 1.5 s^{-1}) results in a native ellipticity at 222 nm, and in the regain of about 75% of the native ellipticity in the near UV. It thus corresponds to the formation of an intermediate with some tightly packed tertiary structure. The last, slow phase (rate constant 0.1 s^{-1}) results in the native protein. It shows essentially a further increase by about 25% of the near UV ellipticity, that coïncides with the last 50% of the tertiary protons becoming protected.

Though the image that emerges from these studies for the complete sequence of folding events appears quite complex, it is clear that, for cytochrome c, a molten globule intermediate is present early on the folding pathway.

How much of the secondary structure present at this early stage is really "native-like"? Attempts to answer that question have been made by trying to model the far UV CD spectrum of the 4 msec intermediate either by a linear combination of the unfolded and native spectra, or by a linear combination of the unfolded spectrum with that predicted for the native peptide backbone on the basis of the known 3D structure of cytochrome c and of the reference spectra for α-helices, β-strands, turns and coils described by Brahms & Brahms (1980). While the fit is obviously much better with a combination of the unfolded spectrum (71%) with the spectrum predicted for the native backbone (29%), we feel that no firm conclusion should be made at this stage (unpublished results).

HEN EGG WHITE LYSOZYME (HEWL)

HEWL WITH INTACT DISULFIDE BONDS

Hen egg white lysozyme consists in a 129 residue polypeptide chain that, in its native state, is crosslinked by four specific disulfide bonds. These disulfide bonds create covalent loops that are known to strongly protect lysozyme against thermal denaturation. HEWL with its disulfide bonds intact can be denatured by guanidine, or urea, or low pH, and refolds very efficiently and rapidly upon removal (by dilution for instance) of the denaturing agent. HEWL can also be refolded after reduction of its disulfide bonds, but the renaturation is not as efficient and occurs much more slowly. We therefore investigated in detail the regain of the far and near UV ellipticity during the renaturation of HEWL with its disulfide bonds intact. For that purpose, we used a combination of flow and stopped-flow CD studies that permitted us to construct with good precision the far UV CD spectrum of a series of folding intermediate, including that present after only 4 msec of folding (Chaffotte et al., 1992b). These studies showed that the refolding proceeds through three distinct phases:
-a "burst" phase, that takes place within the 4 msec dead time of the stopped-flow. It leads to a folded structure with no near UV ellipticity and a far UV CD spectrum indistinguishable from that predicted from the secondary structure contents of native HEWL and the reference spectra of Brahms and Brahms (1980). Thus, this 4 msec intermediate appears as a "specific molten globule" with a secondary structure approximating well that of native HEWL.
-a "rapid" phase, with a rate constant of about 50 s^{-1}, that shows an overshoot of far UV ellipticity. Such overshoots had been interpreted in the past as reflecting the formation of excess secondary structure. In the case of HEWL, it in fact arises from a strong CD contribution from aminoacid side chains, very likely to be mainly S-S bonds between cystine residues.
-a "slow" phase, with a rate constant of 2.5 s^{-1}, leading to the native state and corresponding to the tight packing of aromatic side chains.

Thus, for lysozyme, it appears that the folding proceeds through a "specific molten globule" intermediate, with a native like secondary structure, that is formed very rapidly, i.e. in less than 4 msec.

REDUCED HEWL

The presence of four native disulfide bonds in denatured non-reduced HEWL creates some conformational constraints that may affect the properties of the folding pathway. To investigate the folding in the absence of these constraints, reduced denatured HEWL was prepared according to a classical procedure: HEWL was unfolded in 6M GuHCl and its disulfide bonds were reduced with dithiothreitol at slightly alkaline pH. The reduced protein was then dialysed against acetic acid and lyophilized. Just before use, aliquots were dissolved in 6M GuHCl at acidic pH. The unfolded reduced protein was then diluted 80 fold in phosphate buffer at pH 7.8 in the stopped-flow for refolding studies. However, reduced unfolded HEWL forms aggregates upon dilution in renaturation buffer at the protein concentrations used for CD measurements. This absolutely prohibits stopped-flow studies because aggregates generate artificial dichroic signals, and are likely to clog the stopped-flow lines. To circumvent this difficulty, we performed flow, rather than stopped-flow, CD studies: during half a second, a continuous flow of reduced unfolded HEWL mixed in a 1/80 ratio with renaturation buffer was maintained, and the protein ellipticity was recorded. Then, injection of HEWL was stopped, but the buffer flow was kept for 250 msec to wash away the refolding protein. Thus, the protein in refolding buffer stayed in the lines and observation cell of the stopped-flow for no more than a few milliseconds, a time too short for important aggregation to occur. Such an experiment made it possible to record the ellipticity during the phase of injection of HEWL, while protein that had been refolding for only about 4 msec was constantly flowing through the observation cell. The result of such an experiment is depicted in Figure 1.

The striking observation was that reduced HEWL did not at all regain its native backbone far UV ellipticity during the 4 msec dead time of the experiment. On the contrary, when the same experiment was reproduced with non-reduced HEWL (HEWL with its disulfide bonds intact), the native backbone ellipticity was, as expected, recovered in less than 4 msec (see Figure 1).

Thus, the native disulfide bonds appear to play a crucial role in accelerating the formation of native-like secondary structure in the very early phases of the folding process. This leads one to wonder whether or not proteins that contain native disulfide bonds should be considered as good models for investigating the early events of protein folding.

DISCUSSION

The studies on early folding intermediates summarized above indicate that three distinct polypeptide chains, though of similar sizes, behave in very different ways. In all three cases, the earliest species that could be identified exhibit properties that are characteristic of a molten globule: an important far UV CD signal and no near UV CD signal. But these molten globules have clearly different properties. In the case of the F_2 fragment isolated from the β chain of *E. coli* tryptophan synthase, it seems to be a non-specific molten globule that contains significant amounts of non-native secondary structure. Conversely, for non-reduced HEWL, the molten globule that could be observed seems specific, since its far UV CD spectrum is very close to that predicted from the secondary structure of native HEWL. In the case of cytochrome c, it has not been

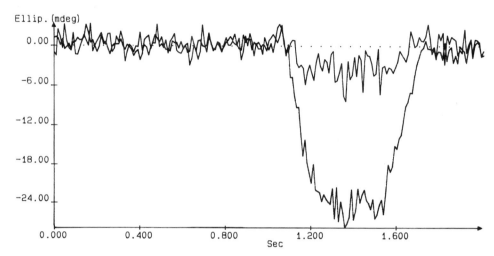

Figure 1 - *Continuous flow kinetics showing the ellipticity at 228 nm of the 4 msec intermediate during the refolding of reduced and of non reduced HEWL.* The kinetics for reduced (upper trace) and non reduced (lower trace) HEWL were obtained in a CD6 dichrograph (Jobin-Yvon - Longjumeau - France) equipped with a SFM3 (BioLogic - Claix - France) stopped-flow accessory as described in Chaffotte et al., 1992b. The sequence of injections was as follows. Recording of the ellipticity was triggered with the observation cell filled with buffer (10 mM potassium phospahte pH 7.8). 1 sec later, a first injection was started: during 500 msec each of the two large syringes injected 1972 μl of buffer and the small syringe injected 57 μl of reduced HEWL (7 mg/ml) in 6 M guanidine pH 3. This resulted in a total flow rate of 8 ml/sec and a dead time of 4 msec. A second injection phase followed immediately (1 ml of buffer from each large syringe in 250 msec) to wash away the protein before it aggregated in the stopped flow. Data points were collected at 10 msec intervals, with a filtering time constant of 10 msec. The traces shown are the averages of 48 successive experiments.

25

possible to clearly establish whether the molten globule observed at 4 msec is specific or not.

It should be pointed out, however, that the three polypeptide chains that we studied are very different in terms of structure and stability. Thus, the isolated F_2 fragment is unable to establish most of the tertiary contacts it normaly makes with the rest of the molecule in the native β_2 subunit of tryptophan synthase, and therefore can not progress further towards its tightly packed native conformation. Hence, for *isolated* F_2, the molten globule we observed is an end product of the folding. It is however not a "dead end" product, since the native protein can be reconstituted by mixing isolated F_2 with a stoichiometric amount of the complementary F_1 N-terminal fragment (Högberg-Raibaud & Goldberg, 1977). On the contrary, in the cases of HEWL and cytochrome c, the molten globules we observed were short lived transient folding intermediates, likely to be on the direct productive pathway of folding of the polypeptide chain.

An other important difference is that in *isolated* F_2, most of the native tertiary interactions are absent throughout the folding process, while for HEWL and cytochrome c tertiary contacts are present early in the process. Thus, for *non-reduced* HEWL, the four native disulfide bonds are present in the denatured state, and therefore also at the very beginning of the folding process. That no detectable formation of a molten globule-like structure can be detected after 4 msec of folding of *reduced* HEWL demonstrates that the four native disulfide bonds in *non-reduced* HEWL play a crucial role in the kinetics of formation of the molten globule we detected at 4 msec for the non-reduced protein. Studies on disulfide bonded model peptides of Bovine Pancreatic Trypsin Inhibitor (BPTI) have nicely illustrated how S-S bonds can speed up the folding of BPTI either by stabilizing folding intermediates (Oas and Kim, 1988) or by kinetically favouring the next steps in the folding (Staley and Kim, 1990). Whether the disulfide bonds of HEWL act kinetically as nucleation centers, or by stabilizing labile early folding intermediates, can not be concluded at present. In the case of cytochrome c, similar effects (nucleation center or stabilization of early intermediates) might be due to interactions of the covalently bound heme with its ligands. Indeed, it has been reported that apo-cytochrome c fails to refold under conditions where the holo-protein does (Stellwagen & Rysavy, 1972; Fisher et al., 1973), suggesting that the heme might be involved in the folding process. Moreover, dichroic signals in the heme absorption region, present at very early stages of holo-cytochrome c folding (unpublished results), strongly suggest that the heme might establish tertiary contacts involved in the first steps of the folding process.

An important observation that emerged from these studies was that, for the three proteins investigated, ellipticity in the far UV revealed the presence at early stages of the folding process of important amounts of secondary structure that could not be detected by the pulsed amide proton exchange method. This was particularly clear for the isolated F_2 fragment of tryptophan synthase for which a large far UV CD signal is present while no protected proton could be detected even in the folded state at equilibrium (Chaffotte et al., 1991). Similarly, the number of protected amide protons present in cytochrome c after 4 msec of folding was by far smaller than the number of residues involved in secondary structure elements, as judged by the far CD spectrum of the 4 msec intermediate (Elöve et al., 1992). Finally, in the case of HEWL, the ellipticity at 225 nm was also shown to be regained considerably faster than the protection of the amide protons (Radford et al. ,1992). How can one explain this quantitative difference between the two methods ? We would like to propose that the early dichroic signals in the far UV are ellicited by regions of the polypeptide chain having conformations that are neither an authentic α helix nor an authentic β strand, i.e. in which the Φ and Ψ angles are not exactly those present in authentic α helices or β strands. In such structures, the amide protons would not be protected. Indeed, the formation of stable hydrogen bonds is highly dependent on the exact distances and on the colinearity of the three atoms involved: the hydrogen, the donor and the acceptor. Thus, the amide proton involved in the hydrogen bonds in a segment of α helix are likely to be effectively protected against exchange only if all the Φ and Ψ angles in this segment are very close to the proper values. The far UV ellipticity also depends on the values of the Φ and Ψ angles, but certainly much less critically than the stability of hydrogen bonds. Thus, if non hydrogen bonded helical or extended structures were formed early during the folding of a protein, one could expect to detect significant far UV CD signals, but no amide proton protection. Such helical and extended structures, stabilized only by hydrophobic interactions, have been predicted by Dill (1985) to appear very early in the folding, and to play a major role in the initial reduction of the accessible conformational space.

The results of our CD stopped flow studies thus lead us to the following view of the early events of protein folding:
- the unfolded polypeptide chain undergoes very rapid folding steps leading to secondary structure elements. These secondary structure elements can be detected by their characteristic far UV ellipticity, but the regular patterns of hydrogen bonds present in α-helices or β-strands are not yet formed (no protection against hydrogen exchange).
- these very early secondary structure elements would be intrinsicaly quite unstable, but could be stabilized by supersecondary or tertiary long range interactions: hydrophobic contacts within the core of the isolated F2 domain, disulfide bonds for HEWL, interactions with the heme and hydrophobic contacts for cytochrome c.
- hydrogen bonds involving amide protons are formed at a later stage. This would bring about both a further stabilization of these secondary structures and their adjustment to the precise patterns of Φ and Ψ angles characteristic of α-helices or β-strands. At that stage, a "specific" molten globule would be formed.
- the later, rate limiting steps, would then correspond - as observed for several proteins (Ptitsyn et al., 1990) - to the tight packing of side chains resulting in the native conformation.

This schematic view of protein folding should be considered at present as tentative only. It has however two virtues. One is to reconcile the observations made by CD stopped flow and pulsed amide proton exchange experiments, and to offer a unifying model to interpret the different behaviours of the three model proteins we investigated in detail. The second is to prompt straightforward experiments for testing the model: investigating, by means of site directed mutagenesis or changes in the folding conditions, the effects of long range interactions on the kinetics of appearance of the very early secondary structure motifs should lead to either justify or discard the essential feature of this model, i.e. the coupling between long range interactions and secondary structure formation through a linked selective process. Experiments along this line are in progress in our laboratory.

REFERENCES

Blond, S. & Goldberg, M.E. (1986) *Proteins: Struct. Funct. & Genet.* 1, 247-255.

Blond-Elguindi, S. & Goldberg, M. E. (1990) *Biochemistry,* 29, 2409-2417.

Brahms, S. & Brahms, J. (1980) *J. Mol. Biol.,* 138, 149-178.

Briggs, M. S. & Roder, H. (1992) *Proc. Natl. Acad. Sci. USA,* 89, 2017-2021.

Bycroft, M., Matouschek, A., Kellis, J. T. Jr., Serrano, L. 1 Fersht, A. R. (1990) *Nature,* 346, 488-490.

Chaffotte, A. Guillou, Y., Delepierre, M., Hinz, H. J. & Goldberg, M. E. (1991) *Biochemistry,* 30, 8067-8074.

Chaffotte, A., Cadieux, C., Guillou, Y. & Goldberg, M.E. (1992 a) *Biochemistry,* 31, 4303-4308.

Chaffotte, A. F., Guillou, Y. & Goldberg, M. E. (1992 b) *Biochemistry,* 31, 9694-9702.

Dill, K. A. (1985) *Biochemistry,* 24, 1501-1509.

Elöve, G. A. & Roder, H. (1991) in *Protein Refolding* (Georgiou, G. Ed.) ACS Symposium Series, No 470, pp 50-63, American Chemical Society, Washington, DC.

Elöve, G. A., Chaffotte, A. F., Roder, H. & Goldberg, M. E. (1992) *Biochemistry,* 31, 6876-6883.

Fisher, W.R., Taniuchi, H. & Anfinsen, C. B. (1973) *J. Biol. Chem.,* 248, 3188-3195.

Friguet, B., Djavadi-Ohaniance, L. & Goldberg, M. E. (1989) *Res. Immunol.,* 140, 355-376.

Högberg-Raibaud, A. & Goldberg, M. E. (1977) *Biochemistry,* 16, 4014-4020.

Kaufmann, M., Schwartz, T., Jaenicke, R., Schnackerz, K. D., Meyer, H. E. & Bartholmes, P. (1991) *Biochemistry,* 30, 4173-4179.

Kuwajima, K. (1989) *Proteins: Struct. Funct. & Genet.,* 6, 87-103.

Kuwajima, K., Hiraoka, Y., Ikeguchi, M. & Sugai, S. (1985) *Biochemistry,* 24, 874-881.

Kuwajima, K., Yamaha, H., Miwa, S., Sugai, S. & Nagamura, T. (1987) *FEBS Lett.,* 221, 115-118.

Kuwajima, K., Garvey, E. P., Finn, B. E., Matthews, C. B. & Sugai, S. (1991) *Biochemistry,* 30, 7693-7703.

Miranker, A., Radford, S. E., Karplus, M. & Dobson, C. M. (1991) *Nature,* 349, 633-636.

Murry-Brelier, A. & Goldberg, M. E. (1988) *Biochemistry,* 27, 7633-7640.

Murry-Brelier, A. & Goldberg, M.E. (1989) *Proteins: Struct. Funct. & Genet.,* 6, 395-404.

Oas, T. G. & Kim, P. S. (1988) *Nature,* 336, 42-48.

Pflumm, M., Luchins, J. & Beychok, S. (1986) *Methods Enzymol.,* 130, 519-534.

Ptitsyn, O. B. (1973) *Vestn. Akad. Nauk. SSSR,* 5, 57-68.

Ptitsyn, O. B., Pain, R. H., Semisotnov, G. V., Zerovnik, E. & Razgulyaev, O. I. (1990) *FEBS Lett.,* 262, 20-24.

Radford, S. E., Dobson, C. M. & Evans, P. A. (1992) *Nature*, 358, 302-307.

Staley, J. P. & Kim, P. S. (1990) *Nature*, 344, 685-688.

Stellwagen, E. & Rysavy, R. (1972) *J. Biol. Chem.*, 247, 8074-8077.

Udgaonkar, J. E. & Baldwin, R.L. (1988) *Nature*, 335, 694-695.

Roder, H. , Elöve, G. A. & Englander, S. W. (1988) *Nature*, 335, 700-704.

Tsong, T. Y. (1976) *Biochemistry*, 15, 5467-5473.

Udgaonkar, J.E. & Baldwin, R. L. (1990) *Proc. Natl. Acad. Sci. USA*, 87, 8197-8201.

NATIVE-LIKE INTERMEDIATES IN PROTEIN FOLDING

Wolfgang Pfeil

Max-Delbrück-Center for Molecular Medicine
Robert-Rössle-Street 10
D-13125 Berlin-Buch, Federal Republic Germany

ABSTRACT

Apocytochromes may represent intermediates in *in vivo* folding of the holo proteins. Therefore, the question arises, whether apocytochromes have co-operative structure. In the present paper, the holo and apocytochromes b_5 and $P450_{cam}$ were studied by differential scanning calorimetry (DSC), limited proteolysis, second derivative spectroscopy, circular dichroism, and size exclusion chromatography.

Apocytochrome b_5 is able to undergo a reversible two-state thermal transition. However, transition temperature, denaturational enthalpy and heat capacity change are reduced compared with the holo protein. According to the spectral data, the co-operative structure is mainly based on the core region formed by the residues 1-35 and 79-90. This finding is in full agreement with NMR data (Moore and Lecomte, 1993).

Holo cytochrome $P450_{cam}$ shows in DSC three folding units (domains). Compared with the holo protein, apocytochrome $P450_{cam}$ melting is a two-state process. However, the enthalpy change at thermal melting is reduced from 980 kJ/mol for the holo protein to 135 kJ/mol for the apo form, and the apocytochrome $P450_{cam}$ stability amounts to $\Delta G = 7.5$ kJ/mol.

The apocytochromes are regarded as native-like proteins with extremely low protein stability of about 3 kT, thus showing at the same time some intermediate-like properties. Parts of the molecule produce enhanced partial specific heat capacity and seem not to contribute to the enthalpy and heat capacity change at unfolding. The importance of the properties at *in vivo folding* are discussed.

INTRODUCTION

Currently, there is much interest in intermediates in protein folding. Unfortunately, intermediates need not to accumulate during folding to such an extent to enable direct experimental investigations under equilibrium conditions. Therefore, it is convenient to enrich intermediate forms for studies on structure and stability using extreme pH, certain

Statistical Mechanics, Protein Structure, and Protein Substrate Interactions
Edited by S. Doniach, Plenum Press, New York, 1994

31

denaturant concentration, or even high pressure. As an example, it could be referred to the acid form of α-lactalbumin (obtained at pH 2) which is usually regarded as the prototype of the molten globule state (Ptitsyn, 1992).

In the present paper an attempt will be made, to study proteins that correspond to naturally occurring intermediates. On some heme proteins it was observed that the incorporation of the prosthetic group proceeds several hours after the synthesis of the apo protein (e.g., cytochrome $P450_{cam}$, cytochrome b_5 according to DuBois and Waterman, 1979; Shawver et al., 1984). Iso-1-cytochrome c is supplied with the prosthetic group even after passage of the mitochondrial membrane (Dumont et al., 1991).

Generally, apoprotein is less stable than the corresponding holo protein. The decrease of protein stability on removal of the prosthetic group was shown on the example of myoglobin (Griko et al., 1988). Apocytochrome c was found to be completely unfolded (Privalov et al., 1989). Nevertheless, it remains open at which stage of folding structure occurs that is able to undergo a co-operative structural transition and in which properties this structure differs from the native protein.

MATERIALS AND METHODS

The tryptic fragment of cytochrome b_5 (residues 1-90) was prepared from rabbit liver as described by Pfeil and Bendzko (1980). Recombinant $P450_{cam}$ was isolated from E. coli strain TB1 according to the procedure described in detail by Jung et al. (1992). The apo proteins were prepared by the butanone extraction method following the protocol of Wagner et al. (1981).

Scanning calorimetric measurements were performed on two instruments: the MicroCal MC-2D scanning calorimeter (MicroCal Inc., Northampton, MA) with the DA-2 data acquisition system; and the DASM 1M microcalorimeter (SKB Puschino, USSR). The measurements were performed at a scan rate of about 1K/min using a protein concentration ranging from 1.2 to 4.5 mg/ml.

For the high precision measurement of protein partial specific heat, the DASM 1M instrument was interfaced with a PC-AT via the DAS8-PGA analog-digital board (Keithley Metrabyte Co., Taunton, MA). Special software was written in Turbo Pascal for data registration and treatment as well as for file transfer to the DA-2 and Origin software packages (MicroCal Inc., Northampton, MA).

Further details of the experimental procedure are described by Pfeil et al. (1993) and Pfeil (1993).

RESULTS

The following experimental study concerns both cytochrome b_5 and cytochrome $P450_{cam}$ for which the apo forms really represent intermediates under *in vivo* conditions.

Cytochrome b_5 is a membrane protein. However, most biophysical studies including X-ray structure analysis were performed on the protein devoid of its membrane anchor, i.e. the tryptic fragment (residues 1-90) (Mathews et al., 1972; Pfeil and Bendzko, 1980; Moore et al, 1991). In scanning calorimetry, both the apo and holo protein show reversible thermal transitions (Figure 1). However, transition temperature, enthalpy and heat capacity change are considerably reduced for the apo protein. Moreover, the heat capacity of the protein below the transition region is enhanced and nearly in between the values for the holo protein and the expected value for the completely unfolded protein (calculated according to Privalov and Makhatadze, 1990). In accordance with Privalov and Makhatadze (1990, 1992), it can be concluded that nonpolar parts of the molecule are exposed to the solvent. Gibbs energy change at unfolding amounts to $\Delta G = (7\pm1)$ kJ/mol for apocytochrome b_5 compared with $\Delta G = 25$ kJ/mol for the holo protein at neutral pH and 25°C.

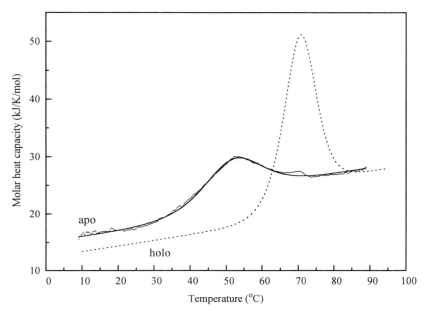

Figure 1. Scanning calorimetric curves of holo and apocytochrome b_5 at pH 7.4. Broken line: holo cytochrome b_5 (data from Pfeil and Bendzko, 1980). Raw data: apocytochrome b_5. Solid line: Two-state fit with $T_{trs} = (48.5\pm0.7)$°C, $\Delta H = (148\pm5)$ kJ/mol, and $\Delta Cp = (4.2\pm0.5)$ kJ/K/mol.

Table 1. Thermodynamic quantities at holo and apocytochrome b_5 unfolding

		holo	apo	extrapol.	Remarks
T_{trs}	(°C)	69.5	48.5±0.5	69.5	
ΔH	(kJ/mol)	329	149.3±6.7	238	at T_{trs}
ΔCp	(J/K/g)	6.0	4.2±0.6		1
ΔCp	(J/K/g)	6.0	4.2±0.5		2
ΔG	(kJ/mol)	25	7.1±1.1		at 25°C
Cp	(J/K/g)	1.32	1.57±0.12		at 20°C

1 From ΔH versus T_{trs}, single measurements at pH 6-9
2 Average from single calorimetric recordings on 7 independently prepared samples.

At the same time, apocytochrome b_5 is more sensitive to thermolysin digestion, in particular at elevated temperature (data not shown). According to the circular dichroism spectrum, the secondary structure content is reduced (Figure 2). The circular dichroism, spectrum in the near UV region shows still existing tertiary structure (Figure 2, inset). Furthermore, the Stokes radius remains unchanged compared with holo protein (19±1 Å).

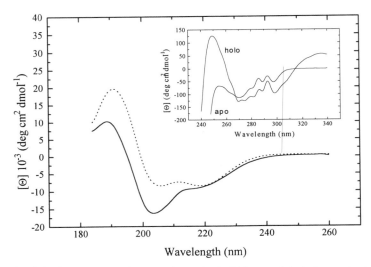

Figure 2. Circular dichroism spectrum of apocytochrome b$_5$ (solid line) and the holo protein (broken line) in the far UV region, pH 7.4. Inset: Circular dichroism spectrum of apocytochrome b$_5$ and the holo protein in the near UV region, pH 7.4.

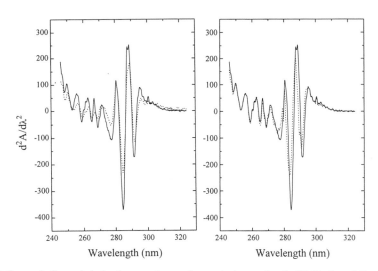

Figure 3, left panel: Second derivative spectrum of apocytochrome b$_5$ (solid line) and the holo protein (broken line) in potassium phosphate buffer, pH 7.4. Right panel: Second derivative spectrum of apo-cytochrome b$_5$ in 50 mM potassium phosphate pH 7.4 (solid line) and in 6 molar guanidine hydrochloride (broken line).

Of some interest are the second derivative spectra of apo and holo cytochrome b_5 since the spectra are similar in the region of Trp and Tyr but not in the region of Phe (Figure 3, left panel). The spectrum in the region of the Phe residues, however, is similar to that of apocytochrome b_5 in 6 molar guanidine hydrochloride (Figure 3, right panel).

The X-ray structure of (holo) cytochrome b_5 provides a reasonable explanation of the findings (Figure 4). The sequence of rabbit cytochrome b_5 (Ozols, 1989) was aligned here into the PDB file 3B5C (Mathews and Durley, 1991) referring to oxidized bovine cytochrome b_5. The heme group is embedded in a four helix bundle with three adjacent Phe residues. These residues are affected by removal of the prosthetic group, in contrast to the three Tyr and the single Trp.

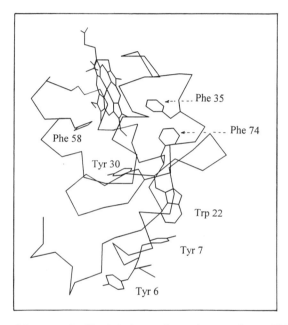

Figure 4. The location of the aromatic side chain in cytochrome b_5 according to PDB file 3B5C. Tyr27 is replaced by His in the rabbit cytochrome b_5 (Ozols, 1989). The sketch was prepared using KINEMAGE (Richardson and Richardson, 1992).

It can be concluded that the four helix bundle which contains the Phe residues is disturbed by removal of the heme. Thus, the helix content decreases and the partial specific heat capacity increases as a consequence of the exposure of nonpolar groups. As indicated by the Stokes radius, the heme binding regions seems to remain in a relatively compact state. At the same time a core region which includes the Tyr and Trp residues remains intact. The core region was recently identified by NMR studies of Moore et al. (1991), Moore and Lecomte, (1990, 1993).

Therefore, apocytochrome b_5 can be considered as a native-like protein. However, parts of the molecule, forming the four-helix bundle in the holo protein, seem not to take part in the co-operative thermal unfolding transition.

On the next example, briefly a similar study will be shown on a larger protein (Pfeil et al., 1993). The protein here is cytochrome P450$_{cam}$ from *Pseudomonas putida*, a water soluble protein consisting of 414 residues and the heme group.

In scanning calorimetry, the protein shows a more complicated transition which can be resolved by deconvolution into three domains (Figure 5A, left panel). The transition at about 42°C is related to the heme group as it can be shown by optically monitored melting curves (Nölting et al., 1992; Pfeil et al., 1993). For comparison, the heat capacity function of apocytochrome P450$_{cam}$ is quite different (Figure 5B, right panel). Again, Cp is found in between the heat capacity of the holo protein and the heat capacity of the completely unfolded protein, as calculated according to Privalov and Makhatadze (1990). Nevertheless, a thermal transition of the apo protein can be identified. Approximation by a two state transition with heat capacity change gives the following result (Figure 5B, inset). The

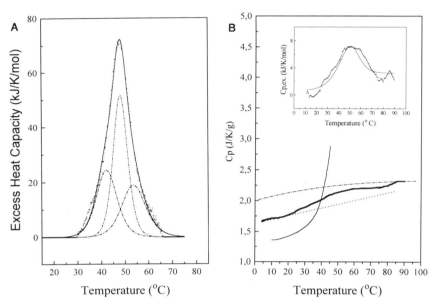

Figure 5A, left panel: Deconvolution of the excess heat capacity function of holo-P450$_{cam}$ measured in the presence of KCN. The results of deconvolution are summarized in Tab II.

Figure 5B, right panel: Partial specific heat of apo-P450$_{cam}$ at pH 8.0 (raw data) and holo cytochrome P450 at pH 8.0 (straight line) compared with the heat capacity function of unfolded apo-P450$_{cam}$ (dashed line). Baseline treatment was done using linear extrapolation (dotted line).

Figure 5B, inset: Two-state fit of the molar heat capacity function of apo-P450$_{cam}$ at pH 8.0. Results see Table 2.

relation between thermodynamic quantities of the holo and apo protein are summarised in the following table (Table 2) showing in particular the enthalpy change reduced from about 980 kJ/mol for the holo protein to about 135 kJ/mol for the apo form. At the same time, ΔG amounts to about 7 kJ/mol. The apo protein is much more sensitive to proteolytic digestion than the holo protein (data not shown). The occurrence of additional electrophoretic bands at tryptic digestion show that cleavage site are accessible that are buried in the holo protein. Other properties of apocytochrome b$_5$ will be mentioned briefly: the helix content is reduced from about 55 to about 35%, the derivative spectrum shows both native-like and unfolded-like band positions, and the Stokes radius remains almost unchanged (5% increase compared with the holo protein).

Table 2. Thermodynamic quantitites at holo and apo cytochrome $P450_{cam}$ unfolding

	holo P450	apo P450
C_p (J/K/g)	1.36±0.06	1.71±0.06 [1]
T_{trs1} (°C)	41.9±1.0	50±2
ΔH_1 (kJ/mol)	296±15	135±10
ΔC_p (kJ/K/mol)	n.d.	3.2±0.5
T_{trs2} (°C)	47.8±0.7	---
ΔH_2 (kJ/mol)	405±38	---
T_{trs3} (°C)	54.3±0.6	---
ΔH_3 (kJ/mol)	279±34	---
ΔG (kJ/mol) at 25°C	n.d.	7.5±1.0
mean of n measurements	n= 6	n= 5

[1] The partial specific heat capacity is given for pH 8.0 and 25°C. Cp of apo-$P450_{cam}$ amounts to C_p = (1.78±0.08) J/K/g at pH 2.5. C_p = 2.07 J/K/g is expected for the fully solvated polypeptide chain of the protein calculated according to Privalov and Makhatadze, 1992).

DISCUSSION

The apo proteins of the cytochrome b_5 and $P450_{cam}$ are distinctly different from the molten globule state. The proteins are able to undergo a thermal transition with a measurable enthalpy and heat capacity change. The proteins remain compact and show significant secondary and tertiary structure. However, parts of the molecule seem not to contribute to the unfolding process as judged by the thermodynamic quantities. In cytochrome b_5 the still existing core region could be localised. In apocytochrome $P450_{cam}$ one of the originally three domains remains as a co-operative unit. However, the stability of the two apo proteins is reduced to about $\Delta G \approx 7$ kJ/mol, i.e. no more than about 3 kT. Furthermore, the enhanced partial specific heat capacity and the susceptibility to proteolytic cleavage indicate exposure of nonpolar groups.

The two apoproteins represent native-like forms. However, they are characterised by the loss of co-operative structure due to the removal of the prosthetic group. The apocytochromes b_5 and $P450_{cam}$ bear resemblance in this respect to apomyoglobin (Griko et al., 1988; Cocco and Lecomte, 1993) and the more recent findings made by Eder et al. (1993) on a folding intermediate of subtilisin BPN'.

On the other hand the apocytochromes b_5 and $P450_{cam}$ occur much earlier in folding under *in vivo* conditions than the holo proteins. Therefore the question arises how the highly protease and aggregation-sensitive proteins persist in living cell. Since no denaturant or other extreme condition are necessary to study the two apoproteins, they could be of some interest in folding studies including chaperones (Ellis and v.d. Vies, 1991; Wickner et al., 1991; Gething and Sambrook, 1992).

Acknowledgements

The paper was supported by Deutsche Forschungsgemeinschaft grant Pf. 251/1-1.

REFERENCES

Cocco, M.J., and Lecomte, J.T.J., 1990, Characterization of hydrophobic cores in apomyoglobin: a proton NMR spectroscopic study, *Biochemistry* 29: 11067-11072.

DuBois, R.N., and Waterman, M.R., 1979, Effect of phenobarbital administration on the level of the in vitro synthesis of cytochrome P-450 directed by total rat liver RNA, *Biochem. Biophys. Res. Commun.* 90: 150-157.

Dumont, K.E., Cardillo, T.S., Hayes, M.K., and Sherman, F., 1991, Role of cytochrome c heme lyase in mitochondrial import and accumulation of cytochrome c in Saccharomyces cerevisiae, *Molec. Cell. Biol.* 11: 5487-5496.

Eder, J., Rheinecker, M., and Fersht, A.R., 1993, Folding of subtilisin BPN': characterization of a folding intermediate, *Biochemistry* 32: 18-26.

Ellis, R. J., and van der Vies, S. N., 1991, Molecular chaperones, *Annu. Rev. Biochem.* 60: 321-347.

Gething, M.-J., and Sambrook, J., 1992, Protein folding in the cell, *Nature* 355: 33-45.

Griko, Yu.V., Privalov, P.L., Venyaminov, S.Yu., and Kutyshenko, V.P., 1988, Thermodynamic study of the apomyoglobin structure, *J. Mol. Biol.* 202: 127-138.

Jung, C., Hui Bon Hoa, G., Schröder, K.-L., Simon, M., and Doucet, J. P., 1992, Substrate analogue induced change of the CO stretching mode in the cytochrome P450cam-carbon monoxide complex, *Biochemistry* 31: 12855-12862.

Mathews, F.S., Levine, M., and Argos, P., 1972, Three-dimensional Fourier synthesis of calf liver cytochrome b_5 at 2.8 Å resolution, *J. Mol. Biol.* 64: 449-464.

Mathews, F.S., and Durley, R.C.E., 1991, Brookhaven Protein Data Bank, file 3B5C.

Moore, C.D., and Lecomte, J.T.J., 1990, Structural properties of apocytochrome b_5: presence of a stable native core, *Biochemistry* 29: 1984-1989.

Moore, C.D., Al-Misky, O.N., and Lecomte, J.T.J., 1991, Similarities in structure between holocytochrome b_5 and apocytochrome b_5: NMR studies of the histidine residues, *Biochemistry* 30: 8357-8365.

Moore, C.D., and Lecomte, J.T.J., 1993, Characterization of an independent structural unit in apocytochrome b_5, *Biochemistry* 32: 199-207.

Nölting, B., Jung, C., and Snatzke, G., 1992, Multichannel circular dichroism investigations of the structural stability of bacterial cytochrome P-450, *Biochim. Biophys. Acta* 1100: 171-176.

Ozols, J., 1989, Structure of cytochrome b_5 and its topology in the microsomal membrane. *Biochim. Biophys. Acta* 997: 121-130.

Pfeil, W., and Bendzko, P., 1980, Thermodynamic investigations of cytochrome b_5. I. The tryptic fragment of cytochrome b_5, *Biochim. Biophys. Acta* 626: 73-78.

Pfeil, W., Nölting, B.O., and Jung, C., 1993, Apocytochrome $P450_{cam}$ is a native protein with some intermediate-like properties, *Biochemistry*, in press.

Pfeil, W., 1993, Thermodynamics of apocytochrome b_5 unfolding, *Protein Science*, submitted.

Privalov, P.L., Tiktopulo, E.I., Venyaminov, S.Yu., Griko, Yu.V., Makhatadze, G.I., and Khechinashvili, N.N., 1989, Heat capacity and conformation of proteins in the denatured state, *J. Mol. Biol.* 205: 737-750.

Privalov, P.L., and Makhatadze, G.I., 1990, Heat capacity of proteins. II. Partial molar heat capacity of the unfolded polypeptide chain of proteins: protein unfolding effects, *J. Mol. Biol.* 213: 385-391.

Privalov, P.L., and Makhatadze, G.I., 1992, Contribution of hydration and non-covalent interactions to the heat capacity effect on protein unfolding, *J. Mol. Biol.* 224: 715-723.

Ptitsyn, O.B., 1992, The molten globule state, *in:* "Protein Folding," Creighton, T.E., ed., pp. 243-300, W.H. Freeman and Co., New York.

Richardson, D.C., and Richardson, J.S., 1992, The kinemage: a tool for scientific communication, *Protein Science* 1: 3-9.

Shawver, L.K., Seidel, S.L., Krieter, P.A., and Shires, T.K., 1984, An enzyme-linked immunoadsorbent assay for measuring cytochrome b_5 and NADPH-cytochrome P-450 reductase in rat liver microsomal fractions, *Biochem. J.* 217: 623-632.

Wagner, G. C., Perez, M., Toscano Jr., W. A., and Gunsalus, I. C., 1981, Apoprotein formation and heme reconstitution of cytochrome $P450_{cam}$. *J. Biol. Chem.* 256: 6262-6265.

Wickner, W., Driessen, A. J. M., and Hartl, F.-U., 1991, The Enzymology of protein translocation across the Echerichia coli plasma membrane, *Annu. Rev. Biochem.* 60: 101-124.

A FIRST-ORDER PHASE TRANSITION BETWEEN A COMPACT DENATURED STATE AND A RANDOM COIL STATE IN STAPHYLOCOCCAL NUCLEASE

Apostolos G. Gittis[1], Wesley E. Stites[2,3] and Eaton E. Lattman[1]

[1]Department of Biophysics and Biophysical Chemistry
[2]Department of Biological Chemistry
School of Medicine, Johns Hopkins University
Baltimore MD 21205-2185 USA

[3]Present Address:
Department of Chemistry and Biochemistry
University of Arkansas
Fayetteville AR 72701-1201

INTRODUCTION

It is well established that many small, globular proteins undergo a two-state unfolding transition that is closely similar to first-order phase transitions in macroscopic systems. In such transitions the protein exists in just two detectable states, native and denatured, separated by an eneregy barrier; all the intermediate states are unstable and only transiently populated. The two-state model has been extraordinarily successful in accounting for the equilibrium denaturation behavior of proteins. However, tantalizing suggestions have appeared that this model is insufficient. Experimental evidence for residual structure in the denatured state has been accumulating for many years. Recently, for instance, changes in the amount and type of residual structure in large fragments of staphylococccal nuclease (nuclease) have been observed upon mutation or varying solvent conditions (Shortle & Meeker, 1989). Acid denatured nuclease also appears to have residual structure (Nakano & Fink, 1990). The importance and relevance of these observations to the stability and structure of nuclease, and of proteins in general, has been subject to question.

The Molten Globule

There has also been a great deal of interest in the stable intermediate form of some proteins that has come to be known as the "molten globule". Review articles (Kuwajima, 1992; Ptitsyn, 1992; Dill & Shortle, 1991; Kim & Baldwin, 1990; Kuwajima, 1989) have summarized much of the work in this area. Although it is non-native, the molten globule is much more compact than a random coil and has significant amounts of secondary structure as judged by CD spectra. The molten globule state is usually found under conditions of fairly high ionic strength and low pH. Whether or not molten globule states are important in protein unfolding under physiological conditions is an important question. Dill, for example, has suggested that under physiological conditions, when the concentration of denatured state in equilibrium with the folded form is unobservably low, the denatured form of the protein is compact (Dill, personal communication).

The nature of the transition between the molten globule state and the fully unfolded random coil is an open question (Karplus & Shakhnovich, 1992; Ptitsyn, 1992). It is well

Statistical Mechanics, Protein Structure, and Protein Substrate Interactions
Edited by S. Doniach, Plenum Press, New York, 1994

established that the breakdown of the native state is a first-order transition, i.e. there are two (or more) distinct populations with an energy barrier to their interconversion. There is no clear cut reason to believe that the breakdown of the molten globule is first or second-order. (Authors in this field have generally used "second-order transition" to mean a transition in which there is a single, perhaps shallow, energy minimum at any temperature, but that the value of the order parameter X at which the minimumm occurs is a function of temperature.) Theoretical studies have drawn mixed conclusions. Some studies have concluded that the transition should be second-order (Finkelstein and Shakhnovich, 1989; Shakhnovich and Finkelstein, 1989), while others conclude that a second-order transition is more likely but a first-order transition is possible (Chan and Dill, 1991; Alonso et al., 1991).

It is important that the molten globule state be thermodynamically characterized to determine both its energetics and the order of its transition to the fully denatured state (Ptitsyn, 1992; Kuwajima, 1992). As pointed out in Karplus and Shakhnovich (1992), there is no experimental evidence with the proper thermodynamic analysis for a well-defined transition between denatured states regardless of the conditions used to cause the denaturation. As discussed in Ikegami (1977, 1981) and Ptitsyn (1992) the least ambiguous way to prove the cooperativity (first-order phase transition) between two states is to show a bimodal distribution of the protein between the two states. Unfortunately, this is not a trivial task as many methods of probing protein structure return only information about the average state. The order of the phase transition may also be determined directly by calorimetric methods. However, the observations on the thermal melting of the molten globule state are controversial (Ptitsyn, 1992). Calorimetric methods run into difficulty because the measured excess heat capacity change upon unfolding of the compact denatured state is very small or zero (Kuwajima, 1992, 1989; Ptitsyn, 1992). Recent work (Yutani et. al, 1992) on the energetics of the molten globule state of apo-α-lactalbumin has shown that the enthalpy difference between the molten globule and the fully unfolded state upon heating was almost zero. It was inferred that the molten globule does not exhibit any cooperative transition upon heating, but no analysis is performed to determine the order of the transition. A method for determining whether the thermal transition between the molten globule and the fully unfolded state is continuous (second-order) or not would be most useful.

NUCLEASE DENATURATION

As mentioned above, large fragments of staphylococcal nuclease have been explored as models of the denatured state (Shortle & Meeker, 1989; Flanagan et al., 1992). The native state of these fragments has been rendered unstable under physiological conditions by the removal of thirteen amino acids from the carboxy terminus of the protein. Although these fragments are clearly not in the native state, it has been shown by CD that they contain substantial amounts of residual structure. The fragments are also much more compact, as assessed by gel exclusion chromatography, than true random coils. These characteristics are broadly similar to those found in the molten globule states of other proteins, although the CD signal of these fragments is perhaps less than typical for a molten globule. The structure in these fragments decreased and their apparent hydrodynamic radius increased when guanidinium hydrochloride (GuHCl) was added. Changes in the amount and type of structure were observed when different mutations were placed in the fragments. Whether or not this fragment system truly corresponds to a molten globule state remains to be resolved. However, unlike most molten globules, this simple model system can be examined under physiologically relevant conditions of pH and ionic strength.

Fluorescence Studies

In this work we examine three substitution mutants of full-length staphylococcal nuclease, with a single residue in each replaced by a tryptophan, which serves as a fluorescence probe. These same mutations were also examined in the 1-136 fragment. In these fragments we find evidence for the first-order, cooperative breakdown of structure similar to that seen in the full-length protein.

The plot of tryptophan fluorescence intensity, I, versus [GuHCl] for full-length wild type and for the full-length mutants valine 66→tryptophan (V66W), lysine 70→tryptophan (K70W) and glycine 88→tryptophan (G88W) is depicted in Figure 1a. In Figure 1b is the corresponding plot of $\log(K_{app})$ versus [GuHCl], where K_{app} is the apparent equilibrium constant between the folded and unfolded states. Wild type displays the classic sigmoidal transition of the two-state model in the first plot and in the second plot the data has near perfect linearity, which has a sstrong empirical correlation with two-state behavior. The mutants K70W and G88W also fit the two-state model well. In marked contrast, V66W does not have a sigmoidal transition and the plot of log(K_{app}) against [GuHCl] is not a straight line. Clearly, in the case of V66W, the two-state model fails to provide an adequate description of the process of protein denaturation. A clear possibility is that the two tryptophan residues in V66W are sampling different environments.

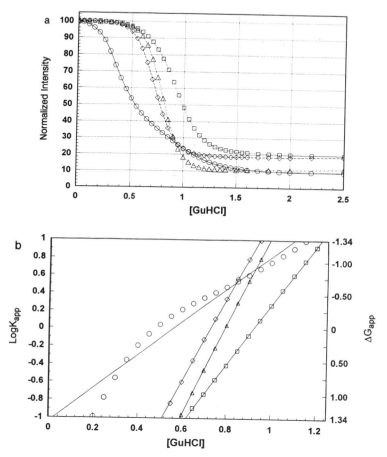

Figure 1. GuHCl denaturation of full-length and fragment 1-136 proteins. Note that the sole tryptophan in wild-type protein, W140, is removed in the fragments. The full-length mutant proteins each have two tryptophan residues while wild type has one. At zero denaturant concentration, the emission wavelength maxima of the fragment proteins relative to full-length wild-type protein (data not shown) is blue shifted in the case of V66W and red shifted in the cases G88W and K70W. This is consistent with the degree of solvent exposure inferred from the crystallographic data. **a.** (upper) Plot of tryptophan fluorescence intensity, excitation at 295 nm, emission at 325 nm, versus concentration of guanidine hydrochloride for full-length proteins. The initial fluorescence intensity for each protein, wild type Δ, V66W O, K70W ◊, and G88W ☐ is normalized to 100. The baselines at high denaturant concentrations are not identical because the initial fluorescence intensity of tryptophans 70 and 88 is much lower than tryptophans 66 and 140. **b.** (lower) Plot of the logarithm of the apparent equilibrium constant, K_{app}, versus concentration of GuHCl for full-length proteins. The apparent equilibrium constant, K_{app}, was determined at various GuHCl concentrations using the equation: Kapp = In - I/I - Id, where I is the fluorescence intensity of the sample, In is the value of native state fluorescence, and Id is the value of the denatured state fluorescence. Lines are the least squares fit to the data over the range of log(K_{app}) = 1 to -1.

We also followed circular dichroism intensity at 222 nm versus [GuHCl] for the full-length wild-type, V66W, K70W and G88W proteins. These experiments confirm these findings. Compared to tryptophan fluorescence, the circular dichroism baseline is poor and the signal is weak. Despite the additional uncertainty caused by these factors, it is clear that wild-type, K70W and G88W give clear sigmoids upon denaturation, while V66W does not.

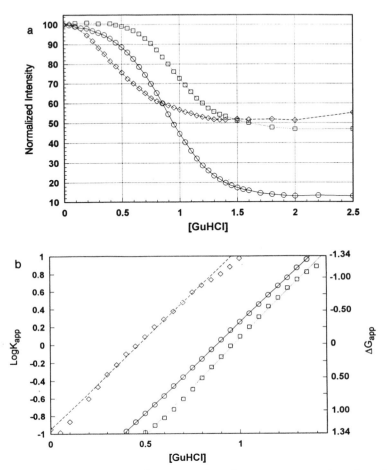

Figure 2. a. (upper) Plot of tryptophan fluorescence intensity versus concentration of guanidine hydrochloride. The initial fluorescence intensity for each fragment protein, V66W O, K70W ◊, and G88W ▢, is normalized to 100. The fragment of V66W shows almost a tenfold change in intensity. This is consistent with a change from a largely apolar environment to a largely polar environment. The fragments of G88W and K70W show only about a twofold change in intensity. This implies that the change in environmental polarity for these residues is much less marked, consistent with their expected higher solvent exposure. This lower intensity change also means there is a worse signal to noise ratio than for V66W, contributing to the overall poorer fit of the G88W and K70W fragments. In K70W there is not a clear initial baseline. The value of the initial fluorescence used in calculating the apparent equilibrium constant, Kapp, was estimated as previously described Shortle et al., 1990. **b.** (lower) Plot of the logarithm of the apparent equilibrium constant, Kapp, versus concentration of guanidine hydrochloride. Lines are the least squares fit to the data over the range of $\log(K_{app}) = 1$ to -1.

The three tryptophan mutants were also placed in the 1-136 fragment of staphylococcal nuclease previously studied (Shortle & Meeker, 1989). This fragment lacks the last thirteen amino acids of nuclease, including the tryptophan at position 140 (W140). The removal of these amino acids greatly destabilizes the native state. Nonetheless, as found

Table 1. Results of guanidine hydrochloride denaturation.

	Full length 1-149			Fragment 1-136		
	m_{GuHcl}[1]	C_m[2]	ΔG_{H_2O}[3]	m_{GuHcl}[1]	C_m[2]	ΔG_{H_2O}[3]
V66W						
Tryptophan Flourescence	6.60[4]	0.35	2.32	2.73	0.88	2.40
	2.45	0.74	1.66			
CD 222 nm	biphasic[5]			1.6[6]	0.65	1.0
Wild type						
Tryptophan Fluorescence	6.72	0.79	5.33	no fluorophore		
CD 222 nm	6.15	0.80	5.02	2.8[6]	0.25	0.7
K70W						
Tryptophan Fluorescence	6.17	0.73	4.49	2.7[7]	0.46	1.3
CD 222 nm	6.49	0.75	4.87	1.9[6]	0.19	0.3
G88W						
Tryptophan Fluorescence	4.46	0.92	4.11	2.9[8]	0.97	2.8
CD 222 nm	4.60	0.95	4.33	2.8[6]	0.84	2.4

[1] Units of kcal/(mol.M). Unless otherwise noted estimated error is ± 0.14.

[2] Denaturation midpoint concentration of GuHCl in molar. Unless otherwise noted estimated error is ± 0.02.

[3] Units of kcal/mol. Unless otherwise noted estimated error is ± 0.1.

[4] Denaturations and data analysis were performed as previously described (Shortle et al., 1990). Briefly, the plot of $\log(K_{app})$ versus the concentration of guanidine hydrochloride was fit by least squares over the range of $\log(K_{app})$ = 1 to -1. The slope of this line, mGuHCl, is listed above in units of kcal/(mol.M). The line was extrapolated back to zero denaturant concentration to determine K , and hence, the apparent free energy of denaturation in the absence of guanidine hydrochloride, . The exceptions to this are the two sets of values listed for full-length V66W which were derived by non-linear regression (Marquardt-Levenberg algorithm, Sigma Plot, version 4.02, Jandel Scientific) assuming a three state model for denaturation. We assumed that the reaction could be represented as N↔D1↔D2, where N represents the native state, D1 the compact denatured state and D2 the random coil state. The individual equilibria linking these states were assumed to be two state. This model fits the experimental data well. The first set of numbers are the values derived for the N↔D1 transition, the second set are those for the D1↔D2 transition. Especially notable is the fact that the second set of numbers matches the values found for the denaturation of the 1-136 fragment of V66W. This strongly supports the argument that the fragments are a good model of the transitions occuring in the denatured state of the full-length protein.

[5] CD data showed similar behavior to the tryptophan fluorescence with a distinct broadening of the sigmoidal transition at higher concentrations of guanidine hydrochloride. Because of the sloping baselines a reliable non-linear regression could not be carried out to resolve the two transitions of the biphasic denaturation curve.

[6] Because of the low signal strength and difficulty in determining baselines, the numbers returned from analysis of the CD data from the fragments are considered less reliable than those derived from tryptophan fluorescence. Estimated error in m is ± 0.2, in C is ± 0.09, ΔG_2 is ± 0.3.

[7] Estimated error in m is ± 0.2, in C is ± 0.04, ΔG_2 is ± 0.2.

[8] The slight, but distinct, increase in fluorescence intensity as low concentrations of guanidine hydrochloride are added makes the choice of a native baseline problematic. Therefore, neither fluorescence or CD returns very accurate values for the fragment of G88W. Estimated error in m is ± 0.2; in C is ± 0.05,; ΔG_2 is ± 0.2. Nevertheless, the magnitude of the figures is consistent with results from other fragments.

from the previous work, the CD spectra of these fragments show considerable residual structure. The plot of tryptophan fluorescence intensity, I, versus [GuHCl] for the 1-136 fragments of V66W, K70W and G88W proteins is depicted in Figure 2a along with the corresponding plot of $\log(K_{app})$ versus [GuHCl] in Figure 2b. The 1-136 fragments of V66W, K70W and G88W display the sigmoidal loss of fluorescence intensity and the linear variation of $\log(K_{app})$ versus [GuHCl] indicative of a two-state transition. These transitions are much broader than those of the full-length proteins.

Again we examined circular dichroism intensity at 222 nm versus [GuHCl] for the 1-136 fragments of wild type, V66W, K70W and G88W. In the fragments the circular dichroism signal is extremely weak and estimation of the correct slope for the beginning baseline is quite problematic. The results obtained from CD are in good qualitative agreement with those from the tryptophan fluorescence data. The transitions observed in the fragments are much broader than those in the full-length proteins.

Fluorescence Results

The results of all these experiments are summarized in Table I. In the full-length wild type, K70W and G88W, tryptophan loses its high fluorescence intensity in a fairly narrow transition. The behavior of full-length V66W is more complex. In contrast, the fragments of wild type and the other two mutants, as well as the fragment of V66W, all show a common broad transition as residual structure breaks down. A simple explanation of the V66W data is that native state breaks down cooperatively into an intermediate, compact, denatured state that in turn breaks down cooperatively into another denatured state that more closely resembles a true random coil. By extension, this behavior should exist in the other mutants and in wild type, but with the stability of the intermediate state much reduced.

MEAN FIELD THEORY ANALYSIS

Thermal denaturation

The suggestion is that the denaturation of 1-136 fragment mimics the breakdown of an intermediate species in the full-length protein. Using this model system we wished to address the question of the order of the phase transition of this breakdown. Accordingly we carried out thermal denaturations of the 1-136 fragments of these three mutants, monitoring tryptophan fluorescence. This data were analyzed as previously published (Shortle et al., 1988) to determine T_m and van't Hoff enthalpy. It has been shown (Privalov, 1979) that the thermodynamic parameters of denaturation for many proteins are the same regardless of method of denaturation, e.g. heat, pH or GuHCl. The order of the thermally induced phase transition may not necessarily be the same as that of the guanidine hydrochloride induced phase transition in the fragment system we study here but the precedent established by normal native to denatured transitions (Privalov, 1979) would imply that this is so.

Critical Exponent β

We carried out an analysis of the transition based on a mean field approach (Ikegami, 1977). In such an analysis it is necessary to account properly for the microscopic nature of the systems involved, and to justify the choice of the order parameter X upon which the analysis is based. The underlying assumption is that the tryptophan fluorescence signal, I, reflects the corresponding "local structure" of the protein. If we now introduce according to Ikagami (Ikegami, 1981; Ikegami, 1977; Pathria, 1984) an order parameter, X, derived from the fluorescence intensity, the order of the transition can be established from the temperature variation of X. At a second-order phase transition, the temperature dependence of the order parameter near the transition temperature, T_m, the temperature where X = 0, is described by an exponent β (called a critical exponent). This relationship takes the form

$$X = c(1 - T/Tm)^{\beta}$$

where c is a proportionality constant. The value of the exponent β depends on the universality class of the system. The finite size effects at a second-order phase transition are determined by the universality class of the system, but do not affect the value of the critical exponents (Barber, 1983). Thus β is not affected by the size of the system elements. Mean field theory predicts that, for a second-order phase transition, β will have a value of 1/2, which represents an upper limit on the values expected (Amit, 1984). For a first-order transition, the temperature variation of the order parameter is discontinuous, due to phase coexistence (Landau, 1990). In a finite system the discontinuity of the order parameter at a first-order phase transition is not sharp (Landau, 1990) and its temperature variation can become similar to that at a second-order transition.

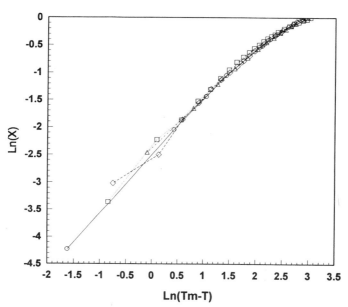

Figure 3. Log-log plot of the order parameter, X, versus $(T-T_m)$ in degrees Kelvin. These four consecutive thermal denaturations were all performed on the same sample of V66W 1-136 fragment. The reversibility of the system and the ruggedness of staphylococcal nuclease is obvious from the virtual superposition of the data after repeatedly heating the sample to over 80°C. Only the last run shows significant change with an apparent increase in T_m of approximately 3°C. The order parameter, X, is derived from the fluorescence intensity using the equation: $X = 1 - 2Ic - I/Ic - Ir$, where Ic is the value of the compact denatured state fluorescence, and Ir is the value of the random coil state fluorescence. The slope of this line is equal to β. A least squares fit of the last four data points was performed and then the values thus obtained were averaged. The same procedure was followed for the other fragments and for full-length wild type. The fourth denaturation of the V66W 1-136 fragment is plotted here also. Despite the increase in apparent T_m there is little change in β.

In order to apply the analysis of finite size effects at a first-order phase transition and directly determine the order of the transition, one needs thermodynamic data, such as specific heat (Challa et al., 1986; Landau, 1990). Thus the value of the critical exponent β can determine whether or not the phase transition is second-order. When the transition is not second-order one has to resort to comparison with a system that is known to exhibit a first-order phase transition. For this purpose we use the full-length (1-149) wild-type staphylococcal nuclease.

Results

In Figure 3 is shown the plot of log(X) versus $\log(T_m - T)$ for four separate thermal denaturations of V66W 1-136 fragment. The value of β, the slope of this line near the melting temperature, was determined by least squares fit of the last four data points. The results of the thermal denaturations of the 1-136 fragments of V66W, G88W, and K70W are summarized in Table II along with data from the thermal denaturation of full-length (1-149) nuclease. The values of β found for the fragments are clearly larger than the upper limit value of 0.5 expected for a second-order phase transition and are comparable to that found for the full-length protein. This is strong evidence that the transition in the fragments is first-order in character. Also interesting is the observation that melting temperatures for the fragments are comparable to that found for the full-length protein while van't Hoff enthalpy has approximately halved.

Table II. Results of thermal denaturation

	Number of determinations[a]	T_m (°K)	van't Hoff ΔH (kcal/mole)	β
V66W 1-136	3	322.5 ± 0.5	37.0 ± 0.5	0.92 ± 0.04
K70W 1-136	2	323.0 ± 0.9	30.3 ± 2.4	1.03 ± 0.24
G88W 1-136	1	331.2	37.5	1.06
Wild type (full-length)	2	326.4 ± 0.3	76.1	0.82 ± 0.02

[a] Each thermal denaturation was independently analyzed and the results averaged. These average results and standard deviations are shown. The data shown for V66W 1-136 does not include a fourth run with decreased T_m. Inclusion of these data slightly increases the averages.

The mutant V66W provides vital evidence to link the behavior of the fragments to that of the full-length protein. There are clearly two transitions present in guanidine hydrochloride denaturation of full-length V66W and, equally clearly, the character of the second transition is reflected in the guanidine hydrochloride denaturation of fragment V66W. It seems likely that the second transition in the full-length protein is first-order just as the analogous transition is in the fragment system. Presumably similar transitions from the compact denatured state to the random coil state are also occurring in the full-length G88W, K70W and wild-type proteins. The presence of this second transition is obscured in these proteins by the fact that it is occurring over the same denaturant range as the transition from native to the compact denatured state. Only in V66W are the two transitions well separated and thus discernible. However, the fragment system provides unambiguous evidence for a second transition in all nuclease mutants examined.

ACKNOWLEDGMENTS

This work was supported by NIH grants GM36358 and GM34171 to E. E. Lattman and D. Shortle. W.E.S. was supported by a NIH postdoctoral fellowship. We thank David Ssshsortle, Neil Clarke, Jeremy Berg and Steve Quirk for careful reading of the manuscript and discussion.

A account of this work has appeared elsewhere (Gittis et al., 1993).

REFERENCES

Alonso, D. O. V., Dill, K. A., Stigter, D. 1991. The three states of globular proteins: acid denaturation. Biopolym. 31:1631-1649.

Amit, D. J. 1984. "Field Theory, the Renormalization Group, and Critical Phenomena," Revised 2nd Ed., World Scientific Publishing, Singapore.

Barber, M. M. 1983. Finite-Size Scaling, in:: "Phase Transitions and Critical Phenomena," Vol. VII, Domb, C. & Lebowitz, J. L., eds., Academic Press, New York, 146-266.

Challa, M. S. S., Landau, D. P. & Binder, K. 1986. Finite-size effects at temperature-driven first-order transitions. Phys. Rev. B 34:1841-1852.

Chan, H. S. & Dill, K. A. 1991. Polymer principles in protein structure and stability. Annu. Rev. Biophys. Biophys. Chem. 20:447-490.

Dill, K. A. & Shortle, D. 1991. Denatured states of proteins. Annu. Rev. Biochem. 60:795-825.

Finkelstein, A. V. & Shakhnovich, E. I. 1989. Theory of cooperative transitions in protein molecules. II. Phase diagram for a protein molecule in solution. Biopolym. 28:1681-1694.

Flanagan, J. M., Kataoka, M., Shortle, D. & Engelman, D. 1992. Truncated staphylococcal nuclease is compact but disordered. Proc. Nat. Acad. Sci. 89:748-752.

Gittis, G.G., Stites, W. & Lattman, E.E. A first-order phase transition between a compact denatured state and a random coil state in staphylococcal nuclease. Biochemistry. In the press.

Hynes, T. R. & Fox, R. O. 1991. The crystal structure of staphylococcal nuclease refined at 1.7 Å resolution. Prot. Struct. Funct. Genet. 10:92-105.

Ikegami, A. 1977. Structural changes and fluctuations of proteins. I A statistical thermodynamic model. Biophys. Chem. 6:117-130.

Ikegami, A. 1981. Statistical thermodynamics of proteins and protein denaturation. Adv. Chem. Phys. 46:363-413.

Karplus, M. & Shakhnovich, E. 1992. Protein Folding:Theoretical Studies of Thermodynamics and Dynamics, in : "Protein Folding," Creighton, T. E., ed., W. H. Freeman, New York, 127-195.

Kim, P. S. & Baldwin, R. L. 1990. Intermediates in the folding reactions of small proteins. Annu. Rev. Biochem. 59:631-660.

Kuwajima, K. 1992. Protein folding in vitro. Curr. Opinion in Biotechnology 3:462-467.

Kuwajima, K. 1989. The molten globule as a clue for understanding the folding and cooperativity of globular-protein structure. Proteins:Struct., Funct., Genet. 6:87-103.

Landau, D. P. 1990. Monte Carlo Studies Of Finite Size Effects At First And Second-Order Phase Transitions, in: "Finite Size Scaling and Numerical Simulations of Statistical Systems," Privman, V., ed. World Scientific Publishing, Singapore, 223-260.

Loll, P. J. & Lattman, E. E. 1989. The crystal structure of the ternary complex of staphylococcal nuclease, Ca^{2+}, and the inhibitor pdTp, refined at 1.65 Å. Proteins, Struct., Funct., Genet. 5:183-201.

Nakano, T. & Fink, A. L. 1990. The folding of staphylococcal nuclease in the presence of methanol or guanidine thiocyanate. J. Biol. Chem. 265:12356-12362.

Pathria, R. K. 1984. "Statistical Mechanics," Peragamon Press, New York.

Privalov, P. L. 1979. Stability of proteins. Small globular proteins. Adv. Prot. Chem. 33:167-241.

Ptitsyn, O. G. 1992. The Molten Globule State, in: "Protein Folding," Creighton, T. E., ed., W. H. Freeman, New York, 243-300.

Shakhnovich, E. I. & Finkelstein, A. V. 1989. Theory of cooperative transitions in protein molecules. I. Why denaturation of globular protein is a first-order phase transition. Biopolym. 28:1667-1680.

Shortle, D., Meeker, A. K., Freire, E. 1988. Stability mutants of staphylococcal nuclease: large compensating enthalpy-entropy changes for the reversible denaturation reaction. biochem. 27:4761-4768.

Shortle, D. & Meeker, A. K. 1989. Residual structure in large fragments of staphylococcal nuclease: effects of amino acid substitutions. Biochem. 28:936-944.

Shortle, D., Stites, W. E. & Meeker, A. K. 1990. Contributions of the large hydrophobic amino acids to the stability of staphylococcal nuclease. Biochem. 29:8033-8041.

Stites, W. E., Gittis, A. G., Lattman, E. E., Shortle, D. 1991. In a staphylococcal nuclease mutant the side-chain of a lysine replacing valine 66 is fully buried in the hydrophobic core. J. Mol. Biol. 221:7-14.

Yutani, K., Ogasahara, K., Kuwajima, K. 1992. Absence of the thermal transition in apo-α-lactalbumin in the molten globule state. J. Mol. Biol. 228,:47-350.

STUDIES ON "HYPERSTABLE" PROTEINS: CRYSTALLINS FROM THE EYE-LENS AND ENZYMES FROM THERMOPHILIC BACTERIA

Rainer Jaenicke

Institut für Biophysik und Physikalische Biochemie
Universität Regensburg
D-93040 Regensburg
Germany

Proteins exhibit marginal stabilities, equivalent to only a few weak intermolecular interactions. Extreme environments require either molecular adaptation in terms of local structural changes or stabilization by "extrinsic factors" not encoded in the amino acid sequence. Such factors are, among others, specific metabolites, cofactors, ions and chaperones. No general strategies of stabilization have yet been established. However, certain incremental contributions to stability have been elucidated, analyzing extremely stable proteins, e.g., from "extremophiles", on one hand, and temperature-sensitive mutants, on the other. Stabilization may occur at the level of 1. local packing of the polypeptide chain in terms of secondary or supersecondary structural elements, 2. domains and domain interactions, and 3. subunit assembly. Experimental approaches and representative examples are: ad 1. fragmentation of single-chain domain proteins, and comparison of homologous proteins from extremophilic and mesophilic organisms (thermolysin, lysozyme of bacteriophage T4, NAD-dependent dehydrogenases from *Thermotoga maritima* and other sources); ad 2. "nicking" of domain proteins and alteration of connecting peptides between domains (gammaB-crystallin); ad 3. dissociation reassociation experiments using oligomeric proteins (lactate dehydrogenase).

"Gene sharing" in the eye-lens offers a whole repertoire of proteins with anomalous stability properties. The discussion will focus on recent work on recombinant gammaB- and betaB2-crystallins, as well as comparative studies on NAD-dependent dehydrogenases and chaperone proteins.

Introduction

Biological systems have developed various ways to cope with extreme conditions in their environment. Their components may either exhibit high intrinsic stability, or the cells may use "extrinsic factors" (such as compatible solvent components, ligands, chaperones, etc.), or increased turnover in order to compensate for destruction. In the following article, I shall focus on *protein* stability, selecting two examples in order to illustrate the different levels at which intrinsic stabilization may be accomplished in nature: eye-lens proteins and proteins from hyperthermophilic microorganisms. Whether they are representative, is

Statistical Mechanics, Protein Structure, and Protein Substrate Interactions
Edited by S. Doniach, Plenum Press, New York, 1994

unclear, since general "strategies of stabilization" of biomolecules have still to be uncovered (if they exist). What makes the two examples interesting is the fact that, on the one hand, eye-lens proteins do not undergo any turnover during the lifetime of an organism; on the other hand, intrinsically stable proteins from hyperthermophiles, e.g., *Thermotoga maritima, Pyrodictium occultum* or *Pyrococcus furiosus,* are perfectly stable at temperatures beyond the boiling point of water. Bulk of the data refer to "hyperstable proteins" of the latter type. In connection with proteins from the eye lens, I shall restrict myself to a monomeric model system, gammaB-crystallin, and to some constructs probing the mutual stabilization of its domains.

There is life on the entire globe. The limits of viability for the biologically relevant variables are: for temperature $\leq 110^{\circ}$C, for hydrostatic pressure ≤ 120 MPa, for water activity ≥ 0.6 (corresponding to salinities up to 6 M), and for pH, values below 2 and beyond 10. Organisms exposed to extreme conditions may respond either by "avoidance", or they have to adapt their cellular components. In the first case, the stress parameter is excluded or compensated, for example, by neutralizing the anomalous external pH, or by levelling a salt gradient through isosmotic concentrations of compatible cytoplasmic components; in the second, mutations have resulted in increased intrinsic stability, so that all cellular components are resistant against the specific variable. Evidently, structural elements facing the outside world cannot avoid stress parameters. In the case of extremes of temperature and pressure (because of the isothermal and isobaric conditions in a given habitat), the entire cell inventory is expected to show enhanced intrinsic stability. However, there are extrinsic effectors, like specific ligands or chaperone proteins, which may assist cellular components in overcoming stress (*Jaenicke, 1991a; Jaenicke et al., 1988; Jaenicke & Buchner, 1993*).

Organisms which have evolved such that they *require* extreme conditions for their whole life cycle are called "extremophiles", or, more specific for extremes of temperature, pressure, salinity and pH: *thermo-* or *psychrophiles, barophiles, halophiles* and *acido-* or *alkalophiles.* The present paper is focused on the upper limit of temperature, i.e., in the case of *"hyperthermophilic"* bacteria, to temperatures close to the boiling point of water. As shown by mRNA sequence analysis, these organisms are close to the root of the phylogenetic tree (**Fig. 1**), suggesting that they preceded their mesophilic counterparts in the time course of evolution (*Woese & Fox, 1991; Woese et al., 1990; Stetter, 1992*) The adaptive strategies allowing full metabolic and reproductive activity under conditions close to the limits of hydrothermal degradation (*Bernhardt et al., 1984*) are still unresolved. In the following, possible mechanisms of stabilization will be discussed, based on model systems such as domains and subdomains, on one hand, and selected examples of thermostable pointmutants or proteins from hyperthermophilic microorganisms, on the other. It is obvious that the solution of the stability problem has biological and technological implications in connection with the adaptive response to physiological stress and applications in protein design.

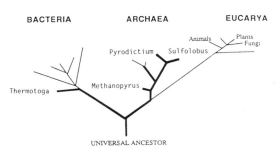

Figure 1. Universal phylogenetic tree showing the three kingdoms of life. Hyperthermophiles, including the universal ancester, indicated by heavy line.

Contributions to Protein Stability

Protein stability is accomplished by the cumulative effect of non-covalent interactions at many locations within a given molecule. In considering the contributions of local structural elements, limited proteolysis may be used in order to determine the minimum length of a polypeptide chain that is required to still form an intrinsically stable native-like structure. NMR analysis has shown that oligopeptides down to six residues do form stable (non-random) conformations, supporting the idea that local structures may serve as "seeds" in the folding process (*Wright et al., 1988; Jaenicke, 1991b*). However, there is evidence that the short fragments do not necessarily adopt the same conformation in unrelated protein structures; e.g., reverse turn motifs observed in small peptides seem to be absent in the known three-dimensional structures of proteins containing these sequences (*Creighton, 1988*). With regard to the stability of protein fragments, it has been known for a long time that proteins are cooperative structures showing mutual stabilization of structural elements. To find out at which fragments size native-like structure is no longer formed, thermolysin was used as a model. The N-terminal portion of the enzyme is found to stabilize the all-helical C-terminal domain. This may be shortened to the 62-residues three-helix bundle without loosing much of its native (secondary) structure. In shortening the polypeptide chain, the free energy of stabilization drops steadily to half its value, while the temperature limit of denaturation is shifted from 87 to 64°C only. At a minimum size of 20 residues, no residual structure is left, "wrong interactions" causing ag-gregation (*Vita et al., 1989*).

The incremental stabilization observed for fragments or subdomains holds also at the higher levels of the hierarchy of protein structure. A striking example illustrating the mutual stabilization of *domains* is the eye-lens protein gammaB-crystallin. As one would predict, the two-domain protein shows a bimodal denaturation transition, with the second phase coinciding with the unfolding of the isolated N-terminal domain (*Rudolph et al., 1990*). As mentioned, the complete molecule is known to undergo no degradation during the whole lifetime of an organism. The N-terminal domain shares this extreme intrinsic stability, whereas the C-terminal half is surprisingly unstable, thus proving that domain interactions must contribute significantly to the overall stability of the protein (**Fig.2A**). The reason why in the present example two topologically closely related domains show such drastic deviations in their stability properties is unknown. The importance of the domain interactions becomes clear, when gammaB- and betaB2-crystallin are compared. The two proteins (from bovine eye-lens) are closely related, distinct mainly by the linker peptide connecting the two domains in each polypeptide chain (*Bax et al., 1990*).

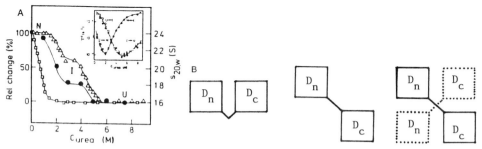

Figure 2. Equilibrium transition of a two-domain protein: gammaB-crystallin. **A:** Urea-dependent unfolding/folding of bovine gammaB-crystallin at pH 2. N,U,I: native, fully and partially unfolded protein; θ_{222} ($\triangle$), $s_{20,w}$ ($\bullet$). Insert: Urea-dependence of unfolding/folding rate constants; data for the N-terminal fragment parallels the I $\rightleftharpoons$ U transition ($\bullet$) (*Rudolph et al., 1990*); recombinant C-terminal domain ($\square$) (E. Mayr, unpublished data). **B:** Alternative models depicting the possible connectivities and domain interactions for the gammaBbeta-mutant of gammaB-crystallin.

However, betaB2-crystallin forms a dimer whereas gammaB-crystallin is monomeric, even at extreme concentrations (*R. Jaenicke, unpublished*). In the homodimer (betaB2), the extended conformation of the linker peptide between the domains has been suggested to force the molecule to favor intermolecular domain interactions, compared with intramolecular contacts in gammaB. This agrees well with the decreased stability of the C-terminal domain when it is isolated from its N-terminal counterpart. Exchanging the linker of gammaB-crystallin against the one of betaB2 by cloning techniques, yields a *monomeric* protein with stability characteristics *identical to gammaB*.This indicates clearly that the domain interface rather than the connecting peptide determines the mode of domain association. In the gammaBbeta construct, the connecting peptide must be twisted to a turn similar to the one in natural gammaB-crystallin (**Fig.2B**) (*E.-M. Mayr, R. Jaenicke, R. Glockshuber, unpublished*).

This result shows that the interdomain contacts are responsible for the overall stability of gammaB-crystallin, independent of the type of the connecting peptide. It is interesting to note that mixing the N- and C-terminal domains at a 1:1 ratio does not yield a gammaB-like "nicked gammaB". Obviously, the covalent linkage provides the high local concentration that is required for the mutual domain stabilization in the natural protein, consistent with its monomeric state, and confirming the evolutionary principle of duplication and fusion of crystallin domains (*Wistow & Piatigorsky, 1988*).

What has been discussed here for the mutual stabilization of domains in multidomain proteins holds also for the interactions of subunits in oligomeric or multimeric proteins. Their stabilizing effect is clearly demonstrated in the case of porcine muscle lactate dehydrogenase where the stability decreases steadily from the native tetramer down to domain fragments. The tetramer, as a dimer of dimers, is highly stable, with 2M guanidinium-chloride as the limiting concentration ($c_{1/2}$) where deactivation occurs. The "proteolytic dimer" (with the N-terminal decapeptide cleaved off) requires structure-making salts to exhibit activity; $c_{1/2}$ is found to be shifted to $\leq$ 1M GdmCl (**Fig. 3A**). The monomer is inactive under any condition and only "structured" as a short-lived folding intermediate on the pathway of reconstitution. The separate NAD- and substrate-binding domains (after "nicking") are unstable, but still sufficiently well-defined in their conformation to recognize eachother and to pair correctly: in joint reconstitution experiments, they lead to partial renaturation exhibiting a mutual "chaperone effect" (*Opitz et al., 1987*).

There are examples where stabilization due to subunit association goes to the extreme (viruses, chromatin, ferritin, bacterial surface layers etc.). In all these cases, the monomers have average size, stability and flexibility. This way, the complex structures

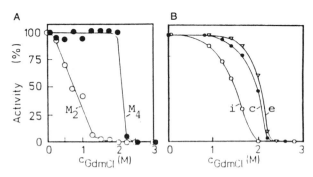

Figure 3. Equilibrium transitions of oligomers and glycoproteins. **A:** GdmCl-dependent deactivation of pig muscle LDH (●) and the proteolytic dimer lacking the N-terminal decapeptide (○); the "proteolytic dimer" exhibits enzymatic activity in the presence of 1M $(NH_4)_2SO_4$ (*Opitz et al., 1987*). **B:** GdmCl-induced deactivation of internal (○), core-glycosylated (●) and external (▽) yeast invertase (with 0, 34, 65% carbohydrate) (*Kern et al., 1992*).

exhibit their characteristic low turnover, while the capsomers or subunits may still perform their essential morphopoietic or catalytic functions, without being inhibited with respect to translocation, targeting, processing, etc.

The increments of protein stability discussed so far originate from local structures, fragments, domains and subunits, i.e., from the intrinsic properties of a given amino acid sequence. However, additional factors *not encoded in the sequence* may be crucial in both protein stability and folding. It is well-known that ligands such as ions, cofactors, metabolites, and components covalently linked to proteins may contribute to protein stability. As an example, **Fig. 3B** illustrates the effect of glycosylation on the thermal stability of invertase from yeast (*Kern et al., 1992*).

Limits of Growth *versus* Limits of Stability of the Covalent Structure

The protein inventory of the above mentioned hyperthermophilic bacteria commonly exhibits intrinsic stability (for a detailed discussion, see below). As shown by recombinant DNA techniques, mesophilic hosts may be applied to express such proteins in active form. To give an example, GAPDH from *Thermotoga maritima* still forms its native three-dimensional structure when its gene is expressed in *Escherichia coli*, i.e., about 60 degrees below its optimum temperature. Obviously, the acquisition of the native structure in the heterologous host does not require the physiological temperature conditions of the guest. This is not trivial since the weak intermolecular interactions which stabilize the native structure of proteins show characteristic temperature dependences which, far below the optimum temperature, may cause cold denaturation (*Privalov & Gill, 1989; Jaenicke, 1990*). Actually, upon *reconstitution* at 0°C, the enzyme forms an inactive low-temperature intermediate which may be explained by the involvement of hydrophobic interactions in the acquisition of the native three-dimensional structure (*Schultes & Jaenicke, 1991; Rehaber & Jaenicke, 1992, 1993*).

Apart from temperature-induced perturbations of the native three-dimensional structure, common Arrhenius behavior leads to a more or less complete loss of catalytic function at low temperature. In the case of hyperthermophilic micro-organisms, significant cell growth requires temperatures beyond 50°C (*Stetter, 1992; Adams & Kelly, 1992*). Considering the temperature effect on the metabolic network, it is obvious that temperature adaptation not only requires directed alterations of the intrinsic stability of cell constituents, but also tuning of the kinetics. Depending on the activation energies of the reactions involved, shifts in the optimum temperature from mesophiles to thermophiles may cause dramatic kinetic dislocations (*Table I*).

On the high-temperature extreme, one may ask why there seems to be a definite limit of growth and reproduction at temperatures around 120°C. Considering this limit and the reversible deactivation of enzymes at water contents below the normal degree of hydration of proteins (≤ 0.25 g H_2O/g protein), the ultimate requirement for life seems to be the presence of liquid water. In testing this hypothesis, stabilization of the liquid state of water at high hydrostatic pressure proves that the temperature limit of viability (T_{max}) cannot be shifted significantly (*Bernhardt et al., 1988*). Life under "Black Smoker" conditions (26 MPa and 250°C) must be science fiction: the susceptibility of the covalent

Table I. Alterations of relative reaction rates normalized to 20°C

T°C	Activation Energy (kJ/mol)		
	16	32	64
20	1	1	1
60	3	10	90
100	14	186	30 000

structure of the polypeptide chain toward hydrolysis, and the hydrothermal degradation of essential biomolecules (*Bernhardt et al., 1984; White, 1984*) would require compensatory "anaplerotic reactions" which are incompatible with the relatively low metabolic activity of most extremophilic organisms.

It is worth mentioning that the hydrophobic hydration also seems to vanish close to the above temperature limit (*Baldwin, 1986; Privalov, 1990, 1992*). Therefore, one may assume that both the covalent structure of the protein and the weak interactions involved in its three-dimensional structure determine the upper limit of protein stability. It is important to note that the building blocks of proteins from extremophiles (including hyperthermophiles) are exclusively the canonical 20 natural amino acids. Thus, molecular adaptation to altered environmental conditions can merely depend on the mutative change in the local and global distribution of the amino acids along the polypeptide chain. From studies on synthetic polypeptides it became clear that the natural amino acids basically allow to build protein molecules with stabilities beyond those commonly observed for natural proteins. The conclusion we may draw from this observation is that molecular adaptation obviously results in optimum protein flexibility rather than maximum stability.

Conformational Stability of Proteins

The central issue in the adaptation of biomolecules to extreme conditions is the conservation of their functional state which is characterized by a well-balanced compromise of stability and flexibility (*Tsou, 1988; Jaenicke, 1991a*). Basic mechanisms of molecular adaptation are changes in packing density, charge distribution, hydrophobic surface area, and in the ratio of polar/non-polar or acidic/basic residues. Evidently, dramatic changes in stabilization energy do not occur, even in the most extreme cases. In general, the overall free energy of stabilization of globular proteins in their natural solvent environment is of the order of 40-50 kJ/mol, independent of both the size of the protein and the mode of denaturation. Determining the stability increment per residue, the result is one order of magnitude below the thermal energy (kT). This proves, that the overall stability of a polypeptide chain must involve cooperativity, because the addition of stability increments per amino acid residue in the process of structure formation would not allow to overcome the thermal energy (*Dill, 1990*). In cases where the molecular mass of a protein is too small to provide the necessary size of a cooperative unit, the structure may be stabilized by covalent crosslinking or ligand binding. Both improve the stability of the entire molecule by decreasing the entropy of unfolding.

The requirement for flexibility is fulfilled by the subtle compensation of attractive and repulsive interactions. Their contributions to the free energy are exceedingly large; however, their superposition yields a small difference between big numbers. Considering the energy contributions of the relevant weak intermolecular forces involved in protein stabilization, on balance, stability is attributable to the equivalent of a few hydrogen bonds, ion pairs or hydrophobic interactions (*Jaenicke, 1991a*).

Extremophilic proteins do not exhibit properties qualitatively different from non-extremophilic ones. The essential adaptive alteration tends to shift the normal characteristics to the respective extreme in the sense that under the mutual physiological conditions the molecular properties are comparable; with other words, adaptation to extremes of physical conditions tends to maintain "corresponding states" regarding overall topology, flexibility and solvation. Considering the parabolic profile of the temperature dependence of the free energy of stabilization (*Privalov & Gill, 1988, 1989*), it is evident that enhanced thermal stability may be accomplished by a variety of mechanisms. As depicted schematically in Fig. 4A, the profile of the mesophilic wild-type protein could either be shifted to higher temperature, or it could be lowered, or flattened; in all three

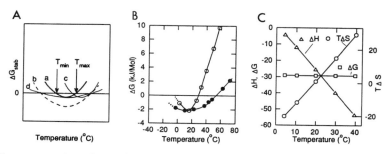

Figure 4. Temperature dependence of the free energy of stabilization and ligand binding. **A:** Hypothetical $\triangle G_{stab}$ profiles for meso- (a) and thermo- philic (b-d) proteins. **B:** $\triangle G_{stab}$ for phosphoglycerate kinase from yeast (O) and *Thermus thermophilus* (●), in the presence of 0.5 M (O) and 2 M GdmCl (●) **C:** Entropy/enthalpy compensation upon binding of NADH to pig muscle LDH-M$_4$.

cases, T_m, the "melting temperature" of the protein, would be increased to the higher value T_m'. Experimental evidence seems to indicate that the third alternative holds true (Fig. 4B). Thus, the temperature dependence of $\triangle G_{stab}$ of thermophilic proteins is less pronounced than the one of the corresponding wildtype protein. A similar independence of temperature has been found for the free energy of ligand binding to enzymes; in this case, entropy-enthalpy compensation renders thermal adaptation unnecessary (Fig. 4C).

In trying to quantify the thermodynamics of thermophilic adaptation, even for *hyper*thermophiles the changes in $\triangle\triangle G_{stab}$ do not exceed the equivalent of a few additional weak interactions. At present, no comparative X-ray analysis of any protein from a mesophile, thermophile and hyperthermophile has been put forward at sufficiently high resolution to pin down specific residues responsible for the gradual increase in protein stability. The most thorough studies in this context refer to X-ray crystallographic and thermodynamic studies on pointmutants of bacteriophage T4 lysozyme (*Matthews, 1987, 1991; Alber, 1989; Jaenicke, 1991a*). The results illustrate how subtle the different weak interactions are balanced in native globular proteins, and how much, even marginal local strain in the three-dimensional structure may affect protein stability. From the point of view of protein engineering and the attempt to provide deeper insight into the mechanisms of thermophilic adaptation, the results are a test case to prove that presently no unequivocal predictions or general strategies of protein stabilization can be given. One problem in this context is the fact that high-resolution X-ray structures may not be adequate to detect the real structural changes accompanying "isomorphous single point mutations", because the lattice forces in protein crystals may freeze or shift positions of the polypeptide backbone which are not representative for the relevant *local* solution structure.

Selected Enzymes from Hyperthermophiles

Adaptation of proteins occurs at the DNA level leading to specific local alterations in structural genes and, subsequently, amino-acid sequences. In the case of thermal adaptation, for a decade, such changes have been assumed to be manifested in certain well-defined shifts in the amino acid *composition* (*Argos et al., 1979*). The significance of the proposed "rules" is ambiguous (*Böhm, 1992; Böhm & Jaenicke, 1992, 1993*), for obvious reasons: The low values of the extra free energy of stabilization ($\triangle\triangle G_{stab}$) required for the increase in stability can be provided by an astronomically large number of subtle changes in local weak interactions. Evidently, summarizing these changes in terms of preferred amino acid exchanges, minimization of crevices, or enhanced ion-pair formation etc. is an oversimplification. What seems to be well-established are the following observations: 1. Proteins from thermophiles commonly exhibit high *intrinsic* stability which becomes marginal at their respective physiological temperatures. At optimum temperature, homologous enzymes show similar conformational flexibility. The increase in rigidity of thermophilic proteins at room-temperature is evident from decreased exchange rates of amide protons, as well as increased resistance against proteolysis and denaturation using mesophilic homologs as references. 2. Available structural data prove that "thermophilic proteins" are closely similar to their mesophilic counterparts regarding their basic topologies and enzyme mechanisms. Correspondingly, virtually all the amino acids constituting their active sites are conserved. 3. The correlation of sequence alterations with changes in thermal stability is expected to be highly complex. It is closely related to the protein folding problem, as increased stability is equivalent to the difference in minimum energy. So far, no general strategy of thermal stabilization is available. To give a few representative examples, amylases, GAPDH and LDH from the hyperthermophilic eubacterium *Thermotoga maritima* will be discussed regarding their solution properties and self-organization.

Amylases. *Thermotoga maritima* can be grown on minimal medium with starch as carbon source. The uptake of glucose is not accomplished by secreting amylases from the cytosol or the periplasm into the outside medium. Thus, starch as a nutrient must be degraded on the cell surface in order to be able to pass the outer sheath and the periplasm. Actually, Percoll gradient centrifugation clearly proves that more than 85% of the total α-, β- and glucoamylase activity are associated with the "toga". The expression of the three isoenzymes is too low to isolate the proteins in quantities required for a detailed characterization. Compared with α-amylase from *Bacillus licheniformis* (T_{max}= 75°C), the amylases from *Thermotoga* show high intrinsic stability with upper temperature limits beyond 95°C; significant turnover of the enzyme requires physiological temperatures ($\geq$ 70°C). The molecular masses and the enzymological properties of the homologous enzymes are similar (*Schumann et al., 1991*).

D-Glyceraldehyde-3-phosphate Dehydrogenase. Among the cytosolic enzymes which have been screened, GAPDH shows the highest expression level. Since homologous GAPDH's from various sources (including thermophiles) have been investigated in detail, this enzyme was chosen for a detailed analysis. The enzyme has been purified to homogeneity (*Wrba et al., 1990b*), its amino-acid sequence has been determined at the protein (*Schultes et al., 1990*) and DNA level (*Tomschy et al., 1993*), and its physicochemical and enzymological properties have been compared with the enzyme from mesophiles and other thermophiles (*Rehaber & Jaenicke, 1992*). Considering the amino acid compositions, it is obvious that only the exchanges Lys $\rightarrow$ Arg, Ser $\rightarrow$ Thr and Val $\rightarrow$ Ile agree with the results of previous statistical analyses (*Argos et al., 1979*). The primary structure is highly homologous with the enzyme from other thermophilic bacteria (*Table II*): comparing the sequences of the enzyme from *Thermotoga, Bac. stearothermophilus* and *Thermus aquaticus*, 63 and 59% identity are observed, only 8% of the exchanges are non-conservative; taking all known structures together, about 33% of the residues are identical. Homology modelling yields equal topologies (*Böhm, 1992*), in accordance with the observation that the secondary and quaternary structure are closely similar and that the essential residues in the active site are conserved in all species. Marked differences are: (a) enhanced intrinsic stability, (b) decreased exchange rates in H-D exchange experiments, (c) activation of the enzyme at low denaturant concentrations, (d) a relatively small change in hydrodynamic volume upon coenzyme binding, (e) a low-temperature folding intermediate upon reconstitution (*Table III. Fig. 5 ; Wrba et al., 1990b; Rehaber & Jaenicke, 1992*).

The given characteristics are indicative for low flexibility, suggesting high packing density and strong subunit interactions of the Thermotoga enzyme at low temperature.

Table II. GAPDH sequences from (**1**) *Thermotoga*, (**2**) *Bac. stearothermophilus*, (**3**) *Thermus*, (**4**) yeast, and (**5**) lobster. Numbering refers to the Thermotoga sequence; (.) identical amino acids, (-) deletions (*Schultes et al., 1990*).

```
          1                                          70                                                        140

1  ARVAINGFGRIGRLVYRIIYERKNPDIEVVAINDLT-DTKTLAHLLKYDSVHKKFPGKVEYTENSLIVDG  KEIKVFAEPDPSKLPWKDLGVDFVIESTGVFRNREKAELHLQAGAKKVIITAPAKGEDITVVIGCNEDQL KP-EHTIISCASCTTNSIAPIVKVLHE

2  AVKVG.........N.F.AAL--KNPDIE.VAV..LT-NADGLAHLL....VHGRLDAE.VVNDGDVSVN..E.IVKAERN.ENLA.GEIGVDIVVE...R.TKRED.AK.LEA.....I.S..AKVENITV.M.VNQDKY.DPKAHHVI.N......CL..FA..LHQ

3  MKVG.........Q.F.I-L--HSRGVE.ALI..LT-NDKTLAHLL....IYHRFPGE.AYDDQYLYYD..A.RATAVKD.KEIP.AEAGVGVVIE...V.TDADK.KA.LEG.....I.T..AKGEDITL.M.VNHEAY.DPSRHHII.N......SL..VM..LEE

4  VRVA.........L.M.IAL--SRPNVE.VAS..PFIBLDYAAYMF....THGRYAGE.SHDDKHIIVD..K.ATYQERD.ANLP.SSGDSVIAID...V.KELDT.QK.IDA.....V.T..S-STAPMF.M.VDSEKY-SDLKIV.N......CL..LA..IND

5  AcSKIG.........L.L.AAL--SCG-AQ.VAV..PFIALEYMVYMF....THGVFKGE.KMEDGALVVD..K.TVFNEMK.ENIP.SKAGAEYIVE...V.TTIEK.SA.FKG.....V.S..S-ADAPMF.C.VNLEKYS-KDMTVV.N......CL..VA..LHE
```

```
            210                                        280

1  KFGIVSGMLTTVHSYTNDQRVLDLP-HKDLRRARAAAVNIIPTTTGAAKAVALVVPEVKGKLDGMAIRVPTPDGSITDLTVLVEKETTVEEVNAVMKEATEGRLKGIIGYNDEPIVSSDIIGTTFSGIFDATITNVIGGKLVKVASWYDNEYGYSNRVVDTLELLLKM

2  E.GIVRGMM....SY.NN.RIL.L.-H..L.GA.A.AES...TT......VAL.L.ELK.KLN...M....PNV.VV.LVAELEK.VTVEEVNAAL.E.A..E.K.ILA.SEEPL.SRNYNGSTV..TI..LSTMVDIGKM..VVS.....T..SH..V.LAAYINAKGL

3  A.GVEKALM....SY.NN.RLL.L.-H..L.RA.A.AIN...TT......TAL.L.SLK.RFD...L....ATG.IS.ITALLKR.VTAEEVNAAL.A.A..P.K.ILA.TEDEI.LZBIVMDPH..IV..KLTKALGNMX..VFA.....W..AN..A.LVELVLRKGV

4  A.GIEEGLM....SL.AT.KTV.G.SH..W.GG.T.SGN...SS......VGK.L.ELQ.KLT...F....VBV.VV.LTVKLDK.TTYDEIKKVV.A.A..K.K.VLG.TEBAV.SSBFLGSBH..IF..SATKALGNMX..LVS.....Y..ST..V.VEHVAKA

5  N.EIVEGLM....AV.AT.KTV.G.SA..W.GG.G.AQN...SS......VGK.I.ELD.KLT...F....PDV.VV.LTVRLGK.CSYDDIKAAM.T.S..P.Q.FLG.TEDDV.SSDFIGDNR..IF..KAGIQLSKTF..VVS.....F..SQ..I.LLKHMQKVDSA
```

Table III. Properties of homologous mesophilic and thermophilic GAPDHs and LDHs

Property	GAPDH		LDH	
	Thermotoga	Yeast	*Thermotoga*	Pig (H_4)
Molecular mass (M_4) (kDa)	145	144	144	140
Subunit mass (M_1) (kDa)	37	36	36	35
Change in $s_{20.w}$ (holo minus apo)	3.5%	5.3%	n.d.	7.0%
Thermal denaturation (T_m holo °C)	109	40	91	42
Denaturation in GdmCl ($c_{1/2}$ 20°C)	2.1M	0.5M	1.8M	0.5M
Activation at 0.5 M GdmCl (70°C)	300%	0%	n.d.	n.d.
Rel. H-D exchange rate ($ß_{rel}$ 25°C)	64	100	n.d.	n.d.
A_{SP} (U/mg at °C)	200 (85)	70 (20)	≈1000 (55)	400 (25)
K_M µM substrate (°C)	400 (60)	160 (25)	60 (55)	76 (25)
K_M µM NAD$^+$/*NADH*	79 (60)	44 (25)	*27 (55)*	*10* (25)
Opt. reactivation at 0°C/≥30°C	0%/85%	35%/85%	n.d.	50%/95%

High-resolution X-ray analysis will provide insight into the detailed mechanism; work in this direction is in progress.

Lactate Dehydrogenase. Thermotoga LDH shows less than 1% of the expression observed for GAPDH, thus causing extreme difficulties in the purification and cloning of the enzyme (*Wrba et al., 1990a; Ostendorp et al., 1993*). The DNA sequence has been elucidated, again proving the enzyme to be highly homologous to LDH's from other thermophilic bacteria. Correspondingly, the structural properties of the enzyme do not offer much surprise, except that (i) the co-enzyme NAD$^+$ and the activator fructose-1.6-bisphosphate have additive stabilizing effects (*Wrba et al., 1990a*), and (ii) that under physiological conditions, the enzyme does not operate with maximum efficiency, since K_M for NADH and pyruvate *increase* with increasing temperature (*Hecht et al.,1989*).

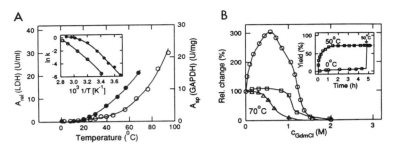

Figure 5. Temperature- and GdmCl-dependence and denaturation/renaturation of GAPDH and LDH from *Thermotoga maritima*. **A:** Activity under standard conditions. **Insert:** Arrhenius plot. **B:** GdmCl-dependent denaturation of GAPDH monitored by activity (○), fluorescence (□) and circular dichroism (△). **Insert:** Reactivation after denaturation (6 M GdmCl) at 0 and 50°C; after final state was reached, T was shifted to 50°C *(Rehaber & Jaenicke, 1992)*.

Thermostable Pointmutants

It has been mentioned that the excess free energy of stabilization for thermostable proteins does not involve more than a few percent of the total number of weak interactions involved in the secondary, tertiary and quaternary contacts within the molecule. Considering the three GAPDH's from *Thermotoga maritima*, *Methanothermus fervidus* and *Bacillus stearothermophilus* with their respective T_m-values 110, 80 and 68°C, about 100 out of the 330 amino acids per subunit differ between the three enzymes; obviously, the answer to the question of which of these is responsible for the increment in thermal stability is not trivial. One way out of this dilemma would be to investigate point mutations yielding proteins of known structures with differing thermal stabilities. Experiments along these lines made use of creatinase from *Pseudomonas putida*, a homodimer of known crystal structure, and pyruvate oxidase from *Lactobacillus plantarum*, a homotetrameric flavoprotein with strong binding sites for FAD, TPP and a divalent cation.

Creatinase. Creatine amidinohydrolase is a homodimer of 91 kD molecular mass the amino acid sequence and 1.9 Å X-ray structure of which are known *(Hoeffken et al., 1988)*. Each subunit consists of two domains with central antiparallel ß-pleated sheets surrounded by α-helices on the outside. In order to unravel the relationship of their structure, function and stability, single-, double- and triple-mutants with known three-dimensional structure and exchanges in the periphery and in the core of the molecule were characterized regarding their biochemical and physicochemical properties. As shown in *Table IV*, wild-type CR and its mutant proteins do not exhibit significant differences in their enzymological characteristics. The V182I exchange refers to the starting point of the helix connecting the two domains; positioned in the periphery of the molecule, it has only a marginal stabilizing effect. A109V and V355M increase tertiary interactions reflected by the first denaturation transition, leaving the overall backbone stability unaltered. Interestingly, extrinsic stabilization of the enzyme (e.g., by polyols) exhibits the same "localized" protecting effect (Schumann et al., 1993a). V355M improves the packing density within the core of the C-terminal domain. The effects of A109V and V182I cannot be easily explained: the increase in hydrophobicity must either decrease the entropy of the denatured state or it must be compensated by readjustments of the global structure. Secondary or quaternary interactions are not affected. In going from single to double and triple mutants, to a first approximation, the increments of stabilization are additive, in agreement with the above idea that $\triangle G_{stab}$ is the sum of local stability increments *(Schumann et al., 1993)*.

Table IV. Physicochemical and enzymological properties of point mutants of creatinase from *Pseudomonas putida* (*Schumann, 1992*)[*]

Creatinase	Exchanges	K_M(mM)	$G_{1/2}$ A_{sp}	$t_{1/2}$ Aggregation	$R_{90,rel}$	T_m(°C) ΔC_p	A_{sp}	$10^4 k_2$ A_{sp}	$U_{1/2}$ CD
Wildtype	-	14.3	0.30	13	100	48.4	40.5	21.0	5.0
M 1	109V	12.5	0.47	24	46	52.0	43.5	5.9	5.0
M 2	355M	13.9	0.47	32	37	50.1	44.0	9.0	5.0
M 3	182I	14.7	0.33	19	74	48.6	40.5	18.0	5.0
M 1+2	A109V+V355M	16.1	0.56	39	22	53.5	46.8	1.9	5.0
M1+2+3	A109V+V355M+V182I	15.6	0.58	67	12	54.7	47.5	0.6	5.0

[*]Wildtype and mutant CR show identical molecular masses (91kD), elution properties, specific activities (13±2 U/mg) and activation energies (41.5±2.5 kJ/Mol). K_M for creatine; $G_{1/2}$, c_{GdmCl} (M) at 50% transition; $t_{1/2}$,$R_{90,rel}$, half-time of aggregation (min), and maximum rel. light scattering (400nm) at 0.5M GdmCl, pH 8, 25°C; T_m, transition temperature from calorimetry (DSC) and deactivation; k_2, 1st-order rate constant of thermal deactivation at 47°C; $U_{1/2}$, c_{urea} (M) where 50% denaturation are monitored by far-UV CD (Θ_{222nm}).

In many related analyses no clear-cut correlations between the increase in stability and alterations in the three-dimensional structure can be defined because normally the crystallographically determined differences between mutants (rms 0.2 Å) are not focused at the exchanged residues, but spread over the whole molecule. Making use of correlation functions, improved thermal stability is found to be connected with enhanced hydrophobicity, reduced surface area, and increased packing density of the interior core of the molecule (*Böhm & Jaenicke, 1992*).

Pyruvate oxidase. In contrast to creatinase (where point mutants do not show a significant influence on the quaternary structure of the enzyme), pyruvate oxidase gains thermal stability from improved FAD binding and subunit association (*Table V; Risse et al., 1992*). Presently, only the high-resolution structure of the wild-type enzyme is available (*Muller & Schulz, 1993*); the detailed interpretation of the mutant data in terms of the involvement of exchanged amino acid residues in quaternary contacts requires further crystallographic analysis.

Table V. Stabilizing effect of point mutations of pyruvate oxidase from *Lactobacillus plantarum* (*Risse et al., 1992*)[*]

Pyruvate oxidase	Exchanges	Midpoint of deactivation T°C	pH	c_{urea}(M)	Effect of ligands on T_m apo	+TPP	+FAD	holo	% M_4
Wildtype	-	42.0	6.70	0.75	48.5	48.5	50.0	48.5	<5
M 1	P178S	50.0	7.30	1.65	47.5	47.5	53.5	50.0	30
M 2	S188N	48.5	7.15	1.35	47.5	47.5	52.0	49.0	15
M 3	A458V	49.0	7.55	1.85	46.0	48.5	56.0	49.0	5
M 1+2+3		51.5	7.85	2.15	46.0	47.5	57.0	51.0	5

[*]Wildtype and mutant PO show identical specific activities (5.3±0.2 U/mg), K_M (phosphate) (2.0±0.3 mM), K_M (pyruvate) (0.36±0.03 mM), fluorescence emission and far-UV CD; T_m (°C): the TPP binary complex and the holoenzyme are monitored in the presence of 1 mM Mn^{2+}/TPP, the FAD binary complex in the presence of 1 mM Mn^{2+}/0.1 mM FAD; % M_4, relative amount of tetramers in the $4M = 2M_2 = M_4$ equilibrium from size exclusion HPLC at 1.5 mg/ml, 20°C.

Extrinsic Effectors and Chaperone Proteins

Metabolites as "extrinsic effectors" have briefly been mentioned. One example is cyclic 2.3 diphosphoglycerate which has been shown to enable non-thermostable proteins to gain thermal stability (*Hensel et al., 1988; Huber et al., 1989*). Because of their functional analogy to *"compatible solutes"* and anti-freeze compounds, it should be remembered that similar cellular components allow halotolerant organisms, as well as plants and fish, to survive salt and low-temperature stress without halophilic or psychrophilic adaptation. More specific protein-ligand interactions, like those involved in the formation of holoenzymes from their respective apoenzymes plus *cofactors* or *coenzymes*, may be crucial in the stabilization of both the native state and intermediates on the pathway of folding and association. The effects of Zn^{2+}/NAD or Mn^{2+}/TPP/FAD on the reconstitution of horse-liver alcohol dehydrogenase or pyruvate oxidase may serve as examples (*Jaenicke, 1987; Risse et al., 1992*).

Further stabilizing factors involving the nascent rather than the native protein, refer to chaperones assisting in structure formation or counteracting misfolding and misassembly (*Jaenicke, 1991b; Jaenicke & Buchner, 1993*). How this "assistance" works, and what is the intermediate state of the nascent or refolding polypeptide chain is still unresolved. The phenomenon was originally detected as *heat-shock response*, devised to counteract misaggregation as the major consequence of protein heat denaturation. The occurrence of heat-shock proteins in *all* cells, and their high degree of conservation during evolution led to the view that they must have a fundamental function as "chaperones" for the nascent or infant polypeptide chain (*Fischer & Schmid, 1990; Ellis & van der Vies, 1991; Jaenicke, 1991b; Seckler & Jaenicke, 1992*). The idea that binding of early folding intermediates to hydrophobic surfaces competes with misfolding, and that additional heat-shock components and/or ATP-hydrolysis lead to conformational changes and subsequent release of the polypeptide chain is a reasonable working hypothesis for future experiments. That such two-step mechanisms may be meaningful has been observed in a number of cases, where reconstitution of thermophilic proteins was arrested at low temperature; after switching to high temperature, folding proceeds to high yields at greatly enhanced rate (*Schultes & Jaenicke, 1991; Rehaber & Jaenicke, 1992*). Since aggregation and association are endothermic, low temperature allows the folding polypeptide chain to make the correct tertiary contacts, thus transforming the "collapsed polypeptide chain" (as an early intermediate) into the native three-dimensional structure.

In connection with thermal stress, one may ask whether (hyper)thermophiles at the upper temperature limit of viability show heat shock response. Evidently, this is the case, as *Sulfolobus shibatae* ($T_{max} = 80°C$) and *Pyrodictium occultum* ($T_{max} = 105°C$) and other archaea contain (and overexpress) chaperone-like, extremely thermostable ATPases. These enzymes resemble proposed cytosolic eukaryotic chaperones and have the capacity to bind folding intermediates and prevent their irreversible denaturation (*Phipps et al., 1991; Sparrer, 1992*).

Conclusions

As has been shown, proteins, due to the delicate balance of stabilizing and destabilizing interactions, are only marginally stable. Contributions to the net free energy of stabilization range from local interactions at the level of elements of secondary structure and subdomains to interactions between domains and subunits. Enhanced intrinsic stability requires only minute local structural changes so that general strategies of stabilization cannot be established. This holds especially because, apart from mutative changes of amino acid sequences, nature provides other ways of stress adaptation. On one hand, a wide variety of specific or unspecific extrinsic factors are known to improve protein stability; on the other, cellular components such as chaperones may be involved in protective mechanisms. General limits of viability refer to the susceptibility of the covalent structure of the polypeptide chain toward hydrothermal degradation, the dissociation of weak interactions involved in protein stabilization, perturbations of protein selforganization, and dislocation of metabolic pathways. How life succeeded in "multiplying and replenishing the earth" to the extent that practically the whole surface of our planet is *"biosphere"* is still enigmatic and will remain a challenge for generations of biochemists to come.

Acknowledgments

I should like to thank the John E. Fogarty International Center for Advances Study, NIH, Bethesda, for support and hospitality. Fruitful discussions with G. Böhm, R. Huber, G. Kern, R. Ostendorp, P.L. Privalov, V. Rehaber, B. Risse, R. Rudolph, J. Schumann, H. Schurig, R. Seckler, K.O. Stetter, A. Wrba and P. Závodszky are gratefully acknowledged. Work performed in the author's laboratory was generously supported by grants of the Deutsche Forschungsgemeinschaft, the Fonds der Chemischen Industrie and the European Community.

References

Adams, M.W.W. & Kelly, R.M. (Eds.) (1992) *Biocatalysis at Extreme Temperatures* ACS Symp. Ser. **498**, 215 p.

Alber, T. (1989) *Annu. Rev. Biochem.* **58**, 765-798.

Argos, P., Rossmann, M.G., Grau, U.M., Zuber, H., Frank, G. & Tratschin, J.D. (1979) *Biochemistry* **18**, 5698-5703.

Baldwin, R.L. (1986) *Proc. Natl Acad. Sci. USA* **83**, 8069-8072.

Bax, B., Lapatto, R., Nalini, V., Driessen, H., Lindley, P.F., Mahadevan, M., Blundell, T.L. & Slingsby, C. (1990) Nature 347, 776-780.

Bernhardt, G., Lüdemann, H.-D., Jaenicke, R., König, H. & Stetter, K.O. (1984) *Naturwissenschaften* **71**, 583-586.

Bernhardt, G., Jaenicke, R., Lüdemann, H.-D., König, H. & Stetter, K.O. (1988) *Appl. Environ. Microbiol.* **54**, 1250-1261.

Böhm,G. (1992) Dissertation, University of Regensburg.

Böhm, G. & Jaenicke, R. (1992) *Protein Sci.* **1**, 1269-1278.

Böhm, G. & Jaenicke, R. (1993) *Int. J. Peptide Protein Res.*, **43**, 97-106.

Creighton, T.E. (1988) *Biophys. Chem.* **31**, 155-162.

Dill, K.A. (1990) *Biochemistry* **29**, 7133-7155.

Ellis, R.J. & van der Vies, S.M. (1991) *Annu.Rev.Biochem.* **60**, 321-347.

Fischer, G. & Schmid, F.X. (1990) *Biochemistry* **29**, 2205-2212.

Hecht, K., Wrba, A. & Jaenicke, R. (1989) *Eur.J.Biochem.* **183**, 69-74.

Hensel, R. & König, H. (1988) *FEMS Microbiol.Lett.* **49**, 75-79.

Hoeffken, H.W., Knof, S.H., Bartlett, P.A., Huber, R., Moellering, H. & Schumacher, G. (1990) *J. Mol. Biol.* **204**, 417-433.

Huber, R., Kurr, M., Jannasch, H.W. & Stetter, K.O. (1989) *Nature* **342**, 833-834

Jaenicke, R. (1987) *Progr.Biophys.Mol.Biol.* **49**, 117-237.

Jaenicke, R. (1990) *Phil.Trans.R.Soc.* **B326**, 535-553.

Jaenicke, R. (1991a) *Eur.J.Biochem.* **202**, 715-728.

Jaenicke, R. (1991b) *Biochemistry* **30**, 3147-3161.

Jaenicke, R. & Buchner, J. (1993) *Chemtracts: Biochem. Mol. Biol.*, **4**, 1-21.

Jaenicke, R., Bernhardt, G., Lüdemann, H.-D. & Stetter, K.O. (1988) *Appl. Environ. Microbiol.* **54**, 2375-2380.

Kern, G., Schülke, N., Schmid, F.X. & Jaenicke, R. (1992) *Prot. Sci* **1**, 120-131

Matthews, B.W. (1987) *Biochemistry* **26**, 6885-6888.

Matthews, B.W. (1991) *Curr.Opin.Struct.Biol.* **1**, 17-21.

Muller, I. & Schulz, G.E. (1993) *Science*, in press.

Opitz, U., Rudolph, R., Jaenicke, R., Ericsson, L. & Neurath, H. (1987) *Biochemistry* **26**, 1399-1406.

Ostendorp, R., Liebl, W., Schurig, H. & Jaenicke, R. (1993) Eur. J. Biochem., **216**, 709-715.

Phipps, B.M., Hofmann, A., Stetter, K.O. & Baumeister, W. (1991) *EMBO J.* **10**, 1711-1722.

Privalov, P.L. (1990) *CRC Crit.Rev.Biochem.Mol.Biol.* **25**, 281-305.

Privalov, P.L. (1992) In: *Protein Folding* (Creighton, T.E., Ed.), Freeman: New York, 83-

Privalov, P.L. & Gill, S.J. (1988) *Adv.Prot.Chem.* **39**, 193-231.

Privalov, P.L. & Gill, S.J. (1989) *Pure Appl.Chem.* **61**, 1197-1104.

Rehaber, V. & Jaenicke, R. (1992) *J. Biol. Chem.* **267**, 10999-11006.

Rehaber, V. & Jaenicke, R. (1993) *FEBS Lett.*, **317**, 163-166.

Risse, B., Stempfer, G., Rudolph, R., Möllering, H. & Jaenicke, R. (1992) *Prot. Sci.*, **1**, 1699-1709.

Risse, B., Stempfer, G., Rudolph, R., Schumacher, G. & Jaenicke, R. (1992) *Prot. Sci.*, **1**, 1710-1718.

Rudolph, R., Siebendritt, R., Nesslauer, G., Sharma, A.K. & Jaenicke, R. (1990) *Proc. Natl Acad. Sci.* **87**, 4625-4629.

Schultes, V. & Jaenicke, R. (1991) *FEBS Lett.* **290**, 235-238.

Schultes, V. & Deutzmann, R.& Jaenicke, R. (1990) *Eur.J.Biochem.* **192**, 25-31.

Schumann, J. (1992) Dissertation, University of Regensburg.

Schumann, J., Möllering, H. & Jaenicke, R. (1993a) Biol. Chem. Hoppe-Seyler, **374**, 427-434.

Schumann, J., Böhm, G., Schumacher, G., Rudolph, R. & Jaenicke, R. (1993b) Prot. Sci., **2**, 1612-1620.

Schumann, J. & Wrba, A., Jaenicke, R. & Stetter, K.O. (1991) *FEBS Lett.* **282**, 122-126.

Seckler, R. & Jaenicke, R. (1992) *FASEB J.* **6**, 2545-2552.

Sparrer, H. (1992) Thesis, University of Regensburg.

Stetter, K.O. (1992) Proc. 3rd Rencontre de Blois, Frontiers of Life.

Tomschy, A., Glockshuber, R. & Jaenicke, R. (1993) *Eur. J. Biochem.*, **214**, 43-50.

Tsou, C.-L. (1988) *Biochemistry* **27**, 1809-1812.

Vita, C., Fontana, A. & Jaenicke, R. (1989) *Eur. J. Biochem.* **183**, 513-518.

White, R.H. (1984) *Nature* **310**, 430-432.

Wistow, G. & Piatigorsky, J. (1988) Annu. Rev. Biochem. 57, 479-504.

Woese, C.R. & Fox, G.E. (1977) *Proc. Natl Acad. Sci. USA.* **74**, 5088-5090.

Woese, C.H., Kandler, O. & Wheelis, M.L. (1990) *Proc. Natl Acad. Sci. USA.* **87**, 4576-4579.

Wrba, A., Jaenicke, R. & Huber, R. & Stetter, K.O. (1990a) *Eur. J. Biochem.* **188**, 195-201.

Wrba, A., Schweiger, A., Schultes, V., Jaenicke, R. & Závodszky, P. (1990b) *Biochemistry* **29**, 7584-7592.

Wright, P.E., Dyson, H.-J. & Lerner, R.A. (1988) *Biochemistry* **27**, 7167-7175.

ORIGINS OF MUTATION INDUCED STABILITY CHANGES IN BARNASE: AN ANALYSIS BASED ON FREE ENERGY CALCULATIONS

Shoshana J. Wodak and Martine Prévost

Unité de Conformation de Macromolécules Biologiques
Université Libre de Bruxelles
CP160/16, avenue Paul Héger, P2
B–1050 Brussels (Belgium)

ABSTRACT

The thermodynamic stability of proteins, measured by their denaturation free energy, is in general not very large. The observed values, which are in the range of 10 kcal/mol, are thought to result from a delicate balance between different types of interactions. Among those, hydrophobic interactions are believed to play a major role. Site directed mutagenesis combined with thermal and spectroscopic stability measurements have provided valuable insights into the contributions to protein stability from individual amino-acids. But the information they provide on the physical origins of these contributions is often incomplete. Theoretical analyses based on a detailed microscopic description of the molecular systems should therefore be helpful in gaining a fuller understanding of the physical phenomena. Methods for computing free energy changes by using molecular dynamics simulation techniques are particularly well suited for this purpose, as they evaluate thermodynamic quantities that can be directly compared with experimental values. These methods are however still not routine and the reliability of the results obtained, especially for complex systems such as proteins, depends critically on the validity of the basic assumption that underlie the simulations. An advantage of the free energy simulation methods is that they offer the possibility of decomposing the computed values into contributions from different parts of the system (protein versus solvent) as well as from individual energy terms such as non–bonded and vibrational terms. This decomposition is important because it contains more information than the experimental results. However, unlike the overall free energy change, which is formally independent of the integration path, the individual components have a path dependence, so that concepts that arise from such an analysis must be confirmed by supplementary calculations.

The free energy simulation study of two hydrophobic mutations in barnase, namely replacement of Ile at position 96 by Ala and Val respectively, is used here to illustrate how issues concerning basic assumptions such as adequate sampling, convergence of the simulations, or definition of the denatured state, can be addressed. We furthermore show that the analysis of the overall free energy change in terms of its specific contributions, can provide valuable insights, which are supported in part by independent calculations performed either on the native protein or on simple model systems.

Statistical Mechanics, Protein Structure, and Protein Substrate Interactions
Edited by S. Doniach, Plenum Press, New York, 1994

In particular, we find that the major contributions to the decrease in stability produced by the Ile→Ala mutations arise from bonding terms involving degrees of freedom of the mutated side chain, and from non–bonded interactions of the side chain with its environment in the folded protein. In contrast, hydration effects are seen to play a minor role in the overall free energy balance for both the Ile→Ala and Ile→Val transformations. These results have implications for our understanding of the hydrophobic effect.

INTRODUCTION

This paper describes the application of the free energy simulation method to two hydrophobic substitutions where Ile at position 96 in barnase is replaced by Val and Ala respectively. Barnase is an extra–cellular ribonuclease from B. Amyloliquefaciens. This small enzyme (110 residues) is of particular interest because its stability and folding mechanism is being extensively studied (Kellis et al., 1988, 1989; Fersht et al., 1992). Its crystal structure is known to 2.0 Å resolution (Mauguen et al., 1982) and the structure of several mutants is being determined, (Fersht, personal communication). This one–domain protein undergoes reversible thermal and urea induced denaturation which closely approximates a two–state equilibrium (Hartley, 1968). It contains significant secondary structure, including a β–sheet (50–55, 70–75, 85–91, 94–101, 106–108) in which Ile96 is situated. This residue is fully buried and participates in a closely packed hydrophobic interface with the first α–helix (residues 6–18).

We report an analysis of the simulations of the alchemical transformations Ile96→Val and Ile96→Ala in the native solvated protein and in an extended conformation used as a model for the denatured state in water. For comparison, the same transformations are also simulated with the extended conformation in the gas phase. A more complete account of this study can be found in Prévost et al. (1991, 1993).

METHODOLOGY

Details of the computational methodology have been given elsewhere (Prévost et al., 1991; 1993). Only a brief summary of the calculation procedure is given here.

To compute the free energy difference, ΔG, between two states A and B representing respectively, the wild type and mutant proteins, a "computer alchemy" operation is undertaken where one amino acid is transformed into another (Gao et al., 1989). This is achieved using a hybrid potential:

$$V(\mathbf{r}^N, \lambda) = (1 - \lambda)V_A(\mathbf{r}^N) + \lambda\, V_B(\mathbf{r}^N) \qquad (1)$$

This potential is a linear combination of $V_A(\mathbf{r}^N)$, and $V_B(\mathbf{r}^N)$, the empirical potentials describing wild type and the mutant proteins respectively, with λ being a coupling parameter, varied from 0 to 1. $\mathbf{r}^N$ are the atomic coordinates of the system encompassing here the protein and solvent molecules.

The free energy difference ΔG between the two states A (wild type) and B (mutant) has been obtained from either of two formally exact expressions, the so called 'exponential formula' (EF) (Zwanzig, 1954):

$$\Delta G = - k_B T \sum_i \ln \left\langle e^{- \frac{V(\mathbf{r}^N, \lambda_{i+1}) - V(\mathbf{r}^N, \lambda_i)}{k_B T}} \right\rangle_{\lambda_i} = - k_B T \sum_i \ln \left\langle e^{- \frac{\Delta V \Delta \lambda_i}{k_B T}} \right\rangle_{\lambda_i} \qquad (2)$$

where $\Delta V = V_B(\mathbf{r}^N) - V_A(\mathbf{r}^N)$, $\Delta \lambda_i = \lambda_{i+1} - \lambda_i$, k_B the Boltzmann constant and T the absolute temperature; the angle brackets represent an ensemble average obtained using the potential $V(\mathbf{r}^N, \lambda_i)$ from eq (1) to represent the system.

The other equivalent formulation used here, is called thermodynamic integration (TI). It uses the following expression (Kirkwood, 1935):

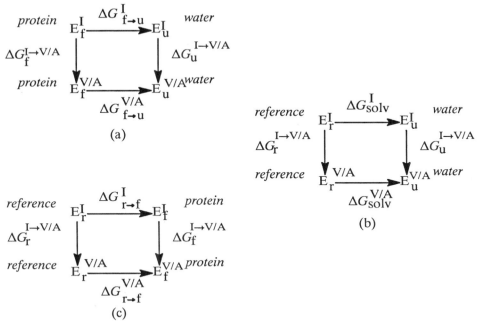

Fig. 1. The thermodynamic cycles in the calculations and the experiments.

(a) The thermodynamic cycle for the unfolding process of barnase. The vertical direction concerns alchemical processes corresponding to the Ile→Val or Ile→Ala transformation. On the left hand side the transformation occurs in the folded protein, yielding the free energy difference $\Delta G_f^{I \to V/A}$. On the right hand side, the transformation occurs in the unfolded state, modeled in our computation by an extended heptapeptide, yielding the free energy difference $\Delta G_u^{I \to V/A}$. The horizontal direction concerns chemical steps of the unfolding reactions. Unfolding of the wild type protein in presence of Ile is shown on top; it corresponds to the free energy difference $\Delta G^I_{f \to u}$. Unfolding of the Val or Ala containing mutant is shown on the bottom, with the corresponding change in free energy of $\Delta G^{V/A}_{f \to u}$.

(b) The thermodynamic cycle for the solvation process of the unfolded protein (the extended heptapeptide), e.g. transfer from the reference phase (gas phase) to water. The vertical direction concerns alchemical processes corresponding to the Ile→Val or Ile→Ala transformation. On the left hand side, the transformation occurs in the gas phase, yielding the free energy difference $\Delta G_r^{I \to V/A}$. On the right hand side, the reaction occurs in water, yielding the free energy difference $\Delta G_u^{I \to V/A}$. The horizontal direction concerns the solvation process for the unfolded protein, whose free energy is ΔG_{solv}. The process of taking the Ile containing unfolded protein from the gas phase to water is shown on top. The solvation process for the Val or Ala containing unfolded protein is shown on the bottom.

(c) The thermodynamic cycle for a folding/solvation process which brings the unfolded protein in the reference (gas phase) to the solvated folded state. The vertical direction concerns alchemical processes corresponding to the Ile→Val or Ala transformation. On the left hand side, the transformation occurs in the gas phase, yielding the free energy difference $\Delta G_r^{I \to V/A}$. The right hand side shows the same transformation in the folded protein, which yields the free energy difference $\Delta G_f^{I \to V/A}$. The horizontal direction concerns the folding/solvation reaction. The process of transferring the Ile containing unfolded protein from the gas phase to the Ile containing folded state is shown on top. It yields the free energy difference $\Delta G^I_{r \to f}$. The same process in presence of Val or Ala is shown on the bottom, with the corresponding change in free energy of $\Delta G^{V/A}_{r \to f}$.

$$\Delta G = \int_0^1 \langle \partial V / \partial \lambda \rangle_\lambda d\lambda \approx \sum_i \langle \Delta V \rangle_{\lambda_i} \Delta \lambda \qquad (3)$$

where the canonical average $\left\langle \frac{\partial V}{\partial \lambda} \right\rangle_{\lambda_i}$ is equal to $\langle \Delta V \rangle_\lambda$ provided the hybrid potential energy function $V(\mathbf{r}^N, \lambda_i)$ is linear in λ. The integration over λ, approximated here by a summation over a finite number of $\Delta \lambda$ "windows", then yields the required free energy value.

Differences due to changes in the number of particles may arise due to the kinetic energy contribution to ΔG. These differences cancel out however when identical alchemical transformations are considered in two different states, such as the native and unfolded ones described in the thermodynamic cycle of Fig. 1a.

To ensure convergence of the calculations which involved small non–polar transformations, an elaboration of the classical procedure was used. Several intermediate states were defined along a pathway from A (wild type) to B (mutant) as illustrated in Fig. 2.

This procedure was applied to the folded state starting with the high resolution crystallographic coordinates of barnase (Mauguen et al., 1982). Unlike the folded state, the unfolded state is rather ill defined. However, following experimental evidence that the unfolded state of barnase contains an appreciable amount of β structure (Matouschek et al., 1989), the unfolded state was modeled as an extended heptapeptide in water, comprising in addition to residue 96 the three residues on each side of the mutated position. Further details about the computation are given in legend of Fig. 2.

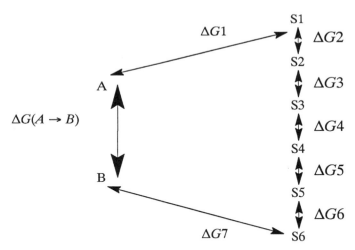

Fig. 2 Schematic representation of the pathway used for the alchemical transformation of Ile into Ala.

The pathway between the end states A and B is defined by six intermediate states labelled S1–S6. These states are generated by gradually modifying van der Waals parameters and bond lengths of the side chain representing one given end–state towards the side chain representing the other end–state. The overall free energy difference is computed as the sum of the free energy differences between successive states along this pathway (ΔG_j), yielding $\Delta G(A \rightarrow B) = \sum_{j=1}^{7} \Delta G_j$. A similar pathway between Ile and Val was taken with two intermediate states.

In both the folded and unfolded states, the ensemble averages over configuration space required for eq (2) and (3) were computed at room temperature (300 K) using Stochastic Boundary Molecular Dynamics (SBMD) (Brunger et al., 1985). The simulated system for the folded protein included 472 protein atoms and 21 water molecules while that of the unfolded state included 76 protein atoms and 145 water molecules. All the free energy calculations were performed with the CHARMM program (Brooks et al., 1983). Long–range interactions were smoothly truncated at 8.5 Å with a shifting function for the electrostatic interaction and a switching function for van der Waals interaction. A dielectric constant of unity was used.

Calculations of ΔG for a given λ consist of an equilibration simulation of 5 ps followed by an averaging period of 10 ps. For Ile→Ala each alchemical transformation involved 105 ps of equilibration and 210 ps of averaging. For Ile→Val each alchemical transformation involved 45 ps of equilibration and 90 ps of averaging. With an integration step of 1 fs, a 5 ps simulation required about 30 minutes of CPU time on the Cyber 205 computer at the former John von Neuman Center.

COMPUTED CHANGES IN PROTEIN STABILITY FOR THE ILE96 SUBSTITUTIONS INTO ALA AND VAL

Table 1 lists the computed free energy values obtained by the exponential formula (EF) and thermodynamic integration (TI) procedures for the Ile96 into Ala and Val alchemical transformation respectively. These are, $\Delta G_f^{I \to A/V}$ for the solvated native protein, $\Delta G_u^{I \to A/V}$ for the solvated unfolded state, and $\Delta G_r^{I \to A/V}$ for the gas phase reference state. By use of the thermodynamic cycle (Fig. 1a), the corresponding difference in

Table 1a Computed free energy changes (in kcal/mol) for the Ile→Val mutations in barnase.

Contribution	$\Delta G_f^{I \to V}$ protein	$\Delta G_u^{I \to V}$ water	$\Delta G_r^{I \to V}$ reference	$\Delta\Delta G_{f \to u}$	$\Delta\Delta G_{r \to u}$	$\Delta\Delta G_{r \to f}$
TI	−2.77	−3.93	−2.06	−1.16	−1.87	−0.71
EF	−2.63	−3.69	−2.18	−1.06	−1.51	−0.45
EXP				−0.9[a];−1.1[b]	−0.16[c]	

Table 1b Computed free energy changes (in kcal/mol) for the Ile→Ala mutations in barnase.

Contribution	$\Delta G_f^{I \to A}$ protein	$\Delta G_u^{I \to A}$ water	$\Delta G_r^{I \to A}$ reference	$\Delta\Delta G_{f \to u}$	$\Delta\Delta G_{r \to u}$	$\Delta\Delta G_{r \to f}$
TI	−3.09	−8.3	−7.15	−5.21	−1.15	4.06
EF	−3.39	−6.81	−6.56	−3.42	−0.25	3.17
EXP				−3.3[a];−4.0[b]	−0.21[c]	

Negative values in $\Delta\Delta G$ correspond to contributions in which the wild type (Ile) is stabilized relative to the mutant (Val or Ala). ΔG_f, ΔG_u and ΔG_r are the free energies for the alchemical transformation Ile→Val or Ile→Ala, in the folded state *(protein)*, the unfolded state *(water)*, and the unfolded state in the gas phase *(reference)* respectively. $\Delta\Delta G_{f \to u}$ is the unfolding free energy difference for the alchemical transformation. $\Delta\Delta G_{r \to u}$ and $\Delta\Delta G_{r \to f}$ are respectively, the solvation free energy differences for Ile versus Val or Ala in water and in the protein interior. The corresponding $\Delta\Delta G$ values are derived from the corresponding ΔG values as explained in the text. EF, TI and EXP stand for, the exponential formula (eq (2)), the thermodynamic integration procedure (eq (3)), and experimental results respectively. The superscripts a, b and c indicate that values are taken from Kellis et al., 1988, 1989 and Wolfenden et al., 1981 respectively.

unfolding free energy, $\Delta\Delta G_{f\rightarrow u}$, is obtained from the following expression (Wong & McCammon, 1986):

$$\Delta\Delta G_{f\rightarrow u} = \Delta G_u^{I\rightarrow V/A} - \Delta G_f^{I\rightarrow V/A} = \Delta G_{f\rightarrow u}^{V/A} - \Delta G_{f\rightarrow u}^{I} \qquad (4)$$

The calculated $\Delta\Delta G_{f\rightarrow u}$ values are $-1.06/-1.16$ kcal/mol for Val mutation and $-3..42/-5.2$ kcal/mol for the Ala mutation (successive values correspond to those obtained using the EF and TI procedures respectively). These values agree in sign and magnitude with the experimental measures which are $-0.9/-1.1$ kcal/mol for Val and $-3.3/-4.0$ kcal/mol for Ala. The negative values indicate that unfolding the mutant is energetically more favorable than unfolding the wild type, and hence that the wild type folded state is more stable than the mutant folded state.

Using another thermodynamic cycle the solvation free energy difference $\Delta\Delta G_{r\rightarrow u}$, is given by (Fig. 1b):

$$\Delta\Delta G_{r\rightarrow u} = \Delta G_u^{I\rightarrow A} - \Delta G_r^{I\rightarrow A} = \Delta G_{solv}^{A} - \Delta G_{solv}^{I} \qquad (5)$$

The $\Delta\Delta G_{r\rightarrow u}$ values calculated by the EF and TI procedures are $-0.25/-1.15$ kcal/mol for the mutation into Ala and $-1.51/-1.87$ kcal/mol for the mutation into Val. The corresponding differences obtained from experimental measures are -0.21 kcal/mol for Ile versus Ala, and -0.16 kcal/mol for Ile versus Val. Although the correct sign is obtained in all cases, the calculated solvation free energy differences tend to be larger than the experimental values. However these changes being very small, the computed values are within the error margin of the calculations (see below).

In addition, the free energy change for going from the gas phase to the folded protein was also calculated by use of the relation (Fig. 1c).

$$\Delta\Delta G_{r\rightarrow f} = \Delta G_f^{I\rightarrow A} - \Delta G_r^{I\rightarrow A} \qquad (6)$$

The values obtained are $-0.45/-0.71$ kcal/mol and $+3.17/+4.06$ kcal/mol for the mutations into Val and Ala respectively. No direct measurement of this quantity is available from experiment. It can however be obtained by difference from the other experimental data given in Table 1. This yields values of $0.71/0.94$ kcal/mol for the Val mutation and $+3.1/+3.8$ kcal/mol for the Ala mutation. The two values for each mutant correspond to the two different experimental measures for $\Delta\Delta G_{f\rightarrow u}$. We thus see that the simulation and experiments agree that the essential effect of the mutation into Ala is a change in the stability of the folded state rather than the differential solvation of Ile and Ala in the unfolded state. This is less clear for the mutation into Val where the small change in free energy is within the error margin of the computations.

ERROR ESTIMATION

The free energy values computed from ensemble averages of complex systems such as the one considered here, may be subject to errors from various origins.

One obvious source of error would be incomplete convergence of the simulations (Beveridge & DiCapua, 1989), which can be assessed from estimates of their statistical precision. Computations based on the standard deviation of individual ΔG_j free energy values obtained by the TI procedure, estimate the precision of the overall free energy differences calculated here to be about 1 kcal/mol. The precision of the corresponding $\Delta\Delta G$ values is thus of about 2 kcal/mol. Also, our computations have been performed with two formally equivalent procedures (the EF and TI methods) that would give identical results if full convergence had been achieved. The discrepancy between the free energy differences for Ile versus Ala obtained by these methods is 1.8 for denaturation and 0.9 kcal/mol for solvation. In the case of Ile→Val, the discrepancy between the $\Delta\Delta G$ values obtained by these methods is smaller. It is 0.1 and 0.35 kcal/mol for unfolding and solvation respectively.

To evaluate whether the trajectories used to compute the ensemble averages were long enough to allow the system to relax in response to the perturbations, water radial distribution functions (rdf's) around atoms of the Val and Ile side chains were computed near the endpoints of the transformation pathways in the solvated unfolded state where each of the side chains is fully present respectively. Results showed that these rdf's displayed the expected shapes, indicating that the correct water structure was formed around the mutated side chains during the perturbation procedure (Prévost et al., 1993).

In addition to random errors, systematic errors may also arise. Due to the severe limitation in computer time, sampling of configuration space is often incomplete. As a result, contributions from multiple rotational isomeric states are neglected. Assuming that only the rotational isomers of the mutated side chains contribute significantly to the free energy difference, the procedure of Straatsma and McCammon (1989) was used to evaluate the correction that one would need to apply to the computed free energy change for the Ile→Ala mutation to adequately account for rotational isomer sampling. The value of this correction was found to be about 0.5 kcal/mol (Prévost et al., 1993), in accord with results obtained previously when computing binding free energies of antiviral drugs to human rhinovirus (Wade and McCammon, 1992).

A further source of systematic error may arise from the model used to represent the unfolded state about which little information is available. To assess its magnitude, the Ile→Val transformation was recomputed in the gas phase reference state using an α–helical structure for the heptapeptide instead of the extended one described above. The $\Delta G_r^{I→V}$ values differed by less than 0.5 kcal/mol between the two conformations (Prévost et al., 1993).

In conclusion we evaluate the statistical imprecision of the present calculations to be between 1–2 kcal/mol. Several of the systematic errors are evaluated to be about 0.5 kcal/mol and hence to lie well within the overall statistical error of the calculation. Systematic errors stemming from shortcoming of the potential function may also be of consequence, but have not been evaluated here.

SPECIFIC CONTRIBUTIONS TO THE FREE ENERGY DIFFERENCE

Decomposition of the computed free energy difference

An advantage of eq (3) is that the total free energy change can be expressed as a sum of various contributions involving the interactions of the mutated side chain (Gao et al., 1989; Tidor & Karplus, 1991). It is thus possible to separate contributions from the solvent and the protein, or to determine the contributions of different types of terms in the energy function. These contributions are however not state functions and their value may vary with the particular pathway taken between the initial and final states. These values are therefore subject to a larger uncertainty than the overall free energy differences and concepts that arise from their analysis must therefore be confirmed by independent computations whenever possible.

Here we discuss the decomposition results for the larger Ile96→Ala change. In this case, we chose to write ΔG as a sum of four terms:

$$\Delta G = \Delta G_{cs} + \Delta G_{nbs} + \Delta G_{ci} + \Delta G_{nbi} \tag{7}$$

Each of these terms arises from specific contributions to ΔV, which are illustrated in Fig. 3.

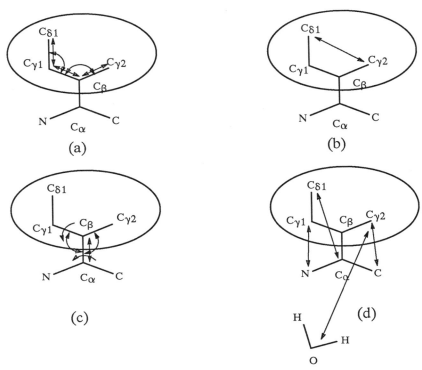

Fig. 3 Illustration of the terms included in the different free energy components of eq (7).

(a) contributions to ΔG_{cs} which comprise covalent terms consisting of bond stretching, valence angle bending. The specific bonds and angles involved are depicted by arrows. (b) contributions to ΔG_{nbs}, the intra-sidechain non bonded term, which are limited to the van der Waals interaction between the $C_{\delta 1}$ and $C_{\gamma 2}$ atoms. (c) and (d) contributions to the last two terms in eq (7). They concern interactions of atoms belonging to the Ile sidechain with the atoms of the remainder of the protein and/or the solvent. Terms included in ΔG_{ci} are illustrated in (c). These comprise covalent components – bond stretching, and valence angle bending – as well as one torsion, as depicted by the arrows. (d) depicts the non bonded terms that contribute to ΔG_{nbi}, which are limited to van der Waals interaction. Equivalent drawings could be made for the Val and Ala sidechains with the exclusion of (a) for Ala since it contains only a C_β atom and lacks the self–energy components.

Table 2 lists the different components of the computed free energy differences in the folded, unfolded and reference states. Note that the free energy components considered here being limited to non–bonded and vibrational terms, contain no Coulombic contributions. The latter were shown to have a much larger path dependence than the former two terms in calculations performed by Shi et al. (1993).

Table 2 Contributions to the computed free energy changes (in kcal/mol) for the Ile→Ala mutations in barnase from different terms of the potential energy.

Contribution	$\Delta G_f^{I\rightarrow A}$ protein	$\Delta G_u^{I\rightarrow A}$ water	$\Delta G_r^{I\rightarrow A}$ reference	$\Delta\Delta G_{f\rightarrow u}$	$\Delta\Delta G_{r\rightarrow u}$	$\Delta\Delta G_{r\rightarrow f}$
c s	−3.28	−4.73	−3.18	−1.45	−1.55	−0.10
c i	−1.40	−3.09	−5.07	−1.69	1.98	3.67
nbs	−0.97	−0.79	−0.24	0.18	−0.55	−0.71
nbi	2.56	0.31	1.35	−2.25	−1.04	1.21
TI(total)	−3.09	−8.3	−7.15	−5.21	−1.15	4.06

The different free energy contributions considered are listed in column 1. 'cs' stands for contributions from side chain covalent terms, 'ci' for covalent interaction terms. 'nbs' represents contributions from non bonded terms that involve only the mutated side chain atoms, and 'nbi' represents non bonded interactions between atoms of the side chain and those of the protein or solvent environments. A more detailed description of the potential energy terms included in each of these contributions is illustrated in Fig. 3. For other details see legend of Table 1.

We find that important contributions to the computed free energy difference $\Delta\Delta G_{f\rightarrow u}$ between the wild type (Ile) and mutant (Ala) proteins arise from the *non-bonded interaction* term $\Delta\Delta G_{nbi}$ (−2.25 kcal/mol) and from the covalent terms, $\Delta\Delta G_{ci}$ (−1.45 kcal/mol) and $\Delta\Delta G_{cs}$ (−1.65 kcal/mol), while that of the *non-bonded side chain* term $\Delta\Delta G_{nbs}$ (−0.18 kcal/mol), is negligible.

The computed contributions of the *non bonded interaction term*, ΔG_{nbi}, which in this case are entirely due to van der Waals interactions, are unfavorable for the mutant relative to the wild type, in both the folded and unfolded states (columns 1 and 2 in Table 2). However, the unfavorable effect in the unfolded state is marginal, leading to an overall negative contribution to $\Delta\Delta G_{nbi}$ which is favorable to Ile in the native state.

To gain further insight into the origins of these contributions, control calculations were performed, and several additional aspects of the Ile→Ala simulations were analyzed (see Prévost et al, 1991 and 1993 for details). These suggested that the replacement of Ile96 by Ala in the core of barnase results in a substantial loss of non-bonded enthalpy, which is partially compensated by a gain in conformational entropy. A central argument to these suggestions has been the finding that a cavity is formed around residue 96 in the Ala containing simulated mutant structure. As a result, the packing around this residue should be loosened, causing some loss in non-bonded interactions in the mutant protein relative to the wild-type, while leading, at the same time, to favorable entropic contributions due to increased local flexibility.

The other term contributing substantially to the computed free energy change is the *side chain covalent* term ($\Delta\Delta G_{cs}$). ΔG_{cs} is large and favors the mutant in both the folded and unfolded states. This is expected since the covalent motional contributions are always positive (destabilizing). To obtain an independent check on these results, an estimate was made of the expected free energy due to intra-sidechain vibrational terms. This was done using a classical harmonic model for 1-butane in the gas phase, taken to represent the Ile side chain (the value for the single atom representing the Ala side chain being zero). It yielded a value of −3.66 kcal/mol for the Ile side chain, representing essentially the vibrational enthalpy of the six classical internal degrees of freedom. The sum of ΔG_{cs} and ΔG_{nbs} in the Ile→Ala transformation in the gas phase reference state computed in the present simulations yields a very similar value (–3.42 kcal/mol), an indication that in the gas phase extended heptapeptide, the decomposition yields reasonable results for these contributions and that the intrinsic vibrational properties of the Ile side chain are well preserved in this phase. In the folded and solvated unfolded states on the other hand, the same sum is somewhat more negative. The origin of these differences, which make significant contributions to $\Delta\Delta G_{f\rightarrow u}$ (see Table 2) is not clear.

The free energy contributions from the *covalent interaction* term ΔG_{ci}, also favor the mutant in both the unfolded and folded states. But this favorable effect is smaller in the folded state, leading here too to a negative $\Delta\Delta G_{ci}$ contribution (see Table 2). Somewhat surprisingly, the ΔG_{ci} computed for Ile→Ala in the gas phase reference state is particularly destabilizing for Ile. Tentative rationalizations of these results have been proposed (Prévost et al., 1991, 1993). They invoked strain caused by water molecules surrounding the larger non–polar side chain in the unfolded state, and by strong side–chain– backbone interactions in the gas phase. Uncertainties due to the qualitative nature of the decomposition results may however also be at play.

DISCUSSION

The simulations of the substitution of Ile96 by Ala and Val in barnase, yield free energy changes that are in satisfactory agreement with experimental measures of protein stability changes. They also yield fair estimates of the difference of solvation free energy between Ile and Ala or Val, though here the computed values are systematically larger than the experimental measures.

Analysis of these results in terms of individual contributions to the computed free energy changes provides insight into the origins of the observed effects. It shows that covalent intra–sidechain terms, and non–bonded interaction terms provide major contributions to the free energy change associated with the alchemical transformation in all three considered states, the folded protein, the unfolded solvated peptide and the gas phase reference state. More importantly even, it shows that these contributions differ in different environments and hence contribute to the overall free energy balance. Hence, non bonded interactions, and possibly also vibrational terms seem to contribute to what is commonly referred to as hydrophobic stabilization of proteins.

In previous evaluations of free energy differences (Singh et al., 1987), the contributions from covalent terms within the group undergoing transformation was neglected on the basis that such contributions do not change in different environments (e.g. the protein and water) and therefore would cancel out when applying the thermodynamic cycle. More recent studies (Gao et al., 1989; Pearlman & Kollman, 1991) have on the other hand recognized the importance of these terms, in agreement with the results described here. Due to the qualitative nature of the decomposition results, and to computational uncertainties, the actual values of the covalent free energy contributions discussed here, may be in error, but the findings that they do play a role are likely to stand. Calculations on model systems using different solvent environment and perturbation pathways, combined with experimental analyses of their vibrational properties, should be helpful in obtaining a more quantitative picture.

The importance of the contributions from terms representing non–bonded interactions of the side chain with its environment has also been emphasized by Kellis et al. (1988), in interpreting their experimental results on the hydrophobic mutations of barnase. Assuming that experimental solvation free energies (which corresponds to

$\Delta\Delta G_{r\to u}$, computed here) represent contributions from non bonded interactions in the unfolded state, and that covalent terms can be neglected (because they cancel out), these authors concluded that the non bonded contribution from the alchemical transformation in the folded protein is the dominant term for this case, while that from the alchemical transformation in the unfolded state is negligible. This conclusion fits well with the calculation results described here. Note however that the latter contend that only part of $\Delta\Delta G_{f\to u}$ arises from non bonded interactions, while the remainder is due to covalent terms. Here too the computed values may be in error, for reasons stated above, and further analysis is needed to obtain a more quantitative picture. Such analysis will, among other things, require detailed knowledge of the 3D structure of the mutant protein, which is presently unavailable.

ACKNOWLEDGEMENTS

This work was carried out in collaboration with Martin Karplus (Chemistry Department, Harvard University, USA) and Bruce Tidor (now at Whitehead Institute, Nine Cambridge Center, USA). Martine Prévost is a research associate of the National Fund for Scientific Research (Belgium). This study was supported in part by the European Communities BRIDGE programme (project BIOT–CT91–0270).

REFERENCES

Beveridge, D.L., and DiCapua, F.M., 1989, *Ann. Rev. Biophys. Chem.* 18:431–492.

Brooks, B. R., Bruccoleri, R. E., Olafson, D., States, D., Swaminathan, S., and Karplus M., 1983, *J. Comp. Chem.* 4:187–217.

Brunger, A.T., Brooks III, C.L., and Karplus M., 1985, *Proc. Natl. Acad. Sci. U.S.A.* 82:8458–8462.

Fersht, A.R., Matouschek, A., and Serrano, L., 1992, *J. Mol. Biol.* 224:771–782.

Gao, J., Kuczera, K, Tidor, B., and Karplus, M., 1989, *Science* 244:1069–1072.

Hartley, R. W., 1968, *Biochemistry* 7:2401–2408.

Jorgensen, W. L., Chandrasekhar, J., Madura, J. D., Impey, R. W., and Klein, M. L., 1983, *J. Chem. Phys.* 79:926–935.

Kellis Jr, J.T., Nyberg, K., Sali, D., and Ferhst, A.R., 1988, *Nature* 333:784–786.

Kellis Jr., J.T., Nyberg, K., and Fersht, A.R., 1989, *Biochemistry* 28:4914–4922.

Kirkwood, J. G., 1935, *J. Chem. Phys.* 3:300–313.

Matouschek, A., Kellis Jr, J. T., Serrano L., and Fersht, A. R., 1989, *Nature* 340:122–126.

Mauguen, Y., Hartley, R.W., Dodson, G.G., Dodson, E.J., Bricogne, G., Chothia, C., Jack, A., 1982, *Nature* 297:162–164.

Pearlman, D. A., and Kollman, P. A., 1991, *J. Chem. Phys.* 94:4532–4545.

Prévost, M., Wodak, S.J., Tidor, B., and Karplus, M., 1991, *Proc. Natl. Acad. Scie. USA* 88:10880–10884.

Prévost, M., Wodak, S.J., Tidor, B., and Karplus, M., 1993 (submitted)

Shi, Y.-y., Mark A. E., Cun–Xin W., Fuhua H., Berendsen H. J. C., and van Gunsteren W. F., 1993, *Protein Engineering* 6:289–295.

Singh, U. C., Brown, F. K., Bash, P. A., and Kollman, P. A., 1987, *J. Am. Chem. Soc.* 109:1607–1614.

Straatsma, T. P., and McCammon, J. A., 1989, *J. Chem. Phys.* 90:3300–3304.

Tidor, B., and Karplus, M., 1991, *Biochemistry* 30:3217–3228.

Wade, R. C., and McCammon, J. A., 1992, *J. Mol. Biol.* 225:679–696 and 697–712.

Wolfenden, R., Andersson, L., Cullis, P. M., and Southgate, C. C. B., 1981, *Biochemistry* 20:849–855.

Wong, F. C., and McCammon, J. A., 1986, *in* "Structure, Dynamics and Function of Biomolecules", A. Ehrenberg, A. et al., eds., Springer, Berlin, pp. 51–55.

Zwanzig, R. W., 1954, *J. Chem. Phys* 22:1420–1426.

MONTE CARLO SIMULATION OF ELECTROSTATIC INTERACTIONS IN BIOMOLECULES

Bo Jönsson and Bo Svensson

Physical Chemistry 2
Chemical Centre
POB 124 S-22100 Lund, Sweden

INTRODUCTION

The function of a protein invariably involves the binding or interaction with a second molecule or aggregate. Regardless of whether the interaction is specific or not, it constitutes an essential part of protein function. An accurate description and a deeper understanding of any such process is of great value. In biology, the type of interaction we ultimately want to describe covers a wide range from large protein-DNA complexes to simple titration curves of a single amino acid residue. The Monte Carlo (MC) simulation approach we will describe here is applicable to a wide range of such protein complexes, but it is limited to the electrostatic component of the interaction. Here we will use the binding of small atomic ions as illustrative examples of the technique. The simulation technique is, due to its simplicity and efficiency, also applicable to electrostatic interactions in larger and more complex systems.

The binding of a charged substrate to a protein in solution clearly also involves non-electrostatic interactions, which unfortunately requires a detailed knowledge of the atomic interactions. One way to circumvent this problem, at the expense of only treating the long range Coulombic interactions, is by approximating the solvent with a dielectric continuum and then scale all electrostatic interactions with the appropriate dielectric constant. This, so-called *primitive model*[1], has long been the main approach in electrolyte theory and also in more

Statistical Mechanics, Protein Structure, and Protein Substrate Interactions
Edited by S. Doniach, Plenum Press, New York, 1994

complex systems found in colloid and surface science - i.e. electrical double layers[2-3]. The primitive model has also been applied to proteins in solution, but in order to be tractable it had to include different types of approximations[4-10].

Our aim is to show that the primitive model, with a discrete representation of all protein atoms and net charges, combined with MC simulations provide an accurate and very efficient route for probing electrostatic interactions in proteins and biomolecules in general. It is more applicable to systems where the dominant interaction is of electrostatic nature - in general, it will be relatively more accurate the stronger the electrostatic interactions are.

MODEL

As the protein molecule will exclude solvent molecules from its interior, the interesting question arises as to what one should choose for the permittivity of the interior region. A seemingly viable choice is the electronic permittivity[11]. However, this may lead to an underestimation as the assumption of a fixed protein structure precludes dielectric response from nuclear motions. Furthermore, it is not apparent where one should locate the dielectric boundary between protein and solvent. The presence of a discontinuity severely complicates the electrostatic interactions and, in order to preserve simplicity, we have assumed no discontinuity in the model. That is, the interior of the biomolecule is assumed to have the same dielectric constant as the solvent.

The explicit particles are divided into two categories: The first contains those which are fixed, while the second consists of mobile species. Mobility implies that a thermal average is to be calculated over these particles. Typically, the first includes the atoms of a macromolecule with fixed structure, while counterions and added salt constitute the second set. Only electrostatic and hard core interactions are explicitly considered while the solvent is treated as a structureless continuum, scaling the electrostatic interactions with its dielectric permittivity, ε_r. Each atom or particle is assigned a hard core diameter, σ_i and a net charge, q_i. Thus the interaction between two species i and j is,

$$u^{el}(r_{ij}) = q_i q_j / 4\pi\varepsilon_0\varepsilon_r r_{ij} \qquad r_{ij} \geq (\sigma_i + \sigma_j)/2 \qquad (1a)$$

$$u^{hc}(r_{ij}) = \infty \qquad r_{ij} < (\sigma_i + \sigma_j)/2 \qquad (1b)$$

where ε_0 is the permittivity of vacuum and r_{ij} the distance between particles i and j.

Figure 1 shows the entire system enclosed within a spherical cavity or cell. Typically the number of mobile ions needed to simulate a protein of 100 residues at physiological ionic strength is in the order of 100. We estimate that a simulation of a biomolecule consisting of 100,000 atoms would take a day. With an explicit evaluation of all interactions the

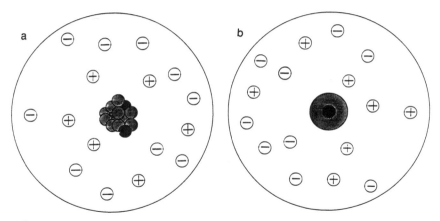

Figure 1. a) Protein model used in the MC simulations and in the mean field theory. Salt particles are shown as open spheres with plus or minus sign and protein atoms are shaded. Only a fraction of the protein atoms included in the calculations are shown. The binding sites are indicated as black spheres. b) Same as a) but for the chelators 5,5'-Br₂BAPTA and quin-2. The chelator is assumed spherical and is shown as a shaded sphere with the binding site at its centre(black).

computation time only increases linearly with the number of aggregate atoms. Partially treating the long range interaction via a multipole expansion would further reduce the size dependence in the computing time.

In the protein simulations we used coordinates from the Brookhaven data bank, which were obtained from x-ray diffraction studies of crystalline Ca^{2+} loaded calbindin[12-13] and calmodulin[14-16], respectively. The total number of protein atoms included in the calbindin and calmodulin simulations were 722 and 1166, respectively. The calmodulin fragments included amino acid residues 1 to 74 and 78 to 148, respectively and their coordinates were taken to be the same as for the intact protein. Each protein atom was represented as a hard sphere, impenetrable to any solvent ions. The atomic diameters in calbindin were chosen as 3.6, 2.0, 3.3 and 3.2 Å for C, H, N and O, respectively, whereas in calmodulin and its fragments all diameters (no hydrogens were included) were 4 Å.

Several different mutants of calbindin were studied. In each of these, one or more of the negatively charged amino acids glutamate (E) and aspartate (D) were replaced by the neutral glutamine (Q) and asparagine (N), respectively (with this notation E17Q means that at residue number 17, a negatively charged glutamate has been replaced by a neutral glutamine). Negatively charged carboxylic oxygens were given a charge of -0.5, while positively charged lysine residues carried a positive unit charge on the ζ-nitrogen. All other atoms were kept neutral. In this model a mutation of glutamate and aspartate was done in a straightforward way by simply adding +0.5 unit charge to the negatively charged oxygen atoms in order to change the carboxylic group into an "amide group".

When simulating the small chelators, 5,5'-Br₂BAPTA and quin-2, an even simpler model was used. The chelator was modelled as one single charged hard sphere with a radius of 14 Å and a net charge of -4e located at the binding site in the center.

The protein or aggregate was placed in the centre of a spherical cell, to which counterions and salt ions was added. These were treated as mobile charged hard spheres confined in the cell. The hard core diameter of the ions, including calcium, was either 4.25 Å (calbindin) or 4.0 Å (calmodulin). All simulations were performed in the canonical ensemble using the algorithm due to Metropolis et al.[17].

THEORY

Consider the binding of a charged ligand M^+ to a protein P. The stoichiometric binding constants for the processes,

$$Pr + M^+ \rightarrow PrM^+ \quad \text{or} \quad Pr + nM^+ \rightarrow PrM_n^{n+}$$

are given by,

$$K_1 = \frac{[PrM^+]}{[Pr]\,[M^+]} \quad \text{and} \quad K_n = \frac{[PrM_n^{n+}]}{[Pr]\,[M^+]^n} \tag{2}$$

where $[Pr]$ and $[M^+]$ are the concentrations of free protein and ligand respectively, and $[PrM^+]$ and $[PrM_n^{n+}]$ are the concentrations of protein with one and n bound ligands, respectively. The binding constant can also be represented here as,

$$pK = -^{10}\log K = \beta \Delta G / \ln 10 \tag{3}$$

where $\beta = 1/k_B T$ and ΔG is the change in free energy for the binding process.

There are many contributions to the free energy change associated with the binding process. In this work we are concerned only with the electrostatic contribution to it, denoted by ΔG_{el}. The hypothesis is that this is the dominating contribution to $\Delta\Delta G$ and that other contributions, including structural changes, are either small or independent of added salt, mutations or other charge changes in the protein. This permits us to calculate *relative* changes in the binding constants with respect to a chosen reference state using,

$$K/K^{ref} = \exp[-\beta(\Delta G_{el} - \Delta G_{el}^{ref})] \tag{4}$$

We can write the electrostatic free energy difference as

$$\Delta G_{el} = \mu^{ex}(M^+, nB) - n\mu^{ex}(M^+, F) \tag{5}$$

where $\mu^{ex}(M^+, nB)$ is the excess chemical potential of the n bound ligands M^+ and, $\mu^{ex}(M^+, F)$, the same quantity for a free M^+. Thus, our aim is to calculate ΔG_{el} via MC simulations for several salt concentrations and for a range of differently charged proteins. The latter can be obtained experimentally by mutations or by changing pH.

We use Monte Carlo simulations to obtain these excess chemical potentials, assuming a fixed protein or aggregate structure. This should be a reasonable approximation if the structural fluctuations of the bound protein do not lead to large variations in the Coulombic potential.

The excess chemical potential of a bound ion was obtained by introducing the test particle at the binding site, as specified in the crystal structure. Widom's method is a convenient way to calculate chemical potentials in a canonical MC simulation, which works well for systems that are not too dense[18-19]. The excess free energy, or excess chemical potential, associated with a perturbing charge at a given position reads,

$$\beta\mu^{ex}(q, \mathbf{r}) = -\ln< \exp[-\beta \{q\Phi(\mathbf{r}) + u^{hc}(\mathbf{r})\}] > \qquad (6)$$

$\Phi(\mathbf{r})$ is the electrostatic potential at position $\mathbf{r}$, q is the valency of the hypothetical charge and $u^{hc}(\mathbf{r})$ takes care of any possible hard core overlap. The angular brackets denote an ensemble average over the unperturbed system. Both the positions, $\mathbf{r}$, and the valency q are input parameters. It is also possible to use more complicated perturbation potentials, corresponding to the simultaneous insertion of several charged particles.

RESULTS AND DISCUSSION

Calcium Binding Chelators

Table 1 summarizes the binding constant shifts obtained from experiment and theory on the two chelators. The experimental shifts for the two chelators are identical within the uncertainties, which indicate that from an electrostatic point of view they behave very similarly, although their covalent structure differs. These results are in agreement with Pethig et al.[20], who compiled salt shifts in the Mg^{2+} binding of a dozen different BAPTA analogues, which all showed very similar salt behaviour. The theoretical and experimental shifts are in excellent agreement, despite the simplicity of the model.

The excellent agreement between experiment and theory depends of course on the choice of the different model parameters. However, the results are surprisingly insensitive to the choice of radii for the small ions, the calcium ion and the chelator. For example, ΔpK between 2 and 150 mM KCl changes from 1.29 to 1.14 if the chelator diameter is increased from 14 to 20 Å. Decreasing the diameter to 10 Å increases the shift to 1.48 pK units. Changing the chelator diameter may also to some degree be compensated by a change in the size of the

Table 1. Theoretical and experimental binding constant shifts for the calcium chelators. The chelator in 2 mM KCl has been taken as reference. [a] From reference [21].

c_{KCl} (mM)	MC	Exp.[a] BAPTA	quin-2
2	0.00	0.00	0.00
10	0.32	0.26	-
25	0.60	0.64	-
50	0.85	0.89	0.88
150	1.29	1.36	1.35
500	1.79	1.77	1.72
700	1.93	1.86	1.82
1000	2.05	1.97	1.94

calcium ion. Keeping the chelator diameter at 14 Å and increasing the diameter of Ca^{2+} to unrealistical 8 Å gives a shift of 1.15 pK units. Similarly, a change of the salt ion radii would hardly be noticeable, unless they were chosen to be very small or very large. One reason for the insensitivity is that the results are dominated by the long range electrostatic interactions and the hard core or excluded volume effects play only minor roles. This is also likely to be the reason for the nearly identical salt dependence of $5,5'\text{-Br}_2\text{BAPTA}$ and quin-2 seen in the experiment. It should be pointed out that the chemical potentials and pK shifts becomes more model dependent at increased salt concentration.

In the above calculations we have assumed a uniform dielectric permittivity equal to that of pure water. We have investigated what effect a low dielectric permittivity, $\varepsilon_r = 2$ inside the chelator, would have on the shift. In this case the electrostatic interactions become more complicated than eq.(1) due to the reaction field contribution from the low dielectric sphere. Details about simulations of electrolytes in such systems has been given elsewhere[22-23]. Although a rather low value of the interior dielectric constant was chosen, the effect on the shift was found to be very small. For a chelator with a diameter of 14 Å in 0.15 M KCl, we find a shift of 1.27 relative to the 2 mM reference solution, which should be compared to 1.29 for a uniform dielectric permittivity. This is a typical result as long as the charges are far from the dielectric discontinuity.

Calbindin

We now turn to the more complex situation with two calcium ions binding to the protein calbindin, for which the apo form has a net charge of -8e. There is a wealth of experimental data available for the binding constant of many different mutants both at varying salt concentration and salt valency. Table 2 shows the binding constant shifts obtained from MC simulations together with available experimental results. In the presence of a univalent salt such as KCl, we see a uniformly good agreement. The largest absolute deviation is half a pK unit. The available experimental data for K_2SO_4 solution show an almost equally good agreement, see Table 3.

Table 2. Theoretical and experimental binding constant shifts for different mutants of calbindin at various KCl concentrations. The wild-type protein in 2 mM KCl has been taken as reference. [a] From references [24-25].

c_{KCl} (mM)	MC	Exp.[a]	
2	-		wild-type
50	2.41	2.9	
100	3.08	3.6	
150	3.42	4.0	
2	1.15	1.2	E17Q
50	3.25	3.4	
100	3.87	4.0	
150	4.22	4.2	
2	2.51	2.4	E17Q+D19N
50	4.25	4.1	
100	4.79	4.6	
150	5.08	4.9	
2	3.24	3.4	E17Q+D19N+E26Q
50	4.68	4.8	
100	5.16	5.3	
150	5.42	5.4	

Table 3. Theoretical and experimental binding constant shifts for different mutants of calbindin at various K_2SO_4 concentrations. The wild-type protein in 2 mM KCl has been taken as reference. [a] From references [26].

c_{Salt} (mM)	MC	Exp.[a]	
1	1.52	-	E17Q
10	2.76	3.31	
50	4.20	4.56	
100	4.83	5.35	
1	3.53	-	E17Q+D19N+E26Q
10	4.43	4.86	
50	5.49	5.80	
100	5.98	6.13	

Simpler Models - Screened Coulomb Potential

Sometimes one can simplify the primitive model further by performing a partial average over the freely moving salt particles. This is particularly relevant in a situation where highly charged aggregates are interacting in a solution of small salt particles. In this case the average potential between aggregates often takes the form of a screened Coulomb interaction, greatly simplifying the description. It is of some interest to study the result a simple screened Coulomb interaction would predict in the present case, in particular as it has been employed for similar studies in the past[27]. Thus let us assume that the Ca^{2+}-protein and Ca^{2+}-Ca^{2+} interaction is described by,

$$u(r_{ij}) = q_i q_j e^2 \exp(-\kappa r_{ij})/4\pi\varepsilon_0\varepsilon_r r_{ij} \qquad\qquad (10)$$

where the mobile ions now only enters through κ, the Debye-Hückel inverse screening length. For mutant E17Q in a 0.05 M salt solution eq.(10) gives a pK shift of 4.5, which is about 1.6 units larger than experimentally observed. In 0.15 M KCl solution the discrepancy becomes even larger, 2.6 units. Obviously, the screening predicted by eq.(10) is much too large and by applying the simple screened Coulomb interaction one has introduced too severe an approximation. The failure is probably due to the fact that it is derived under the assumption that salt particles may penetrate the protein interior, that is, the screening is assumed to be isotropic. In reality the protein excludes salt particles and the screening between protein charges and Ca^{2+} ions is not as effective.

Simpler Models - Multipole Expansion

As was mentioned above, the detailed and simplified charge distributions on calbindin lead to similar pK shifts. From a computational point of view it is of interest to see to what extent the protein shape and charge distribution can be further simplified without affecting the binding free energies. In the first step we replaced the protein charge distribution with a net charge at its centre. This changed the calculated pK shifts by almost one pK unit and is obviously a too crude model to be used. In the next attempt we also added a dipole moment at the protein centre while keeping the protein shape defined by the uncharged atoms. In the final step we also replaced the protein with an equivalent sphere of the same volume with a net charge and a dipole moment at its centre. These two latter models changed the pK shifts by less than 0.2 units, which is well within the experimental uncertainties in most systems.

These test simulations demonstrate the possibility for simplifications of the protein model. This would allow us to calculate free energy changes at a modest expense for very large biomolecules.

Calmodulin and its Fragments

The experimental calcium binding constants for the two tryptic fragments of calmodulin have been determined by Linse et al.[28] using the same technique as for calbindin. The net charge of apo calmodulin is -23e, while that of the fragments is -12e and -13e, respectively. Both fragments show a high affinity for calcium ions and the binding constants are sensitive to addition of KCl. Table 4 shows a comparison of experimental and theoretical binding constant shifts using the 2 mM KCl solution as reference. There is a good agreement between experiment and theory for the second fragment, which happens to be where the experimental results are most accurate. The comparison for the first fragment is less favourable, although there is agreement within the combined experimental and theoretical uncertainties. The salt concentration for one of the data points given by Linse et al. is only specified as "low". The actual concentration is most likely less than 10 mM (Linse, personal communication). Thus,

Table 4. Theoretical and experimental binding constant shifts for the two calmodulin fragments TR_1C and TR_2C at a concentration of 20 µM. The binding constants at 2 mM are taken as references. The experimental uncertainty is ±0.6 and ±0.2 for TR_1C and TR_2C, respectively. The theoretical numbers are significant to within 0.1 of a pK unit.[a] From references [34]. [b] The lowest salt concentration is not accurately known in the experiments, because of the contributions from buffer and residual salt in the protein solution.

c_{KCl} (mM)	TR_1C		TR_2C	
	Theory	Exp.[a]	Theory	Exp.[a]
150	4.0	3.5	3.9	3.7
100	3.5	3.3	3.5	3.2
50	2.7	2.9	2.7	2.6
10	1.0	-	1.1	-
2 [b]	0.0	0.0	0.0	0.0
0	-0.4	-	-0.3	-

the overall agreement going from 150 mM to "low" KCl concentration is fairly good, being off by a few tenths of a pK unit. To specify the salt concentration at the millimolar level in a biomolecular solution is a notorious problem, which hampers the comparison between theory and experiment.

Table 5 contains a similar comparison for intact calmodulin. Calmodulin has a very high net charge, which makes it a really severe test on theory. There is a qualitative agreement and a few points are in very good agreement, but there is a discrepancy in shifts when going from 50 to 100 mM. At present we would be inclined to ascribe this difference as being due to titrating groups on the protein, which is not taken into account in our canonical simulations.

Table 5. Theoretical and experimental binding constant shifts for calmodulin at a concentration of 20 µM. The binding constant at 2 mM is taken as reference. The experimental uncertainty is approximately ±0.5 and the theoretical numbers are significant to within 0.1 of a pK unit. [a] From references [29]. [b] The lowest salt concentration is not accurately known in the experiments, because of the contributions from buffer and residual salt in the protein solution.

c_{KCl} (mM)	ΔpK	
	Theory	Exp.[a]
150	7.1	6.7
100	6.1	6.1
50	4.3	5.3
25	2.8	3.5
10	1.2	1.9
5	0.3	-
2 [b]	0.0	0.0
0	0.4	-

Note that the binding constant initially goes up when the salt concentration increases from 0 to 2 mM. This is due to a change of the free calcium chemical potential[29] and experimentally it has also been seen in DNA solution[30].

CONCLUSIONS

In the present model we have assumed a uniform dielectric permittivity, which is obviously incorrect - the protein interior does not have the same dielectric response as pure water. Still we see a good agreement between theory and experiment in systems involving substantial changes in the electrostatic interactions - e.g. in calmodulin the binding process involves a charge change of 8e and in calbindin a change of 4e. The agreement might be fortuitous, but in the case of calbindin the comparison has been extended to a number of mutants and salt of differrent valency with the same result. The model with a uniform dielectric permittivity does also seem to work for small chelators and here the experimental results of Pethig et al.[20] gives further support to the model. With a dielectric boundary, the results will be extremely sensitive to the exact location of the boundary between the low and high permittivity region. Experimentally we have no guidance to where to place this boundary. It is also obvious that the low dielectric region will only play a role as long as its permittivity is much smaller than that of water, i.e. $<<80$. We conclude that the relevant parts of the protein show a reasonably high dielectric response and it is only the core of the protein that may have a low permittivity.

Recently, the Poisson-Boltzmann (PB) and linearized PB approximations have been numerically solved for protein models similar to that described here. Application of these approximations to biomolecular systems has led some authors to emphasize the importance of the low dielectric interior of biomolecules. It must be remembered, however, that the PB equation is an approximation and its accuracy detoriates when ionic correlations increase, as they are likely to do when dielectric discontinuities are introduced.

Monte Carlo simulations using the present model are extremely fast and reliable numbers are obtained from simulations requiring less than a minute of computer time on an ordinary work station. Further advantage comes from the perturbation procedure, where any number of perturbations can be calculated during one single simulation at marginal cost. That is, one simulation is enough to obtain the chemical potential for a bound ion for a large number of mutants.

REFERENCES

1. Friedman, H. L. (1962) Ionic Solution Theory, Interscience Publishers, New York.
2. Derjaguin, B. V.; Landau, L., Acta Phys. Chim URSS, 14 (1941) 633.
3. Verwey, E. J. W.; Overbeek, J. Th. G. (1948) Theory of the Stability of Lyophobic Colloids; Elsevier: Amsterdam.
4. Tanford, C. & Kirkwood, J. G., J. Am. Chem. Soc., 79 (1957) 5333.
5. Tanford, C., J. Am. Chem. Soc. 79, 5340.
6. Marcus, R. A. (1955) J. Chem. Phys., 23 (1957) 1057.
7. Gilson, M. K. & Honig, B. H., Nature, 330 (1987) 84-86.
8. Juffer, A. H., Botta, E. F. F., van Keulen, B. A. H., van der Ploeg, A. & Berendsen, H. J., Comp. Phys., 97 (1991) 144.
9. Sharp, K. & Honig, B., Chem. Scr., 29A (1989) 71; and references therein.
10. Bashford, D. & Karplus, M., Biochemistry, 29 (1990) 10219.
11. Harvey, S. C., Proteins, 5 (1989) 78.
12. Szebenyi, D.M.E., Obendorf, S. K. & Moffat, K., Nature, 294 (1981) 327.

13. Szebenyi, D.M.E. & Moffat, K., J. Biol. Chem., 261 (1986) 8761.
14. Babu, Y. S., Sack, J. S., Greenhough, T. G., Bugg, C. E., Means, A. R. & Cook, W. J., Nature, 315 (1985) 37.
15. Babu, Y. S., Bugg, C. E. & Cook, W. J., J. Mol. Biol., 204 (1988) 191.
16. Kretsinger, R. H. & Weissman, L. J., J. Inorg. Chem., 28 (1986) 289.
17. Metropolis, N., Rosenbluth, A.W., Rosenbluth, M.N., Teller, A.H., & Teller, E., J. Chem. Phys., 21 (1953) 1087.
18. Widom, B., J. Chem. Phys., 39 (1963) 2808.
19. Svensson, B., & Woodward C., Molec. Phys., 64 (1988) 247.
20. Pethig, R.; Kuhn, M.; Payne, R.; Adler, E.; Chen, T.-H.; Jaffe, L. F., Cell Calcium, 10 (1989) 491.
21. Svensson, B., Jönsson, Bo, Fushiki, M. & Linse, S., J. Phys. Chem., 96 (1992) 3135.
22. Friedman, H. L., Mol. Phys., 29 (1975) 1533.
23. Linse, P., J. Phys. Chem., 90 (1986) 6821.
24. Linse, S., Brodin, P., Johansson, C., Thulin, E., Grundström, T. & Forsén, S., Nature, 335 (1988) 651.
25. Linse, S., Johansson, C., Brodin, P., Grundström, T., Drakenberg, T., Forsén, S., Biochemistry, 30 (1991) 154.
26. Svensson, B., Jönsson, Bo, & Woodward, C. E. & Linse, S., Biochemistry, 30 (1991) 5209.
27. Waugh, D. F., Slattery, C. W. & Creamer, L. K., Biochemistry, 10 (1971) 817.
28. Linse, S., Helmersson, A. & Forsén, S., J. Biol. Chem., 266 (1991) 8050.
29. Svensson, B., Jönsson, Bo, Woodward, C. E. & Thulin, E., Biochemistry, 32 (1993) 2828.
30. Honig, B., Personal communication.

THE MOLECULAR ORIGIN OF THE LARGE

ENTROPIES OF HYDROPHOBIC HYDRATION

Themis Lazaridis[#] and Michael E. Paulaitis [*]

Center for Molecular and Engineering Thermodynamics
Department of Chemical Engineering
University of Delaware
Newark, DE 19716 (USA)

[#]Present address: Department of Chemistry
Harvard University
Cambridge, MA 02138
[*]Author to whom correspondence should be addressed

ABSTRACT

A rigorous statistical mechanical expansion for the entropy in terms of multiparticle correlation functions is used to identify the major contributions to the entropy of hydration of simple, hydrophobic solutes. With a factorization assumption for the solute-water correlation function we have been able to separate the contributions to the entropy of hydration due to translational and orientational solute-water correlations. The required correlation functions are obtained by Monte Carlo simulation. This approach is applied to infinitely dilute solutions of methane, inert gases, and model Lennard-Jones particles in water in order to examine the dependence of the solute size and the solute-water pair interaction energy on the hydration entropy. Solute-water orientational correlations are found to decay linearly with intermolecular distance and virtually vanish in the second hydration shell. The orientational and translational contributions are comparable in magnitude. The primary factor determining the solute-water correlation entropy is found to be solute size, with the pair interaction energy playing a secondary, yet significant role. The dependence of the orientational entropy on the curvature of the solute surface is discussed in connection with enthalpy-entropy compensation phenomena. We find that the large entropies and heat capacities of hydrophobic hydration are well accounted for by solute-water correlations alone and that large perturbations in water structure are not required to explain hydrophobic behavior at room temperature.

INTRODUCTION

The role of water in stabilizing the structure and mediating the interactions of biological macromolecules is now well appreciated. The interaction of protein groups with water is

Statistical Mechanics, Protein Structure, and Protein Substrate Interactions
Edited by S. Doniach, Plenum Press, New York, 1994

87

commonly studied through thermodynamic measurements on the dissolution of small molecules in water. However, the molecular interpretation of these thermodynamic measurements is still at a rather primitive stage, with the vague concept of "water structure" frequently invoked to explain the observed enthalpy and entropy changes. Yet, understanding the molecular origin of the observed thermodynamic behavior is essential for a detailed analysis of the complex, cooperative processes that take place in protein folding and molecular recognition. Here we attempt to provide a rigorous interpretation of the large negative entropies that characterize the hydration of nonpolar groups in terms of well-defined statistical mechanical distribution functions.

At room temperature dissolution of nonpolar compounds in water is slightly favored by enthalpy, but strongly opposed by a large, negative change in entropy [Frank and Evans, 1945; Privalov and Gill, 1988]. However, and given that the heat capacity of solution is very large and positive, the situation is reversed as the temperature is increased [Privalov and Gill, 1988]. To explain these observations Frank and Evans proposed that the structure of water around nonpolar solutes is more "ordered" than bulk water. This could occur through strengthening or through an increase in the number of hydrogen bonds between water molecules in the hydration shell. Another explanation was based simply on the directional, hydrogen bonding properties of water, which rearranges around the nonpolar solute in order to maintain the maximum possible number of hydrogen bonds [Glew, 1962]. In this view, hydration water need not be different from bulk water, only its orientational freedom is restricted due to the inability of the solute to participate in hydrogen bonds, which makes certain orientations of the adjacent water molecules energetically unfavorable [Stillinger, 1977]. While orientational preferences have been observed in computer simulations [Geiger et al., 1979; Rossky and Karplus, 1979; Pangali et al., 1979; Postma et al., 1982; Alagona and Tani, 1980], no quantitative assessment of the contribution of this factor to the entropy has been made due to the lack of a direct, theoretical connection between the entropy of solution and angular probability distributions.

An entirely different point of view has emerged from Pierotti's application of Scaled Particle Theory (SPT) to the calculation of thermodynamic properties of solutes in nonpolar liquids [Pierotti, 1963] and water [Pierotti, 1965]. In this work water is viewed essentially as a simple fluid endowed with the actual density and thermal expansivity of water. The success of SPT suggested that the microscopic structure of water is not directly implicated in the solution thermodynamics. Stillinger improved SPT by considering the effect of the hydrogen bonding structure of water on the solute-water radial distribution function [Stillinger, 1973]. Similarly, in the integral equation theory of Pratt and Chandler [1977], the hydrogen bonding structure of water is taken into account only indirectly through the use of the experimental O-O radial distribution function and density of water at different temperatures. Here, as in the SPTs, solute-water orientational correlations are not evaluated as a separate contribution to the entropy. Despite that, their prediction for the entropy of hydration of methane at room temperature was in excellent agreement with the experimental value. As recognized by several authors, the large negative entropies of solution calculated by these methods for water compared to nonpolar solvents are due to the smaller size of water, and not directly to its hydrogen bonding properties [Lucas, 1976; Lee, 1985].

In this work a rigorous statistical mechanical expression for the entropy of solution is used to quantitatively evaluate the contribution to the entropy of translational and orientational pair correlations between the solute and the water molecules. The hydration entropy of methane as a function of temperature and of inert gases at room temperature is calculated and compared to experimental data. The dependence of the entropy on solute size and the solute-water interaction energy is also examined by simulations of model Lennard-Jones particles in water.

CORRELATION EXPANSION FOR THE ENTROPY

In a recent publication Wallace derived an expansion for the entropy of a fluid in terms of one-, two-, etc. particle correlation functions in the canonical ensemble [Wallace, 1987]. Similar expansions had been derived earlier in the grand canonical ensemble by Nettleton and Green [1958], Morita and Hiroike [1961], and Raveché [1971]. Baranyai and Evans also recently demonstrated the equivalence of the two expressions [1989]. The development of the correlation expansion for the case of hydrophobic hydration has been given elsewhere [Lazaridis and Paulaitis, 1992], with some corrections and revisions appearing in more recent communications [Lazaridis and Paulaitis, 1993a; 1993b]. The formulas for the multiparticle correlation function expansion of the entropy used in this work are briefly presented below.

Neglecting intramolecular degrees of freedom, the entropy of a polyatomic fluid in the canonical ensemble can be expressed as follows [Lazaridis and Paulaitis, 1992]:

$$S_N = -\frac{kN}{\rho}\frac{\Omega}{\sigma} \int f_N^{(1)} \ln (h^s f_N^{(1)}) \, dp_1 dJ_1 - \frac{k\rho^2}{2\,\Omega^2} \int g^{(2)} \ln g^{(2)} \, dr^2 d\omega^2$$

$$- \frac{k\rho^3}{3!\Omega^3} \int g^{(3)} \ln\delta g^{(3)} \, dr^3 d\omega^3 - \; ... \tag{1}$$

where k is Boltzmann's constant, h is Planck's constant, ρ is the number density (N/V), r^N the Cartesian coordinates of the N molecules, p^N the linear momenta, ω^N the Euler angles, J^N the angular momenta, $\Omega = \int d\omega$, σ the symmetry number of the molecules (2 for water), s the number of external degrees of freedom per molecule (6 for nonlinear molecules), $f_N^{(1)}$ the one-particle distribution function, and $g^{(n)}$ the n-particle correlation function. The quantity $\delta g^{(3)}$ is defined by

$$g^{(3)}(1,2,3) = g^{(2)}(1,2) \cdot g^{(2)}(1,3) \cdot g^{(2)}(2,3) \cdot \delta g^{(3)}(1,2,3) \tag{2}$$

and similarly for $\delta g^{(n)}$, n=4,...,N. The above expression is derived by a Kirkwood factorization of the N-particle correlation function and substitution into the Gibbs expression for the entropy [Wallace, 1987]. The first term, the one-particle contribution, reflects the ability of each particle to visit any point in the volume of the system. The second term accounts for pair correlations in the positions and orientations of the molecules, while the higher order terms represent correlations that cannot be accounted for by the superposition approximation ($\delta g^{(3)} = 1$ in eq. 2).

The entropy of a mixture can be similarly expressed in terms of multiparticle correlation functions. In the case of an aqueous mixture of a monatomic solute the expression for the entropy is

$$S = N_w\left(\frac{6}{2}k - k\ln(\rho_w\Lambda_w^3 q_w^{-1})\right) + N_s\left(\frac{3}{2}k - k\ln(\rho_s\Lambda_s^3)\right)$$

$$-N_w^2\frac{k}{2V\Omega^2}\int g_{ww}\ln g_{ww} \, dr d\omega^2 - N_s^2\frac{k}{2V}\int g_{ss}\ln g_{ss} \, dr$$

$$- N_s N_w \frac{k}{V\Omega}\int g_{sw}\ln g_{sw} \, dr d\omega + \text{higher order terms} \tag{3}$$

where subscripts s, w denote solute and water, respectively; N_w, N_s are the number of molecules; ρ_w, ρ_s are number densities; q_w is the rotational partition function of water; and g_{ww}, g_{sw}, g_{ss} are pair correlation functions.

The partial molar entropy of the solute is obtained by differentiating eq. 3 with respect to N_s at constant T, P, and N_w. At infinite dilution this partial molar entropy is given by

$$\bar{S}_s^\infty = \left(\tfrac{3}{2}k - kln(\rho_s\Lambda_s^3)\right) - k\left(1-\rho_w\bar{v}_s^\infty\right) - k\frac{\rho_w}{\Omega}\int g_{sw}ln g_{sw} d\underline{r}d\omega +$$

$$\frac{k}{2}\frac{\rho_w^2}{\Omega^2}\bar{v}_s^\infty\int g_{ww}ln g_{ww}d\underline{r}d\omega^2 - \frac{k}{2}\frac{\rho_w}{\Omega^2}\left(\frac{\partial}{\partial x_s}\int g_{ww}ln g_{ww}d\underline{r}d\omega^2\right)_{T,P}$$

$$+ \text{ higher order terms} \tag{4}$$

The term, $k(1-\rho_w\bar{v}_s^\infty)$, arises from the derivative of the first two terms in eq. 3 and has been recently derived [Sharp et al., 1991] on the basis of Flory-Huggins theory. We will refer to this term as the "volume entropy."

The two terms containing the water-water pair correlation function account for "solvent reorganization" upon solute insertion at constant temperature and pressure. The calculation of these terms, as well as of the higher order terms, is very difficult to perform by simulation. Therefore, in what follows we concentrate on the two-particle solute-water contribution, which can be expressed as

$$S_{sw} = - \frac{k(N-1)}{V\Omega} \int g_{sw}^{(2)} ln g_{sw}^{(2)} d\underline{r} d\omega \tag{5}$$

By performing the following factorization of the pair correlation function [Lazaridis and Paulaitis, 1992],

$$g_{sw}^{(2)}(r,\theta,\phi,\chi) = g_{sw}^r(r) \cdot g(\theta,\chi,\phi) \qquad \text{for } r \leq r_{sh} \tag{6}$$

$$g_{sw}^{(2)}(r,\theta,\phi,\chi) = g_{sw}^r(r) \qquad \text{for } r \geq r_{sh}$$

where g_{sw}^r is the orientationally averaged radial distribution function (RDF) and r_{sh} is defined as the distance to the first minimum in the RDF, we can separate the solute-water pair correlation entropy into a translational and an orientational contribution:

$$S_{sw} = - \frac{k(N-1)}{V} \int g_{sw}^r ln g_{sw}^r d\underline{r} \qquad \text{(translational)}$$

$$- \frac{k(N-1)V_i}{V\Omega} \int g(\theta,\chi,\phi) \cdot ln\{g(\theta,\chi,\phi)\} d\omega \qquad \text{(orientational)} \tag{7}$$

where $V_i = \int_0^{r_{sh}} g_{sw}^r d\underline{r}$. For spherically symmetric solutes, the orientational correlation function will be uniform with respect to the third Euler angle, ϕ. The angle θ is defined as that between the dipole vector of water and the solute-water oxygen axis, while χ describes rotation around the water dipole vector (fig. 1).

The first integral in eq. 7 runs over the entire volume of the fluid. However, in simulations g_{sw}^r can only be calculated up to a certain distance, usually less than 15 Å. This is not a problem if this RDF is short-ranged—i.e., exactly unity at larger separations. However, in the canonical ensemble g_{sw}^r does not become exactly 1 at any distance from the solute. In other words, the first integral in eq. 7 is not "local" [Baranyai and Evans, 1989] and consideration must be given to the long-range behavior of the RDF. It can be shown [Lazaridis and Paulaitis, 1993b] that the correction due to this long-range behavior is simply

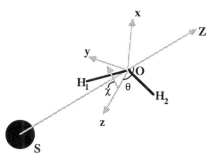

Figure 1. Definition of the angles used to describe the orientation of water molecules with respect to the solute S. xyz is the body-fixed coordinate system.

$$S_{corr} = k\rho \int_0^R (g_{sw}^r - 1) \, d\underline{r}$$
(8)

where R is the distance up to which the RDF is calculated in the simulation. Similar terms appear in the entropy expansion in the grand canonical ensemble [Nettleton and Green, 1958; Morita and Hiroike, 1961; Raveché, 1971]. We found this correction to be very small when the solute is about the same size as water, but it can become comparable to the $g_{sw}^r \ln g_{sw}^r$ integral in eq. 7 for large solutes.

To compare with experiment we need to establish the connection between the partial molar entropy (eq. 4) and the reported standard entropies of solvation. Ben-Naim's standard solvation entropy [Ben-Naim, 1978] is defined as the temperature-derivative of the residual chemical potential

$$\Delta S_s^* = -\frac{\partial \mu_s^{res}}{\partial T}$$
(9)

where

$$\mu_s^{res} = \mu_s - kT \ln \rho_s \Lambda_s^3$$
(10)

The partial molar entropy is

$$\bar{S}_s = -\frac{\partial \mu_s}{\partial T} = -\frac{\partial \mu_s^{res}}{\partial T} - k \ln \rho_s \Lambda_s^3 + kT\alpha + \frac{3}{2}k$$
$$= \Delta S_s^* - k \ln \rho_s \Lambda_s^3 + kT\alpha + \frac{3}{2}k$$
(11)

where α is the thermal expansion coefficient of the solution. Comparing eq. 11 in the limit of infinite dilution with eq. 4 gives

$$\Delta S_s^{*\infty} = -kT\alpha^\circ - k\left(1 - \rho_w \bar{v}_s^\infty\right) - \frac{k(N-1)}{V} \int g_{sw}^r \ln g_{sw}^r \, d\underline{r}$$

$$- \frac{k(N-1)V_i}{V\Omega} \int g(\theta,\chi,\phi) \cdot \ln\{g(\theta,\chi,\phi)\} \, d\omega + \dots \tag{12}$$

where α° is the thermal expansion coefficient of pure water.

METHOD OF CALCULATION

The RDF and the orientational distribution function are calculated by Monte Carlo simulation using Jorgensen's BOSS program [Jorgensen, 1989]. For these simulations one particle is inserted in a box of 216 water molecules pre-equilibrated at 25°C. Then, depending on solute size, one or two water molecules exhibiting the highest interaction energy with the solute are deleted to approximate particle dissolution at constant pressure. In the case of methane, we used NPT simulation data at a number of temperatures, whereas in all other cases NVT simulations were performed at 25°C. NPT and NVT simulations were found to give no detectable difference in the calculated entropy beyond statistical uncertainty. Each simulation consisted of 5 million or more configurations for equilibration and from 20 to 60 million configurations for sampling. The TIP4P model for water [Jorgensen et al., 1983] was used in all simulations. The solute-water pair correlation entropy was calculated from eq. 7 with the RDF numerically integrated up to 8 Å for the smaller solutes and up to 10.5 Å for the larger ones. Due to spherical symmetry, the orientational distribution function is uniform with respect to ϕ and can be set equal to $p(\theta,\chi) \cdot c$, where c is a constant determined from the normalization condition for $g_{sw}^{(2)}$. For the calculation of $p(\theta,\chi)$ a two-dimensional histogram was prepared with data taken from the first hydration shell. Details of the numerical integration are given in Lazaridis and Paulaitis [1992].

HYDRATION OF METHANE AS A FUNCTION OF TEMPERATURE

Hydration of methane was studied at a number of temperatures between 5 and 65°C. Methane was represented by a simple Lennard-Jones particle (σ=3.73 Å, ε=0.294 Kcal/mol). The calculated orientational distribution function $p(\theta,\chi)$ at 25°C is shown in fig. 2. The letters A-D denote characteristic orientations of the water molecules around the solute. From this figure it is concluded that the most probable orientations are those with one of the vertices of the tetrahedral coordination structure of water pointing outwards (B or C), while those orientations where one of the vertices points toward the solute (A or D) are least probable. This is an expected result and has been observed in many previous simulations [Geiger et al., 1979; Rossky and Karplus, 1979; Pangali et al., 1979; Postma et al., 1982; Alagona and Tani, 1980]. The calculated orientational distribution was similar at all temperatures, although the peaks of the maxima became less pronounced at higher temperatures.

The results for the translational and orientational contributions to the entropy are listed in table 1. The error bars were estimated as the maximum deviation between the result from the full 20 million data points and those from different subsets of 15 million data points. As expected, both contributions diminish with increasing temperature.

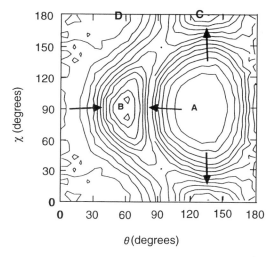

Figure 2. A contour plot of the calculated orientational probability distribution $p(\theta,\chi)$ of water molecules in the first hydration shell of methane at 25°C (from Lazaridis and Paulaitis [1992]). Arrows indicate the direction of increasing probability. Letters denote characteristic orientations. A: hydrogen radially inwards ($\theta=125,\chi=90°$) ; B: hydrogen radially outwards ($\theta=55,\chi=90°$); C: lone pair radially outwards ($\theta=\sim125,\chi=0°$) ; D: lone pair radially inwards ($\theta=\sim55,\chi=0°$).

Table 1. Translational and orientational correlation contributions to the entropy of solution of methane in water at several temperatures calculated by Monte Carlo simulations (from Lazaridis and Paulaitis [1992]). All values in e.u. (1 e.u. = 1 cal/mol·K).

T (°C)	Translational	Orientational	Total
5	-10.9 ± 1.2	-8.3 ± 1.2	-19.2 ± 1.7
15	-10.5 ±0.2	-8.2 ±0.6	-18.7 ± 0.6
25	-8.4 ± 0.3	-6.6 ±0.5	-15.0 ± 0.6
35	-7.8 ± 0.3	-6.0 ± 0.4	-13.8 ± 0.5
45	-8.1 ± 0.3	-5.7 ± 0.2	-13.8 ± 0.4
55	-7.0 ±0.4	-4.6 ± 0.3	-11.6 ± 0.5
65	-7.1 ± 0.2	-4.4 ± 0.5	-11.5 ± 0.5

Using the experimental partial molar volume of methane in water [Masterton, 1954], the volume entropy contribution is calculated to be +2.1 e.u. at room temperature. For the $kT\alpha°$ term the experimental thermal expansion coefficient of pure water is used ($\alpha°=257.21\cdot10^{-6}K^{-1}$) [CRC Handbook, 1983]. At room temperature this term is about -0.2 e.u. With these values, the sum of the four terms in eq. 12 is equal to -13.1 e.u., which is smaller than the experimental value of -15.94 e.u. [Ben-Naim and Marcus, 1984].

The heat capacity will be determined primarily by the temperature dependence of the translational and orientational correlation entropy. Therefore, to calculate the heat capacity, the sum of the translational and orientational entropy was expanded in a Taylor series and a least squares fit was performed, giving a value of 52.2 cal/mol/K for the heat capacity of hydration at room temperature. This value is very close to the experimental values: 56.6 [Rettich et al., 1981]; 50.0 cal/mol/K [Naghibi et al., 1986]. The temperature dependence of this heat capacity is, however, overestimated [Lazaridis and Paulaitis, 1992].

Eq. 6 assumes that the orientational distribution function is independent of distance within the first hydration shell and that it is uniform in the second and higher shells. These assumptions have been tested by calculating the orientational distribution function in three subshells of the first hydration shell as well as in the second hydration shell. The first subshell extends up to 3.8 Å, the second between 3.8 and 4.6 Å, and the third between 4.6 and 5.4 Å. The second hydration shell extends from 5.4 to 7.4 Å. Since the subshell calculations demand better statistics than a calculation sampling over the whole first hydration shell, the simulation was extended to 100 million configurations. The orientational distribution in the three subshells was found to be qualitatively similar; only in the third subshell did we observe a shift in the position of the maxima. However, the distribution becomes progressively less pronounced with increasing distance from the solute. In the second hydration shell the distribution is more nearly uniform although some very weak orientational preferences are still evident. Notably, these orientational preferences are opposite to those in the first shell; i.e., on average, water molecules in the second shell tend to point one of their tetrahedral bonding vectors toward the solute, apparently to form hydrogen bonds with the water molecules in the first shell, which are oriented in the opposite direction.

The second integral in eq. 7, divided by Ω, gives the orientational entropy per water molecule and is a measure of the extent of orientational correlations. This quantity is observed to decay almost linearly with distance from the solute. The total orientational contribution to the hydration entropy per methane molecule is given by a sum of terms identical to those in eq. 7, one term for each hydration shell or subshell, where $(N-1)V_i/V$ is replaced by the average number of water molecules in that shell or subshell. This quantity is given in table 2. These contributions sum to -8.6 e.u., which is more negative than that obtained above (-6.6 e.u.), but is in reasonable agreement given the statistical uncertainties inherent in this approach for calculating the entropy. It should be noted that orientational correlations are almost nonexistent beyond the first hydration shell and thus appear to make a very small contribution to the entropy. This is consistent with neutron and X-ray diffraction measurements for pure water [Narten, 1972] showing that orientational correlations decay much faster than positional correlations. This additional, negative contribution brings the result for the total hydration entropy of methane in better agreement with experiment (-15.1 vs. -15.94 e.u. from experiment at room temperature [Ben-Naim and Marcus, 1984]).

Therefore, while the assumption of decoupled orientational and translational correlations (eq. 6) is not strictly correct, the results obtained based on this assumption are quite close to those obtained through the more rigorous subshell calculations. Since the computational requirements of performing rigorous subshell calculations are quite large, it will be expedient to continue to apply eq. 6, knowing, though, that this approximation may underestimate the magnitude of the orientational correlation entropy by 1 or 2 e.u.

Table 2. Orientational Entropy Contributions for First Hydration Subshells and the Second Hydration Shell (from Lazaridis and Paulaitis [1993b]).

Hydration Shell or Subshell	Average Number of Water Molecules	Orientational Entropy Contribution (e.u.)
First Hydration shell		
1st subshell	5.6	-3.2
2nd subshell	8	-3.4
3rd subshell	7	-1.3
Second hydration shell	25.7	-0.7

Table 3. Lennard-Jones Parameters for the Inert Gases.

Component	σ (Å)	ε (kcal/mol)
He	2.55	0.02
Ne	2.78	0.069
Ar	3.401	0.2339
Kr	3.624	0.317
Xe	3.935	0.433
Rn	4.36	0.576

HYDRATION OF INERT GASES

The inert gases are modeled as Lennard-Jones particles with the parameters given in table 3 (from Lazaridis and Paulaitis [1993b]). The calculated sum of the four terms in eq. 12 is reported in table 4 and compared to experimental data. These calculations are based on a total of 60 million configurations for each system. To obtain the volume entropy term we use the experimental partial molar volume of argon [Tiepel and Gubbins, 1972] and estimate the partial molar volumes of the remaining gases from a corresponding states correlation based on Kirkwood-Buff solution theory [Brelvi and O'Connell, 1972]. We take the characteristic volume in this correlation to be the atomic volume ($\pi\sigma^3/6$) for all the gases. From table 4 it can be seen that the magnitude of the hydration entropy is underpredicted by approximately 2-3 e.u. for all inert gases, except radon. Based on the results of the previous section, this discrepancy may be due to averaging the orientational distribution over the whole first hydration shell and to neglecting the small contribution of orientational correlations in the second hydration shell. If we assume that, as in the case of methane, these contributions amount to -2 e.u., the results would agree quite well with experiment. Overall, the results appear to reasonably reproduce the observed increase in the magnitude of the hydration entropy with increasing size of the molecule.

Table 4. Simulation results for the hydration entropy of inert gases at 25°C (from Lazaridis and Paulaitis [1993b]). All values in e.u. (1 e.u. = 1 cal/mol·K). The $kT\alpha°$ term is included in the total calculated entropy.

	total calculated	experimental[a]		
helium	-5.0	-8.0		-7.8
neon	-7.3	-10.1	-10.8	-10.0
argon	-12.6	-14.7	-15.6	-14.4
krypton	-13.7	-16.3	-16.8	-16.0
xenon	-15.5	-17.4	-19.6	-17.8
radon	-18.5	-18.2		

a: first and second columns from Ben-Naim, A. *Solvation thermodynamics*, Plenum:New York, 1987; third column from Krause, D. Jr., Benson, B.B. *J. Soln. Chem.*. **1989**, 18, 823 after change of standard states.

DEPENDENCE OF THE ENTROPY ON SIZE AND INTERACTION ENERGY

We have also studied a number of model Lennard-Jones solutes in water to examine separately the effect of interaction energy and solute size on the entropy. As a measure of molecular size we use the thermal diameter σ_{th} for solute-water interactions, defined as the intermolecular distance where the interaction energy is equal to the thermal energy (0.6 kcal/mol at 300K). With a very small ε the Lennard-Jones potential will closely emulate a hard sphere potential. Simulations of purely repulsive solutes are performed to facilitate a comparison of the present method with analytical theories, such as Scaled Particle Theory. Based on the thermal diameter we define an effective solute diameter as

$$\sigma_s^{eff} = 2\sigma_{th} - \sigma_w \tag{13}$$

where σ_w is the Lennard-Jones parameter for water.

Table 5 lists the model solutes studied. The first four systems emulate a point cavity with different values of the interaction energy. Similarly for the next six systems, which in this case, are for a solute of finite size. The remaining systems correspond to purely repulsive solutes of different size.

The translational and orientational correlation terms in eq. 12 were calculated and are reported in table 5. These results reveal that the size of the solute is the dominating factor in determining both the translational and orientational correlation entropy, with the solute-water interaction energy playing a less significant role. Fig. 3 shows the rapid increase in magnitude of the translational correlation entropy with increasing effective solute diameter. This behavior can be rationalized by a free volume argument [Lee, 1983]: as the size of the solute increases, the water molecules can pack around the solute more effectively. This restricts the translational freedom of the solute, since any motion of the solute must occur in concert with the motion of a large number of water molecules.

The translational correlation entropy is negative for all solutes except the point cavities. It should be noted that the positive values are obtained only after the long-range correction terms are added. The uncorrected integral in eq. 7 up to 8 Å gives a negative value for the translational contribution. The volume entropy term should make a negative contribution to the entropy. The upper bound for the magnitude of this contribution, obtained by setting

Table 5. Lennard-Jones systems studied and results obtained (from Lazaridis and Paulaitis [1993b]). The units are Å for σ, kcal/mol for ε, and entropy units (cal/mol·K) for the entropies. N_C is the number of water molecules in the first hydration shell.

σ_s	ε_s	ε	σ_{th}	σ_s^{eff}	translational correlation S	orient. correl. S per water molecule	N_c
1.252	0.001	0.01233	1.575	0.0	+1.53	-0.082	7.4
0.962	0.066	0.1	1.575	0.0	+1.38	-0.084	5.3
0.901	0.263	0.2	1.575	0.0	+1.20	-0.132	5.3
0.874	0.592	0.3	1.575	0.0	+1.02	-0.120	7.6
4.773	0.001	0.01233	3.075	3.0	-5.6	-0.310	19.2
3.667	0.066	0.1	3.075	3.0	-6.7	-0.320	18.2
3.435	0.263	0.2	3.075	3.0	-7.7	-0.366	19.6
3.330	0.592	0.3	3.075	3.0	-8.4	-0.388	19.9
3.266	1.052	0.4	3.075	3.0	-9.0	-0.340	18.7
3.226	1.644	0.5	3.075	3.0	-9.5	-0.360	19.4
2.019	0.001	0.01233	2.0	0.85	+0.35	-0.150	9.5
3.155	0.001	0.01233	2.5	1.85	-1.84	-0.214	13
6.184	0.001	0.01233	3.5	3.85	-9.4	-0.374	23
8.078	0.001	0.01233	4.0	4.85	-13.9	-0.316	28
10.222	0.001	0.01233	4.5	5.85	-20.7	-0.350	35
12.622	0.001	0.01233	5.0	6.85	-28.2	-0.306	42
15.269	0.001	0.01233	5.5	7.85	-36.5	-0.322	51
18.174	0.001	0.01233	6.0	8.85	-47.4	-0.272	58.5

$\bar{v}_s^\infty = 0$, is -2 e.u., which would make the entropy slightly negative. For comparison, Scaled Particle Theory predicts a small, negative value (-0.8 e.u.) for the entropy of solution of a point cavity in water [Pierotti, 1965].

In fig. 4 we plot the orientational correlation entropy per water molecule in the first hydration shell as a function of solute size. The magnitude of this quantity increases with solute size up to approximately 4 Å and then decreases slightly for larger solutes. Although there is some statistical noise in the data, this trend is clear. The initial increase and subsequent decrease in the magnitude of the orientational correlations can be understood as follows. In the limit of a point cavity, water molecules are positionally restricted but orientationally free to form hydrogen bonds in all directions. Thus, one would expect the magnitude of the orientational entropy in this limit to be very small. As the solute size increases and the curvature of its surface decreases, water molecules are forced to adopt more restricted orientations in order to form the maximum number of hydrogen bonds. As the curvature decreases further, water molecules near the surface are not able to form four linear hydrogen bonds no matter how they orient. Thus the energetic advantage of "straddling" configurations is diminished, and the orientational configurational space for water molecules around the solute becomes more uniform energetically. This leads to a

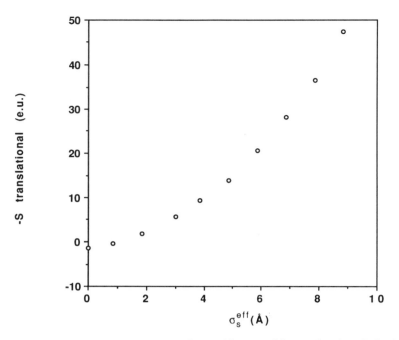

Figure 3. Translational correlation entropy of the Lennard-Jones particles as a function of effective solute diameter (from Lazaridis and Paulaitis [1993b]).

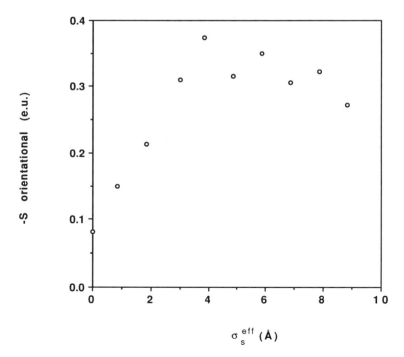

Figure 4. Orientational correlation entropy per water molecule as a function of effective solute diameter (from Lazaridis and Paulaitis [1993b]).

broadened orientational distribution and a higher (less negative) orientational entropy. An interesting connection can be made to simulations of water next to a hydrophobic surface [Lee et al., 1984], where it was found that water molecules tend to point one of their hydrogens or lone pairs *toward* the surface, thus sacrificing on average one hydrogen bond. Apparently this configuration becomes energetically favored over the "straddling" configuration next to a flat surface. It follows that the magnitude of the orientational entropy may exhibit a minimum with respect to surface curvature at the point where these two configurations become energetically equivalent.

The effect of ε on the orientational entropy is small and appears almost constant for σ_s^{eff} = 3Å over the entire range of ε. The effect on the translational entropy is more substantial, but much smaller than that for solute size over the usual range of Lennard-Jones parameters for small solutes. The increase in the magnitude of the translational entropy with ε is not difficult to understand, since stronger solute-water interactions naturally lead to more sharply peaked correlations. These results clearly show a significant entropic effect associated with turning on solute-water interactions that can lead to contributions of up to 4 e.u. in the hydration entropy.

CONCLUSIONS

From the results presented above two important conclusions can be drawn about the molecular origin of the large hydrophobic entropies of hydration:

1) a significant portion of the entropy of solution is due to reduced orientational options for water molecules interacting with the solute. It is important to note, however, that statements such as "hydration water has lower entropy than bulk water" are not strictly correct. The orientational entropy defined here refers to *correlations* between solute and water molecules and cannot be "assigned" to water alone (see also Ben-Naim and Marcus, 1984; p. 2025). The existence of orientational preferences does not necessarily imply that hydration water is less "mobile" or more "ordered" than bulk water, as suggested elsewhere [Glew, 1962]. Water molecules in the hydration shell of *another* water molecule will also exhibit strong orientational preferences, presumably the opposite of those observed next to nonpolar solutes.

2) the contribution of translational correlations seems to be higher than in nonpolar solvents. For example, the standard entropy of solution of methane in CCl_4 is -1.7 e.u. [Wilhelm and Battino, 1973, after change of standard states according to Ben-Naim and Marcus, 1984] and that of xenon in a great number of organic solvents is on average -4.1 ± 0.5 e.u. [Pollack, 1991]. Apparently, the small size of water is involved in increasing the translational contribution. Scaled Particle Theory would be expected to predict this effect with sufficient accuracy, since the translational term in the present formulation corresponds to the *entropy of solution in a fluid that has the density and packing fraction of water but no orientational degrees of freedom*. However, the fact that this theory has given entropies of solution as high as the experimental ones, while totally neglecting the orientational contribution, suggests that the translational contribution must be overestimated by SPT.

The observed temperature dependence of solute-water orientational correlations suggests a mechanism for the change in the nature of hydrophobic hydration from entropic to enthalpic as the temperature is raised. If water molecules oriented randomly with respect to the solute, the orientational contribution to the entropy of solution would vanish and the enthalpy of solution would be positive due to the loss of hydrogen bonding interactions. In actuality, water will seek an orientational distribution which corresponds to a state of minimum free energy. At low temperatures, enthalpic contributions to the free energy are more important and water molecules sacrifice orientational freedom in order to minimize interaction energy. This is consistent with Shinoda's statement that, at lower temperatures, water reorganization ("iceberg formation") enhances solubility [Shinoda, 1977]. As temperature is increased, however, entropic contributions to the free energy prevail and the orientational distribution becomes more and more uniform at the expense of enthalpy. Thus, the hydrophobic effect may appear entropic in some instances and enthalpic in others. Therefore, it is incorrect to ascribe the large enthalpies of hydrophobic hydration observed at high temperatures entirely to van der Waals interactions [Privalov and Gill, 1988].

The contributions from the terms omitted in the entropy expansion in eq. 4 should be further examined. Particularly interesting would be the change in water-water pair correlations and the three-particle solute-water-water terms, which should contain any "water-structure enhancement" effects. From the results of the present work, water structure enhancement is not required to explain the large entropies of hydration, in accord with Stillinger's view [Stillinger, 1977]. However, the magnitude of these terms should be estimated independently, to rule out the possibility of large cancellation of errors in the entropy calculation, as has been suggested [Smith et al., 1993].

One of the most interesting findings of the present work is that the magnitude of the orientational entropy per water molecule in the first hydration shell exhibits a maximum at an effective solute diameter of 4 Å. Differences in the characteristics of hydration of small solutes versus flat hydrophobic surfaces have long been recognized [Lee et al., 1984]. However, the thermodynamic consequences of these differences have not been examined in detail. The sacrifice of a hydrogen bond next to a flat surface will necessarily manifest itself thermodynamically in an increase in the enthalpic cost of the interaction. This may be one reason for the observation of enthalpy-driven association processes in water [Hinz, 1983; Smithrud et al., 1991]. While the enthalpy must become progressively more positive with decreasing surface curvature, the entropy probably changes in a more complex way. It is unknown, however, how the *net free energy* of the interaction will change with surface curvature. If we accept that the reorientational relaxation of water to a state of lower entropy and energy is accompanied by a favorable free energy (which should be true, since this process occurs spontaneously next to small solutes at room temperature), it is natural to hypothesize that when this process cannot occur (e.g., next to a flat surface), the free energy will be higher.

This picture of hydrophobic hydration is in qualitative agreement with an empirical geometric model proposed recently for the dependence of the hydrophobic free energy on the curvature of the solute-water interface [Sharp et al., 1991; Nicholls et al., 1991]. In this model it is assumed that the free energy is proportional to the solvent accessible surface of *water* molecules in the hydration shell of the solute. It is hoped that the present work will help provide a quantitative, molecular foundation to such methods.

Extension of the present method of calculating the entropy to more complex systems, such as molecular, polar, or ionic solutes, is highly desirable, but not trivial. For instance, in the case of ions, the perturbation of water-water pair correlations next to the ion is significant and must be taken into account in the entropy calculation. However, water-water pair correlation functions will be of higher dimensionality than those for the monatomic, uncharged solutes studied here and cannot be obtained easily by simulation unless inordinately long simulation times are employed. The study of these systems will probably require thoughtful approximations and perhaps some mathematical ingenuity.

ACKNOWLEDGMENTS

Prof. W. L. Jorgensen is gratefully acknowledged for making the program BOSS available to us. We also thank Prof. R. H. Wood for numerous helpful discussions. This work was supported by the National Science Foundation (grant CPE8351228). Partial support was also provided by Union Carbide, Merck and Exxon.

REFERENCES

Alagona, G., Tani, A. *J. Chem. Phys.* **1980**, 72, 580.

Baranyai, A., Evans, D.J., *Phys. Rev. A* **1989**, 40, 3817.

Ben-Naim, A., *J. Phys. Chem.* **1978**, 82,792.

Ben-Naim, A., Marcus, Y., *J. Chem. Phys.* **1984**, 81, 2016.

Brelvi, S.W., O'Connell, J.P., *AIChE. J.* **1972**, 18, 1239.

CRC Handbook of Chemistry and Physics, 64th Ed., CRC Press, Boca Raton, 1983.

Frank, H.S., Evans, M.W. *J. Chem. Phys.* **1945**, 13, 507.

Geiger, A., Rahman, A., Stillinger, F.H., *J. Chem. Phys.* **1979**, 70, 263.

Glew, D.N. *J. Phys. Chem.* **1962**, 66, 605.

Hinz, H.-J., *Ann. Rev. Biophys. Bioeng.* **1983**, 12, 285.

Jorgensen, W.L. BOSS version 2.8, Yale University, New Haven, Connecticut, 1989.

Jorgensen, W.L., Chandrasekhar, J., Madura, J.D., Impey, R.W., Klein, M.L. *J. Chem. Phys.* **1983**, 79, 926.

Lazaridis, T., Paulaitis, M.E., *J. Phys. Chem.* **1992**, 96, 3847.

Lazaridis, T., Paulaitis, M.E., *J. Phys. Chem..* **1993a**, 97, 5789.

Lazaridis, T., Paulaitis, M.E., *J. Phys. Chem..* **1993b**, submitted for publication.

Lee, B. *J. Phys. Chem.* **1983**, 87, 112.

Lee, B., *Biopolymers* **1985**, 24, 813.

Lee, C.Y., McCammon, J.A., Rossky, P.J. *J. Chem. Phys.* **1984**, 80, 4448.

Lucas, M., *J. Phys. Chem.* **1976**, 80, 359.

Masterton, W.L., *J. Chem. Phys.* **1954**, 22, 1830.

Morita, T., Hiroike, K., *Prog. Theor. Phys.* **1961**, 25, 537.

Naghibi, H., Dec, S.F., Gill, S.J. *J. Phys. Chem.* **1986**, 90, 4621.

Narten, A.H. *J. Chem. Phys.* **1972**, 56, 5681.

Nettleton, R.E., Green, M.S., *J. Chem. Phys.* **1958**, 29, 1365.

Nicholls, A., Sharp, K.A., Honig, B. *Proteins* **1991**, 11, 281.

Pangali, C., Rao, M., Berne B.J. *J. Chem. Phys.* **1979**, 71, 2982.

Pierotti, R.A., *J. Phys. Chem.* **1963**, 67, 1840.

Pierotti, R.A., *J. Phys. Chem.* **1965**, 69, 281.

Pollack, G.L. *Science* **1991**, 251, 1323.

Postma, J.P.M., Berendsen, H.J.C., Haak, J.R., *Faraday Symp. Chem. Soc.* **1982**, 17, 55.

Pratt, L.R., Chandler, D., *J. Chem. Phys.* **1977**, 67, 3683.

Privalov, P.L., Gill, S.J. *Adv. Protein Chem.* **1988**, 39, 191

Raveché, H.J., *J. Chem. Phys.* **1971**, 55, 2242.

Rettich, T.R., Handa, Y.P., Battino, R., Wilhelm, E. *J. Phys. Chem.* **1981**, 85, 3230.

Rossky, P.J., Karplus, M., *J. Am. Chem. Soc.* **1979**, 101, 1913.

Sharp, K.A., Nicholls, A., Fine, R.F., Honig, B. *Science* **1991**, 252, 106.

Sharp, K.A., Nicholls, A., Friedman, R., Honig, B., *Biochemistry* **1991**, 30, 9686.

Shinoda, K. *J. Phys. Chem.* **1977**, 81, 1300.

Smith, D.E., Laird, B.B., Haymet, A.D.J., *J. Phys. Chem .* **1993**, 97, 5788.

Smithrud, D.B., Wyman, T.B., Diederich, F., *J. Am. Chem. Soc.* **1991**, 113, 5420.

Stillinger, F.H. *J. Sol. Chem.* **1973**, 2, 141.

Stillinger, F.H. *Phil. Trans. R. Soc. Lond. B* **1977**, 278, 97.

Tiepel, E.W., Gubbins, K.E., *J. Phys. Chem.* **1972**, 76, 3044.

Wallace, D.C. *J. Chem. Phys.* **1987**, 87, 2282.

Wilhelm, E., and Battino, R., *Chem. Rev.* **1973**, 73, 1.

STATISTICAL MECHANICS OF SECONDARY STRUCTURES IN PROTEINS: Characterization of a molten globule–like state

J.Bascle, S.Doniach[1],T.Garel and H.Orland[2]

Service de Physique Théorique, CE-Saclay
F-91191 Gif-sur-Yvette Cedex, France

ABSTRACT

In order to obtain a simple characterization of the molten globule state of proteins, we re–examine a simple model for the formation and stability of secondary structures in proteins. We consider chains on a cubic lattice; each monomer can be considered to reperesent a helical turn, and we assign a Boltzmann weight $\exp(\epsilon_H/T)$ when two monomers are aligned, and 1 when they make a turn. Here, ϵ_H is positive and represents the energy gain due to hydrogen bonds (H-bonds). In addition, we include an attractive nearest neighbour monomer–monomer interaction to represent the hydrophobic effect which drives the collapse of the chains. In the thermodynamic limit (and in the mean-field approximation), the system undergoes two transitions: first, a second order collapse transition at high temperature, from random coil to globular structures, similar to the usual theta point of polymers. Then, at lower temperature, the system undergoes a first order freezing transition, from a high temperature phase where the helix–like secondary structures are present but mobile in the system, to a frozen phase where secondary structures invade the whole globule, and make their turns only on the outside surface. We have checked these results by extensive Monte Carlo simulations on finite chains, and have indeed observed the theta point at higher temperature, and the freezing transition at lower temperature. The structure factor in this frozen phase exhibits more structure than in the non-frozen phase, and the number of co–linear bonds is maximized. Above the freezing transition, the number of H-bonds drops sharply, but the system still exhibits significant secondary structure, and we observe a slight increase of the radius of gyration. The collapsed phase above the freezing point thus provides a simplified representation of the molten globule phase of real proteins.

---

[1] also at Department of Applied Physics, Stanford University, Stanford, CA94305, USA

[2] also at Groupe de Physique Statistique, Université de Cergy Pontoise, 95806 Cergy Pontoise Cedex, France

1. Introduction

Understanding the nature of the compact state of macromolecules is of interest from both physical and biological standpoints. In a physical context, the influence of short range order on a collapsed polymer is a relatively unexplored problem. For biology, the partially folded states of proteins may be important as kinetic intermediates in the general protein folding pathway, and in their role in trans–membrane transport. In order to elucidate the possible partially foded states of proteins we re–examine a highly simplified model of a protein previousely introduced by Kolinski, Skolnick and Yaris [1].In this model the protein is represented on a (cubic) lattice, with the following characteristics: (i) to take into account the steric repulsion of the atoms, the peptide chain is represented by a Self-Avoiding Walk (SAW) on the lattice; (ii) the hydrophobic effect [2], which is widely believed to be the main driving force for the collapse of proteins, is modelled by a short range attraction; (iii) finally, the possibility of hydrogen-bonding leading to secondary structure is represented by a term which favors local alignment of the chain. This model therefore mimicks the competition between global compactness and local one-dimensional order, which for instance occurs in an α -helical protein. Models, which are more appropriate to the existence of (quasi) two-dimensional β sheets in folded proteins, can be similarly defined. The model is studied analytically at high and low temperatures. We find three distinct thermodynamic regimes:

  1) a fully folded regime at low temperatures;
  2) a compact but disordered state reminiscent of the "molten globule" state characterized by Ptitsyn [3], at intermediate temperatures;
  3) a swollen (non-compact) state at high temperatures.

To check these predictions, we have done extensive Monte Carlo simulations on chains of 126 monomers, using the Monte Carlo chain growth method developed in refs [4,5]. The layout of this paper is as follows: the model is defined in section 2; the analytical approach is presented in section 3; we briefly comment on our numerical simulations in section 4, and examine the possible implications of the model for the molten globule state of proteins. A more detailed version of the present work can be found in ref [6].

2. The Model

We consider a simple three dimensional cubic lattice, with lattice spacing a. The protein is represented by a chain of n links on the lattice. Only SAW's are considered, which automatically take steric effects into account. The hydrophobic effect is modelled by a short range attraction. Finally, we will take a link to represent a helical turn (figure 1(a)). We assign an energy $-\varepsilon_H$, if two successive links are aligned, to mimic the energy of formation of hydrogen bonds between neighbouring helical turns (figure 1(b)) . Otherwise, the assigned energy is zero (figure 1(c)).

With this crude picture of helices, links and nodes lose their natural interpretation as amino-acids, but we will keep this denomination. In particular, one could use an energy $\varepsilon_H(i)$, where i would be the type of amino-acid. For instance, an alanine–like amino–acid might have a large ε_H (good helix former), whereas a glycine–like amino–acid might have $\varepsilon_H = 0$.

The partition function of the model may be written

$$Z = \sum_{\text{allchains}} \exp\left(-\beta \sum_{0<i<j<n} v(\vec{r}_i - \vec{r}_j) + \beta \sum_i^n (\vec{r}_{i+1} - \vec{r}_i) \cdot (\vec{r}_i - \vec{r}_{i-1})\varepsilon_H/a^2\right) \quad (1)$$

Figure 1. (a) Definition of the model: one lattice link represents a helical turn. (b) Two aligned links allow for the formation of an H-bond (of energy $-\epsilon_H$). (c) Two unaligned links have zero energy.

where $\{\vec{r}_0, \vec{r}_1, ..., \vec{r}_n\}$ are the vectors of the $(n+1)$ nodes of the chain, and β denotes the inverse of the thermodynamic temperature $1/k_BT$, where k_B is the Boltzmann constant. In (1) the summation is over all chains on the lattice (whether self-avoiding or not). The self-avoidance and the short distance attraction representing the hydrophobic effect is contained in the definition of $v(\vec{r})$:

$$
\begin{aligned}
v(\vec{r}) &= +\infty & \text{if} && \vec{r} = 0, \\
&= -v_0 & \text{if} && |\vec{r}| = a, \\
&= 0 & \text{if} && |\vec{r}| > a.
\end{aligned}
$$

The phase diagram of this model can be inferred from analytic approximations, that we now present for the case of a homopolymer in the large n limit.

3. Analytical results

3.1 High temperature regime: the θ-point

At sufficiently high temperatures, following the original arguments of Flory [7,8], the attractive interaction v_0 will no longer be able to cause the chain to collapse, and the ensemble will tend to that of self-avoiding random walks on the lattice. We further assume that $\varepsilon_H < v_0$, and that ε_H can be neglected at high temperature. In this limit, a virial expansion can be made in

$$
f(\vec{r}) = e^{-\beta v(\vec{r})} - 1 \tag{2}
$$

leading , to lowest order in f, to

$$
Z = Z_0 (1 + (6\,(e^{\beta v_0} - 1) - 1)N \int \frac{d^d q}{(2\pi)^d} S_G(\vec{q}) + ...) \tag{3}
$$

where $S_G(q)$ is the structure factor (Debye) in the limit $v_0 = 0$, i.e., the structure factor of a free chain on the lattice. The second virial coefficient may now be deduced from (3) to be

$$
B_2(T) = 1 - 6\,(e^{\beta v_0} - 1) \tag{4}
$$

As pointed out by Flory, the condition $B_2(T) = 0$ defines a temperature θ above which the chain is no longer compact and follows a self avoiding random walk. In the limit $T \gg \theta$, the radius of gyration then satisfies

$$
r_g \simeq n^\nu \tag{5}
$$

with $\nu \simeq 3/(d+2) = 0.6$ using Flory's estimate.

105

At low temperature, $B_2(T) < 0$, and the attraction wins over the entropy. The chain is collapsed with an exponent $\nu = \frac{1}{3}$. The temperature θ is given by

$$\theta/v_0 = \frac{1}{\ln(\frac{7}{6})} \simeq 6.49 \tag{6}$$

At this temperature, the chain is almost ideal ($\nu = \frac{1}{2}$). Note that the real θ is expected to be lower than the estimate of eq.(6), since , for instance, we have neglected the helical rigidity ε_H.

3.2 Low temperature regime: the Hamiltonian walk approach

At low temperature, as mentioned above, the chain is collapsed with very few vacancies in the bulk. In this regime, it is reasonable to represent the chain as a Hamiltonian path on the lattice (i.e. a self-avoiding walk passing through each point of the lattice once and only once). Such Hamiltonian walks have frequently been used to model polymer chains in a bad solvent [9].

In this restricted model, the partition function reads:

$$Z = \sum_{\mathcal{H}} \exp(\beta \varepsilon_H N_H(\mathcal{H})) \tag{7}$$

where the sum over $\mathcal{H}$ means a sum over all Hamiltonian paths and $N_H(\mathcal{H})$ denotes the number of aligned pairs of links. This model can be studied using mean field theory [10,11,12]. The results are the following. As the temperature is decreased, the system undergoes a first order freezing transition at a temperature, T_m, approximately given by

$$T_m/\varepsilon_H = \frac{1}{\ln(\frac{4}{e-2})} \simeq 0.58. \tag{8}$$

This transition is driven by the the loss of entropy of the gas of corners in the ordered phase and the correspondoing gain in ε_H.

At temperatures immediately above T_m, the system consists of a liquid of corners with free energy per monomer given by

$$F = -T \ln(\frac{2(1 + 2e^{-\beta \varepsilon_H})}{e}) - \varepsilon_H. \tag{9}$$

Below T_m, the system is frozen in its (crystal-like) ground state configuration ($F = -\varepsilon_H$). The ground state consists of fully stretched configurations of the chain, with corners expelled at the surface of the lattice. Below T_m, the entropy per monomer is zero (or equivalently, the chain entropy scales like $n^{\frac{2}{3}}$). Above T_m, the entropy per monomer is finite and reads (see eq.(9))

$$S = \ln(\frac{2(1 + 2e^{-\beta \varepsilon_H})}{e}) + 2\beta \varepsilon_H \frac{e^{-\beta \varepsilon_H}}{1 + 2e^{-\beta \varepsilon_H}} \tag{10}$$

In terms of secondary structure, the system has still a sizeable amount of structure above T_m.For instance, the average length of helices (more precisely the average number l of consecutive aligned links) is given above the transition by

$$l = 1 + \frac{1}{2}e^{\beta \varepsilon_H} \tag{11}$$

and is infinite below. The thermodynamics of the freezing transition at T_m is summarized in figure 2.

For intermediate temperatures $(T_m < T < \theta)$, the chain is therefore collapsed with a mobile, liquid-like amount of secondary structure, and the corresponding configurational entropy. This intermediate phase is somewhat reminiscent of the molten globule phase described in the protein literature [3,13,14]. Of course, the simple model considered here does not include side-chains, so that their contribution to the entropy of formation of the molten-globule state is not taken into account [15].

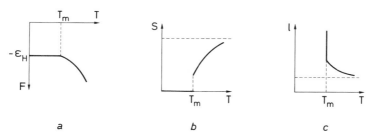

Figure 2. Summary of the melting transition: (a) The free energy (b) The entropy (c) The average "helix" length.

4. Monte Carlo simulations

To check these results, we have performed Monte Carlo (MC) simulations on a model chain. We have not used dynamical MC methods, which, due to the problem of high barriers and rugged landscapes, get rapidly trapped in the collapsed phases. We have used instead, a MC ensemble growth method. In this method an ensemble of chains is grown, group of monomers by group of monomers. The resulting conformations are then replicated (or deleted) according to their Boltzmann weights. The details of the method are described in ref [4,5]. For our purposes, we just note that this growth method lessens the trapping-in-local-minima-problem, and that one may follow various quantities during the growth of the chain (e.g. the radius of gyration), and deduce in one MC run how these quantities vary with the size n of the chain (e.g. the exponent ν of eq (5)).

The simulations were performed for a model chain of $n = 126$ monomers on a cubic lattice. The attractive energy v_0 was 2 kcal/mole and the helical rigidity ε_H was chosen to be 1 kcal/mole in order to represent a range of partially-folded ("swollen-globule" [16]) states. (Following the studies of Kolinski et al. [1] , a larger value of the rigidity parameter would narrow the range between the θ-point and the fully-folded state.) The total population of chains to build the ensemble was $M = 25000$, and five independent runs were made, so as to have error bars on the results. In order to further stabilize the fully folded structure at low temperature, we have worked with a periodic chain, consisting of sequences of four alanine-like links ($\varepsilon_H = 1$ kcal/mole) followed by two glycine-like links ($\varepsilon_H = 0$). A detailed comparison between the simulations and the analytical approximations can be found in ref [6]. Here, we just point out that we did in fact find the three phases (collapsed frozen; collapsed liquid; random coil) as the temperature was increased. The numerical transition points agree well with eqs (6) and (8), as does the exponent ν. A slight increase of the radius of gyration is also observed above the freezing temperature. Finally, we have made a detailed study of the intermediate phase, and have used this to suggest how the molten globule state joins on to a large spectrum of partially folded states below the θ–point(see ref [6] for details).

REFERENCES

[1] Kolinski A; Skolnick J; Yaris R. " Monte Carlo simulations on an equilibrium globular protein folding model." *Proc. Natl. Acad. Sci. USA* 1986 Oct, **83**(19):7267-71.

[2] T.E.Creighton, *Proteins*, W.H.Freeman, New York, (1984).

[3] O.B.Ptitsyn, "Protein Folding: Hypotheses and Experiments", *J. Prot. Chem.* **6**, 273-293 (1987).

[4] T. Garel and H. Orland, "Guided replication of random chains: a new Monte Carlo method", *J. Phys.* **A23**, L621-624 (1990).

[5] B. Velikson, T. Garel, J.C. Niel, H. Orland and J.C. Smith, "Conformational distribution of heptaalanine: analysis using a new Monte Carlo chain growth method", *J. Comp. Chem.* **13**, 1216-1233 (1992).

[6] J.Bascle, S.Doniach, T. Garel and H. Orland revised manuscript in preparation

[7] P.J. Flory, *Principles of Polymer Chemistry*, Cornell University Press, Ithaca, N.Y. (1971).

[8] P.G. de Gennes, *Scaling Concepts in Polymer Physics*, Cornell University Press, Ithaca, N.Y. (1979).

[9] J. des Cloizeaux and G. Jannink, *Polymers Solution, their Modelling and Structure*, Clarendon Press, (1990).

[10] J. Bascle, T. Garel and H. Orland, "Mean-field theory of polymer melting", *J. Phys. A* **25**, L1323-1329 (1992).

[11] J. Bascle, T. Garel and H. Orland, "Formation and stability of secondary structures in globular proteins", *J. Physique (France) II*, **3**, 245-253 (1993).

[12] J. Bascle, T. Garel and H. Orland, "Some physical approaches to protein folding", *J. Physique (France) I*, **3**, 259-275 (1993).

[13] Ptitsyn OB; Pain RH; Semisotnov GV; Zerovnik E; Razgulyaev OI. "Evidence for a molten globule state as a general intermediate in protein folding." *Febs Letters*, 1990 Mar 12, **262**(1):20-4.

[14] V.Dagget and M.Levitt, " A model of the molten globule from molecular dynamics simulations", *Proc. Natl. Acad. Sci. USA* **89**, 5142-5146 (1992).

[15] Shakhnovich EI; Finkelstein AV. "Theory of cooperative transitions in protein molecules. I. Why denaturation of globular protein is a first-order phase transition." *Biopolymers*, 1989 Oct, **28**(10):1667-80.

[16] Finkelstein AV; Shakhnovich EI. "Theory of cooperative transitions in protein molecules. II. Phase diagram for a protein molecule in solution." *Biopolymers*, 1989 Oct,**28**(10):1681-94.

HYDROPHOBIC ZIPPERS: A CONFORMATIONAL SEARCH STRATEGY FOR PROTEINS

Ken A. Dill and Klaus M. Fiebig

Department of Pharmaceutical Chemistry
University of California
San Francisco, California 94143-1204

We describe a computational method for constructing hydrophobically clustered compact conformations of polymer chains having hydrophobic and polar monomers. We call the method "hydrophobic zippers."[1,2] We have used hydrophobic zippers to construct conformational ensembles that resemble the compact denatured and molten globule states of proteins. We have also used hydrophobic zippers to attempt to find global minimum ("native") conformations of simple HP (H: hydrophobic, P: polar) copolymers in lattice models. In addition, we believe that the sequence of serial steps followed by the hydrophobic zippers construction process closely parallels the sequence of kinetic events in the fast stages of the folding processes of globular proteins.

The problem of protein folding kinetics is how a protein can search a vast conformational space on a short (biological) time scale (of the order of seconds - hours) to find the unique native state. The principal difficulty is in assembling the "nonlocal" contacts. We divide intrachain interactions into two types: local, involving near neighbors along the chain sequence, and nonlocal, involving monomers that are spatially adjacent in a given conformation that are not necessarily close together in the sequence. Local interactions, principally helical and turn propensities in proteins, are not the source of the computational difficulty, since these can be computed in a time that scales linearly with the chain length. The greater computational difficulty arises from trying to assemble an optimal hydrophobic core, i.e., trying to optimize the nonlocal interactions, since this is the basis for the combinatorial explosion of conformations. Hence, we view the folding problem principally as a problem of how the protein assembles an optimal hydrophobic core.

Hydrophobic zippers is not itself a kinetic process, nor does it involve "move sets," as Monte Carlo dynamics methods do. Each hydrophobic zipper is an ordered list, i.e., a conditional sequence of events. Each event involves an opportunistic pairing of two hydrophobic (H) monomers together in a way that is consistent with: (a) their connectivity in a chain and (b) other contacts already brought together as earlier events. Thus a hydrophobic zipper is a list of steps in a sequential assembly process, by which a monomer sequence is converted to a spatial conformation of the chain.

Using a simple 2-dimensional lattice model of HP copolymers, Figure 1 shows a hydrophobic zipper that causes the sequence (HPHPPHPHPHPH) to fold up into its globally optimal native conformation[1] (HH interactions are favorable, so the native state is the conformation with a maximum number of HH contacts). The zipper is the sequence of HH contacts in the order (3,6), (3,8), (1,8), (1,10), (1,12).

Statistical Mechanics, Protein Structure, and Protein Substrate Interactions
Edited by S. Doniach, Plenum Press, New York, 1994

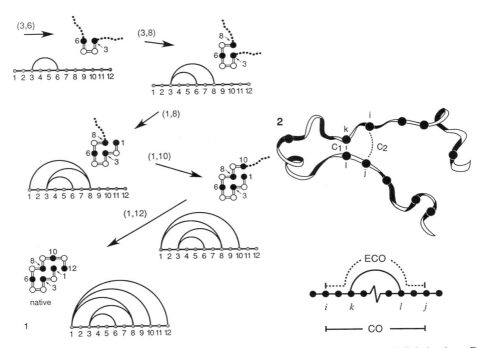

Figure 1. A hydrophobic zipper sequence of steps in folding a chain. Dark beads are H; light beads are P. Shown also is the polymer graph for each conformation: straight lines indicate covalent links between the numbered monomers, and curved links indicate HH contacts.

Figure 2. Effective contact order (ECO) is defined as a shortest path on the polymer graph (defined in the caption of Figure 1). If contact C1 is given, then the formation of contact C2 involves only a small conformational search, as if it were a chain with only $|k-i| + |j-l| + 1$ bonds.

Now we describe the hydrophobic zipper recipe for choosing the sequence of HH contacts. In the simplest implementation of zippers, the first decision is simply based on compiling the set of all pairs of H residues that are spaced $(i, i+3)$ in the sequence. These are the closest in sequence that could form a contact on a 2-dimensional square lattice. No tighter turn is possible on that lattice. From that set of potential HH pairs, we choose randomly, and form the first HH contact. In the example in Figure 1, contact (3,6) was the first chosen. The second decision is made in much the same way. To describe subsequent decisions, we first define two terms. Two monomers at positions i and $i+k$ that form a spatial contact are referred to as having *contact order* k. Figure 2 defines the *effective contact order,* $|k-i| + |j-l| + 1$ of a contact between monomers i and j, given an existing contact (k, l). Effective contact order defines a relationship between two HH contacts. With this terminology, we now describe the second and subsequent decisions in a hydrophobic zippers process. For the second decision, we choose from among the set of all remaining possible HH contacts that are either contact order 3 (i.e., $(i, i+3)$), or have effective contact order (ECO) 3, i.e., contacts that are brought into close spatial proximity by the formation of the preceding HH contact. In Figure 1, (3,8) has ECO 3 if (3, 6) has already been formed. Contact (3,8) was randomly chosen in this case as the second step of the hydrophobic zipper in Figure 1. The third and subsequent decisions follow in the same way: choices for the next HH contact are made randomly from among the set of contacts that have order 3 or ECO 3 at that stage of the process.

In this example, we choose only from among contacts of order 3 or ECO 3, but more generally zippers would also choose from contacts of order 5 or ECO 5, with a smaller probability, and from higher orders with even smaller probabilities, depending on the relative fractions of the conformational ensemble which are consistent with the given contact. The ECO of a contact is readily and unambiguously determined by finding the shortest path on the polymer graph, as indicated in Figure 3. There exist many simple algorithms for finding shortest paths through graphs. One hydrophobic zipper can be generated rapidly on a computer. For 100 monomer chains on 3-dimensional simple cubic lattices, each zipper takes a few seconds on a computer workstation.

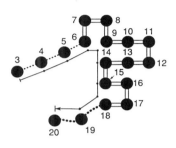

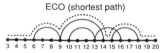

Figure 3. This shows a more general example defining ECO in a more complex conformation.

This recipe zips a sequence of hydrophobic and polar monomer units into a compact conformation having as many HH contacts as the algorithm can find. The zipping process terminates at *endstates*. Endstates are conformations that are either native states or "stuck." Stuck conformations are those for which the computer can find no more HH contacts that can be added to yield a new viable conformation.

Hydrophobic zippers often reach native states.[1,2] This has been determined for short chains in the 2-dimensional HP lattice model, for which all the native states are known by exhaustive enumeration,[3,4] and for longer HP chains on 3-dimensional simple cubic lattices,[5] for which native states are known by a discrete geometry hydrophobic core construction process described elsewhere.[6] In the 2D short chain models, a considerable fraction of all HP sequences can find their native states by hydrophobic zippers processes.[1] For longer 3D chains, computer time becomes a significant limitation.

Hydrophobic zippers is a nonexhaustive process. Why should it reach the unique native conformation so often? The native state is the conformation with the maximum number of HH contacts for a given total loss in conformational entropy (from the open conformations, to the one native state). Said the other way around, to achieve a native state we seek a process that will lose a minimum of conformational entropy per every HH contact formed. The small ECO contacts chosen by hydrophobic zippers are generally those for which the chain loses the least conditional entropy, relative to its prior conformation.[1] In a current implementation, hydrophobic zippers has found global minimum conformations of 50-mer HP chains on 3D cubic lattices.[5]

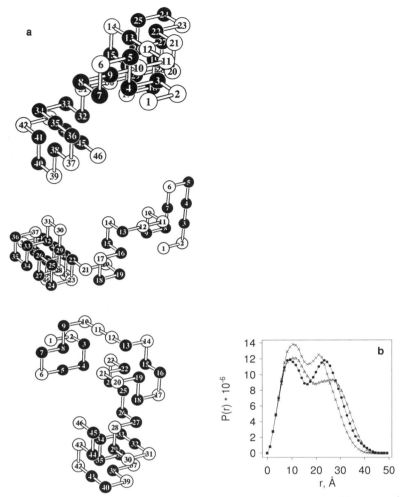

Figure 4a. 3 conformations representative of compact denatured states of a given HP sequence in the 3D cubic lattice model. The sequence and chain length are chosen to mimic the crambin protein.

Figure 4b. The p(r) curves for the 3 conformations in Figure 4a,[8] where $p(r)$ is the distribution of pairwise atomic distances r throughout the molecule.

For each HP sequence, we generate an ensemble of hydrophobic zippers. Most zippers do not reach native conformations. To seek native conformations, we choose the zipper endstates with the largest number of HH contacts. On the other hand, endstates which are not native appear to have properties of the compact denatured states of proteins. The stuck endstates are conformations at which hydrophobic zippers runs into an "entropy catastrophe." That is, at early stages, the number of HH contacts of order 3 or ECO 3 is usually large, provided a chain has a sufficiently large fraction of monomers that are H. However, as zipping proceeds, the pool from which potential HH contacts are chosen diminishes to the point at which the chain can no longer find additional HH contacts without considerable conformational entropy loss.

We believe the same sequence of events (as an ensemble of zippers) describes the fast stages (but probably not the slow stages) of the kinetics of folding of globular proteins. This is supported by a comparison with several lines of experimental evidence, reviewed elsewhere.[2] Zippers involve only forward steps, as would a kinetic process with small reverse rate. The slow stages of protein folding probably involve considerable barrier-climbing, i.e., breaking incorrect HH contacts in order to proceed further toward the native state,[7] whereas the zippers recipe as currently implemented has no provision for such reversals.

We believe the stuck endstates of hydrophobic zippers closely resemble compact denatured states.[8] Stuck zippers endstates tend to be only slightly larger in radius than native conformations; they have multiple hydrophobic clusters but no single well-developed core; a given HP sequence tends to have an ensemble of endstates with considerable similarities in the locations of their helices and turns and their hydrophobic clusters, but otherwise considerable differences in their conformations. Recently small angle x-ray scattering on 3 proteins[9-11] has shown that a bimodal pairwise atomic distribution function is a fingerprint of compact denatured states that distinguishes them from native or highly unfolded states. Figure 4 shows that hydrophobic zippers leads to such bimodal distributions.[8]

We do not believe that hydrophobic zippers are the only computational strategies that could lead to globally optimal states in reasonable computer times. On the contrary, the discrete geometry construction method[6] in its present implementation is faster, and we believe other methods too may be fast. Rather, hydrophobic zippers provides a justification and a physical model for kinetics and the mechanism by which proteins can find their global minima without exhaustive conformational searching, but it also does provide one strategy for quickly generating compact chain conformations that tend toward a high degree of hydrophobic clustering, as occur in native and denatured proteins.

REFERENCES

1. K.M. Fiebig and K.A. Dill, Protein core assembly processes, *J. Chem. Phys.* **98**:3475 (1993).
2. K.A. Dill, K.M. Fiebig, and H.S. Chan, Cooperativity in protein-folding kinetics, *Proc. Nat'l. Acad. Sci. USA* **90**:1942 (1993).
3. K.F. Lau and K.A. Dill, A lattice statistical mechanics model of the conformational and sequence spaces of proteins, *Macromolecules* **22**:3986 (1989).
4. H.S. Chan and K.A. Dill, "Sequence space soup" of proteins and copolymers, *J. Chem. Phys.* **95**:3775 (1991).
5. K.M. Fiebig and K.A. Dill (in preparation).
6. K. Yue and K.A. Dill, Sequence-structure relationships in proteins and copolymers, *Phys. Rev. E* **48**: 2267 (1993).
7. H.S. Chan and K.A. Dill, Energy landscapes and the collapse dynamics of homopolymers, *J. Chem. Phys.* **99**:2116 (1993).
8. E.E. Lattman, K.M. Fiebig, and K.A. Dill, Modeling compact denatured states of proteins, *Biochemistry* (in press).
9. T.R. Sosnick, and J. Trewhilla, Denatured states of ribonuclease A have compact dimensions and residual secondary structure, *Biochemistry* **31**:8329 (1992).
10. J.M. Flanagan, M. Kataoka, D. Shortle, and D.M. Engleman, Truncated staphylococcal nuclease is compact but disordered, *Proc. Natl. Acad. Sci. USA* **89**:748 (1992).
11. M. Kataoka, Y. Hagihara, K. Mihara, and Y. Goto, Molten globule of cytochrome *c* studied by small angle X-ray scattering, *J. Mol. Biol.* **229**:591 (1993).

THEORETICAL PERSPECTIVES ON IN VITRO AND IN VIVO PROTEIN FOLDING

D. Thirumalai

Institute for Physical Science and Technology and
Department of Chemistry
University of Maryland
College Park, Maryland 20742, USA

INTRODUCTION

It is now well established that in order to execute specific biological functions proteins must adopt well defined three dimensional structures.[1,2] The diverse biological demands placed on proteins are probably the reason why one observes seemingly large classes of protein structures. However, the process by which proteins acquire their three dimensional structure remains an unsolved problem in biochemistry and molecular biology. The text book description of the folding of proteins follows the classic work of Anfinsen on the spontaneous *in vitro* refolding of ribonuclease.[3] This work suggests that the protein folding is a self-assembly process which means that all the information needed for folding is contained in the various solvent induced interactions between the amino acids comprising a protein molecule. The self-assembly process does not require additional input of energy or other constraints. Since many proteins have been made to fold *in vitro* it follows that the major principles of protein folding can be formulated without the need to involve extra cellular components. It therefore follows that the protein folding problem reduces to a statistical mechanical description of finite sized branched heteropolymers.

Although the thermodynamic hypothesis of Anfinsen is well accepted this basic tenant has been questioned based largely on the famous Levinthal argument.[4] From a purely biological point of view one can argue that the necessity to carry out various biological functions in a finite time scale (typically ranging from seconds to several minutes) must also provide some constraints on the natural selection of foldable sequences, folding into a global free energy minimum, etc. It may well be that the functional protein conformations are only metastable local minima in the free energy hypersurface.[5] Although, the largely academic issue of whether the folded conformation is a global free energy minimum or belongs to one of the several accessible basin of attraction is not settled, it is clear that in a large number of cases folding kinetics and the prediction of the structure of the native state fall in the realm of physical chemistry.

Recent studies of *in vivo* assembly of nascent proteins into regular structures have shown that extra cellular cofactors and interactions with other proteins are necessary to

Statistical Mechanics, Protein Structure, and Protein Substrate Interactions
Edited by S. Doniach, Plenum Press, New York, 1994

achieve protein folding.[6]   The direct role of certain proteins, called molecular chaperones, has been established in the folding of proteins in both eukaryotes and prokaryotes.   The precise action of the molecular chaperones (MC) has not been established, however.   It is likely that MC prevent incorrect intramolecular interactions and prevent favorable interactions between different proteins.[6,7] Alternatively chaperone molecules may act to severely restrict the size of the conformational space of proteins thus enabling the protein to reach the energetically most favorable basin of attractions in finite times.   Regardless of the precise mechanism of chaperone action it is clear that their role has to be incorporated if a solution of the "second half of the genetic code" is sought.[6]

In this article, I describe our efforts to extract some general guiding principles underlying the kinetics of *in vitro* protein folding by studying several simple model proteins using a variety of simulation methods.   There are several groups who have used this basic strategy as well as other statistical mechanical approaches to understand the thermodynamics and kinetics of protein folding.[8-24]   Although it is not yet clear that this approach has provided many results of interest to experimentalists it is becoming increasingly evident that some basic concepts have emerged from these studies. In addition to the discussion of folding kinetics in simple models of proteins we also present a speculative treatment of the role played by molecular chaperones in the *in vivo* assembly of proteins.   The major purpose of this article is to show that both *in vitro* and *in vivo* assembly of proteins can be eventually understood in physicochemical terms.   The rest of this article is organized as follows: In Section II we present our studies of the folding kinetics of simple α-carbon representations of Crambin like protein using low friction Langevin dynamics.   The kinetic picture that emerges from these and related dynamic lattice Monte Carlo simulations are presented in Section III and the implication of these studies for in vitro folding are discussed.   In Section IV we present tentative schemes for chaperone mediated self assembly of proteins in vivo.   The article is concluded in Section V with additional remarks.

II. LOW FRICTION LANGEVIN SIMULATIONS OF MODEL PROTEINS

In order to extract a few general guidelines for *in vitro* protein folding kinetics we have constructed a model for a mock polypeptide chain (resembling Crambin) and followed the kinetics of approach to the native state starting from a fully denatured conformation.[5,25]   Since side chains are not modelled in our studies the heteropolymer can be thought of as an α-carbon representation of a protein.   We have used a variety of simulation techniques to investigate the dynamics of the unfolding to folding reaction. These methods are based on a combination of molecular dynamics, quench techniques, and simulated annealing techniques.   A summary of the methodology is provided in this section.   The details can be found elsewhere.[25]

A. Model

Our objective has been to propose "minimal" models of polypeptides the ground states of which resemble the various structural motifs found in proteins.   By "minimal" we mean the simplest model that captures the essential interactions in proteins.[26]   The minimal models contain some, but not all, of the features that play a role in imparting stability to globular proteins.   For example, our model for the polypeptide chain does not contain side chains, which are known to be responsible for intramolecular hydrogen bonding. The model does not explicitly take solvent molecules into account.   This may be especially important because recent studies indicate that molecular dynamics simulations that explicitly include the atomistic nature of water provide a far more realistic picture of protein dynamics.[27] Nevertheless since our primary interest is in gleaning certain general guidelines about kinetics of the folding process it is our expectation that the qualitative aspects of the results should not change due to the inherent limitations of the models. However, for the reasons presented above the picture of kinetics we suggest should be viewed as tentative.

The use of simple models to obtain generic features of complex phenomena is not new. This has been used extensively in the context of critical phenomena, studies of localization, spin glasses, etc. The idea of studying the behavior of protein, using coarse grained models has also been introduced quite some time ago.[28,29] However only recently concepts developed in other fields have been used to analyze in detail the behavior of idealized models of proteins. In fact, progress in our fundamental understanding of the protein folding phenomenon is perhaps largely attributable to detailed studies of simplified lattice and off-lattice models. The advantages of this approach is that pathways, landscapes for the folding kinetics can be explicitly constructed.[30] It is for these reasons that we began a series of studies to understand the folding of continuous heteropolymer models into β-barrel[4] and α-helical structures.[31] In this article we discuss our kinetic results for folding to a β-barrel structure. The potential energy functions we have developed are generic so that the native structures that result are not entirely due to the intrinsic bias in the energy surface.

Our models incorporate the following types of forces: (a) Long-range forces such as the hydrophobic interaction (for a review of the importance of hydrophobic effects in protein folding, see Ref. 26). Here "long range" refers to the interactions between hydrophobic residues that may be separated by long distance in sequence space as measured by the length along the backbone of the chains. The range of the potential in coordinate space is relatively short and is typically of the order of a few molecular diameters, i.e., about (4-5) Å. We have chosen a simple functional form to mimic these forces. (b) Simple bond angle potential between two successive bonds. (c) Dihedral angle potential due to the rotation about the peptide bond. The bonds are constrained to be of fixed length. Our model protein contains three different types of "residues" in the backbone of the chains, namely hydrophobic (B), hydrophilic (L), and neutral (N). The potential function describing the interaction between the residues is described below. The sequence we have used for the 46-member chains is

$$B_9N_3(LB)_4N_3B_9N_3(LB)_5L \quad . \tag{1}$$

In Eq. (1), LB indicates that an L-type bead is bonded by a B-type bead. We have done most of our simulation work on the n=46 model and have studied the n=58 case less extensively. In both cases we have established that the low temperature form indeed has a well-defined β-barrel structure. The time required for the atoms in the folded β-barrel states to achieve registry is quite long. It is for this reason the mapping of the various minima for the n=58 has not been thoroughly done. However, the robustness of our model in yielding β-barrels for a larger value of n (the number of beads) has been established. In this article we present results only for the n=46. The above sequence has been constructed so that if one has a folded structure, a three or four fold bend would emerge as long as n is greater than a certain minimum value. From simple energy considerations the minimum value of n turns out to be approximately 30. The neutral residues appear in regions where bends are desired for the native low temperature structure. These residues interact with each other only through a short-range repulsion. In real globular proteins these segments are responsible for loops and turns. It is now recognized that in many instances the enzymatic activity of proteins is predicated by the flexibility in the loop regions. The dihedral angle forces are assumed to be weaker for the bonds involving the neutral residues so that the bend formation is enhanced. This constraint proves to be extremely important and is the main reason for the folding of the chain in finite times. The Hamiltonian describing the model heteropolymer is given by the sum of the three forces. The functional form for each is described.

Long-Range Potential: The hydrophobic attraction, which occurs only between residues both of which are of type B, has the form

$$V_{BB}(r) = \varepsilon_h[(\sigma/r)^{12} - (\sigma/r)^6] \tag{2}$$

where r is the distance between the specified residues. For simplicity, σ is chosen equal to the bond length between two successive residues along the backbone and ε_h, and all lengths are given terms of σ. Thus $\varepsilon_h = \sigma = 1$ in reduced units. The interactions between L-L pairs or L-B pairs are governed by the following potential

$$V_{L\alpha}(r) = 4\varepsilon_L(\sigma/r)^{12} + (\sigma/r)^6] \quad \sigma = L \text{ or } B \tag{3}$$

where $\varepsilon_L = 2/3 \varepsilon_h$. Note that the above form is purely repulsive. The interaction between the N residues and L, B, or N residues takes the form

$$V_{N\alpha}(r) = 4\varepsilon_h(\sigma/r)^{12} \quad \alpha = N, L, \text{ or } B \tag{4}$$

Bond Angle Potential: The bond angle formed between two successive residues (k, k+1) is constrained by a simple harmonic potential,

$$V(\theta) = \frac{k_\theta}{2}(\theta - \theta_0)^2 \tag{5}$$

where $k_\theta = 20\varepsilon_h/(rad)^2$ and $\theta_0 = 105°$ (=1.8326 radians).

Dihedral Angle Potential: The potential for the dihedral angle ϕ is conveniently represented as

$$V(\phi) = A(1+\cos\phi) + B(1+\cos(3\phi)) \tag{6}$$

The dihedral potential has three minima, one at $\phi=0$ corresponding to the trans configuration, and two slightly higher minima at $\phi=\pm\cos^{-1}[(3B-A/12B)^{1/2}]$ corresponding to the two gauche states. For all the residues except those in the bend regions, $A=B=1.2\varepsilon_h$. For this choice of parameters the difference between the energies of the trans and gauche states is $1.75\varepsilon_h$. For the residues in the bend region we chose $A=0$ and $B=0.2\varepsilon_h$. It is clear that the bonds in the bends must adopt the gauche (or approximately gauche) conformations. Our choice of the parameter values in the bend regions (mostly involving the neutral residues) achieves this objective by having two major effects. First, it makes all the three minima shallower than when $A\neq 0$ and B is larger. This enhances the kinetics of bend formation by reducing the barrier between the various states. In addition setting $A=0$ makes the trans and gauche states of equal depth, thereby enhancing the thermodynamics of bend formation. Since the major objective of our investigation is to study the nature of various low temperature minima, we have used the simplest physically plausible potentials to ensure that the chain folds into well-defined β-barrel structures in finite time scale simulations.

B. Simulation Methods

Since our primary interest has been to study kinetics of protein folding we have used noisy molecular dynamics to follow the folding of our model protein from a denatured state to the folded β-barrel structure. Several groups have used Monte Carlo algorithms for inferring folding kinetics. It is well known that the dynamics in Monte Carlo studies can depend on the precise moves used and consequently the kinetics may be far less realistic than Langevin or molecular dynamics simulations. On the other hand the dynamics in MD does not suffer from similar criticisms and may represent actual *in vitro* kinetics more closely. (In defense of MC simulations it should be pointed out that Rey and Skolinck compared the dynamics in Monte Carlo and Langevin simulations in the context of protein folding and find very similar results.[32]) For these reasons we have used a modified molecular dynamics method based on the velocity form of the Verlet algorithm. We have also used a damping term with an appropriately chosen friction coefficient. A Gaussian random force is also added to compensate for energy loss due to friction so that fluctuation-dissipation theorem is satisfied. This ensures that these equations of motion lead to the correct equilibrium state at long times. There are several advantages in using noisy molecular dynamics in these simulations.[25] The most important one is that the use of friction and stochastic terms enhances the rate of exploration of the allowed conformation space. The main purpose of the random force is to control the temperature of the system in a physically reasonable way. This is particularly important because we are dealing with finite systems.

For simplicity we present the equations of motion for one Cartesian component of the position and velocity of a single residue. The constraints in our model, namely the fixed bond length between the beads, is enforced using the RATTLE algorithm.[33] The mass m of the particle and the Boltzmann constant k_B are set to unity. The equation of motion is

$$\ddot{r} = F_c + \Gamma - \zeta\dot{r} \equiv F \qquad (7)$$

where F_c is the conformational force given by the negative of the potential energy with respect to the particle position, Γ is the random force with white noise spectrum, and ζ is the friction coefficient. Notice that since the friction coefficient on the bead is taken to be small we retain the inertial term in Eq. (7). We use the velocity form of the Verlet algorithm to numerically integrate the above equation of motion. Since the force depends on the velocity the equations of motion have to be derived self-consistently. We find the following expression for r(t+h) and $\dot{r}$(t+h) through second order in the step size h^2

$$\dot{r}(t+h) = \dot{r}(t)\left[1 - \frac{h\zeta}{2} + (h\zeta)^2/4\right]$$

$$+ 1/2\ h\left[F_c(t) + F_c(t+h) + \Gamma(t+h)\right] \qquad (8a)$$

$$r(t+h) = r(t) + h\dot{r}(t) + \frac{1}{2}h^2\ F(t)) \quad . \qquad (8b)$$

Equation (8) is valid only in the limit of small hζ i.e. in the underdamped limit. In the simulations one uses the discrete version of the equations of motion. The random force which satisfies

$$\langle \Gamma(t)\ \Gamma(t') \rangle = 2\zeta\Gamma\delta(t-t') \tag{9}$$

is sampled at each time step using the standard Box-Muller algorithm.

III. RESULTS FOR KINETICS OF FOLDING

A. Native structure(s)

In order to obtain the ordered low temperature structures we used combinations of slow cooling from high temperatures, simulated annealing,[34] and steepest descent quench methods. The low temperature minimum energy conformations were computed using the following typical strategy. We first demonstrated that for our potential functions the low temperature structure by slow cooling is a well ordered β-barrel confirmation. A typical snapshot of a structure is shown in Fig. (1). The analysis of the β-barrel structures using a crude distance matrix approach indicates that our structures has elements of parallel and antiparallel strands. In order to refine the minima and to explore the existence of other basins of attraction we used a combination of simulated annealing and steepest descent quench methods. A trial β-barrel structure was constructed "manually" such that all bond angles away from the bend regions were set at the equilibrium minimum values and all the bonds away from the bends were made to be in the trans configuration. For convenience, each of the hairpin turns was made planar, resulting in an initially strained bond angles in the bend region. The energy of this structure was minimized by performing "dynamics" with the velocity of all the residues being set to zero after each time step. This is equivalent to the steepest descent quench because the system traverses the path in the free energy landscape to minimize the gradient. The structure that resulted from this had the desired folded conformation which we denote as $\beta1$. To explore other basins of attraction in which the heteropolymer adopts native like structure, we searched for β-barrel-like structures with energies less than that of the $\beta1$ structure. Starting from the $\beta1$ structure, the temperature of the system was raised, keeping it well below the unfolding transition temperature. The system was then repeatedly annealed for lengths of time which varied from 100τ to 300τ. Subsequently, the system was quenched to a local minimum, and the structural and energetic aspects of the minima were determined. This method of searching for the local minima is similar to the simulated annealing Monte Carlo technique. This procedure was repeated several times. The initial structure for a given run was taken to be the previous minimum conformation explored by the chain. In this process we obtained eight β-barrel structures. The various structural characteristics, namely the gross appearance and the radius of gyration, of these eight minima were almost identical. However, they differed in their energies, implying that almost all these are distinct metastable minima. It should be emphasized that we did not by any means exhaust all the minima that can be obtained using this procedure. However, the number of minima corresponding to the folded native like structure is not expected to scale exponentially with the number of residues. We also wish to stress that we have by no means shown that the simulated β-barrel structure with the lowest energy ($-49.57\ \varepsilon_h$) corresponds to a global minimum. The main point is that there appear to be several distinct basins of attraction in which the heteropolymer adopts very similar conformation. It is very likely the size of the basins (the logarithms of which is essentially the entropy) is largest for the native state. Detailed characterization of the energy landscape for lattice model of proteins seems to support this assertion.

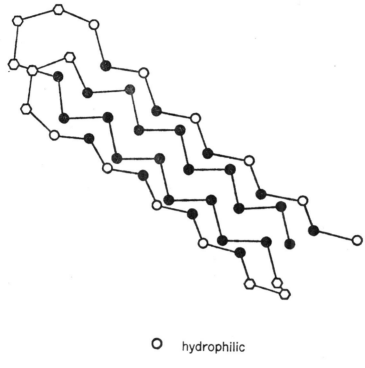

○   hydrophilic

●   hydrophobic

⬡   neutral

Figure 1. Typical beta barrel structure of the heteropolymer chains at low temperatures. The structure was obtained by a slow cooling process. The nature of the various residues for this chain consistingof 46 beads is shown by the various symbols.

B. Kinetics of Approach to the Native State

One of the advantages of using simulations is that we can monitor the detailed kinetics of approach to the native starting from a high temperature random coil configuration. As a part of our studies in the kinetics of folding we have introduced several novel correlation functions whose time dependence provide a microscopic picture of the folding process.[25,30,35] Here we describe the dynamics of one such correlation function which was intended to probe the time scale of approach to a compact conformation similar to that found in the native state. The correlation function is given by[25]

$$\Delta_\alpha(t) = \left\{ \frac{1}{n-1} \sum_{i=1}^{n} \left[\left(X_i^\alpha(t)\right)^2 - \left(X_i^o(t)\right)^2 \right]^2 \right\}^{1/4} \qquad (10)$$

where $X_i^\alpha(t)$ is the coordinate of the ith residue in the αth β-barrel and $X_i^o(t)$ is its corresponding value in the native state. For the latter we chose the coordinates of the conformation with the lowest energy. It would be more desirable to use $X_i^o(0)$ in Eq. (10) where $\{X_i^o(0)\}$ is the coordinate set corresponding the minimum energy conformer. At low enough temperatures we expect $X_i^o(t)$ to differ negligibly from $X_i^o(0)$. We have in fact checked this and do find that $\Delta_\alpha(t)$ for both choices is qualitatively the same. Notice that if $\alpha \neq 0$ and if there is no transition from the minimum labelled α to the lowest energy confirmation during the observational time scale then the longtime value of $\Delta_\alpha(t)$ will be nonzero. Thus the time dependence of $\Delta_\alpha(t)$ is a useful indicator of the kinetics of folding. In Fig. (2) a plot of $\Delta_\alpha(t)$ as a function of t is given. In order to set the relevance of the three temperatures we digress to briefly describe the changes in the structure that takes place in this system as the temperature is lowered. At a temperature $T_\theta(\approx 0.65\varepsilon_h/k_B)$ the heteropolymer collapses from an extended conformation to a compact (collapsed) state. This can be most directly seen from the position of the peak in the specific heat. More recently it has been shown using novel order parameters that the heteropolymer undergoes a first order freezing transition at $T_f(\approx 0.35\varepsilon_h/k_B)$ below which the protein adopts a well defined three dimensional β-barrel structure.[36] Thus the lowest temperature for which $\Delta_\alpha(t)$ is shown is for $T \approx 1.1 T_f$. It is interesting to note that, on the time scale of our simulations, $\Delta_\alpha(t)$ does not decay to zero indicating that there is a bottleneck between the minimum labelled α and the native state that exceeds $2\varepsilon_h$ (typical value of $\varepsilon_h = 1.5$-3.0 kcal/mole based on reasonable estimates of the effective hydrophobic interactions). Perhaps the most interesting observation we have made is that the transition to conformation to the native state occurs in two distinct phases. In the first phase $(t_I \approx (50$-$75)\tau)$ there is a rapid collapse to almost native like structure. A detailed analysis of the structure shows that in this time scale many of the contacts that are found in the native β-barrel is already established. After this fast phase there is a much slower decay in which there are rearrangements of the residues to achieve a conformation (almost) identical to that of the native state. This apparently requires overcoming a barrier or bottleneck implying that the transition state in the unfolding to folding reaction occurs late in the folding process. Similar findings were independently reported by Shakhnovich et al. based on Monte Carlo simulations of lattice models of proteins.[15] More recently Chan and Dill have argued that a two stage kinetics is also found in the dynamics of collapse of homopolymers.[37]

C. Scenario for Folding and Dynamical Scaling Exponent

The qualitative scenario of fast folding phase followed by a slower activated phase is a useful caricature of a multistep kinetic process that is probably characteristic of in the folding process. In order to provide a testable picture of folding Camacho and myself

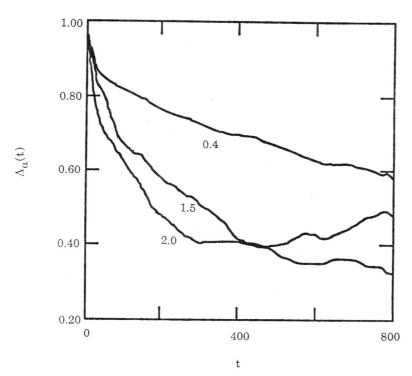

Figure 2. The compactness measure given by Eq. (10) as a function of t (in reduced units) at different temperatures. The numbers adjacent to each curve is the temperature in reduced units ε_h/k_B. The two lowest energy beta barrel structures generated by the simulated annealing method were used to calculate $\Delta_\alpha(t)$.

have studied in detail the kinetics of approach to the native state in a class of lattice models of proteins.[30] One of the models was introduced by Dill and coworkers while the other is a two dimensional version of a related lattice model introduced by Go et al.[38] In addition we introduced a third lattice representation of protein which is a generalization of the original Chan-Dill model.[30] For all these models we have performed exhaustive Monte Carlo simulations to study both thermodynamics and kinetics of folding. The advantage of these lattice models over the off-lattice ones described above is that in the former models enumeration of all allowed conformations for short enough chains is possible. Thus the kinetics and pathways of the folding process can be monitored in terms of the known conformations.[30] Furthermore we have devised numerical methods for computing the energy barriers separating several of the low lying conformations. Thus we have been able to interpret the kinetics of approach to the native state in terms of the underlying rough energy landscape.

The result of this study, the preliminary report of which is given elsewhere,[30] can be summarized as follows. Upon a quench from an infinite temperature to low temperatures we find that the folding process takes place in three distinct stages. In the first stage, labelled the kinetic phase, the heteropolymer chain collapses to a compact conformation perhaps driven by the effective attractive interactions between the hydrophobic residues. The relaxation of the correlation function in this regime is quite complex and may in fact be consistent with a stretched exponential behavior. In the second phase referred to as the kinetic reordering regime, the compact conformation undergoes cooperative rearrangement and the chain attains most of the native like contacts. At the end of this folding stage the molecule has compact almost native like structure. The last stage involves a transition from the almost native like conformation over a barrier to the native conformation. The analysis of several trajectories indicate that there are very few precursors that the molecule visits before making an activated transition to the native conformation. This implies that the kinetic reordering regime invariably terminates at one (or perhaps a few at best) well defined native like conformations. Thus the precursors and the ensuing transition state not only occur late in the folding process but also may be more or less unique. We also find that there are multiple pathways that lead to the precursor state in the unfolding→folding reaction and since this process happens relatively fast in the folding process it follows that experiments that probe the folding kinetics only on long time scales (i.e., greater than several milliseconds) should find that protein folding is invariably an all or none process. The three stage scenario that we have outlined above can be verified by studying the *in vitro* folding process from about 10^{-7} secs to several seconds. The fast folding experiments initiated by Eaton and coworkers will be invaluable in assessing the validity of the theoretical models and proving further guidance.[39]

An estimate of the time scales for the various kinetic regimes is needed for translating the above picture to an experimentally testable phenomenological theory. The first stage of collapse takes place in a time scale that depends on the surface tension γ between the residues and water, the strength of hydrophobic interactions, and N the effective number of amino acid residues. By following the arguments of deGennes[40] one can estimate that

$$\tau_I \approx \tau_o(\gamma, \varepsilon_h) N^2 \qquad (12)$$

where $\tau_o(\gamma, \varepsilon_h)$ is a microscopic time that depends on γ and ε_h the hydrophobic interaction. For N=100 τ_I turns out to be 10^{-6} secs where an estimate of $\tau(\gamma, \varepsilon_h) \approx 10^{-10}$ sec has been made. In our study we had conjectured that the time constant for the kinetic reordering phase scales as[30]

$$\tau_{II} \approx \tau_c N^\varsigma \qquad (13)$$

where we introduced the <u>dynamical folding exponent</u> ζ. We tentatively estimate $\zeta \approx 3$. The estimate in Eq. (13) was based partly on simulations and partly on the picture that the acquisition of native structure from the compact collapsed conformation takes place via a series of local cooperative motion. We wish to emphasize that the value of ζ should be considered as conjectural at this time. More detailed computations are required to obtain a better estimate. The time constant τ_c is extrapolated by comparison with the Monte Carlo simulation data and it is found that $\tau_c \approx 100$ Monte Carlo steps. If one MC step is assumed to correspond to about 10^{-10} sec (which is roughly the time needed to go from one conformation to another "nearby" one in the energy surface) then $\tau_c \approx 10^{-8}$ secs. With this estimate it follows that $\tau_{II} \approx 10^{-2}$ secs for N=100. The transition from the native like state that is reached in the second kinetic reordering regime to the native conformation occurs by overcoming an activation barrier and therefore

$$\tau_{Folding} \approx \tau_o \exp(\Delta F^{\ddagger}/k_B T) \qquad (14)$$

$\Delta F^{\ddagger}$ is the activation barrier. We have no basis to obtain the scaling of $\Delta F^{\ddagger}$ with N, and it is not even clear that this is the correct scaling variable. Since those sequences that

fold do so in finite times we conjecture that $\dfrac{\Delta F^{\ddagger}}{k_B T}$ cannot depend significantly on N.

This dependence can at best be weak and it is possible that the free energy of activation $\Delta F^{\ddagger}$ only grows logarithmically with N, i.e., $\Delta F^{\ddagger} \approx \Delta F_o \ln N$. The scenario that is roughly described by Eqs. (12-14) can in principle be tested experimentally provided the kinetics of the folding reaction can be monitored from 10^{-6} sec time scale.

IV. CHAPERONE MEDIATED ASSEMBLY OF PROTEINS - A TENTATIVE MODEL

The *in vitro* folding studies clearly demonstrate that all the information needed for the acquisition of the three dimensional structure in proteins is contained in the amino acid sequence i.e., protein folding can be viewed as a self-assembly process in a well defined environment (aqueous medium, pH, salt concentration, etc.). However numerous recent *in vivo* folding studies have clearly demonstrated that certain classes of proteins are implicated starting from the translational process till the assembly of the target proteins into multimeric complexes.[6,7,41-43] The action of these cofactors, commonly referred to as "molecular chaperones", is not understood in terms of physiochemical terms. It should be emphasized that the basic premise that the sequence does specify structure has remained unchallenged. The major modification in this traditional picture is that when off-pathway processes compete either for kinetic or thermodynamic reasons protein folding can only occur in the presence of molecular chaperones. It is clear that at the very least the molecular chaperones prevent favorable interproteins interactions that can occur in the concentrated cell environment. These attractive interactions would lead to aggregation and by preventing this outcome the chaperones assist in a "non-invasive" manner in the assembly of proteins in cells.[6] The purpose of this section is to show that one can describe the interaction of chaperones with the evolving polypeptide chains using methods developed in colloid and polymer science.[44] This view point serves to emphasize that even *in vivo* folding of proteins can be addressed in terms of statistical mechanical models. The predictions of the results presented here can be checked either by simulations or by direct experiments.

It has been shown that molecular chaperones belonging to the class of heat shock proteins (Hsps) are involved in the folding of mitochondrial proteins. The two proteins among the family of the Hsps most commonly involved in the assembly process are

Hsp70 and Hsp60.[41] Beckmann et al. showed that cytosolic Hsp70 interacts with the nascent polypeptide chain while it is still in the ribosome.[45] In addition a few studies have also established that the Hsp70 bound to the nascent protein, which appears to be only in a partially folded conformation, is transported across mitochondrial membranes. In the mitochondria the assembly of the partially folded chain is completed in the presence of the chaperonins Hsp60 and perhaps Hsp10 as well.[46,47] These events are assumed to occur in sequence i.e., the Hsp70 protein complex is first formed. This transient complex, whose formation is adenosine triphosphate (ATP) dependent,[46] is then transported to mitochondria in a "translocation competent" state. Here the Hsp70 is released and the subsequent final assembly takes place in the presence of Hsp60 and perhaps other cofactors. Similar findings have been reported for the assembly of proteins in Escherichia coli (E. coli) where the analogues of the Hsp protein, namely Dnak, Draj and GroEl have been identified.[47] Thus there probably is a unified mechanism involved in the chaperone mediated protein assembly. It should be emphasized that except for the initial step in which a binary complex between Hsp70 (or Hsp60) and the target protein is formed the other sequence of reactions outlined above has not been well understood.

We propose a speculative model that rationalizes a few experimental facts concerning the role of chaperones and perhaps gives some clues on the use of simple statistical mechanical ideas in addressing the interactions between chaperone molecules and the folding polypeptide chain. It is believed that one of the functions of Hsps (and their analogues in E. coli) is to prevent proteins from aggregation. This problem is not that relevant in *in vitro* folding because the concentration of proteins can be made to be small enough so that the refolding takes place in "isolation". However due to the concentrated environment in cells aggregation between partially folded proteins is a severe problem. Thus it is believed that one of the roles of chaperone is to prevent the dominant off-pathway reaction leading to aggregation.

We tentatively suggest that the chaperone molecules use the following mechanism to prevent aggregation in the *in vivo* folding process: by suitable interaction (basically by adsorbing onto the exposed hydrophobic patch of the partially folded polypeptide chain) between Hsp70 and the protein aggregation is prevented. The approximate theoretical model which hinges on the following observations gives the criterion for stability of the binary complex. It is known that Hsp70 does not bind to the folded protein. In addition it is known that chaperones bind to a variety of non-native conformations of proteins in E. coli leading to the suggestion that chaperones are highly "promiscuous". Since they interact with a multitude of non-native states of proteins it is most likely that the nature of these interactions are the same. In the process of folding the partially folded conformations expose groups of hydrophobic residues and if the protein concentration is high enough then these exposed hydrophobic particles attract each other leading to aggregation. We suggest that the Hsp70 binds to these hydrophobic patches and serves as "bumpers" so that the net interaction between two Hsp70 bound to partially folded proteins is repulsive. We will refer to the Hsp bonded to the partially folded protein complex as Hsp70-MG to indicate that the target protein is perhaps in a "molten globule" state. In order to provide a quantitative expression of the interaction between two Hsp70-MG molecules we assume that Hsp70-MG is roughly spherical and write

$$V(r) = V_R(r) + V_A(r) \tag{15}$$

where $V_R(r)$ and $V_A(r)$ are the repulsive and attractive contributions to the free energy of interaction $V(r)$. A schematic sketch of interaction between two Hsp-MG is shown schematically in Fig. (3). In order to estimate $V_R(r)$ we assume that Hsp70 is "adsorbed" onto the exposed hydrophobic patch. In this picture the rest of the partially folded

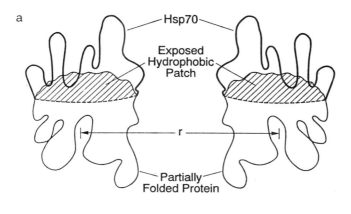

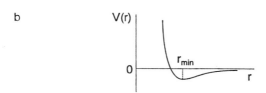

Figure 3. A schematic sketch of the picture of interaction between two molecules formed by the binding of Hsp70 to a target protein. (a) Shows Hsp70 adsorbed onto the exposed hydrophic patch with the rest of the partially folded protein acting as a substrate. Hsp70 acts as a "bumper" thereby preventing aggregation. (b) Sketch of the interaction energy between two complexes shown in (a). Aggregation is prevented provided r_{min} satisfies Eq. (19) and the Hamekar constant satisfies the bound $o \leq A < 12\ k_B T$.

protein is essentially a substrate which for purposes of computing $V_R(r)$ we take to be flat. Then the repulsive interaction between two Hsp70-MG boils down to computing $V_R(r)$ between two flat plates with adsorbed chains. The result can be computed using scaling arguments and one gets[49]

$$V_R(r) \simeq (k_B T/r^2) \lambda_H^2 \qquad (16)$$

where $\lambda_H = \sqrt{a_h}$ with a_h being the area of the exposed hydrophobic surface. The attractive interaction is the van der Waals potential of mean force between two nearly spherical Hsp70-MG and can be taken to be[50]

$$V_A(r) \simeq -AR/12r \qquad (17)$$

where A is the Hamaker constant,[50] R is the size of the spherical Hsp70-MG complex. Thus the net interaction is

$$V(r) \simeq \frac{k_B T \lambda_H^2}{r^2} - \frac{AR}{12r} \qquad (18)$$

which has a minimum at $r_{min} = \left(\frac{24 k_B T}{A}\right) \cdot \left(\frac{\lambda_H^2}{R}\right)$. This analysis suggests that aggregation is prevented if $r_{min} >> R$ which implies

$$\left(\frac{24 k_B T}{A}\right) \left(\frac{\lambda_H^2}{R^2}\right) > 1 \qquad (19)$$

An estimate of $(\lambda_H/R)^2$ can be made by assuming that $\lambda_H \approx f_H N^{1/3} \ell$ where f_H is the fraction of hydrophobic residues, N the number of amino acids and ℓ is the average size of an amino acid residue. This gives a maximum value of $(\lambda_H/R)^2 \approx f_H^2 \approx 1/4$ and the inequality (19) is easily satisfied provided A satisfies the bound $0 \leq A < 12 \, k_B T$. This bound is derived by the following argument. The value of the interaction potential at $r = r_{min}$ is $V_{min} = -\frac{1}{k_B T} (AR/24\lambda_H)^2$. For Hsp70 to be an effective bumper in addition to the requirement in equality given by Eq. (19) we should also have the condition that V_{min} not be very deep. If V_{min} were very attractive then once again strong association between the [Hsp70-MG] complex is possible. This can be prevented if $V_{min}/k_B T \sim 0$ (1) and this leads to the condition that $A < 12 k_B T$. The condition given by Eq. (19) and the limit on the value of A should be satisfied for Hsp70 to function as efficient bumper that prevent protein aggregation. The predictions of the model, namely Eq. (19) and the bound on A, can be checked directly by experiments.

Although we have derived Eqs. (18) and (19) for the special case of Hsp70-MG complex we note that similar potential of mean force is expected for interaction between other chaperones bound to target proteins. In particular we think the same form of the potential should also characterize the interaction between the complex formed by binding chaperonin to the target protein. Of course the parameters, A, R, and λ_H that determine the strength of interaction will vary.

If the condition in Eq. (19) is satisfied the binary complex Hsp70-MG is stabilized and this complex can then be transported to the mitochondria. The release of the Hsp70 from the binary complex would generally require energy and this reaction can therefore occur only by the hydrolysis of ATP. Thus the mechanism in the early stages of *in vivo* folding is roughly the following:

$$[MG] + [Hsp70] \rightarrow [Hsp70\text{-}MG] \overset{nATP}{\rightarrow} MG^* + n(ADP+P_i) \qquad (20)$$

Thus in this model the only role of ATP is to provide energy for release of Hsp70. The second stage of the above reaction is expected to occur inside the mitochondrial matrix. We have denoted the conformation of the target portion upon release by MG^* to indicate that it may be different from MG the conformation that is bound to Hsp70. If the stoichiometry of the moles of ATP hydrolysis is known then the effective (adsorption) binding energy between the Hsp70 and MG can be estimated using the formula[51]

$$\varepsilon = (n\Delta/Nk_BT)^{2/5} \quad . \qquad (21)$$

In Eq. (21) $\varepsilon k_B T$ is the average interaction energy per residue between Hsp70 and MG. In order to derive Eq. (21) we assume that there is an average binding energy per residue ε between the chaperone molecule and the exposed hydrophobic patch of the target protein. Strictly speaking we should consider the free energy gain by the adsorption of Hsp70 on to the rough surface of the protein. In the spirit of obtaining a tractable expression we model the protein as essentially a flat surface and consider the gain in free energy by a protein chain (Hsp70) which gains on an average free energy ε kcal/mole per residue by the adsorption process. This problem of estimating the binding energy can be solved by standard energy balance arguments. The resulting free energy is given by[49]

$$F/k_BT \simeq -\varepsilon\left(\frac{a_R}{D}\right)N + \left(\frac{a_R}{D}\right)^{5/3}N \qquad (22a)$$

where D is the thickness of the adsorbed layer, a_R is the average size of one residue, and N is the total number of residues. The free energy is obtained by minimizing Eq. (21) with respect to D and one finds

$$F/k_BT \simeq -N\varepsilon^{5/2} \quad . \qquad (22b)$$

If n is the stoichiometry of ATP involved in the release of Hsp70 from the [Hsp70-MG] complex and Δ is the available free energy upon hydrolysis ATP then Eq. (21) follows by equaling $n\Delta$ and $N\varepsilon^{5/2}$. If $n = 2$, $\Delta \approx 12$ kcal/mole then Eq. (21) predicts that the effective attractive interaction energy per residue between Hsp70 and the target protein is approximately 0.42 kcal/mole for $N = 100$ at $T = 300$ k.

Upon release of MG^*, the concentration of which is ATP dependent, it is generally accepted that another binary complex between chaperonin 60 and MG^* is formed. The manner in which folding occurs when the target protein is associated with chaperonin 60 is not at all understood. It may even be possible that in this stage other cofactors (such as chaperonin Hsp10) are involved. The current experimental situation is that there appears to be no unifying mechanism to describe the events that take place when MG^* is bound to chaperonin 60 and folding ensues. A tentative scheme for the expected sequence of events has been given by Lorimer et al.[52]

The basic conclusion of our analysis is that the approximate forces between Hsp70 bound to the target protein should have the form given in Eq. (18). Since Hsp70 binds to non-native forms of many proteins it is tempting to suggest that whatever is the form of interaction potential it should be generic. The interaction potential between Hsp70-MG complex can in principle be measured by atomic force microscopy (AFM).[53] This experiment should be feasible because in general chaperone bound to target proteins are very stable. In fact the binary chaperonin 60-rubisco I complex (the structure of which is the same as the complex found *in vivo*) has been isolated by gel filtration chromatography Thus a direct measurement of the potential of interaction between the molecules formed by association of chaperones to the non-native forms of the protein by AFM may help in unraveling the role of chaperones in *in vivo* assembly of proteins.

V. CONCLUDING REMARKS

In this paper I have tried to argue that both *in vitro* and *in vivo* folding can be modeled and analyzed using the principles of statistical mechanical techniques. The argument that traditional theoretical approaches to complex problems can be of use in the study of protein folding is based on the observation that in both cases proteins reach a well defined three dimensional state namely the "native" conformation. This suggests that the vectorial nature of the synthesis of protein in cells (which starts from the N terminal) is of little consequence in the formation of final three dimensional structure. Thus the theoretical ideas developed in the context of *in vitro* protein folding will ultimately be of use in understanding the more complex case of assisted self-assembly of proteins in cellular environment. In this spirit I have suggested possible ways of analyzing protein folding in terms of physiochemical principles.

Perhaps a fruitful way of providing a unifying framework of describing protein folding as a strictly self-assembly process and assisted self-assembly is in terms of the underlying rough energy landscape.[54,55,56] It has been repeatedly argued that many of the complex kinetic behavior observed in proteins is due to the presence of several conformational states separated by barriers ranging from few hundredths kcal/mole to several kcal/mole. This notion forms the basis of the spin glass paradigm of the protein folding problem. Explicit computation of the energy landscape for short heteropolymer models of proteins support this picture to a large extent.[30] In terms of the energy landscape of proteins it is clear that there are many metastable minima. If the volume of the basis of attraction corresponding to these metastable minima is large the folding protein can be kinetically get trapped in one of them. The pathways leading to these would give rise to misfolded structures. However sequences which do find the "native" state in finite times avoid deep glassy traps. Since the occurrence of such glassy states is unavoidable in heterogeneous systems, such as proteins, it follows that in designing foldable sequences one should attempt to minimize the size of the basins of attraction corresponding to glassy configurations. Then the pathways to these unfolded glassy states become effectively blocked, and the protein is more likely to find the native state in finite time scales. If the energy landscape can be quantified in a systematic way (i.e. the factors that govern the size of conformation space corresponding to glassy conformations can be predicted) then the effect of mutations on proteins can be predicted. This would then make protein engineering practical in the de novo design of novel proteins.

Our kinetic scenario embodied in Eqs. (12-14) suggests that unless the structure that develops in the kinetic reordering regime is almost native-like the chances of the protein locating the native conformation becomes small. For those sequences which do fold in finite times the barrier separating the native-like state and the native conformation is not large. We also find that if the energy landscape is smooth with a "large" basin of attraction corresponding to the native state then the folding occurs rapidly. In these situations the misfolded basins of attractions are avoided. These observations, quantified by Eqs.

(12-14) for sequences that reach the native state, are in harmony with the previous discussions based on energy landscape.

The energy landscape picture for describing *in vitro* folding may also be useful in describing chaperone mediated protein assembly especially in those situations that require ATP hydrolysis. It has been conjectured sometime ago that one of the possible roles of stress proteins (of which the chaperones belonging to the heat shock proteins are examples) is to reconstitute already misfolded structures. According to the energy landscape picture the chaperones in the presence of ATP provide the energy and pathway to circumvent the barrier separating the misfolded structure and a structure committed to the formation of native conformation. The barrier height can be directly inferred if the stoichiometry for ATP hydrolysis is known. According to this scenario the ATP dependent role of chaperones should be viewed as catalyst that reverses the dead end reaction leading to the formation of misfolded structures.

The interaction potential between the two binary complexes [Hsp70-MG] that we derived in the previous section is based on the supposition that Hsp70 adsorbs onto the exposed hydrophic patch of the non-native state of the target protein. We wish to emphasize that, although the basic assumption that the primary function of Hsp70 is to prevent aggregation between the protein molecules is likely to be correct, it is entirely possible that the equations we have derived for the stability condition may not be valid. An alternative form of the interaction between Hsp70 (or the chaperonin Hsp60) bound to the MG state of the protein can be obtained by assuming that the ends of the Hsp70 is grafted onto the hydrophobic region of the protein. This situation also predicts stability under certain conditions. However the binary complex, namely [Hsp70-MG], is only metastable and there is a barrier to aggregation.[51] If the size of the barrier exceeds k_BT then Hsp70 is successful in stabilizing the [Hsp70-MG] complex. We do not favor this scenario because grafting implies formation of a covalent bond between the residues of the Hsp70 and the target protein. The breaking of this bond would require considerable energy and therefore this mechanism is unlikely. Although the force law we have derived is based on several tenuous assumptions it is clear that one can describe the key features of the interactions between chaperones and the target protein in terms of concepts developed in the context of polymer and colloid science. It is likely that the experimental determination of the mechanism of stabilization can be inferred by appropriate atomic force microscopy (AFM) measurements.[53]

Acknowledgments

I wish to thank Carlos Camacho, Zhuyan Guo, and J. D. Honeycutt for developing the ideas presented here. I am grateful to Bill Eaton for pointing out relevant references. The work on chaperone mediated assembly of proteins was done during the meeting on proteins in Cargese in June 1993 and I am grateful to several participants for useful discussions. I especially wish to thank Prof. Buzz Baldwin and Prof. Michel Goldberg for giving me invaluable advice. Our work has been supported by a grant from the National Science Foundation.

REFERENCES

1. R. Jaenicke, "Protein Structure and Protein Engineering", (Mossbach Colloq. 39, Springer-Verlag, Berlin, 1988).
2. R. Jaenicke, "Protein Folding: Local Structures, Domains, Subunits, and Assemblies", Biochemistry 30:3147 (1991).

3. C.B. Anfinsen, Principles that Govern the Folding of Protein Chains, <u>Science</u> <u>181</u>:223 (1973).

4. C. Levinthal in "Mossbauer Spectroscopy in Biological Systems" (Proceedings of a meeting held in Allerton House, Monticello, Illinois, University of Illinois, Press, 1969).

5. J.D. Honeycutt and D. Thirumalai, Metastability of the Folded States of Globular Proteins, <u>Proc</u>. <u>Natl</u>. <u>Acad</u>. <u>Sci</u>. (<u>USA</u>) <u>87</u>:3526 (1990).

6. R.J. Ellis and S.M. Van der Vies, "Molecular Chaperones", <u>Ann</u>. <u>Rev</u>. <u>Biochem</u> <u>60</u>: 321 (1991).7. R.B. Freedman in "Protein Folding", Edited by T. E. Creighton 455 (W. H. Freeman and Co., New York, 1992).

8. J.D. Bryngelson and P.G. Wolynes, Spin Glasses and the Statistical Mechanics of Protein Folding, <u>Proc</u>. <u>Natl</u>. <u>Acad</u>. <u>Sci</u>. <u>84</u>:7524 (1987); Intermediates and Barrier Crossing in a Random Energy Model (with Applications to Protein Folding), <u>J</u>. <u>Phys</u>. <u>Chem</u>. <u>93</u>:6902 (1989).

9. M.S. Friedrichs and P.G. Wolynes, Toward Protein Tertiary Structure Recognition by Means of Associative Memory Hamiltonian, <u>Science</u> <u>246</u>:371 (1989).

10. T. Garel and H. Orland, Mean Field Model for Protein Folding, <u>Europhys</u>. <u>Lett</u>. <u>6</u>:307 (1988).

11. T. Garel and H. Orland, Chemical Sequence and Spatial Structure in Simple Models of Biopolymers <u>Europhys</u>. <u>Lett</u>. <u>8</u>:327 (1988).

12. J. Bascle, T. Garel, and H. Orland, Some Physical Approaches to Protein Folding, <u>J</u>. <u>Phys</u>. <u>I</u> (<u>France</u>), 3:259 (1993).

13. E.I. Shakhnovich and A.M. Gutin, Frozen States of Disordered Heteropolymers <u>J</u>. <u>Phys</u>. <u>A22</u>:1647 (1989).

14. E.I. Shakhnovich and A. M. Gutin, Implication of Thermodynamics of Protein Folding for Evolution of Primary Sequences <u>Nature</u> 346:773 (1990).

15. E.J. Shakhnovich, G. Farztdinov, A.M. Gutin, and M. Karplus, Protein Folding Bottlenecks: A Lattice Monte Carlo Simulation, <u>Phys</u>. <u>Rev</u>. <u>Lett</u>. 67:1665 (1991).

16. H.S. Chan and K.A. Dill, Origins of Structure in Globular Proteins, <u>Proc</u>. <u>Natl</u>. <u>Acad</u>. <u>Sci</u>. (<u>USA</u>) 87:6388 (1990).

17. H.S. Chan and K.A. Dill, Compact Polymers <u>Macromolecules</u> 22:4559 (1989).

18. K.M. Fiebig and K.A. Dill, Protein Cone Assembly Processes, <u>J</u>, <u>Chem</u>. <u>Phys</u>. 98:3475 (1993).

19. J. Skolnick and A. Kolinski, Simulations of the Folding of a Globular Protein, <u>Science</u> 250:1121 (1990).

20. J. Skolnick, A. Kolinski, and R. Yaris, Monte Carlo Simulations of the Folding of Beta Barrel Globular Proteins, <u>Proc</u>. <u>Natl</u>. <u>Acad</u>. <u>Sci</u>. (<u>USA</u>), 86:5057 (1988).

21. D.G. Garrett, K. Kastella, and D.M. Ferguson, New Results on Protein Folding From Simulated Annealing, <u>J</u>. <u>Am</u>. <u>Chem</u>. <u>Soc</u>. 114:6555 (1992).

22. P. Leopold, M. Montal, and J.N. Onuchic, Protein Folding Funnels: A Kinetic Approach to the Sequence-Structure Relationship, <u>Proc</u>. <u>Natl</u>. <u>Acad</u>. <u>Sci</u>. (<u>USA</u>) 89:8721 (1991).

23. G. Iori, E. Marinari, and G. Parisi, Random Self-Interacting Chains: A Mechanism for Protein Folding, <u>J</u>. <u>Phys</u>. <u>A24</u>:5439 (1991).

24. D.G. Covell and R.L. Jernigan, Conformations of Folded Proteins in Restricted Spans, <u>Biochemistry</u> 29:3287 (1990).

25. J.D. Honeycutt and D. Thirumalai, The Nature of Folded States of Globular Proteins, <u>Biopolymers</u> 32:695 (1992).

26. K. A. Dill, Dominant Forces in Protein Folding, <u>Biochemistry</u> 29:7133 (1990).

27. M. Levitt and R. Sharon, Accurate Simulation of Protein Dynamics in Solution, <u>Proc</u>. <u>Natl</u>. <u>Acad</u>. <u>Sci</u>. USA, 85:8557 (1988).

28. H. Taketomi, Y. Ueda, and N. Go, Studies on Protein Folding, Unfolding and Fluctuations by Computer Simulation, Int. J. Peptide Protein Res., 7:445 (1975).

29. M. Levitt and A. Warshel, Computer Simulation of Protein Folding, Nature 253:694 (1975).

30. C.J. Camacho and D. Thirumalai, Kinetics and Thermodynamics of Folding in Model Proteins, Proc. Natl. Acad. Sci. (USA) 90:6369 (1993).

31. Z. Guo and D. Thirumalai, Kinetics of Helix Formation, Preprint.

32. A. Rey and J. Skolnick, Comparison of Lattice Monte Carlo Dynamics Brownian Dynamics Folding Pathways of Alpha-Helical Hairpins, Chem. Phys. 158:199 (1991).

33. H.C. Andersen, Rattle: A "Velocity" Version of the Shake Algorithm for Molecular Dynamics Calculations, J. Comp. Phys. 52:24 (1983).

34. S. Kirkpatrick, C. D. Gelatt, and M. P. Vecchi, Optimization by Simulated Annealing, Science 220:671 (1983).

35. Z. Guo, D. Thirumalai, and J.D. Honeycutt, Folding Kinetics of Proteins: A Model Study, J. Chem. Phys. 97:525 (1992).

36. Z. Guo and D. Thirumalai, Kinetics of Protein Folding: Time Scales and Pathways, Preprint.

37. H.S. Chan and K.A. Dill, Energy Landscapes and Collapse Dynamics of Homopolymers, J. Chem. Phys. 99:2116 (1993).

38. Y. Ueda, H. Taketomi, and N. Go, Studies on Protein Folding, Unfolding, and Fluctuations by Computer Simulation II. A Three Dimensional Lattice Model of Lysozyme, Biopolymers 17:1531 (1978).

39. C.M. Jones, E.R. Henry, Y. Hu, C.-K. Chan, S.D.Luck, A. Bhuyan, H. Roder, J. Hofrichter, and W.A. Eaton, Fast Events in Protein Folding Initiated by Nanosecond Laser Photolysis, Preprint.

40. P.G. de Gennes, Kinetics of Collapse for a Flexible Coil, J. Physique Lett. 46:L639 (1985).

41. J.P. Hendrick and F.-U. Hartl, Molecular Chaperone Functions of Heat Shock Proteins, Ann. Rev. Biochem. 62:349 (1993).

42. E.A. Craig, Chaperones: Helpers Along Pathways to Protein Folding, Science, 260:1902 (1993).

43. D.A. Agard, To Fold or Not to Fold .... Science 260:1903 (1993).

44. J.N. Israelachvili in "Intermolecular and Surface Forces", (Academic Press, London, 1991).

45. R.P. Beckman, L.A. Mizzan, and W.J. Welch, Interaction of Hsp70 with Newly Synthesized Proteins: Implications for Protein Folding and Assembly Science 248:850 (1990).

46. M.Y. Cheng, F.-U. Hartl, J. Martin, R.A. Pollock, F. Kalousek, W. Neupert, E.M. Hallberg, R.L. Hallberg, and A.L. Horwich, Mitochondrial Heat-Shock Protein Hsp60 is Essential for Assembly of Proteins Imported into Yeast Mitochondria Nature 337:620 (1989).

47. T. Langer, C. Lu, H. Echols, J. Flanagan, M.K. Hayer, and F.-U. Hartl, Successive Action of Dnak, Dnaj and GroEl along the Pathway of Chaperone-Mediated Protein Folding Nature 356:683 (1992).

48. P.V. Viitanen, A.A. Gatenby, and G.H. Lorimer, Purified Chaperonin 60 (GroEl) Interacts with the Non-Native States of a Multitude of Escherichia coli Proteins, Protein Sci. 1:363 (1992).

49. P.G. de Gennes in "Scaling Concepts in Polymer Physics:, (Cornell University Press, Ithaca, New York, 1985).

50. D.H. Napper in "Polymeric Stabilization of Colloidal Dispersions", (Academic Press, New York, 1983).

51. D. Thirumalai, Interaction Between Chaperone Bound Protein Complexes, Preprint.
52. G.H. Lorimer, M.J. Todd, and P. Viitanen, Chaperonins and Protein Folding: Unity and Disunity of Mechanisms, Phil. Trans. R. Soc. Lond. B 339:297 (1993).
53. D. Sarid in "Scanning Force Microscopy with Applications to Electric, Magnetic, and Atomic Forces", (Oxford University Press, Cambridge, 1993).
54. R.E. Elber and M. Karplus, Multiple Conformational States of Proteins: A Molecular Dynamics Analysis of Proteins, Science 235:318 (1987).
55. H. Frauenfelder, S. Sligar, and P.G. Wolynes, The Energy Landscapes and Motions of Proteins, Science 254:1598 (1991).
56. J.E. Straub and D. Thirumalai, Exploring the Energy Landscape in Proteins, Proc. Natl. Acad. Sci. (USA) 90:809 (1993).

ON THE CONFIGURATIONS ACCESSIBLE TO FOLDED
AND TO DENATURED PROTEINS

J.C. SMITH[1], P. CALMETTES[2], D. DURAND[2], M. DESMADRIL[3],
S. FUROIS-CORBIN[4], G.R. KNELLER[1,5] and B. ROUX[1,6]

[1] Laboratoire de Simulation Moléculaire, SBPM, DBCM,
  C.E. Saclay, 91191 Gif-sur-Yvette Cedex France
[2] Laboratoire Léon Brillouin
  C.E. Saclay, 91191 Gif-sur-Yvette Cedex France
[3] Laboratoire d'Enzymologie Physico-Chimique et Moléculaire
  Université de Paris Sud, 91405 Orsay Cedex, France
[4] IBPC, 13 rue Pierre et Marie Curie, 75005 Paris France
[5] IBM France, 68-76 quai de la Rapée 75012 Paris Cedex, France
[6] Département de Physique, Université de Montréal
  C.P. 6128 succursale A, Montréal Canada H3C 3J7

INTRODUCTION

The determination of the configurations accessible to native and denatured states of globular proteins is required for an understanding of protein folding and function. X-ray crystallography and NMR spectroscopy furnish detailed information on the average structures of folded proteins. The configurations accessible to most folded proteins can be described in terms of fluctuations of the atoms of ~ 1 Å about their average positions. Data on the configurations accessible to denatured proteins are much harder to obtain. An idealised model for a denatured protein is that of the 'random coil' involving a completely unfolded chain randomly sampling conformations limited only by steric requirements. Alternatively, intramolecular protein-protein associations can occur, such that some structure remains in what is traditionally assumed to be a 'fully denatured' state.

In the present paper we review some recent work examining the configurational distributions of folded and denatured proteins. For the folded proteins vibrational and diffusive internal *dynamics* are described. The investigations on the denatured state are restricted to aspects of the time-averaged configurational distribution of the protein chain. In both studies neutron scattering and computer simulations were combined. The neutron is well suited to address problems of the structure and dynamics of molecules[1] as the detectable neutron scattering wavevector transfers (q) and energy

Statistical Mechanics, Protein Structure, and Protein Substrate Interactions
Edited by S. Doniach, Plenum Press, New York, 1994

135

transfers (ω) correspond to the typical distances (Å) and times (ps) involved. The magnitude of the neutron-nucleus interaction depends on the type of nucleus present and is quantified by the scattering length, b. Two forms of neutron scattering are of particular interest: coherent and incoherent. Coherent scattering , which is proportional to the mean square scattering length, $\bar{b}^2$, can give interference effects due to correlations in the positions of different atoms. Incoherent scattering is due to fluctuations from the average b, originating from random nuclear spin orientations in the sample. The incoherent scattering is proportional to the mean square deviation, $\overline{(b - \bar{b})^2}$. It results from self-correlations of the atoms in the system. In what follows we summarise recent work combining computer modelling with small-angle coherent neutron scattering to obtain information on the structural properties of denatured phosophoglycerate kinase and combining molecular dynamics simulation and incoherent inelastic neutron scattering to determine picosecond-timescale motions in folded proteins.

DYNAMICS OF FOLDED PROTEINS

Although dynamical events in folded proteins occur on a wide range of timescales, picosecond (ps) motions make a particularly important contribution to the internal fluctuations of the atoms from their mean positions. These 'low-frequency' protein motions have been extensively investigated using computer simulation techniques that employ empirical energy functions. Incoherent neutron scattering has been able to provide complementary experimental information on the timescales, forms and amplitudes of the low-frequency motions in proteins. Recent progress was made possible by comparing experimentally-derived spectra with those calculated from the results of simulations.[2]

Incoherent Neutron Scattering

The basic, measureable quantity is $S(\vec{q}, \omega)$, the incoherent dynamic structure factor. This is written as

$$S(\vec{q}, \omega) = \frac{1}{2\pi} \int_{-\infty}^{+\infty} dt \ e^{-i\omega t} \mathcal{F}(\vec{q}, t) \tag{1}$$

$$\mathcal{F}(\vec{q}, t) = \frac{1}{N} \sum_{\alpha} b_{\alpha}^2 \langle e^{-i\vec{q}\cdot\mathbf{R}_{\alpha}(0)} e^{i\vec{q}\cdot\mathbf{R}_{\alpha}(t)} \rangle \tag{2}$$

We see from Eq. 1 that the incoherent dynamic structure factor is the time Fourier transform of the incoherent intermediate scattering function, $F(\vec{q}, t)$. The sum in Eq. 2 is over the N atoms, α in the sample. The atomic positions are specified by their time-dependent position vector operators $\mathbf{R}_{\alpha}(t)$. Each isotope has an incoherent scattering length b_{α}. b_{α} for hydrogen is sufficiently large that hydrogen scattering dominates the measured profiles from proteins.

Incoherent scattering probes *self-correlations* in motions of atoms. It can be classified as three types: elastic, quasielastic and inelastic. The inelastic scattering arises from vibrational motion in the sample. Quasielastic scattering is manifest as a broadening of the elastic peak and indicates the presence of nonvibrational, diffusive motions. The elastic scattering gives information on the geometry of the motions involved and the associated atomic mean square displacements in the long time limit.

Low-frequency vibrations in proteins

The simplicity of interpretation and the existence of analytical expressions for basic dynamic and thermodynamic quantities makes the normal mode description of protein dynamics a good starting point for comparison of theoretical and experimental inelastic neutron scattering spectra. In addition, the relatively large amplitude of the low-frequency internal vibrations in proteins makes it of considerable interest to probe them experimentally. Inelastic neutron scattering is well placed for such an investigation because the intensity of the scattering from a vibration depends directly on its *amplitude*; the low-frequency vibrations that dominate the mean-square fluctuations in the harmonic models will also dominate the neutron scattering spectra.

Detailed analyses have been made of the neutron scattering properties expected from harmonic and damped harmonic models of the bovine pancreatic trypsin inhibitor, BPTI.[3-5] Experimental spectra were obtained and it was shown that the amplitudes and frequencies of the low-frequency (10-200 cm^{-1}) modes calculated from harmonic models are in good agreement with experiment. However, the frequencies of the very low frequency modes (< 50 cm^{-1}) were found to be highly sensitive to the method of representation of the long-range electrostatic interactions.

At the available instrumental resolution low-frequency (< 80 cm^{-1}) spectra from the various proteins measured so far (lysozyme, BPTI, myoglobin, cytochrome C, phosphoglycerate kinase etc) have similar forms. The lowest frequency vibrations (< 10 cm^{-1}) do not appear as inelastic scattering in the spectrum. If present, they must be overdamped. This has particular implications for double-lobed proteins such as lysozyme. The characteristic lysozyme hinge-bending motion, if vibrational, would show up as a strong peak in the inelastic neutron scattering spectrum. No such peak is seen[2]; even in the dried protein the hinge-bending mode, if present, is overdamped.

The addition of water molecules to a dry protein can be expected to produce a change in the internal protein dynamics. Neutron experiments on protein powders indicate that hydration increases the picosecond-timescale mean square displacements and the quasielastic scattering from small proteins.[4,6] This may be the result of a water-induced increase in the rate of picosecond-timescale jumps between free energy minima (see below). An alternative model describes the effect of hydration as the addition of low-frequency overdamped modes to the system.[4]

Diffusive motions at room temperature

Vibrational models of isolated proteins neglect several factors likely to be present in real proteins in physiological conditions. One of these is the anharmonicity of the intramolecular potential energy surface. Intriguing data pertaining to the anharmonicity of protein motions have been obtained by determinations of the temperature-dependence of mean square displacements and fluctuations of protein atoms. Neutron scattering[7] and crystallographic studies on myoglobin have indicated that the average mean-square displacement, $< R^2 >$ increases approximately linearly with temperature below $\sim 200K$, in accord with harmonic models for the internal dynamics. At higher temperatures $< R^2 >$ increases more rapidly with temperature, indicating the presence of additional motions. The nonlinear increase in $< R^2 >$ is accompanied by the presence of quasielastic neutron scattering indicating the presence of picosecond- timescale nonvibrational motions in the protein.[7] There is increasing evidence that these additional motions are required for the functioning of some proteins. For example, crystallographic analyses of ribonuclease A have demonstrated that inhibitor molecules will bind only at temperatures above the dynamical transition.[8] The bacteriorhodopsin photocycle works only at temperatures where the additional motions are present.[9]

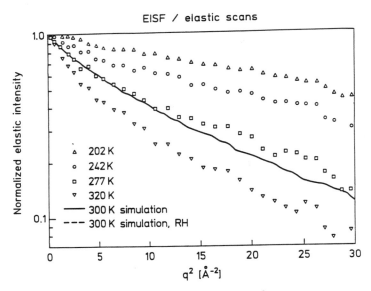

Fig. 1. Log of the elastic incoherent structure factor versus q^2 for myoglobin. The experimental data points are shown for four temperatures (from ref 7.) and the corresponding quantity derived from a molecular dynamics simulation of myoglobin at 300K is given by a solid line.

Information on the geometries of the motions concerned can be obtained by examining the q-dependence of the elastic incoherent structure factor (EISF). $\mathcal{S}(\vec{q}, \omega)$ can be divided into a part arising from purely elastic scattering with $\omega = 0$ and a part with $\omega \neq 0$. Writing

$$\mathcal{F}(\vec{q}, t) = EISF(\vec{q}) + \mathcal{F}'(\vec{q}, t), \tag{3}$$

$$EISF(\vec{q}) = \lim_{t \to \infty} \mathcal{F}(\vec{q}, t), \tag{4}$$

we have

$$\mathcal{S}(\vec{q}, \omega) = EISF(\vec{q})\delta(\omega) + \mathcal{S}'(\vec{q}, \omega). \tag{5}$$

The $EISF$ is determined by the geometry of the volume explored by the atom and is written as

$$EISF(\vec{q}) = \frac{1}{N} \sum_{\alpha} \left| \langle e^{i\vec{q} \cdot \vec{R}_{\alpha}} \rangle \right|^2 . \tag{6}$$

This is effectively the Fourier transform of the average probability distribution of the protons.

In Fig. 1 is shown the elastic scattering as derived experimentally[7] and from a 100 ps molecular dynamics simulation of myoglobin.[10] In both experiment and simulation a nonGaussian form is seen for the elastic scattering indicating that the dynamics cannot be described in terms of the hydrogen atoms all undergoing identical harmonic motion.[11] The calculated elastic incoherent structure factor is in excellent agreement with experiment. The gradient of $log(EISF)$ vs q^2 as $q \to 0$ gives the average mean square displacement of the hydrogen atoms. The low q behaviour in Fig. 1 indicates that the mean square displacements are slightly underestimated in the 100 ps simulation; the low q scattering matches that experimentally obtained at 277K. This can be attributed to the fact that the hydrogen motions contributing to the experimental profile (which has a time resolution of ~ 500 ps) are not quite completely sampled in the simulation. However, the overall agreement indicates that the amplitudes of the vibrational and diffusive motions of myoglobin are well represented in the simulations over the accessible timescale, 0-100 ps.

Two qualitatively different models for the anharmonic dynamics can be considered. In one the atoms jump between potential wells in a confined geometry. Experimental evidence for the presence of multiple minima has lead to the conclusion that at low temperatures proteins are structurally inhomogeneous in a fashion similar to the glassy state.[12] In a given sample different protein molecules explore configurations associated with different potential energy wells. The wells have been termed 'conformational substates'. At 300 K transitions between the substates occur. The 'jumps between substates' picture has been invoked in molecular dynamics simulation analyses.[13,14] It was suggested that for most atoms the harmonic approximation is valid whereas some, mostly side-chain atoms spend some time vibrating in one harmonic well before making a conformational transition into another one.[13] In myoglobin picosecond-timescale jumps between substates were described as rigid-body helix motions accompanied by loop rearrangements.[14]

As the EISF of the simulated dynamics is in good agreement with experiment (Fig. 1) it is of interest to try to identify from the simulation the nature of the nonvibrational motions concerned. The question therefore arises as to whether the full atomic trajectories can be simply described. A model alternative to substate jumps describes the 300 K dynamics as continuous diffusion of collections of atoms in confined volumes specified by the EISF. To investigate the contribution from this type of dynamics we have examined a dynamical quantity closely related to the neutron scattering - the time-dependent mean square displacement, $< \Delta R(t)^2 >$.

In Fig. 2 are presented curves of $< \Delta R(t)^2 >$ per atom for the helices, loops and side chains. After an initial fast increase, the mean square displacements increase linearly with time. This is a clear indication of the presence of *diffusive* motion in the molecule with an effective diffusion constant given by the gradient of the curves. The diffusion constant varies according to the structural element considered. Extending the plots to timescales beyond 50 ps (not shown) leads to a reduction in the gradients although the mean square displacements do not converge in 200 ps, consistent with the presence of characteristic relaxation times of the order of hundreds of ps experimentally.[7] Also shown as dashed lines in Fig. 2 are rigid-body contributions. These were obtained by fitting rigid-body reference structures to the time frames of the simulation. Two types of rigid-body reference structures were fitted; one consisting of the helices (backbone plus side-chains) and one consisting of the side-chains (with the centroid of the rigid body structure pinioned on the C_α atom).

It is clear from Fig. 2 that the rigid-helix contribution to the helix atom mean square displacements is small (about 30%) whereas the side-chain displacements are well represented by rigid-body side-chain motions. These results indicate that the major part of the picosecond-timescale nonvibrational mean-square displacements arises from collision-determined diffusive motions of the side-chains acting as rigid bodies. Torsional jumps of the side-chains do occur, but too rarely to influence the $< \Delta R(t)^2 >$ on the timescale considered. Direct calculations of the neutron scattering properties of the rigid-body trajectories have been performed and indicate that rigid-side chain diffusion dominates the nonvibrational contribution to the measured profiles.[15] A more detailed analysis of the fitted rigid-body trajectories in terms of fluctuations, mean-square displacements, velocity correlation functions and vibrational frequency distributions has been published.[16]

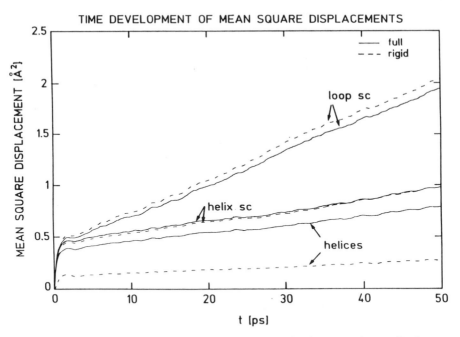

Fig. 2. Time development of the average mean square displacement (normalised per atom) from the molecular dynamics simulation of myoglobin.[16] The dashed lines are rigid side-chain and helix displacements derived from the full trajectory using a quaternion-based method[17] as described in ref. 16

CONFIGURATIONAL DISTRIBUTION OF A DENATURED PROTEIN

A characterisation of the physical properties of unfolded proteins can be expected to provide information pertaining to the stability of native states and to folding pathways. Much recent work has been focussed on intermediate states in protein folding. Here we discuss the configurational distribution of a protein (phosphoglycerate kinase) denatured in 4M guanidine hydrochoride (Gdn-HCl). It has been suggested that the Gdn-HCl denaturant leads to the best approximation to a random coil polypeptide.[18] PGK is a two-domain monomeric enzyme of 44.5 kDa involved in the glycolytic pathway. The folding properties of this enzyme have been well characterised.[19]

In the recent work[20] small-angle coherent neutron scattering data were collected from the native and denatured protein samples. A low-resolution statistical mechanical model for the configurational distribution of the denatured protein was fitted to the data and combined with molecular mechanics calculations to generate plausible atomic detail models for individual configurations of the denatured protein chain.

Small-angle neutron scattering

The structural information on biological macromolecules that can be obtained using small-angle neutron scattering has been reviewed.[21] Small-angle neutron scattering has a distinct advantage over its X-ray counterpart in that, due to the difference in sign of the coherent neutron scattering lengths of deuterium and hydrogen, experimental conditions can be optimised so as to obtain good protein contrast over both its associated denaturant and the solvent; this was achieved in the present experiment by using heavy water and deuterated Gdn-HCl. In general the scattering system can be divided into two regions. One region is that of 'bulk' solvent, characterised by a neutron scattering length density ρ^o. The remaining volume is occupied by material of different average scattering density. In the case of the denatured protein this contains the protein molecules and solvent perturbed from the bulk. For a system at infinite dilution, the coherent scattering arises from correlations within individual protein molecules and associated perturbed solvent. Under these conditions the coherent scattering intensity is given by

$$S(q) = K^2 \langle |\hat{F}(\mathbf{q})|^2 \rangle \tag{7}$$

where $\langle |\hat{F}(\mathbf{q})|^2 \rangle$ is the form factor of the protein molecule plus perturbed solvent and K is its average contrast, given by

$$K = \frac{1}{V} \int_V (\rho(\mathbf{r}) - \rho^o) d\mathbf{r}, \tag{8}$$

where V is the volume giving rise to the contrast and $\rho(\mathbf{r})$ is the coherent neutron scattering length density in the solution at point $\mathbf{r}$.

The average in Eq. 1 is taken over all configurations and orientations of the macromolecule. The form factor amplitude is given by

$$\hat{F}(\mathbf{q}) = C \int_V (\rho(\mathbf{r}) - \rho^o) \, e^{i\mathbf{q} \cdot \mathbf{r}} d\mathbf{r} \tag{9}$$

where C is a constant chosen such that $\hat{F}(0) = 1$. By Fourier transformation of the experimentally-derived scattering function the radial distribution function, $P(r)$, can be derived.

The experimental $I(q)$ and $P(r)$ curves obtained from the recent experiments on denatured PGK are presented in Fig. 3. To interpret the data a statistical mechanical freely-jointed sphere model was constructed. This treats the system with a level of detail equivalent to the experimental resolution. Fits using 1, 10 and 100 spheres are shown in Fig. 3.

Jointed spheres model

In the model the denatured protein is represented using a chain of identical spheres. The spheres consist of scattering material with a constant scattering density; they can contain protein and solvent. They are connected to one another with rigid links such that the distance between the centres of adjacent spheres is equal to their diameter; consecutive spheres are in contact but do not overlap. The chain of spheres is freely-jointed i.e., the chain of spheres is allowed to take any configuration in space subject to the above constraint; excluded volume effects between 1,3 and higher order pairs of spheres are not taken into account. An advantage of the freely-jointed chain model is that most of the scattering expressions can be derived analytically. The model was fitted to the lower resolution scattering (lower values of q, the scattering wavevector) and then compared to the higher q data; the number of spheres and their radius was allowed to vary.

From the experimental data in Fig. 3 an effective radius of gyration of 78 Å was derived for the denatured state; this compares with 23 Å for the native species. As expected a one-sphere model fits the data well at very low q values (below 0.02 Å $^{-1}$) but fails at higher q. This indicates that a single, large uniform sphere can model the configurational distribution only at very low resolution. As the q range increases more spheres are needed and a best-fit is found to have approximately 100 spheres of 8.5 Å radius. For $P(r)$ it is clear that at short distances a model with a small number of spheres fails. The 100-sphere model (sphere radius 8.5 Å) agrees well at short and at long distances, and slightly less well at distances between 10 Å and 70 Å . The need for a large number of spheres indicates that at the resolution level of the instrument no significant persistent structuration is detected and the random coil picture is appropriate. It is interesting to note that the optimal sphere diameter corresponds approximately to the persistence length of a polyalanine chain.

Generation of atomic-detail configurations

The freely-jointed chain represents the low-resolution pathways followed by the protein chain, with roughly four residues per sphere. More detailed information about the chain is not available from the experimental data. From the low-resolution statistical mechanical model it was possible to generate a detailed atomic-resolution model by incorporating further information and making further assumptions. The extra information is that contained in a molecular mechanics force field. The assumption is that the short-range, local conformation of the denatured chain resembles that of the folded molecule. Sections of appropriate length (4 residues) of the native chain were model built into the spheres defining the low-resolution pathway of one of the configurations of the denatured chain. The resulting polypeptide chain was subjected to restrained en-

142

ergy minimisation. Examination of generated atomic-detail configurations shows that the path followed by the backbone of the unfolded protein is quite anisotropic. Such anisotropy is a general property of random chain polymers. Very few chain-crossing contacts were detected in the constructed chains, indicating that protein-protein excluded volume effects are not severe. The RMS deviation of the heavy atoms of unfolded configurations with respect to the the native X-ray structure are ~ 40 Å. This compares with the sequential deviation of segments of four residues within each FJC sphere which is, on average, only 2.1 Å . This illustrates how a denatured chain can exist in which some memory of the native conformation remains over short sequences while over long sequences the global shape is very different. However, the small-angle scattering experiment and the interpretation using a jointed-sphere model do not furnish information as to the conformational properties of the chain within each sphere; the adoption in the modelling of a single configuration in each sphere close to the native one is an additional assumption.

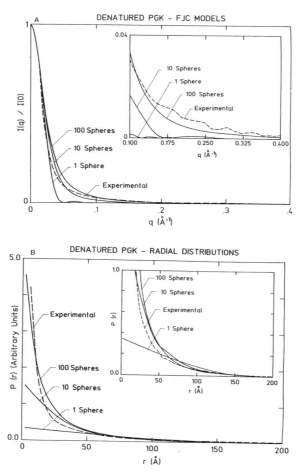

Fig. 3. Scattering function, $I(q)/I(0)$ and radial distribution function, $P(r)$, obtained from small-angle neutron scattering experiment on PGK denatured in 4M Gdn-HCl. Also shown are fits to the data using the jointed spheres models.

CONCLUSIONS

The work presented here illustrates how different forms of neutron scattering can be combined with computer simulations and theoretical modelling to provide information on the accessible configurations of denatured and native proteins.

Incoherent scattering experiments probe the same motions as are simulated using molecular dynamics. This, together with the simplicity of the neutron-nucleus interaction makes the combination of molecular dynamics (or harmonic analysis) and incoherent neutron scattering an excellent strategy for the determination of the picosecond dynamics of folded proteins. A direct comparison between the simulated dynamics and experimental spectra is possible.

It is now clear from comparisons of incoherent inelastic neutron spectra calculated normal modes with experimental results that global, collective, underdamped, low-frequency vibrations do exist in proteins at all temperatures. The vibrational motion dominates the mean-square displacements at low temperatures. At higher temperatures and in solution additional, nonvibrational motions occur. To obtain a simple description of the nonvibrational dynamics molecular dynamics can be used as a stepping stone for interpreting the experiments. The simulations reproduce well the measured neutron scattering profiles for myoglobin. Analysis of the simulations indicates that rigid side-chain diffusive motions make the major contribution to the room temperature mean-square displacements. Similar, combined incoherent scattering/molecular dynamics analyses have been used to decompose and examine in detail methyl group dynamics in crystals of the alanine dipeptide[22] and to describe low-frequency vibrations in crystalline polyacetylene[23].

Coherent inelastic scattering shows considerable promise in the future for the analysis of correlated motions in proteins. An application of this technique, to the determination of phonon dispersion curves in amino-acid crystals, and a complementary theoretical analysis has been recently reported.[24] In the work described here on denatured phosphoglycerate kinase the coherent scattering was measured at small angles and in a spherically and energy-averaged form such that time-averaged radial pair probability distribution functions were obtained. The measured data were interpreted using a statistical mechanical model so as to establish bounds for the overall behaviour of denatured protein chain. This low-resolution information was then combined with a molecular mechanics force field to enable the construction of plausible sample configurations for the denatured protein chain. The incorporation of low-resolution configuration-averaged structural data into an atomic-detail model is a first step towards an analysis of the detailed physical chemistry of a denatured protein.

A further use for the atomic-detail configurations could be as starting points for more extensive computer simulations. The construction of complete model systems with water and denaturant around the freely-jointed chain configurations can now be envisaged. Molecular dynamics simulation could then be used to further characterise the systems. However, the large size of the system would render this simulation computationally expensive. In addition, to fully investigate the denatured protein in this way would require simulation of each of a large number of generated protein configurations. This is impossible computationally at present, but represents a significant goal for the future.

REFERENCES

1. Lovesey, S. (1984). *Theory of Neutron Scattering from Condensed Matter.* International Series of Monographs on Physics, no. 72 Oxford Science Publications. Oxford:Clarendon.

2. Smith, J.C. (1991). *Q. Rev. Biophys.* **24(3)**, 227-291.

3. Smith, J.C., Cusack, S., Pezzeca, U., Brooks, B. & Karplus, M. (1986) *J. Chem. Phys.* **85(6)**, 3636-3654.

4. Smith, J.C., Cusack, S., Tidor, B. and Karplus, M. (1990) *J. Chem. Phys.* **93(5)** 2974-2991.

5. Cusack, S., Smith, J.C., Finney, J.L., Tidor, B. & Karplus, M. (1988) *J. Mol. Biol.* **202**, 903-908.

6. Smith, J.C., Cusack, S. Poole, P. & Finney, J.L. (1987) *J. Biomol. Struct. Dyn.* **4(4)**, 583-587.

7. Doster, W., Cusack, S. & Petry, W. (1989) *Nature* **337**, 754-756.

8. Rasmussen, B.F., Stock, A.M., Ringe, D. & Petsko, G.A. (1992) *Nature* **357** 423-424.

9. Ferrand, M., Dianoux, A.J., Petry, W. & Zaccai. G. (1993) *Proc. Natl. Acad. Sci. (U.S.A.)* In press.

10. Smith, J.C. & Kneller, G.R. *Mol. Simul.* In press.

11. Smith, J.C., Kuczera, K., & Karplus, M. (1990) *Proc. Natl. Acad. Sci.* (U.S.A.) **87**, 1601-1605.

12. Frauenfelder, H., Parak, F. & Young, R.D. (1988). *Ann. Rev. Biophys. Biophys. Chem.* **17**, 451-479.

13. Ichiye, T., & Karplus, M. (1988) *Proteins: Struct. Funct. and Genetics.* **2**, 236-259.

14. Elber, R. & Karplus, M. (1987) *Science* **235**, 318-321.

15. Kneller, G.R. & Smith, J.C. submitted.

16. Furois-Corbin, S., Smith, J.C. & Kneller, G.R. (1993) *Proteins: Struct. Funct. Genetics* (1993) **16(2)** 141-154.

17. Kneller, G.R. *Mol. Simul.* **7**, 113-117.

18. Tanford, C. (1968). *Advan. Prot. Chem.* **23**, 121-275.

19. Yon, J.M., Desmadril, M., Betton, J.M., Minard, P., Ballery, N., Missiakis, D., Gaillard-Martin, S., Perahia, D. & Mouawad, L. (1990) *Biochimie* **72**, 417-429.

20. Calmettes, P., Roux, B., Durand, D., Desmadril, M., & Smith, J.C. (1993) . *J. Mol. Biol..* In Press.

21. Zaccai, G. & Jacrot, B. (1983). *Annu. Rev. Biophys. Bioeng.* **12**, 139-157.

22. Kneller, G.R., Doster, W., Settles,M., Cusack, S. & Smith, J.C. (1992) *J. Chem. Phys.* **97(12)**, 8864-8879.

23. Dianoux, A.J., Kneller, G.R., Sauvajol, J.L. & Smith, J.C. (1993) *J. Chem. Phys.* In press.

24. D. Durand, M. Field, M. Quilichini, & Smith, J.C. (1993) *Biopolymers* **33**, 725-733.

EXPERIENCES WITH DIHEDRAL ANGLE SPACE MONTE CARLO SEARCH FOR SMALL PROTEIN STRUCTURES

Natalya A. Kurochkina, Hong Seok Kang, and B. Lee

Bldg. 37, Room 4B15
Laboratory of Molecular Biology, DCBDC/NCI
National Institutes of Health
Bethesda, MD 20892

INTRODUCTION

In this article, we describe the experiences we had in trying to find the native conformation of polypeptides and a small protein molecule, crambin, by means of the Monte Carlo (MC) sampling procedure.

Many different methods are being used to study how proteins fold (for a recent review, see Chang & Dill, 1993). Many of these are based on a lattice representation of the protein (Covell & Jernigan, 1990; Hinds & Levitt, 1992; Kolinski, Godzik & Skolnick, 1993). The lattice-space MC has important advantages because the conformational search space is very significantly reduced and because it is easy to make local moves. Confining a protein onto a lattice, however, introduces distortions to the structure, with unclear consequences. Wilson & Doniach (1989) studied the folding of crambin using the MC procedure in the dihedral angle space, i.e. the bond lengths and angles were fixed and only the dihedral angles were sampled. We followed this procedure, but with differences in making local moves and in the energy functions used.

METHODS

The calculations were done using a conventional Metropolis MC procedure (Metropolis et al., 1953). It consists of cycles of (1) selecting one or more dihedral angles and changing them, (2) constructing the structure corresponding to the new dihedral angle values, (3) calculating the energy of the new structure, and (4) selecting or rejecting the new structure according to the Metropolis criteria. Since this is a well-known procedure, we describe in the following only those aspects of the method for which we devised a special procedure.

Biased sampling of the main-chain angles

The main-chain dihedral angles were selected according to a set of pre-computed tables of probabilities. The procedure for setting up this probability table has been reported (Kang et al., 1993). Briefly, the probability is set up by counting frequency of occurrences of dihedral angles that are observed in a set of well-determined crystal structures. The effect of

Statistical Mechanics, Protein Structure, and Protein Substrate Interactions
Edited by S. Doniach, Plenum Press, New York, 1994

local sequence, which includes ±4 neighbors, was included by counting conditional occurrences that a residue will have certain $\phi-\psi$ pair of angles when its m-th neighbor is of a given amino acid type. The effect of the local sequence difference can be seen clearly in Figure 1.

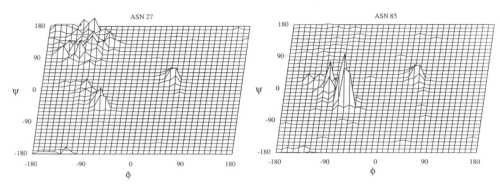

Figure 1. Mesh plot of the $\phi-\psi$ angle probabilities for two specific residues of glutathione peroxidase (1GP1). Note that the same asparagine residue type has a high probability to have the beta-sheet conformation at one position (asn 27, left) and the alpha-helical conformation at another position (asn 85, right). From Kang et al. (1993) with permission.

The procedure used to select the $\phi-\psi$ angles so as to conform to the calculated probabilities was as follows. The two-dimensional space of the $\phi-\psi$ angles of a given residue was divided into 1296 bins, each 10° wide on each side. These bins were serially numbered. A cumulative plot was then made where the x-axis was the bin serial number and the y-axis was the sum of the probabilities in all the previous bins (Figure 2). A random pair of angles ϕ and ψ is selected by selecting a random number on the y-axis of this graph and then by reading off the bin number from the x-axis corresponding to the selected y-axis value.

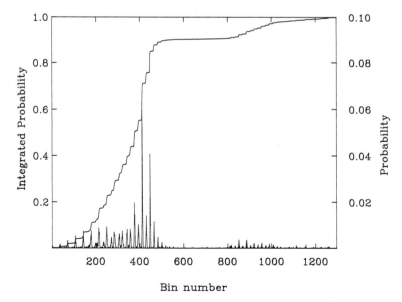

Figure 2. Probability (bottom curve and right scale) and cumulative probability (upper curve and left scale) vs. angle bin number for the trypsin inhibitor.

Making local moves

A dihedral angle space MC procedure suffers from the fact that even a small rotation around a bond in the middle of the chain can produce a large movement towards the ends of the chain. When the chain is compact, large structural changes will generally not be tolerated and the acceptance rate will become small if one attempts to rotate around a single bond at a time. One therefore needs to find conjugated motions, i.e. changing a combination of two or more angles, that will change the structure immediately surrounding the angles changed but which will preserve the position and orientation of the structure outside of this immediate vicinity. Go & Scheraga (1970) showed that this problem is equivalent to the ring-closure problem and gave a general procedure for solving this problem, which involves six angles of the three consecutive peptide groups. This procedure is, however, strictly geometrical and does not take into account the different angle probabilities.

In order to obtain conjugated moves that still reflect the different angle probabilities, we proceeded in a different manner as follows. The geometry of a peptide group is such that the $C\alpha$-N bond is nearly parallel to the C-$C\alpha$ bond. We changed the position of the atoms C and N slightly (by about 0.1 Å) to make these bonds exactly parallel (Figure 3). It can then be shown that the orientational part of the problem can be solved analytically. There are two solutions. One is the trivial solution, corresponding to making a series of three crankshaft motions around each of the peptide groups,

$$\Delta\alpha_1 = \Delta\alpha_2 = \Delta\alpha_3 = 0,$$

where $\alpha_i = \psi_i + \phi_{i+1}$. There is also the non-trivial solution given by

$$\Delta\alpha_1 = \Delta\alpha_2 = -2\delta \text{ and } \Delta\alpha_3 = -2\,\alpha_2,$$

with

$$\tan\delta = \sin\alpha_2 / \cos\theta\,(\cos\alpha_2 - 1)$$

where θ is the bond angle at the $C\alpha$ atom. A detailed derivation of these solutions will presented later elsewhere. The procedure is then to choose either the ϕ or the ψ angle randomly but biased according to the probability of that angle for each peptide group and choose the other angle to conform to one of the above two solutions. The result is that the remainder of the structure translates but does not rotate. Helfand (1971) showed that such sliding motions are not seriously inhibited in dense polymer systems.

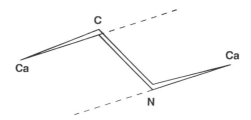

Figure 3. Distortion of the peptide geometry by the change of the position of the C and N atoms such that the $C\alpha$-C and the N-$C\alpha$ bonds become parallel.

In practice, we randomly choose from a repertoire of five different move types: a single rotation around one bond, a small angle move within one angle bin, a crankshaft motion around one peptide group, and a 3-peptide and a 4-peptide Helfand type conjugated moves.

Energy function

We always do the hard sphere bump check. Beyond that we used various combinations of the following energy terms: The hydrogen bond energy, for which we used the electrostatic formulation suggested by Kabsch & Sander (1983), the knowledge-based empirical potential of Miyazawa & Jernigan (1985), and a disulfide bond energy, expressed as a square-well potential. We experimented with the energy term that was proportional to the accessible surface area in order to represent the hydrophobic energy, but had to abandon it because of the excessive computing time demanded by the surface area calculation.

SAMPLE RESULTS

No energy function

The first set of calculations were made without using any energy function at all in order to test the usefulness of the φ-ψ angle probabilities. Results of such calculations are shown in

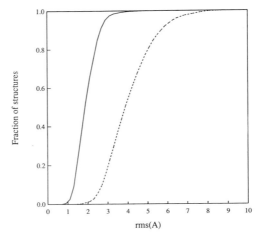

Figure 4. Number of structures versus d-rms deviation from the crystal structure for the C-peptide of ribonuclease A. From Kang et al. (1993) with permission.

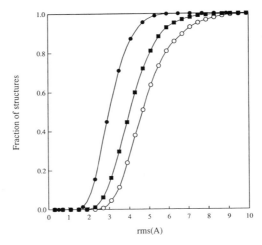

Figure 5. Number of structures versus d-rms deviation from the crystal structure for the C-helix of myohemerithrin. From Kang et al. (1993) with permission.

150

Figure 4 for the C-peptide, which consists of the residues 1 to 13 of ribonuclease A. Shoemaker et al. (1985) has shown that this polypeptide forms a significant amount of α-helical structure in the aqueous solution, as judged by its circular dichroism spectra. It can be seen from Figure 4 that, for more than 90% of the generated structures, the rms deviation of their Cα positions from those in the x-ray structure of ribonuclease A (Protein Data Bank entry 7rsa, Abola et al., 1987; Wlodawer et al., 1988) was less than 3 Å. In contrast, if the φ-ψ angle probabilities were not used, less than 20 % of the generated structures had the corresponding level of resemblence to the x-ray structure.

Figure 5 shows the results of similar calculations for the 18-residue polypeptide corresponding to residues 69 to 87 that make up the C-helix of myohemerythrin (Protein Data Bank entry 2mhr, Abola et al., 1987; Sheriff et al., 1987). This polypeptide was also shown to form some amount of α-helical structure in the aqueous solution although the evidence is much weaker than for the case of the C-peptide (Dyson et al., 1988). The results show that approximately 50% of the generated structures for this peptide had the Cα d-rms deviation less than 3 Å. The corresponding percentages drop to less than 5% and 20%, respectively, if the φ-ψ angle probabilities were not used or used but without the neighbor effect. These results indicate that the φ-ψ angle probabilities are useful for efficient sampling of near native structures. More examples of the use of the φ-ψ angle probabilities can be found in Kang *et al.* (1993).

Hydrogen bond energy alone

In order to see the combined effect of the φ-ψ angle probabilities and the energy function through the use of the Metropolis selection/rejection procedure, we made a number of runs on simple systems using only the hydrogen bonding term in the energy function. In these calculations, the temperature was changed from 900°K to 50°K linearly over 50,000 steps. We found that a 20-residue polyalanine always froms the α-helix as do many other amino acid homopolymers. The transition to all α-helical conformation occurred gradually over the temperature range between approximately 500°K to 100°K. The 20-residue polyglycine was an exception. It stayed in random conformation until approximately 100°K and then abruptly froze into one conformation. The final conformation attained was different for each run. It should be noted that the conformation corresponding to the global energy minimum is the α-helix both for polyalanine and for polyglycine. It is apparent, therefore, that the different behavior is related to the larger conformational space available to polyglycine. Although we have not investigated the behavior of these two molecules under constant temperature MC, it seems clear that the two systems will behave differently at all temperatures. For example, at 300°K, which is about the mid-temperature of the transition for polyalanine, polyalanine will undergo many transitions to and from α-helix whereas polyglycine will remain in random conformation. A more systematic study on the interplay between the effect of the φ-ψ angle probabilities and the energy function will be reported later elsewhere.

Miyazawa-Jernigan (MJ) potential alone

The next set of tests involved similar runs as above, but with the residue-residue empirical potential of Miyazawa and Jernigan (1985) replacing the hydrogen bond energy. These tests were made using crambin, a 46-residue protein from plant seeds, for which a very high resolution x-ray structure is known (Protein Data Bank entry 1crn, Abola et al., 1987; Hendrickson and Teeter, 1981). The same Monte Carlo run was repeated a number of times. The protocol used for each run was to start from the same fully stretched conformation and to change the temperature from 300 K to 100 K on a linear cooling scheme over 100,000 steps (simulated annealing).

The results for a few runs are summarized in Table 1. One finds that all of the runs get trapped in a local energy minimum, as can be seen from the fact that the final energies are significantly higher than that of the native structure. None of the structures obtained had a recognizable degree of secondary structures.

Table 1. The results of Monte Carlo runs with MJ potential.

run id	c-rms[a]	E_{mj}[b]	$\langle E_{mj} \rangle$[c]	$\langle R \rangle$[d]
1	5.9 (7.0)	-113.7	-111.7	10.9
2	10.9(11.1)	-109.0	-108.9	11.3
3	9.7(10.3)	-101.0	-102.8	11.7
4	12.7(12.9)	-103.4	-99.2	11.5
5	12.7(13.1)	-112.1	-105.8	11.3
native		-120.4		9.7

(a) The first numbers are for α-carbon atom, while the numbers in parentheses are for N, H, Ca, C and O atoms of main chain.
(b) Energy of the final structure (kJ/mol).
(c) Average over 5000 steps at the final temperature, 100°K (kJ/mol).
(d) Radius of gyration for all non-hydrogen atoms (Å).

The best run had the c-rms value of 5.9 Å, comparable to the value obtained by Wilson and Doniach (1989). The relation between the MJ energy values and the d-rms values in the ensemble of structures generated in this run is shown in Figure 6. It can be seen that the correlation between energy and the rms is poor when the energy is high, but rather good when the energy is low. In particular, the minimum energy structures do have the smallest deviation from the x-ray structure. Such correlations were, however, not observed for all runs, for example for the run number 28, the final energy value was low but the structure was far from the x-ray structure. This indicates that getting trapped in a local minimum is not the only problem - one must also have an energy function that can better discriminate the good structure from many bad ones.

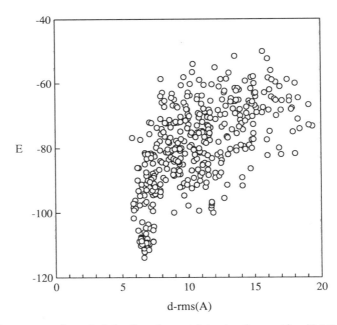

Figure 6. Energy versus d-rms deviation from the crystal structure for crambin with MJ potential alone.

Hydrogen bond energy plus the MJ potential

One obvious way to improve the energy function is to use both the hydrogen bond energy and the MJ potential. This is reasonable since the MJ potential is based on contact frequencies specific to each amino acid pair, which will be dominated by the hydrophobic effect (Miyazava and Jernigan, 1985) and rather insensitive to the hydrogen bonds. The lack of secondary structures in the runs with the MJ potential alone also indicates that an extra hydrogen bonding potential is needed.

The protein used for this test was again crambin. Initially several 60,000-step simulated annealing runs were made with different weights for the hydrogen bonding potential relative to the MJ potential. The optimum weight was chosen by comparing the resulting structures with the x-ray structure, particularly in terms of the location and the length of the helices that develop in the computed structure. Because of the computer time involved, we did not strive to determine the value of the weight to a very high precision. Within the set of computer runs that we made, it was not difficult to determine the optimum weight because the lengths of the helices that develop were found to be rather reliably correlated with the magnitude of the weight.

Once the weight has been determined, the main calculation was done from one long constant temperature run. (This was actually run as 12 separate but consecutive runs.) The temperature chosen was nominally 150 K, which was the mid-point of transition observed in the previous simulated annealing run with the same weight. The total number of steps was approximately 2.4 million, from which every 5,000-th structure was selected for analysis.

Figure 7 shows the energy versus d-rms values for the whole ensemble. The magnified view of the low energy region of this figure is shown in Figure 8. One can see in Figure 8 that there are two sets of structures, one with lower rms than the other. The lowest energy structure belongs to the set with the lower rms, although it does not have the lowest rms. These two sets of structures can also be distinguished by the relative magnitude of the two energy terms, which is shown in Figure 9.

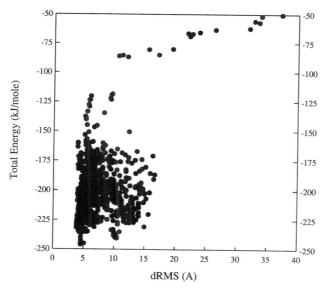

Figure 7. Energy versus d-rms deviation from the crystal structure for crambin obtained from the run with hydrogen bond energy plus the MJ potential.

It can be seen that the MJ and the hydrogen bond energy components are not independent of each other; when the MJ energy is high, hydrogen bond energy is low and vice versa. In structures with low hydrogen bond energies, the helices were found to be longer than those found in the x-ray structure. The lowest rms deviation from the x-ray structure is achieved when these two components are balanced and similar to each other.

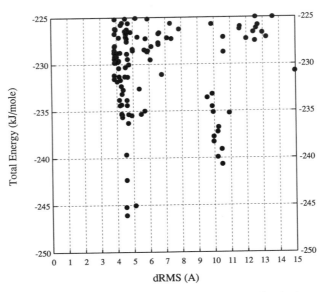

Figure 8. Energy versus d-rms deviation from the crystal structure for the lower energy structures of crambin obtained from the run with hydrogen bond energy plus the MJ potential.

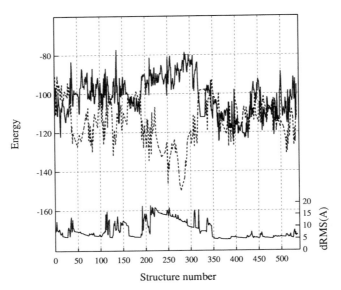

Figure 9. Hydrogen bond energy (broken line) and MJ energy (solid line) for the MC run of crambin with the hydrogen bond energy plus the MJ potential (upper curves and left scale) and the d-rms deviation from the crystal structure (lower curve and right scale).

The rms value of the structure with the lowest energy was not any better than that of the lowest energy structure obtained using only the MJ potential. However, the rms was not uniform throughout the structure, the helical part being much better determined than the rest of the structure. The degree with which the program predicts the secondary structure is indicated in Figure 10, which shows the frequency of obseving a secondary structure for each residue during the whole run. As can be seen, the two helical regions of the structure are rather well predicted. A similar result was also obtained by Wilson & Doniach (1989), who used a somewhat different potential. The β-strands are also well predicted. However, the two strands did not pair up properly to form a β-sheet. The β-sheet in crambin starts with the residues that form a disulfide bond. It is possible that the disulfide bond acts as a seed, and therefore is essential, for the formation of the β-sheet in this structure.

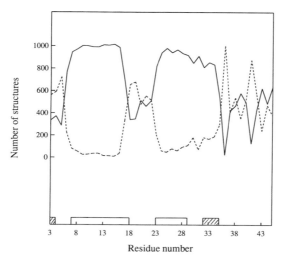

Figure 10. Frequency of every amino acid residue of crambin in α-helical (solid line) or β–sheet (broken line) conformation in the ensemble generated by the MC run with the hydrogen bond energy plus the MJ potential. The assignments for α-helix and β–sheet were made according to Levitt and Greer (1977). Boxes indicate the secondary structures in the crystal structure, filled boxes for β–sheet, open boxes for α-helix.

Hydrogen bond energy, the MJ potential, plus the disulfide bond potential

Since a small protein such as crambin will probably not fold without the disulfide bonds, the next logical step is to add the disulfide bond potential. However, the inclusion of a disulfide bond energy in the form of a square potential made the MC run extremely sluggish. In addition, it was found that if the disulfide bonds were formed ahead of the helices, the helices will not form afterwards. We therefore proceeded as follows.

Figure 11.   Amino-acid sequence of crambin and the arrangement of disulfide bonds.

Since the disulfide bonds in crambin are arranged in nested way as shown on Figure 11, we could search conformations by fragments. The first fragment consisted of residues 7-30, which includes both of the helices of the structure and the 16-26 disulfide bond. The residues that were identified as helical in the previous run without the disulfide bond potential were fixed in the α-helical conformation. The MC run with all three energy terms was then run with only the remaining non-helical regions allowed to vary. As expected, the structure becomes virtually quenched as soon as the disulfide bond is formed. Many such runs were made and several structures for the fragment were selected.

The procedure was then repeated with the next fragment, from residues 4 to 32 inclusive, with the segment 7-29 fixed and with another disulfide bond energy term for the 4-

32 disulfide bond. Finally, the fragment 4-32 was fixed and the remainder of the stucture varied with a third disulfide bond potential for the 3-40 disulfide bond. This procedure resulted in the structures with 3-3.5Å deviation from the crystal structure. One of them is shown in Figure 12. This structure had the Cα c-rms deviation of 3.0 Å from the x-ray structure.

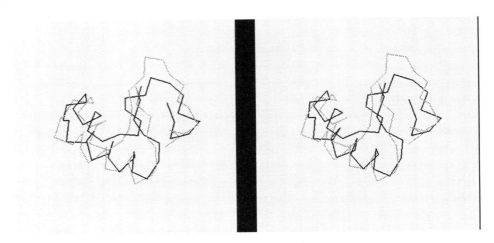

Figure 12. A stereo view of the structure of crambin obtained using the hydrogen bond energy, MJ Energy, and a disulfide bond energy (dark) superimposed on the crystal structure (light). The c-rms deviation between the Cα atoms of the two structures is 3 Å.

SUMMARY

Some locally stable helical structures can be predicted using the φ–ψ probabilites alone without the use of any other energy function (Kang et al. 1993). However, long-range effects can be included only by the use of some additional energy terms. It is not clear at this time what kind of energy terms should be included for this purpose. For instance, Chan & Dill (1990) suggested that hydrophobic force alone may be sufficient for protein folding, while others suggested that the hydrogen bonding and the hydrophobic effects are nearly equally important (Lee, 1991; Privalov & Makhatadze, 1993). Our calculations with crambin with and without the hydrogen bond energy clearly indicate that, at least for a small protein molecule such as crambin, inclusion of a hydrogen bonding potential is essential, without which the helices do not form. Furthermore, the best structure was obtained when the contributions from these two different sources of energy become roughly equal in magnitude.

The best structure that we could obtain was derived by using a disulfide bond potential as well as the hydrogen bonding and the hydrophobic (Miyazawa-Jernigan) terms. The quality of the obtained structure is excellent, from visual inspection and in terms of numerical measures such as the c-rms. This is not a prediction, since the information from the x-ray structure was used in some of the selection steps. However, this experience indicates that a sufficiently good structure can indeed be produced if the correct folding pathway is followed.

ACKNOWLEDGEMENT

We would like to thank Dr. Jong Ryul Kim for supplying us with the graph plotting program and the program that calculates the c-rms values. General assistance with the computer operations provided by Mr. Joseph J. Cammisa is also gratefully acknowledged.

REFERENCES

Abola, E. E., Bernstein, F. C., Bryant, S. H., Koetzle, T. F. & Weng, J. (1987). Protein data bank, *in*: "Crystallographic Databases-Information Content, Software Systems, Scientific Applications," Allen, F. H., Bergerhoff, G. & Sievers, R., eds, Data Commision of the International Union of Crystallography, Bonn/Cambridge/Chester.

Chan, H. S. & Dill, K. A. (1990) Origin of structure in globular proteins. *Proc. Natl. Acad. Sci. USA* 87: 6388.

Chan, H. S. & Dill, K. A. (1993). The protein folding problem. *Physics today* :24.

Covell, D. G., Jernigan, R. L. (1990). Conformations of folded proteins in restricted spaces. *Biochemistry*, 29:3287.

Dyson, H. J., Rance, M., Houghten, R. A., Wright, P. E., Lerner, R. A. (1988). Folding of immunogenic peptide fragments of proteins in water solution. II. The nascent helix. *J. Mol. Biol.* 201:201.

Go, N., Scheraga, H. A. (1970). Ring closure and local conformational deformations of chain molecules. *Macromolecules* 3:178.

Helfand, E. (1971). Theory of the kinetics of conformational transitions in polymers. *J. Chem. Phys.* 54:4651.

Hendrickson, W. A., Teeter, M. M. (1981). Structure of hydrophobic protein crambin determined directly from the anomalous scattering of sulfur. *Nature (London)* 290: 107.

Hinds, D. A., Levitt, M. (1992). A lattice model for protein structure prediction at low resolution. *Proc. Natl. Acad. Sci.* 89:2536.

Kabsh, W. & Sander, C. (1983). Dictionary of protein secondary structure: pattern recognition of hydrogen-bonded and geometrical features. *Biopolymers* 22:2577.

Kang, H. S., Kurochkina, N. A., Lee, B. (1993). Estimation and use of protein backbone angle probabilities. *J. Mol. Biol.* 229:448.

Kolinski, A., Godzik, A., Skolnick, J. (1993). A general method for the prediction of the three dimensional structure and folding pathway of globular proteins. *J. Chem. Phys.* 98:7420.

Lee, B. (1991) Isoenthalpic and isoentropic temperatures and the thermodynamics of protein denaturation. *Proc. Natl. Acad. Sci.* 88:5154.

Levitt, M., Greer, J. (1977) Automatic Identification of secondary structure in globular proteins. *J. Mol. Biol.* 114:181.

Metropolis, N., Rosenbluth, A. W., Rosenbluth, M. N., Teller, A. H. (1953). Equation of state calculations by fast computing machines. *J. Chem., Phys.* 21:1087.

Mijazawa, S. & Jernigan, R. A. (1985). Estimation of effective interresidue contact energies from protein crystal structures: quasi-chemical approximation. *Macromolecules* 18:534.

Privalov, P. L. & Makhatadze, G. I. (1993). Contribution of hydration to protein folding thermodynamics. II. The entropy and Gibbs energy of hydration. *J. Mol. Biol.*, in press.

Rooman, M. J., Kocher, J.-P. A., Wodak, S. J. (1991). Prediction of Protein Backbone Conformation based on seven structure assignments. Influence of local interactions. *J. Mol. Biol.* 221:961.

Sheriff, S., Hendrickson, W. A., Smith, J. L. (1987). Structure of myohemerithrin in the azidomet state at 1.7/1.3 angstrom resolution. *J. Mol. Biol.* 197:273.

Shoemaker, K. R., Kim, P. S., Brems, D. N., Marquesee, S., York, E. J., Chaiken, I. M., Stewart, J. M., Baldwin, R. L. (1985). Nature of the charged-group effect on the stability of the C-peptide helix. *Proc. Natl. Acad. Sci.* 82:2349.

Wilson, C., Doniach S. (1989). A computer model to dynamically simulate protein folding: studies with crambin. *Proteins* 6:93 .

Wlodawer, A., Swensson L. A., Sjolin, L., Gilliland, G. L.(1988) Structure of phosphate-free ribonuclease A refined at 1.26 Å. *Biochemistry* 27: 2705.

CONSTRAINED LANGEVIN DYNAMICS OF POLYPEPTIDE CHAINS

Niels Grønbech-Jensen[1] and Sebastian Doniach
Department of Applied Physics, Stanford University
Stanford, California 94305

Abstract

We have developed an algorithm to compute the trajectory of a polypeptide moving under Langevin Dynamics which incorporates constraints that conserve the the planar geometry of the peptide bonds. We show that for overdamped dynamics the constraints can be implemented at each time step by inversion of a banded matrix. This inversion is computationally very efficient. The forces which are included are those due to the non-bonding interactions between the amino acid atoms, with water put in by a simple dielectric function model. The method can easily be generalized to incorporate explicit water. Because the non-bonding forces are 2-3 orders of magnitude weaker than the bonding forces, we find that the algorithm is stable with time steps on the scale of a thousand times longer than those needed for molecular dynamics simulations. We have tested out the algorithm on simple poly-alanine examples, with extensive runs on 20–alanine. We observe that poly-alanine has rather long-lived meta-stable configurations with partial beta sheet conformations which can live many hundreds of nano-seconds.

1. Introduction

A central problem of the computer simulation of the dynamics of bio–molecules is that of the enormous range of time scales which need to be encompassed. Atomic vibrations occur on sub-pico second time scales while large changes of conformation of even relatively small proteins have time scales ranging anywhere from micro- seconds to seconds.

One approach to overcome some of these problems which has been discussed by a number of authors in the past [1, 2] is to eliminate the very rapid time scales of atomic vibrations by replacing Newtonian dynamics with Langevin dynamics in which the high frequency part of the motion is simply replaced by a noise term with an accompanying friction constant determined by the fluctuation–dissipation theorem. In this work, we have added an additional simplifying step by formulating the dynamics in such a way as to conserve the geometry of the basic peptide bonds in a polypeptide chain, thereby constraining the allowed motion to that of the ϕ and

---

[1]Present Address: Theoretical Division, Los Alamos National Laboratory, Los Alamos, NM 87545

Statistical Mechanics, Protein Structure, and Protein Substrate Interactions
Edited by S. Doniach, Plenum Press, New York, 1994

ψ angles at the C_α hinge points. The formulation is, however, done purely in terms of cartesian coordinates for the individual atoms, with the constraints being put in by a set of forces which orthogonalize the total forces acting on each atom to the constraint directions.

A number of researchers have shown that the slow modes in proteins tend to be overamped [3, 4, 5]. By confining ourselves to over damped Langevin dynamics, we can show that the constraint relations become a set of linear simultaneous equations which we solve at each time step. This is in contrast to the SHAKE algorithm for the application of constraints to under-damped motion in which the presence of the second order acceleration term means that the constraints can only be applied iteratively [6]. We are able to show that the over damped Langevin dynamics can be computed very effectively by inverting a banded matrix at each time step of the simulation. A detailed account of this work may be found in ref [7].

2. Formulation of the Algorithm

Our basic Langevin dynamic equations can be written in the form

$$\begin{aligned}
\mathbf{\Gamma}\frac{d\mathbf{X}}{dt} &= -\nabla E + \mathbf{N} + \mathbf{S} \Rightarrow \\
\frac{d\mathbf{X}}{dt} &= \mathbf{\Gamma}^{-1}\{-\nabla E + \mathbf{N} + \mathbf{S}\}
\end{aligned} \tag{1}$$

where $\mathbf{X} = (\mathbf{x_1}, \mathbf{x_2}...\mathbf{x_n})$ is the cartesian position vector of the atomic coordinates. E represents the non–convalently bonding forces between the atoms of the peptide. In our initial studies, we do not include the effects of water and simulate these in a very approximate way by means of a dielectric screening function [see below]. $\mathbf{\Gamma}$ represents the friction matrix acting on the individual atoms of the peptide. In general, this will include hydrodynamic effects through the Oseen tensor which will produce frictional correlations between different atoms. In the simplest approach, we replace this by a diagonal matrix $\Gamma_{ij} = \gamma\delta_{ij}$ in which γ is the friction constant. $\mathbf{N}$ is a noise-source and $\mathbf{S}$ represents the constraining forces.

The way the constraints are implemented, which is derived by generalizing the approach of Deutsch and Madden ([8]) is as follows: at each time step of the algorithm the total force on each atom resulting from the non–bonding interactions plus noise is computed. Then, that component of the force which will tend to alter a set of constrained bond lengths is projected out analytically through the constraint forces $\mathbf{S}$. Angle constraints are put in by combining a number of distance constraints ,and planar constraints are implemented using dummy atoms located above the plane. If $r_{ij} = |\vec{x}_i - \vec{x}_j|$ is a constrained distance, then by differentiating this with respect to time and inserting Eq. (1) we get:

$$\begin{aligned}
(\vec{x}_k - \vec{x}_l) \cdot (\dot{\vec{x}}_k - \dot{\vec{x}}_l) &= 0 \Rightarrow \\
(\vec{x}_i - \vec{x}_j) \cdot \sum_{ij} \vec{s}_{ij} &= -(\vec{x}_i - \vec{x}_j) \cdot \sum_{j}(-\vec{\nabla}_j E + \vec{\eta}_j)
\end{aligned} \tag{2}$$

where we have used the simplest possible diagonal friction matrix. Given the positions of the atoms, we can now determine the constraint forces given by s_{ij}. This is done by solving the above set of q linear equations (Eq. (2)), where q is the number of constraints in the system. The only s_{ij}'s are for those q pairs of atoms for which the distance $\vec{x}_i - \vec{x}_j$ is constrained to a given value. Thus the number of non-zero off-diagonal elements on one side of the diagonal of this constraint matrix corresponds to the number of constraints. In the case of poly-alanine, the peptide bond geometry can be conserved using 22 constraints. Thus, solution of the simultaneous equations involves inverting a symmetric banded matrix with 22 non-zero off-diagonal elements on each side of the diagonal.

The usual fluctuation-dissipation arguments for a gaussian white noise satisfying

$$\langle \vec{N}_i(t)\vec{N}_j(0)\rangle = \delta_{ij}\frac{k_BT}{\gamma} \tag{3}$$

for the simplest diagonal friction model, show that equation(1) will guarantee a sampling of configurations which satisfies a Boltzmann distribution of energies. [Of course for Langevin dynamics, energy is not conserved since it is pumped in by the noise forces and drained out bythe friction]. We can show that since the constraint forces always act in a sub-space which is orthogonal to the allowed configurational motion, this Einstein relation is not altered by the presence of constraints[7].

Since the dynamics are overdamped, the time scale is set purely by the friction coefficient in equation (1). In order to test the algorithm, we make the simplest possible assumption in which the friction matrix is diagonal. Then, the order of magnitude of the friction coefficient can be computed using Stokes' law for each individual atom. For a sphere of radius a this gives us

$$\gamma = 6\pi\zeta a_i \tag{4}$$

where we have assumed that all atoms have the same effective radius in interaction with the solvent, $a_i = 3\text{Å}$ for simplicity, where ζ is the viscosity. Putting in the viscosity of water, and for energy units measured in kilocalories, this means that a displacement of one Ångström would correspond to a time step of 75 picosecs.

For the test studies reported here we have modelled the non–bonding potentials between the atoms in the peptide using the OPLS pair potentials of Jorgensen and Tirado–Rives[9] with screening of the Coulomb forces to simulate the effects of polarisation of the solvent and of the peptide itself, and a finite cut-off distance. For the screening we have chosen to use the dielectric functional form $\epsilon(r) = exp(r/3)$, where r is the distance measured in Å, suggested in Ref. [10].

3. Testing the algorithm: poly–alanine

For initial tests of the algorithm we used a variable time step so that the average error introduced into the constrainted lengths during one time step would not exceed a fixed amount, δx. For δx of 10^{-3} Ångström, we find that the time steps fall in the range 500 femtosecs to 1 picosec. Then we ran with the fixed time step over many steps and verified that δx remains stable over many thousands of time steps.

We did test runs on a series of poly–alanine models with most work being done on a twenty–alanine model. These runs were set up with initial condition of an α-helix conformation for the poly–alanine. We found that the helix unwinds, as defined by the time at which half the H–bonds are broken, on a time scale of tens of nanosecs, depending on the noise-temperature T. As a function of T, we find that the time for the helix to unwind is approximately exponential. From MD simulations, the unwinding time for helices, is believed to be in the 500 picosec range [11]. We found that we needed to increase the temperature by about a factor of three over room temperature in order to achieve unwinding times of this order. This indicates that the screened OPLS potentials we are using to simulate solvent effects overestimate the strength of the hydrogen bonds in the helix by about a factor of 3. As shown by Daggett and Levitt [13], explicit water is probably needed for a resasonably accurate simulation of helix unwinding.

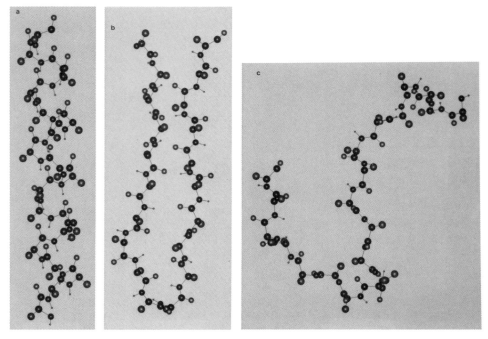

Figure 1. Langevin dynamics of 20–Alanine: (a)initial α helix; (b)temperature=2.0, time=292.5ns; (c)time=592.5ns. (Temperature in units of $T_0 = 505K$).

We ran simulations of up to a million time steps corresponding to approximately a microsecond of simulated time. A striking feature which we observed over many runs was the occurence of a series of metastable intermediate states in which, after the helix unwinds, the two ends of the peptide come together in a beta sheet. [See Figure 1]. These beta sheet intermediates were found to have extremely long life times on the scale of hundreds of nanosecs. Such metastable beta sheet conformations have been found in monte carlo simulations by Velikson *et al* [12].

4. Discussion

Our constrained Langevin dynamics algorithm involves two very fundamental simplifying assumptions: overdamped Langevin dynamics and rigid peptide bond constraints. The resulting motions evolve on a time scale of order a thousand times slower than that needed for molecular dynamics. This is to say that we have replaced virtually all the high-frequency motions in the system by the effective noise term and accompanying friction. Clearly, this is an extreme over–simplification which, in the simplest case, is represented by a single adjustable parameter, the friction constant. The question of how many parameters are needed to reliably simulate the slower motions of large molecular system is completely open [14]. It seems clear that the above assumptions can be refined by modifying the noise spectrum and putting in a more complicated friction matrix. Ultimately, the only reliable test will be to compute various correlation functions such as rotational isomerization times for amino acids in the polypeptide and compare them with experiment. The addition of explicit water to the Langevin algorithm, which can be done computationally efficiently for overdamped dynamics will clearly improve the accuracy of the simulation at the few picosec level. Since the dielectric relaxation time of water is on the scale of five picosec, a time step of one picosec should be a reasonable scale on which to simulate the effect of water on dissolved polypeptides.

The other severe approximation consists in keeping the peptide bonds completely rigid using the matrix inversion scheme. This has consequences of two types: (a) the internal vibration of the peptides have been suppressed so that their effects need to be incorporated into the noise spectrum and friction constants as for the effects of the solvent; (b) because the peptide bonds are rigid, motions of the peptide in which two parts of the molecule come together do not allow for any flexibility other than that of the backbone (ϕ, ψ) angles so that internal distortions of the molecule are suppressed. This will tend to mean that the molecule is "hard" which will be a distortion of the internal dynamics.

Although we have only tested the algorithm so far for poly– alanine, it can be shown that the same approach could be used for any polypeptide where a series of side–chain constraints would be added to the peptide backbone constraints used for poly–alanine. The maximum number of constraints needed for all the amino acids does not exceed 65.

In general, there are two types of application for the kind of algorithm described here. The first is to simulate the kinetics of real molecules, which should be done on the time scale discussed above with inclusion of simulated water molecules. The second application will be for efficient conformational searching for use in protein model building, docking of peptides, etc. In the second application, accurate simulation of the kinetics is not important and one wants to generate a conformational search algorithm which is as fast as possible within the constraints of the peptide bond geometry. The maximum gradient of the potentials determines the convergence of a given discrete algorithm. Because we are only taking into account the non–bonding forces in the molecule these are on the scale of 10^{-3} weaker than the bonding forces. We find that the present algorithm starts to blow up when the time steps exceed a few tens of picosecs, but that it remaine stable for times steps on the scale of 10 picosecs or so. Thus searching conformational space may be done quite efficiently using the present approach.

References

[1] W. Nadler, A.T. Brunger, K. Schulten and M. Karplus
Molecular and stochastic dynamics of proteins
Proc. Natl. Acad. Sci. (USA), 84, 7933 - 7937 (1987)

[2] G. Lamm and A. Szabo
Langevin modes of macromolecules
J. Chem . Phys. 85 (12) 7344 - 7348 (1986)

[3] Kottalam J; Case DA.
"Langevin modes of macromolecules: applications to crambin and DNA hexamers."
Biopolymers. 1990 29(10-11):1409-1422.

[4] Kitao, A.; Hirata, F.; Go, N.
The effects of solvent on the conformation and the collective motions of protein: normal mode analysis and molecular dynamics simulations of melittin in water and in vacuum.
Chemical Physics (15 Dec. 1991) vol.158, no.2-3, p. 447-72.

[5] Smith, J.Cusack, S.; Tidor, B.; Karplus, M.
"Inelastic neutron scattering analysis of low-frequency motions in proteins: harmonic and damped harmonic models of bovine pancreatic trypsin inhibitor"
Journal of Chemical Physics (1 Sept. 1990) vol.93, no.5, p. 2974-91.

[6] W. F van Gunsteren and H. J. C. Berendsen
Algorithms for macromolecular dynamics and constraint dynamics
Mol. Phys. 34, 1311–1327 (1977).

[7] N. Grønbech-Jensen and S. Doniach
Long Time Langevin Dynamics of Molecular Chains
submitted for publication (1993).

[8] Deutsch, J.M.,Madden, T.L.
Theoretical studies of DNA during gel electrophoresis
Journal of Chemical Physics (1989) vol.90, p.2476–2485.

[9] Jorgensen, W.L. Tirado-Rives, J.
The OPLS potential functions for proteins. Energy minimizations for crystals of cyclic peptides and crambin.
Journal of the American Chemical Society (16 March 1988)vol.110, no.6,

[10] M. Pellegrini and S. Doniach
"Computer simulation of antibody binding specificity."
PROTEINS STRUCTURE FUNCTION AND GENETICS. 1993 15(4):436-444.

[11] Soman KV; Karimi A; Case DA.
Unfolding of an alpha-helix in water.
Biopolymers, 1991 Oct 15, 31(12):1351-61.

[12] Velikson, B. Garel, T.; Niel, J.-C.; Orland, H.; Smith, J.C.
Conformational distribution of heptaalanine: analysis using a new Monte Carlo chain growth method.
Journal of Computational Chemistry (Dec. 1992) vol.13, no.10, p. 1216 -33
see also: B. Velikson, J. Bascle, T. Garel, H. Orland,
Structural and ultrametric properties of twenty(L–alanine)
Macromolecules *in press*.

[13] V. Daggett and M. Levitt
Molecular Dynamics Simulations of Helix Denaturation
J. Mol. Biol. (1992) 223, 1121–1138.

[14] Bhattacharya, D.K.; Clementi, E.; Xue, W.
Stochastic dynamic simulation of a protein.
International Journal of Quantum Chemistry (5 June 1992) vol.42, no.5, p. 1397-408.

MOIL: A MOLECULAR DYNAMICS PROGRAM WITH EMPHASIS ON CONFORMATIONAL SEARCHES AND REACTION PATH CALCULATIONS

Ron Elber[1], Adrian Roitberg, Carlos Simmerling,
Robert Goldstein, Gennady Verkhivker, Haiying Li,
and Alex Ulitsky

Dept. of Chemistry, University of Illinois at Chicago
M/C 111 Science and Engineering South Bldg,
845 West Taylor St., Room 4500, Chicago IL 60607-7061
[1]Dept. of Physical Chemistry, The Fritz Haber Research
Center and the Institute of Life Sciences, The Hebrew
Univ. of Jerusalem, Givat Ram, Jerusalem 91904, Israel

INTRODUCTION

The field of computational biology has expanded considerably in
the last few years. Insight to the dynamics of biomolecules,
the design of new drugs and the interactions that lead to
stability of macromolecules has been obtained. Crucial in
bringing these changes was the introduction of "user friendly"
computer programs so that the number of potential users
expanded considerably. It is now possible to visualize complex
molecules and to study their structure and thermodynamics
properties. The strength of this approach and what makes it so
attractive is the possibility of studying the behavior of a
variety of molecules using essentially the same set of tools.
Constructing a large number of molecules was made possible by
the use of a data base of molecular pieces: Different molecules
are described using common fragments. For example, all proteins
are constructed from the same monomers – amino acids. Another
example of repeating fragments is found in the base-pairs of
DNA. The fragment solution is chemically intuitive, however, it
is approximate. In general the intramolecular interactions in
an amino acid are influenced by its neighborhood. For example
the charge distribution is determined not only by the identity
of the amino acid (as is usually assumed) but also by the
solvent, the nearby amino acids and the specific conformation
of the fragment. Nevertheless, this approximate approach has a
number of successes that are documented in the literature[1] and
therefore not covered in this manuscript.

There are (at least) three questions that come to mind
when such a tool is proposed. The first is the one mentioned
earlier. It is related to the assumed transferability of
"fragments" from molecule to molecule, or more generally to the

Statistical Mechanics, Protein Structure, and Protein Substrate Interactions
Edited by S. Doniach, Plenum Press, New York, 1994

165

way in which the interaction potential between all the particles in the system is constructed. This question was discussed in the past[1] and it is not the focus of the research pursued by the authors. Improvement on existing representations and the limitations of this approach are still under investigations. This question is not discussed in this chapter and we refer the interested reader to literature on development of empirical force fields[1].

A second question is associated with the design of such a tool, and how the above idea of generating an arbitrary molecule from a set of fragments is translated into a computer program. We shall discuss in detail the principles of one computer program - MOIL. MOIL (MOlecular dynamics at ILlinois) was developed by the authors. It is of course only one example of many programs for molecular modeling and dynamics. However, the underling structure of MOIL is similar to other computer programs, such as AMBER[2], CHARMM[3] and GROMOS[4]. Therefore understanding the general features of one of these programs is usually sufficient. MOIL (to the best of our knowledge), is the only molecular dynamics program with the full source code in the public domain (**Appendix 1** includes instructions on how to get the code and conditions). This makes MOIL a useful starting point for developing new code, in addition to the direct applications of the program. In this chapter we provide more detailed information than is given in concise scientific publications on the structure of the code. This (we hope) will enable the readers to better appreciate the pitfalls in the present approach and will provide (for those who would like to try) a reasonable basis for writing a program of this type from scratch.

The third question is related to the possible applications of this tool. Here we consider in depth two features available only in MOIL: The calculation of reaction paths in large molecular systems[5] and the use of the LES (Locally Enhanced Sampling) method[6] in conformational searches. This is in addition to the more "common" uses of such a program, such as energy calculation, energy minimization, molecular dynamics and free energy estimates.

This chapter is organized as follows: In the next section we describe the basic ingredients of the program: The typical structure of the data base, how to minimize the space requirements and how to make the code faster. In the third section we consider several simple applications that include the construction of a molecule, energy calculation, energy minimization and dynamics. In section four the LES approach and the technique to calculate reaction paths will be introduced.

THE INTERNAL STRUCTURE OF A MOLECULAR DYNAMICS CODE: THE PROGRAM MOIL

The corner stone of any molecular dynamics program is the energy function. In MOIL the combination of the AMBER force field (covalent part)[2], OPLS parameters for non-bonded interactions[7], and the CHARMM force field (improper torsions) is used[3]. The functional form of the potential is "standard" and it is given by the following formula

$$U(\mathbf{R}) = \Sigma\, K_b\, (b - b_0)^2 \;+\; \Sigma K_\Theta\, (\Theta - \Theta_0)^2 + \Sigma A_\phi \cos\,(n\phi + \delta) \qquad (1)$$

$$+\; \Sigma K_{\phi i}\, (\phi_i - \phi_{io})^2 \;+\; \Sigma\, (A_{ij}/r_{ij}^{12}\; -\; B_{ij}/r_{ij}^{6} \;+\; q_i q_j / \varepsilon r_i$$

$$+\, \Sigma_{1-4}\, v14\,(A_{ij}/r_{ij}^{12}\; -\; B_{ij}/r_{ij}^{6}) + el14\,(q_i q_j / \varepsilon r_{ij})$$

where $\mathbf{R}$ is the coordinate vector of all the atoms in the system, b denotes bonds, Θ angles, ϕ_i the improper torsions, ϕ the proper torsions and r_{ij} the distance between the i-th and the j-th atom (see figure 1 for the definition of the internal coordinates). The different energy terms can be separated into covalent and non-bonded interactions. The covalent part can be further divided to two body, three body and four body terms. The two body terms include the bonds - covalent links between pairs of atoms. The bond energy depends only on the distance between the two atoms of interest - r_{ij} - which is denoted in the formula by b. The three body terms are the angles: Θ-s. They are defined as follows: Let $\mathbf{e}_{ij}$ be a unit vector along the bond between the i-th and the j-th atom and let $\mathbf{e}_{jk}$ be a unit vector along the bond between the j-th and the k-th atoms. The angle Θ_{ijk} between two bonds that share an atom is defined by the formula: $\cos(\Theta_{ijk}) = \mathbf{e}_{ij}\mathbf{e}_{jk}$, with Θ in the range between 0 and 180 degrees. Note that since we do not expect large deviation from ideal values for the bonds or the angles, harmonic energy terms are usually employed. Thus, common empirical potentials that are used in biological molecules are not designed to describe chemical reactivity but rather conformational energy. Some developments in the chemical direction are included in MOIL in which electronic curve crossing, modeled via the Landau Zener formula[8], was already used to investigate the geminate recombination of nitric oxide to myoglobin[8]. The last two terms in the covalent energy (the proper and improper torsions) are four body interactions. The difference between proper and improper torsions is in the ordering of the three bonds (figure 1).

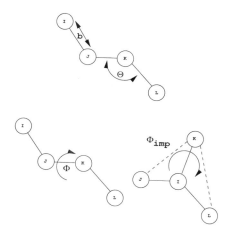

Fig.1 Definition of internal coordinates for two body (bonds), three body (angles) and four body (proper and improper torsions) interactions.

Proper torsions are defined for three sequential bonds between the four atoms i,j,k,l. Let e_{ij}, e_{jk} and e_{kl} be unit vectors along the bonds ij,jk and kl. The proper torsion - ϕ - is the angle between two planes: the plane spanned by the two unit vectors e_{ij}, e_{jk} and the plane spanned by e_{jk} and e_{kl}. $\sin(\phi)$ is given by $e_{ij}(e_{jk} \times e_{kl})$, the cosine of ϕ can be calculated from the normals to the planes:

$$\cos(\phi) = [e_{ij} \times e_{jk}] \, [e_{jk} \times e_{kl}] \qquad (2)$$

An improper torsion is also defined for four atoms linked by three bonds. However, the arrangement of the bonds is not sequential and there is one atom (say i) that is in the center and has three bonds to atoms j,k and l. For example consider atoms with sp^2 type orbitals, (e.g., CH_2O) in which the carbon in the center is bonded to three different atoms. These two types of torsions are topologically distinct. Nevertheless, the improper torsion is computed in an identical way to the proper torsion: It is the angle between the plane defined by the three atoms i,j,k and the plane defined by j,k and l.

The improper torsions were introduced to preserve the planarity or the chirality of different fragments while the regular torsions aim at reproducing experimental barriers for rotations along a bond. We expect improper torsions to have only one minimum, in contrast to regular torsions that typically have two or three minima (rotations along bonds). It is therefore quite common to use different functional forms for the torsion energies: periodic (cosine) functions for proper torsions and harmonic terms for improper torsions. The harmonic term is advantageous for improper torsions since it has a single minimum. Therefore if the molecule is at an angle far from equilibrium the corresponding energy will be high and energy minimization is likely to force the molecule in the right direction. The cosine energy term has multiple energy minima and it is therefore possible that wrong configurations (for improper torsions) have low energies and pass undetected. We use harmonic terms for the improper torsions in MOIL.

To enable convenient calculations of the covalent energy we make extensive use of pointers. The pointers make it possible to formulate the potential in a sufficiently general way such that many different molecules can be described. As an example we consider the bond energy. The identity of the particles linked by bonds is stored in two integer arrays - ib1 and ib2. The number of elements of the two arrays is the number of bonds. The indices of the particles in the molecular system which are connected via the i-th bond are stored as ib1(i) and ib2(i). The coordinates of the first particle (for example) will be given as x(ib1(i)) y(ib1(i)) and z(ib1(i)) where the x,y and z are double precision arrays of the Cartesian components of the coordinates. To calculate the bond energy, it is also required to have easy access to the value of the force constant and equilibrium position. The simplest and fastest solution to this problem is to store the force constants and the equilibrium positions in arrays parallel to ib1 and ib2, e.g. kbond is a double precision array that stores the value of the force constants such that kbond(i) is the force constant of the i-th bond. If memory size is a problem, a more economical choice will be to store the bond parameters according to type, e.g. all N-H bonds in the protein have the same parameters and these parameters can be stored only once. This will require however an additional pointer for each of the bonds - a pointer to the bond type. It is an almost general rule that saving in memory size by adding more pointers and making data extraction less direct reduces the speed of the calculation. Of course one should be careful not to overuse the memory. The situation becomes worse if the arrays become so large that the program requires the use of disk swap space. Overloading the memory is rarely the case with the covalent energy terms but becomes more likely when calculating the nonbonded interactions (see below). Further subtle effects may be found when secondary memory (Cache) is heavily used. This however, may be machine dependent.

In MOIL the arrays ib1, ib2, kbond and req have the same running index - the number of bonds - (req is the equilibrium distance). A similar approach is taken for the other covalent energies: There are three integer pointers to particle locations for the angle energies and four such pointers for the torsions. The different energy parameters are stored according to the index of the internal degree of freedom in the same way as kbond and req. The data is stored internally in the program in a FORTRAN common block that is shared by a large number of programs (the modular structure of the program is discussed later) and is shown in Appendix 2.

The remaining energies correspond to electrostatic and van der Waals forces between atoms that **are not** covalently linked. Covalently linked atoms are separated by (at most) two bonds along the polymer chain. The identity of the covalently linked atoms (which is fixed in most applications) is stored in the exclusion list. The list consists of two integer arrays, one of length of the number of atoms and one of length of the number of exclusions. The first array (excl1) is used to "point" to the last excluded particle in the second array (exc2). Thus, the particles that are "excluded" to the i-th particle are on the second array: exc2(excl1(i-1)+1) to exc2(excl1(i)). Whenever the list of atoms that **interact** via van der Waals and electrostatic interactions is constructed we check each pair to verify that is not in the exclusion list. We also check that it is not in the 1-4 list. The 1-4 list includes the set of atoms separated by three sequential bonds. The non-bonded interactions of the 1-4 type are reduced by constant scaling factors (v14 and el14 in equation (1)) as compared to the regular interactions. The reduction is connected with the empirical observation that if the regular values for the non-bonded interactions are used for the 1-4 pairs then the barriers for torsional transitions become too high compared to the experiment. The 1-4 list is therefore used to reduce the barrier heights for rotations around individual bonds below the experimental value. No precise value is required since the final adjustment will be made by the torsion energy. This is also why a uniform scaling for 1-4 interactions of different atoms can be employed.

The exclusion list is (of course) a guiding tool to compute the major list - the list of atom pairs for which the non-bonded interactions are calculated. The list is generated according to a cutoff criterion and in order to save time it is updated only after a few sequential energy evaluations. For example, after several steps of minimization or several steps of dynamics (typical values vary between 10 and 50 steps) an update is performed. It is done in a hierarchical way: first a list of monomers with their center of mass closer than a cutoff distance (see below) is constructed. Based on the monomer list an atom list is calculated. The distances between atoms of monomers that are in the list are computed. Pairs with distances shorter than a cutoff (typical cutoff distances vary between 8-15Å) are further checked to verify that they are not in the 1-4 or the exclusion lists. The remaining pairs are stored in a similar way to the exclusion list: in two integer arrays. the first array pointx(i) points to the last neighbor of the i-th atom which is stored in the second integer array - listx(pointx(i)). Thus the neighbors of the i-th atoms are stored in listx(pointx(i-1)+1) to listx(pointx(i)). Double counting of pairs is of course avoided, and the list is organized such that the neighbors always have a higher index number. It is therefore not simple to extract all the neighbors of a particle from this list, since this requires a search for neighbors with a lower index.

A significant reduction in computational effort is achieved using the cutoff distance. The basic idea is the following: It is assumed that beyond a certain distance the interaction can be neglected. The computational effort (CE) is proportional to the number of atom pairs whose interaction energies need to be calculated. Without cutoff the Computational Effort is CE=A N^2 where A is a proportionality constant that does not depend on N. For a given cutoff distance the number of neighbors to a particle is approximately constant and independent of the total number of particles in the system. Thus, in the cutoff scheme the computational effort grows linearly with the number of particles. The disadvantages of using cutoff are significant when the system under investigation includes substantial (long range) electrostatic interactions. This is a topic of an ongoing research. Obviously the ideal approach would be no cutoff at all. However, one must bear in mind that some empirical potentials (including OPLS - the potential that is used in MOIL) passed final tuning using cutoff. It is therefore difficult to separate the model potential from the cutoff system that was employed in developing it.

The above discussion of the nonbonded list is a correct but simplified description of what is going on in MOIL. In fact the "x" in pointx and listx is varied from 1 to 3 and we are using three lists instead of one to calculate the non-bonded interactions. We emphasize that since 60-90 percent of the computation time is spend in the non-bonded calculations, it is worth investing the time to tune this part of the code in the spirit described below. There are two points that we should like to use to speed up the computation. The first is that not all particles are charged. It is obviously unnecessary to compute variables required for electrostatic calculations such as distances between particles and finally to multiply the result by zero - the particle charge. The second point is that the range of forces for electrostatic and van der Waals interactions are very different. The Lennard Jones forces are of considerably shorter range compared to the electrostatic. (The van der Waals potential decreases like $1/R^6$ while the electrostatic decreases like $1/R$, where R is the distance between two particles). Therefore it makes sense to use two different cutoffs for the two energy terms. Typical values that we use for van der Waals cutoff is 7-8Å and for electrostatic is 10-15Å. The above two points could have been addressed by "if" statements during the energy calculations. Thus, "if" can be used to check if the particle is charged or if the distance is such that the van der Waals energy is negligible compared to the electrostatic. "if" statements, however, can be costly and may prevent vectorization. This is why we use three lists. The first list is for charged pairs (both particles) that are within the short (van der Waals) cutoff. For these pairs electrostatic interactions and van der Waals energies are calculated. The second list is for pairs within the van der Waals cutoff but for which at least one of the particles is not charged. Therefore there is no need to calculate electrostatic energies. The third list includes charged pairs with distances larger than the van der Waals but smaller than the electrostatic cutoff. Only electrostatic interactions are calculated for these pairs.

The size of the arrays of the non-bonded lists can be very large and in many cases they are a major factor in determining the program size and the memory requirements. Therefore some estimates for memory size are provided below. The lists of non-bonded neighbors for a small protein (around 1000 particles) using cutoffs of the order of 10Å consist of approximately 300,000 neighbors. The identity of the neighbors is stored as integers and the list size is therefore: 4 bytes times 300,000 neighbors = 1.2MB. Since (good) MD programs rarely exceed a few megabytes in memory size, this list can be, and in many cases is, size determining. It is possible to avoid the use of the list by examining all pairs of particles and testing them against the exclusion list, the 1-4 list and the cutoff. However, the additional tests and the fact that all pairs need to be examined will slow the program considerably. It is therefore recommended (unless you simply do not have the space to store the nonbonded list) to use stored neighbors and to avoid as much as possible the above mentioned tests.

In some code people attempt to save space by using short integers for the non-bonded lists and mixing double precision and single precision numbers. This is not recommended on modern architecture which in many cases is slower for short integers. Furthermore, mixing single precision and double precision numbers requires function calls (a major slow down) and may also lead to significant loss of accuracy.

If memory is *not* an issue, it is possible to speed-up the computation further by pursuing some of the calculations on the level of list preparation before the energy is called. One should be quite careful however in deciding that "memory is not an issue", since cache memory is still small and loading from memory to the CPU can be costly too. The only real help we can provide at present is to describe general principles and recommend some trials. An example for use of more memory to reduce actual computation is the following: During the energy calculations we compute the variables A_{ij} (equation 1) by multiplying the square roots of A_{ii} and A_{jj} [7]. For N particles the multiplication requires storage of order N (storing all the A_{ii} and A_{jj}). Storing the A_{ij} requires much more space – the array required is of the same length as that of the neighbor list. If space is available such that one can store two more arrays of A_{ij} and B_{ij}, then the multiplication is required only when the non-bonded list is updated. Furthermore, these long arrays should lead to a better vectorization and cache use. This is since these vectors will have a direct reference in the neighbor loop and the use of pointers will be avoided.

The potential parameters are the major difference between the potential energies of different computer programs. The variations between most of the functional forms employed are rather minor. The parameters include the force constants k_b, k_Θ and k_{imp}, the equilibrium positions for the bonds, angles and improper torsions: b_o, Θ_o and ϕ_{io}, the amplitude, the periodicity and the phase factors for the torsions – A_i, n_i, and δ_i. Other parameters are for the non-bonded interactions and include A_{ij}, B_{ij} for the van der Waals and q_i, q_j and ε for the electrostatic interactions. v14 and el14 are the 1-4 scaling parameters for van der Waals and electrostatic interactions respectively. In MOIL these factors are 0.5 and 0.125 (adopted from OPLS[7]). They are read however as external parameters and therefore can be easily modified.

We comment (as already mentioned above) that the van der Waals (or the Lennard Jones) parameters are stored as single particle properties. The A_{ij} and the B_{ij} are given in terms of the single particle properties – σ_i and ε_i :

$$A_{ij} = 2 \ [\sigma_i^{12} \ \sigma_j^{12} \ \varepsilon_i \ \varepsilon_j]^{1/2} \ ; \quad B_{ij} = 2 \ [\sigma_i^{6} \ \sigma_j^{6} \ \varepsilon_i \ \varepsilon_j]^{1/2} \quad (3)$$

where $\sigma_{i/j}$ is the van der Waals radius for the i/j-th particles and similarly $\varepsilon_{i/j}$ is the well depth. Another common and potentially confusing way of defining σ is as half of the van der Waals radius. In MOIL we use equation (3).

All of the parameters used in MOIL were adopted from other published force fields (covalent structure from AMBER[2], improper torsions from CHARMM[3] and non-bonded interactions from OPLS[7]). The potential energy parameters are stored in an external data-base (external to the FORTRAN code). The data-base is written in a format close to free, which makes it relatively easy to modify or to add to its content. They are kept in a "property file" and are used only when a new molecule is created. Since this file is quite large we did not include it in the present manuscript. However it is also available via anonymous ftp (see Appendix 1). Similarly to the energy function the properties are organized in a hierarchical structure. There are properties of individual particles, of pairs of particles, and of three and four particles. We prefer to use the name "particle" rather than an atom since some of the "particles" that are commonly used are groups of atoms. For example, CH_n is a single particle in the model potential that we use. Each particle has a mass, a van der Waals radius, a van der Waals energy and a charge. More single particle parameters (such as polarizability) can be added (in principle). At present MOIL includes only the properties mentioned above. Assuming transferability of parameters, approximately one hundred different types of particles are required to describe a protein. For example, all the C_α's that are **not** at the beginning of the chain (N terminal) or at the end of the chain (C terminal) have the same "single particle" properties.

Properties of the covalent structure are listed in the property file afterwards. Thus, a list of all possible bonds, angles, torsions and improper torsions for different types of particles is provided. The equilibrium positions and other parameters that are required for energy calculations are listed.

The information provided in the property file is in principle complete. It is possible to construct now any molecule of interest provided that the chemical connectivity of the molecule is defined by the user and all types of needed parameters appear in the property file. At present the MOIL data files support proteins but not DNA or other macromolecules. Generating large molecular systems based on information for one and up to four particles is inconvenient. Therefore an intermediate data structure that includes larger structural segments is employed. In MOIL it is called the monomer file. Similarly to the property file, the monomer file is too large to be included in this manuscript, however, it is available via an anonymous ftp – **Appendix 1**. We make use of the repeating chemical units which are very common in biology, e.g., the amino acids for proteins. The particles in each monomer are assigned a unique name within that monomer and an identification for the look-up table of "single particle properties" – i.e. the property file. The new information in the monomer file is the covalent structure (listing all bonds) of the particles in a monomer. This saves the user a considerable amount of work in defining the connectivity of "standard" pieces. However since the chemical connectivity is usually known, the monomers file is aimed at convenience, not necessity. The database further lists links necessary for polymerization, i.e., "virtual" particles in the PREVious or NEXT monomer that will be bonded to the present monomer. The "virtual" particles become real once the previous and next monomers are specified by the users.

With the help of the monomer file, it is possible to generate the energy function for all macromolecules for which the monomers are defined. This is done by simply listing the sequence of the monomers. The program **con** (the first program in the MOIL package) reads the file of properties, the file of monomers and the file of "polymerization" (the list of monomers that together consist the molecule of interest). For each monomer in the monomer files **con** identifies "virtual particles" – particles that exist in the NEXT or PREVious monomers and are used for the polymerization process. Based on the "virtual particles" and the polymerization list **con** builds the specific molecule of interest. It is worth emphasizing that the short polymerization file is the *only* information that differs from molecule to molecule, since the data on properties and monomers is a general purpose fixed database that was already constructed. Of course, if you have an unusual molecule with monomers that do not appear in the database, adding these monomers to the property and monomer files will be required. **con** further builds the rest of the covalent data that was not defined in the monomer file: (angles, torsions, improper torsions, exclusion and 1-4 lists). This is done automatically, based on the bond structure, in a **maximal way**. For example, the CH3–CH3 molecule has nine torsions of the type H-C-C-H. This is important to remember since care must be used when transferring data between different force fields. For example, some programs use only a single torsion along the C–C bond. The barrier for a single torsion may differ by a factor of nine if a wrong interpretation is given to the data. The output of the con program is a file that defines the energy function for the whole molecule and in MOIL is called the connectivity file. Once it is generated for a given molecule, other more interesting applications can be pursued that use this file. Examples will include energy calculations, energy minimization or dynamics. Each one of the above is a separate program. The programs interact with each other via the common connectivity file and via coordinate files describing the position of the particles in space. From the programmer's point of view the different programs share a large number of common tools, such as interpretation of the command line, routines to read coordinates, connectivity files, and a large set of common databases and FORTRAN common blocks. Furthermore, the same routines for the calculations of the energies and the forces are used by all programs attempting to calculate dynamic or thermodynamic properties.

Splitting of the applications between many separate programs has the advantage that the executing files are more compact. They require relatively small space and memory and are likely to be more efficient. The disadvantage compared to "one program does it all", is that the bookkeeping is more difficult. From the user's point of view, keeping track of many programs can be confusing and time consuming. This problem can be solved however on the shell level. Since we are familiar with the multiple programs, we did not bother to "shell" the different execution files of MOIL. The fact that the executing files remain lean makes it possible to address larger problems than can be investigated using "the one program does it all" approach. The assumption made in MOIL is that of an "educated user", that should be able to find his/her way through the different modules. MOIL is not designed (or aimed) to replace a commercial, mouse driven code.

In the next section we discuss the basic steps that are required to pursue a molecular dynamics simulation using MOIL. The starting point will be the availability of a pdb file. More advanced topics (LES and reaction paths) will be described in the follow up section.

EXAMPLE: MOLECULAR DYNAMICS SIMULATION USING MOIL

In this section we describe the procedure of getting 100 psec simulation time for the protein myoglobin with a partial solvation shell. This section is rather technical and does not contain new scientific results. However it may be useful as a reference point, since essentially all molecular dynamics code must pass through the same steps. It can be useful to understand what is going on in some "black box" procedures even if the procedure is hidden by a nice user interface of a commercial program. This is especially important if problems are encountered.

In MOIL the first step is ALWAYS the generation of the connectivity file. Generating the connectivity file for myoglobin requires six separate files. The monomer file and the property file are already available as a part of the MOIL database. The rest (four files) must be provided by the user. This may seem a lot but as will be demonstrated below these files are quite short, and only the counting to four may pose a problem. One of the files is for polymerization, and a second file provides directions to add bonds. An example for a bond addition that is going beyond what is already available in the monomer file is the bond between the histidine and the heme iron. Another example is of a disulfide bridge. Further modifications to the covalent structure when necessary are listed in the third file. These changes are required only for "non-standard" groups when the automatic procedure for generating angles and torsions is "overdoing it". Finally an input directing the program **con** to the location of the above mentioned files is requested. In many cases the file to add bonds or the file to edit the covalent structure are not required. Here is the **con** input with directions to the location of other files:

```
~ a comment line starts with a "~"
~ below is the address of the pre-prepared monomer file
~ "file" states that this is an i/o line, "mono" is the file
~ type, "name" says it all, "read" - the file is open for read
~ only. Other options to open a file (in addition to "read")
~ is "writ" in which the file must not exist, and "wovr"
~ in which an existing file is written over.
file mono name=(ALL.MONO) read
~
~ ALL.PROP is the name of the pre-prepared property file
file prop name=(ALL.PROP) read
~
~ polymerization file is called below
file poly name=(poly.mb) read
~
~ add bond between iron and histidine
file ubon name=(addb.mb) read
~
~ remove unnecessary angles in the heme planes that
~ were generated by the automated procedure
file uedi name=(edit.mb) read
~
~ output (the connectivity file) will be written on wcon.mb
file wcon name=(wcon.mb) wovr
~
~ execute
action
```

The polymerization file is listed below:

```
MOLC=(MYOG)  #mon=238
NTER MET VAL LEU SER GLU GLY GLU TRP GLN LEU VAL LEU HIS VAL
TRP ALA LYS VAL GLU ALA ASP VAL ALA GLY HIS GLY GLN
ASP ILE LEU ILE ARG LEU PHE LYS SER HIS PRO GLU THR
LEU GLU LYS PHE ASP ARG PHE LYS HIS LEU LYS THR GLU
ALA GLU MET LYS ALA SER GLU ASP LEU LYS LYS HIS GLY
VAL THR VAL LEU THR ALA LEU GLY ALA ILE LEU LYS LYS
LYS GLY HIS HIS GLU ALA GLU LEU LYS PRO LEU ALA GLN
SER HIS ALA THR LYS HIS LYS ILE PRO ILE LYS TYR LEU
GLU PHE ILE SER GLU ALA ILE ILE HIS VAL LEU HIS SER
ARG HIS PRO GLY ASN PHE GLY ALA ASP ALA GLN GLY ALA
```

```
MET ASN LYS ALA LEU GLU LEU PHE ARG LYS ASP ILE ALA
ALA LYS TYR LYS GLU LEU GLY TYR GLN GLY CTRG
HEM1
CO
TIP3 TIP3 TIP3 TIP3 TIP3 TIP3 TIP3 TIP3 TIP3 TIP3
TIP3 TIP3 TIP3 TIP3 TIP3 TIP3 TIP3 TIP3 TIP3 TIP3
TIP3 TIP3 TIP3 TIP3 TIP3 TIP3 TIP3 TIP3 TIP3 TIP3
TIP3 TIP3 TIP3 TIP3 TIP3 TIP3 TIP3 TIP3 TIP3 TIP3
TIP3 TIP3 TIP3 TIP3 TIP3 TIP3 TIP3 TIP3 TIP3 TIP3
TIP3 TIP3 TIP3 TIP3 TIP3 TIP3 TIP3 TIP3 TIP3 TIP3
TIP3 TIP3 TIP3 TIP3 TIP3 TIP3 TIP3 TIP3 TIP3 TIP3
TIP3 TIP3 TIP3 TIP3 TIP3 TIP3 TIP3 TIP3 TIP3 TIP3
*EOD
```

This file is almost self explanatory. It includes the sequence of the protein myoglobin, with the terminators of the peptide chain (non-pdb) – NTER and CTRG – which are defined as monomers in MOIL. Also provided are a monomer HEME (HEM1), a ligand (CO) and a small droplet of water (a few water molecules – TIP3). This file (in contrast to the property and the monomer files) must be constructed by the user for the molecule of interest. More details on the construction of this file can be found in the standard documentation of the program and we do not provide them here. After reading the polymerization file **con** creates the list of all the particles in the molecular system, lists of all the properties of the single particles and all the bonds that are referred to in the "standard" data base. After this is done **con** looks for additional bonds that cannot be included in the general polymerization procedure. An example which is relevant to the previously presented sequence of myoglobin is of bond formation between the proximal histidine and the heme iron. The input for the last is found in the addb.mb file and is

```
bond chem HIS 95 NE2 HEM1 157 FE
*EOD
```

It directs the program to add a bond between the $N_{\varepsilon 2}$ atom of the residue His 95 (the proximal histidine) and the iron at the center of the heme group. Note that the addition of an N terminal residue in MOIL changes the reference number of the proximal histidine from the common 94 in mutant myoglobin to the above 95. This should complete the list of all the bonds in the molecule and the **con** program proceeds to compute the list of all angles, torsions, improper torsions as well as the exclusion and the 1-4 lists. As discussed above the angles and the torsions are generated in a maximal way. This is working as expected in most of the cases, and specifically for all amino acids but not in all other cases. A well known counter example is the case of the heme in which the pyrrole nitrogens and the central iron form some desired and some undesired angles. NA-Fe-NB is a 90° angle that is kept, however NA-Fe-NC is 180° angle that should be removed from the list. It is necessary at this point to tell **con** that some internal degrees of freedom require elimination which is done in the file edit.mb :

```
remo angl chem HEM1 157 NA HEM1 157 FE HEM1 157 NC
remo angl chem HEM1 157 NB HEM1 157 FE HEM1 157 ND
*EOD
```

It should be emphasized that such operations are quite rare. If "special" groups (like the heme) are not present "editing" is not required. More common however is the action of bond addition due to the presence of sulfur bridges.

At this point **con** is all done. All information that required for energy calculations was produced and it is written to the file wcon.mb as the final act of the above program.

Essentially all of the MOIL programs depend on the existence of the connectivity file (which we named here as wcon.mb). The next step in our attempt to run molecular dynamics of myoglobin is to produce a coordinate file readable to the dynamics program from the pdb (protein data bank) coordinates. We are using the format of the CHARMM coordinate files. This is done by the program **puth**. **puth** also places at covalently reasonable positions all the missing hydrogens. In protein crystallography the resolution is insufficient to place hydrogens with certainty. Nevertheless, the polar hydrogens are needed in the model potential that we use. Given the pdb coordinates and the connectivity file, **puth** provides an output coordinate file that is acceptable for all other MOIL modules. Some work is required in preparing the pdb file for a successful manipulation by **puth** which is described in the documentation. As one example we note that multiple positions of side chains (that are sometimes found in experiments and therefore also in pdb files) must be removed before **puth** can be used. A "typical" **puth** input is given below

```
~ Here come the necessary connectivity file
file conn name=(wcon.ery) read
~ This is a pdb file to be read by the program
file rcrd name=(erythro.co) read ctyp=(pdb)
~ This is the new coordinate file - CHARMm style
~ that is used everywhere in MOIL
file wcrd name=(ery.CRD) wovr ctyp=(CHARM)
action
```

The hydrogens are placed by **puth** using covalent geometry only. If more than one possibility exists a random choice is made. It is therefore possible that the position of the hydrogens are not optimal for van der Waals and electrostatic interactions. Short minimization after **puth** is highly recommended. There are two minimizers in MOIL: one is based on conjugate gradient algorithm (**mini_pwl**) and the second on a truncated Newton-Raphson procedure (**mini_tn**). It is surprising but nevertheless true that in all our studies the conjugate gradient was doing better that the truncated Newton-Raphson algorithm. Below is a sample input

```
file rcon name=(wcon.mb10co)   read
file rcrd name=(MB10CO.CRD)   read
file wcrd name=(MB10CO.MIN.CRD)   wovr
file wmin name=(mb10co.min)       wovr
epsi=1.  cdie v14f=8. e14f=2.
tolf=0.01 mistep=50    list=10 rvmx=7. relx=10.
action
```

The organization of the minimization input file is similar to the corresponding input file for **puth** with a few additional instructions. First, the (already existing) connectivity and coordinate files are identified to the program, the output files include the coordinates of the final minimized structure and the minimization history (wmin). In the next two lines some parameters for the energy and the minimization algorithm are defined: **epsi** is the dielectric constant, **cdie** means that constant dielectric will be used, **v14f** and **e14f** are 1-4 scaling parameters (as suggested for the OPLS force field). **tolf** is the maximum force gradient that is acceptable as a converged solution and **mistep** is the maximum number of minimization step that is allowed in the present calculations. **list** is the number of minimization steps between updates (recalculation) of the list of interacting particles according to the cutoff criterion. **rvmx/relx** are cutoff distances for the van der Waals and electrostatic interactions respectively. We comment that the input files of all MOIL modules are similar and consist first of a list of relevant files followed by assignment of relevant parameters. None of the input files require FORTRAN formatting and this includes also the property and monomer files. The line interpreter in MOIL picks only the first four characters of a given name, for example **mistep** is interpreted internally as **mist,** the additional characters are ignored. The order of the assigned parameters is not important and an instruction line can be extended to the next line by adding a dash ("-") at the end of the line. The interpreter breaks the line to individual expressions using the spaces in the line. The use of alternatives (like tabs) is not supported. After the "line break" is completed the interpreter "echoes" the line to the standard output. Thus the appearance of the line in the standard output does not imply that all the parameters were correctly read by the program. The expressions are searched later to find parameters that the program needs.

Once the minimized structure is obtained (MB10CO.MIN.CRD) we can continue with the dynamics. This is the last example closing this "technical" session.

```
file conn name=(wcon.mb10co) read
file rcrd name=(MB10CO.MIN.CRD) read
file wcrd name=(MB10CO.DCD) bina  wovr
file wvel name=(MB10CO.VCD) bina  wovr
#ste=50000  step=0.002 info=100 #equ=10000  #crd=10 #vel=10
#lis=10 #scl=10  #tem=1  rand=-3451187 tmpi=10 tmpf=300
rmax=10. epsi=1.  cdie v14f=8.  e14f=2. shkb
action
```

Similarly to the minimization we provide the program with a connectivity file (first line), a starting coordinate file (second line) and we assign output coordinates and velocities files which are written in a compressed form (wcrd and wvel). History of the dynamics (i.e. regular reports along the trajectory such as average temperature, energies etc.) are written on the standard output. After the list of files some additional instructions are provided. **#ste** is the number of steps to be calculated. The **step** size is 0.002 psec in our example, **info**rmation on the trajectory is written on the standard output each 100 steps of dynamics. 10,000 equilibration steps (**#equ**) are calculated and coordinate sets (**#crd**) are written to the dynamics coordinate file each 10 steps. The non-bonded lists (**#lis**) are re-calculated each ten steps and the velocities are also scaled each ten steps to preserve the kinetic temperature. There is only one temperature in the example that we considered (**#tem**). Multiple temperatures in the system can be useful in a number of cases. For example, consider a conformational transition of a solvated peptide. In order to explore alternative configurations of the peptide in shorter periods it is possible to simulate the peptide at higher temperature than the solvent. **rand** provides a seed for a random number generator that samples the velocities from a Boltzmann distribution at the beginning of the trajectory. **tmpi** and **tmpf** are the initial and final temperatures in the equilibration process, **rmax** is a cutoff distance which sets both **rvmx** and **relx** equal to the same number and it is used for compatibility reasons. **epsi, cdie, e14f,** and **v14f** were described in the minimization step. **shkb** directs the program to fix the length of all the bonds in the system using the velocity SHAKE algorithm (RATTLE[10]). More parameters and options can be found in the documentation of the program.

The protocol required for producing dynamics from protein data bank files is therefore quite long. However it is rather straightforward and the input for the different programs is quite short. Essentially the same steps are followed by all molecular dynamics programs. As a summary we list again the programs one needs to run and the main result of each of the runs:

1. **con** – generating the connectivity file
2. **puth** – generating from the protein data bank coordinates another coordinate file that includes hydrogens.
3. **mini_pwl** – minimized the output of **puth** to obtain energetically more reasonable starting point
4. **dyna** – run a molecular dynamics trajectory. Main result: coordinates as a function of time.

SPECIAL TOPICS: THE LOCALLY ENHANCED SAMPLING (LES) METHOD AND THE SELF PENALTY WALK (SPW) ALGORITHM

In this section we describe two relatively new computational methods that are unique to MOIL. To the best of our knowledge, these computational methods, while published and established were not implemented yet in other generally accessible codes.

The first of the techniques is called LES (locally enhanced sampling). LES is a mean field approach that we introduced to the simulations of proteins. The words "mean field" can be misleading, since some of our applications of LES are exact, i.e., free energy calculations[16] and structure determination by energy optimization[6]. Applications of LES to time dependent problems are approximate. Nevertheless, LES made it possible for the first time to investigate ligand diffusion through a flexible protein matrix in a statistically meaningful way[12,14,15]. The application of LES in conjunction with experiment identified transient binding sites in the myoglobin matrix[15]. Recently, we significantly improved the accuracy of the application of LES to dynamics by use of a binary collision correction[11]. Detailed derivation of the LES protocol was given in the past[6,11,12], and here we describe the main results and how LES is translated to the computer code. In LES we selectively enhance the sampling of a part of the system that is of prime interest. As an example we consider the problem of side chain modeling in which LES is employed to enhance the sampling of the side chain configurations. Modeling of side chains is useful in refinement of low resolution X-ray and NMR structures in which only the protein backbone can be traced. Positioning of side chains is also important in homology modeling in which the backbone structure of the modeled protein is adopted from a known protein structure with a similar sequence. One way of determining the side chain structure is via energy minimization (energy like the one defined in MOIL). Direct energy minimization using **mini_pwl** is unlikely to work, since **mini_pwl** finds a local energy minimum that is the closest to the initial guess. More global optimization that searches for minima beyond the local neighborhood is therefore required. Simulated annealing is one such approach[21]. In simulated annealing we heat the system and slowly cool it down. The kinetic energy makes it possible to overcome barriers that separate minima. If the cooling is done infinitely slow we are guaranteed to find the global energy minimum (the lowest energy minimum of all available minima). Practical (finite) cooling can give a minimum which is not the lowest. Nevertheless typical results are considerably better than direct minimization. The LES protocol is an addition to the simulated annealing that dynamically modifies the target function – the energy – that we optimize. Pictorially what is done is the following: for each of the $C\alpha$ of the different residues we attached multiple copies of the same side chain. The multiple copies of the side chain are set in a special way. They do not see the other copies of the same residue and they interact with the protein backbone and the rest of the copies in an average way. We now anneal the new system. The basic idea in LES is to average the potential over a distribution that is determined from a mean field condition[6,11]. The distribution is given by a Hartree product of the part that we do not want to enhance and the part that we do. Here, the side chains are the part that we enhance. The averages over the Hartree products considerably reduce the barrier heights and therefore make the optimization easier. Of course, one should be careful not to make the potential too flat by using distribution functions that are too broad. Without barriers there is no more than one minimum which is not necessarily the correct one. Nevertheless, the requirement that the potential will not be over-cooked does not mean that we need to "eat" it raw! The shape of the distribution used in the averages is determined in LES by the temperature of the system and by the equations of motion. More formally, let $\mathbf{R}_r$ be the coordinate vector of the "rest of the system" and $\mathbf{r}_i$ be the coordinate vector of the i-th side chain that its position we should like to optimize. Then the probability density of coordinates – ρ - is approximated by $\rho = \rho_r(\mathbf{R}_r) \prod \rho_i(\mathbf{r}_i)$ where ρ_r and ρ_i's are given by:

$$\rho_r = \delta(\mathbf{R}_r - \mathbf{R}^o(t)) \qquad \rho_i = \Sigma \, \omega_k \, \delta_k(\mathbf{r}_i - \mathbf{r}^o(t))$$

The vectors $\mathbf{R}^o(t)$ and $\mathbf{r}^o(t)$ contain time dependent parameters for which we solve equations of motion:

$$\mathbf{M} d^2 \mathbf{R}^o(t)/dt^2 = -\Sigma \ \omega_i \nabla_\mathbf{R} U(\mathbf{R},r_i) \qquad (4)$$
$$\omega_i m_i \ d^2 r_i(t)/dt^2 = -\omega_i \nabla_{r_i} U(\mathbf{R},r_i)$$

where $\mathbf{M}$ and $\mathbf{m}_i$ are the mass matrices and ω is a diagonal weight matrix for the multiple copies. If all the copies of the different side chains have the same weight, as is usually the case, the diagonal elements of ω are equal to $1/N_i$ where N_i is the number of copies of the i-th side chain. Note that the forces (i.e. $-\nabla_\mathbf{R} U(\mathbf{R},r_i)$; $-\nabla_{r_i} U(\mathbf{R},r_i)$) are averaged over the current distribution that is determined in a self-consistent way from the equations of motion. This average reduces barrier heights and increases the relative energies of the minima. We refer the reader to an earlier publication for a proof[6]. The problem of annealing the density is reduced to solving ordinary trajectories (initial value differential equations) for a different system that include multiple copies of side chains.

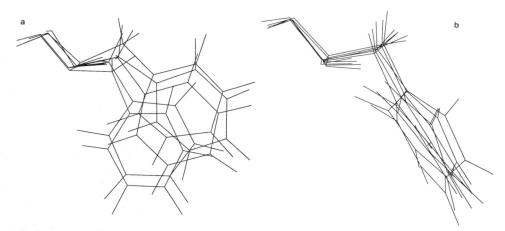

Fig.2 Snap shots in time for the distribution of the phenylalanine 43 side chain during the annealing of 45 side chains in myoglobin[23]. (a) High temperature structure, (b) Room temperature structure. Note that in the lower temperature structure the spatial distribution of the copies of the side chain is much narrower, as it should.

The results of an annealing using the LES protocol are exact. We proved[6] that the global energy minimum of the original and the mean field systems is the same and therefore the annealing of the LES energy with lower barriers is clearly advantageous to the regular annealing. We obtain the correct answer with better and easier exploration of alternative conformations. More details on the methodology and other applications of LES can be found elsewhere[6,11,12,14,15]. Here we should like to discuss several programming issues.

MOIL was built with LES in mind, therefore defining the mean field system is done in a relatively easy (?!) way. All that the user needs to do is to **mult**iply the desired monomers during the creation of the connectivity file and to provide multiple coordinates for the same monomers. The only new feature can be found in the poly file:

```
~
MOLC=(PEP3) #mon=8
NTER GLY mult ALA 4 PHE CTER
*EOD
```

We define a molecule that in "real life" should have five monomers NTER-GLY-ALA-PHE-CTER (including the terminus residues NTER and CTER). We request the multiplication of the alanine residue by a factor of 4 (mult ALA 4). The four copies of the alanine residue do not see each other and they are all connected to the GLY on the left and to the PHE on the right. The interaction energies with the copies are all normalized by a factor of 4. The program considers the system to have 8 monomers with four of them having the special properties described below. The special properties - the normalization of the forces, the normalization of the masses (multiplying the mass by the weight ω) and the exclusion of the interactions between copies of the same residues - are all taken care of during the creation of the connectivity file. LES particles are treated in an identical way to real particles, which makes programming and management of the code simple and straightforward. One can run or program dynamics applications, energy minimization and watch "movies" of multiple side chain copies (figure 2). For all practical purposes one can forget that these are not real particles but rather parameters describing the evolution of approximate probability density. At present LES is regularly used for searching plausible diffusion pathways[11,12,14,15], for optimization[6] and for free energy calculations[16]. This demonstrates the versatility and the usefulness of the new protocol for very different computational problems.

There is one disadvantage from the user point of view to the above set-up of monomer multiplication. It is not possible to multiply as a single segment several residues, unless the user will define them as a single residue (a process which can be time consuming). A facility that enables the multiplication of any group of atoms will be available in future releases of MOIL[13]. There is another subtle computational point that is worth discussing and is related to the calculations of the non-bonded energy in LES. The non-bonded parameters are kept in the connectivity file as single particle parameters. For each LES particle we "renormalize" the interactions by dividing the parameter by the number of copies (e.g. A_{ii} B_{ii} , q_i are translated to A_{ii}/N B_{ii}/N and q_i/N). This gives the correct result for an interaction of any LES particle with any real particle, or between any LES particle of a multiplied monomer with multiplied monomers of other residues. However, it does not give the correct answer for interactions within a LES monomer. For example, if we multiply a lysine, the "tail" of the side chain will interact incorrectly (using the "renormalized" parameters) with the backbone of the same residue. According to the protocol described above they are divided by N^2 (one N for each particle) rather than by N as required by formula (4). In the energy code we test for such pairs and we multiply them by N for re-renormalization. Since

182

the number of such pairs is usually small the computational cost of the re-renormalization is not large. Nevertheless, this is not an optimal solution. A more efficient approach will be to assign a special list for these interactions and to avoid the check mentioned above. The new list is likely to appear in future releases of MOIL.

Another unique piece of code in MOIL is **chmin**. It is a program to calculate reaction coordinates in large molecular systems. In contrast to LES, **chmin** is a separate module that uses ordinary connectivity files. It is based on the SPW (Self Penalty Walk) algorithm that we developed[5,22]. In the SPW approach the reaction coordinate is approximated by a discrete set of intermediates that interpolate between the known reactants and the known products. The intermediates resemble a "polymer" in which each of the monomers is a complete copy of the whole molecule(!). The "polymer" is hooked up at its two end points. One end is connected to the reactants and one end to the products. A minimum energy configuration for the "polymer" is a minimum energy path. Therefore, the basic calculation in the SPW approach is an energy minimization for the "polymer". Yet another view of the reaction path is of a walk between the reactant and the product, a walk in which there is a penalty for the polymer on crossing itself. This is why the algorithm is called SPW - Self Penalty Walk. The SPW can easily handle more than 100 particles and is capable of calculating continuous and well defined low energy paths even if a reasonable adiabatic coordinate is not available. This makes it quite unique compare to alternative approaches such as the reference coordinate approach[17] or mode following[18] adopted from codes of small molecules. Its advantage compared to algorithms used for small molecules is that it does not require the second derivative matrix. This makes it possible to calculate reaction coordinates in large molecular systems in a stable and reasonably economic way. Second derivative matrices are difficult to manipulate in large molecules for two reasons. First, the memory requirement is growing like N^2 where N is the number of particles. Second, the manipulation of these matrices and the extraction of relevant modes becomes exceedingly difficult when the density of the modes is high, as is the case for proteins. The polymer in **chmin** is larger than the original system. However it is optimized using first derivatives only. This makes the space requirement relatively small compared to mode following (see however the discussion below). The minimization of the energy of the polymer is also more stable than searches for "saddle points" pursued by the mode following techniques.

Some computational limitations are discussed next: An initial guess for the complete path may be provided by the user. If a reasonable guess is available this option is highly recommended. Alternatively the program starts from a linear interpolation between the reactant and the product. The interpolation is done in Cartesian space. This is a straightforward algorithm that always produces something. However, the linear interpolation usually gives paths with high energy barriers and therefore requires considerable computer (and human) time for refinement. Because the code works simultaneously on all the path structures, memory is an issue. Consider for example a system with 5,000 particles. Each of the particles has three degrees of freedom corresponding to motion along X, Y or Z axes. To specify the positions of the particles in the system 15,000 numbers (double precision) or 120Kbytes are needed. This is not too bad. However, if a path that consists of 100 intermediate structures is considered, the size increases to 12MB which is a significantly larger number and on workstations may create considerable difficulties. We therefore usually work with a smaller number of grid points and if there are additional intermediate minima between the reactants and the products we connect them in independent runs[5]. Another memory "issue" is the non-bonded lists. For 5,000 particles we anticipate lists of total size of 1MB per structure. Generating individual lists for every intermediate requires 100MB of memory. Even if possible on some workstations, significant degradation in performance due to swapping and cache misses are expected. In MOIL we made the approximation of using one set of lists for the whole set of structures. The lists are generated according to the intermediate that has the middle index. For example, examining a path with 101 structures, structure 50 is used to update the lists. A single set of lists may create difficulties especially if the difference between the two end structures is large and the cutoff distance used is small. A possible solution to the problem is to use a larger cutoff distance. Another solution is to change the code to include more than one set of lists (for example, one set for the reactant half of the path and one set for the half of the path of the product). Yet another solution is to generate arrays of differences between the lists, in addition to one common set of lists. The difference arrays are expected to be much smaller than individual lists. No..., programming the last modification to the code is not high on our agenda but the talented reader is encouraged to do so.

For the record we provide a typical input to the **chmin** program which is similar in spirit to what we saw previously

```
file conn name=(VAL.WCON) read
file rcrd unit=5 read
file wcrd name=(VALMIN.PTH) bina wovr
#ste=10000 #pri=10 #wcr=1000 list=10000 repl=1000.
lmbd=2. gama=20. grid=5
rmax=9999. epsi=1. cdie v14f=8. el14=2. cini
action
file name=(MOIL1.CRD) read
file name=(MOIL60.CRD) read
```

The first line direct the **chmin** program to the location of the connectivity file. In the second line the program is informed that the reactants and the products files will be identified after the "action". This is what the unit=5 (standard input in FORTRAN) means. MOIL1.CRD includes the coordinates of the reactants and MOIL60.CRD the coordinates of the products. Similarly to minimization and dynamics **#ste** is the number of optimization steps for the whole reaction path. **repl lmbd** and **gama** are energy parameters for intermonomer interaction in the SPW "polymer", **grid** is the number of path structures and **cini** directs the program to use a linear interpolation between the two structures. **chmin** overlaps first the reactant and the product structures such that the rms will be a minimum and then calculates Cartesian averages between the structures. The path coordinates are written on VALMIN.PTH, optimization history is written to the standard output.

In our experience we found that the **chmin** code is very robust, this is in molecular systems that vary from several tens of particles to thousands of particles. We did not find a molecular system on which our code was not able to find a reasonable solution. Further calculations of free energy path along the discrete set of points provided by **chmin** is also possible by (for example) **umbr** or **freee**. **umbr** pursues umbrella sampling calculation along the reaction coordinate[19] and **freee** – free energy perturbation[20]. This completes the required code for calculations of rates for chemical reactions and conformational transitions, within the limits of the transition state theory.

This also concludes the present paragraph describing some advanced features of MOIL.

FINAL REMARKS

In this chapter we discussed the general structure of a molecular dynamics program called MOIL. MOIL is designed as a flexible simulation package for users and programmers. For example, using a database of amino acids that is provided with MOIL it is possible to investigate computationally any protein. Furthermore, the large set of common subroutines and tools that is available via an anonymous ftp makes the code useful to new programmers in the field. We further provided a number of "tips" to those who designed their own computer experiments or planned to do programming on similar codes, "tips" that are hard to find in usual scientific publications. We discussed the design and typical steps taken in common molecular dynamics applications as well during LESs common applications that are unique to MOIL such as the LES mean field simulations and the **chmin** program.

ACKNOWLEDGMENTS

This research was supported by grants from the NIH and the DOE. RE thanks Alan Hinds for very useful discussions on code optimization. We are also grateful for the constructive suggestions and comments we received from many MOIL users.

APPENDIX 1: ANNOUNCEMENT OF MOIL RELEASE AND CONDITIONS

RE-RELEASE OF A PUBLIC DOMAIN MOLECULAR DYNAMICS PROGRAM
(An electronic mail message broadcasted in the computational chemistry list)

We are announcing a new release of moil - moil.006 . moil is a public domain molecular dynamics program that is available via anonymous ftp. At the end of the file we attached instructions of how to get the code. The code was developed by Ron Elber, Adrian Roitberg, Carlos Simmerling Haiying Li, Gennady Verkhivker, Robert Goldstein and Alex Ulitsky. Other contributors that helped in testing, developing parameter sets and writing documentation: Danuta Rojewska and Chyung Choi.

Compare to previous releases (003 and 004) the present code has a number
of bug fixes and new features. The most major addition
is the visualization program for Silicon Graphics (moil-view ,
author: Carlos Simmerling). Other features include considerable
speed up of most intensive computations (up to a factor of 3) and
undoubtedly a number of unknown bugs. Please note that old connectivity
files will not run on 006 and it is necessary to regenerate them.

At present existing modules are:
 con   : creating connectivity files
 con.specl: creating special connectivity files for curve crossing
calculations
 con.specll: need also to run con.specll to get the curve crossing to
work.
 energy   : good old vanilla energy calculation
 dyna   : good old vanilla dynamics using velocity Verlet with RATTLE
(velocity SHAKE), also possible to use periodic (cubic) boundary
conditions and also possible to use LES (Locally Enhanced Sampling
Protocol). We worked on optimizing the dynamics code, The previous
version of RATTLE was improved and the use of double cutoffs (for van der
Waals and electrostatic) also speeds-up things. While more can be
undoubtedly done, at present we are doing better than anyone else we are
compared to.

 chain or **chmin**: Reaction path algorithms that are based on optimization
                of the complete path.
                (R. Czerminski and R. Elber IJQC 24,167(1990))
 ccrd   : Converting CooRDinates between different formats
 boat   : provide values of BOnds Angles and Torsions. I.e.
                internal coordinates for a set of structures.
 rms_resd : rms calculations for those who care
 fluc   : fluctuations for those who are not at a minimum (from MD)
 umbr   : Umbrella sampling calculations.
 freee   : free energy calculations along reaction paths by perturbation
or thermodynamic integration
 mini_tn : minimization using truncated Newton Raphson
 mini_pwl : Conjugate gradient algorithm for energy minimization.
 puth   : add hydrogens to protein data bank coordinates to obtain
acceptable MOIL coordinates for energy calculations. Done according to
geometric criteria (no fancy stuff)
 therm   : free energy calculations where the potential parameters are
changing. Alchemistry.
 contacts : analyzing dynamics to find collisions between specific
residues
 rgyr   : calculate radius of gyration
 torstat : compact torsion statistics (as compared to boat), give
phi.psi.chil values for all residues
 mult   : multiply coordinates of desired residues for LES calculations.

AND FINALLY:

 moil-view : (creator - Carlos Simmerling)
 A visualization program (not modeling!) for Silicon Graphics (only).
Easy to use and well interfaced with the features of moil. Dynamics of
shaded spheres, sticks or mixtures. Easy recording to video tapes.
Note that this program does NOT require the use of moil but can
also use other file formats such as PDB.

The code is in the public domain. We are giving up any right to the code.
You may include it in your own code and distribute it further. You may
transfer it "as is" to other users without asking us for permission.
However, we request that this message will be kept. In case that changes
are made to the code we request that a list of changes will be provided
to the new users. We shall not be responsible for other people's bugs, we
can hardly handle our own.

186

The code is given with the friendly understanding that you will not try to use this code as part of a commercial software package. However, we put no restrictions on other type of products that may emerge using this code. Reference should be made to the authors of the programs as listed in the makefile to the extent that this program helps/distracts you in your work. Contact Ron Elber (ron@pap.chem.uic.edu) for proper references.

BUGS/PROBLEMS OR OTHER TYPES OF DISCOMFORT ASSOCIATED WITH MOIL CAN BE REPORTED TO: ron@pap.chem.uic.edu
We shall make significant effort to answer inquiries and especially to fix bugs. Note however (as is quite clear from the price tag on this package) that we are not making our living selling this code. Therefore we do not promise to provide satisfactory answers to all questions. On what is free there is no guarantee!

TO GET THE CODE:

```
ftp  128.248.186.70
user-name:anonymous
passwd:  your  e-mail
cd dist
get  moil.006.tar.Z (source code only)
get  moil.dat.tar.Z (documentation and parameters/property and monomer
                 files, in the documentation you made find useful
                 the file tutorial)
get  moil.test.tar.Z (test files - input and output - quite large and
                 may be a real pain to ftp, useful also as a starting
                 point to construct your own input).
get  view.tar.Z (the moil-view visualization program of Carlos
                 Simmerling
quit
```

```
On your machine
uncompress  moil.006.tar.Z
tar  xvf  moil.006.tar
 tar  xvf  moil.006.tar
repeat the same procedure for the other files.
```

```
cd  moil.src.006
```

edit the makefile and change the variables FC and FFLAGS to fit your machine.
then type "make" and you should be all set.

A good place to start is the tutorial in moil.doc, please note however that some of the docu were prepared with the old versions, so things may have slightly different output than you expect. Old input should work.
To compile moil-view go to the view.source directory and type make. The machine MUST be a Silicon Graphics. Look at the file view.doc to get started.

APPENDIX II: THE FORTRAN COMMON BLOCK THAT INCLUDES ALL THE CONNECTIVITY DATA (CONNECT.BLOCK)

C DEFINITION OF THE MOLECULE TOPOLOGY AND POTENTIAL ENERGY TERMS

```
C--------------------------------------------------------------------
```

```
c *** vectors of length npt (number of particles in the system)
c                npt < maxpt
c poimon  - integer vector with monomer pointer
c                (poimon(ipt) = the # of the monomer of the ipt particle)
c ptid    - integer vector with id of particle types
c ptnm    - (char*4) name of the particle as a function of the particle
number
c ptms    - (double) mass of the particle                    "
c invms   - (double) 1/mass of particle, added by carlos for shake speed
c ptchg   - (double) charge of particle                      "
c epsgm6  - (double) twice square root of van der Waals energy times
sigma^6
c epsgm12 - (double) twice square root of van der Waals energy times
sigma^12
c ptwei   - (double) particle special weight used in LES calculations
c lesid   - integer id of LES residue. If normal residue set to 0 if les
c              residue all the copies under a single MULT are set to the
c              same number
c flgchr  - is a flag charge. It is set to true of the particle is
charged
c              and to false if it is not.
c
        integer poimon(maxpt),ptid(maxpt),lesid(maxpt)
        logical flagchr(maxpt)
        character*4 ptnm(maxpt)
        double precision ptms(maxpt),ptchg(maxpt),epsgm6(maxpt)
        double precision epsgm12(maxpt),ptwei(maxpt),invms(maxpt)
c------------------------------------------------------------------------


c------------------------------------------------------------------------
c BULK - character vector with the name of BULK (molecular systems)
c assembled
c pbulk - pointer to the last particle of the i-th BULK
c
        character*4 BULK(MAXBULK)
        integer pbulk(MAXBULK)
c------------------------------------------------------------------------


c------------------------------------------------------------------------
c *** vector of length imono (number of monomers in the molecule)
c              imono < maxmono
c poipt   - From monomers to particles poipt(i) is the last particle
c              of monomer i
c
        character*4 moname(maxmono)
        integer poipt(0:maxmono)
c------------------------------------------------------------------------


c------------------------------------------------------------------------
c *** vectors of length nb (number of bonds) nb < maxbond
c ib1 ib2  - pointers to particles that are bonded
c              ib1(k) is the pointer to particle k.
c kbond    - (double) force constant for harmonic energy of bonds
c req      - (double) equilibrium position for harmonic energy of bonds
c imb1 imb2 - pointers to morse bonds
c rmeq     - equilibrium distance for morse potential
c D,alpha  - parameters of Morse potential
c Arep     - parameter of repulsion energy function
c Brep     - parameter of repulsion energy function
c beta1    - parameter of repulsion energy function
c
        integer ib1(maxbond),ib2(maxbond),imb1(maxmorsb),imb2(maxmorsb)
        double precision kbond(maxbond),req(maxbond),rmeq(maxmorsb)
```

```
        double precision D(maxmorsb),alpha(maxmorsb),Arep
        double precision beta1(maxmorsb),Brep
c--------------------------------------------------------------------


c--------------------------------------------------------------------
c *** vectors of length nangl (number of angles)
c              nangl < maxangl
c iangl1 iangl2 iangl3 - pointers to particles in angles
c kangl    - (double) force constant for harmonic energy of angles
c angleq   - (double) equilibrium position for harmonic energy of angles
c
        integer iangl1(maxangl),iangl2(maxangl),iangl3(maxangl)
        double precision kangl(maxangl),angleq(maxangl)
c--------------------------------------------------------------------


c--------------------------------------------------------------------
c *** vectors of length ntors (number of torsions)
c          nunmer of torsion =< maxtors
c itor1 itor2 itor3 itor4 - pointers to particles in torsions
c ktors[1-3]   - (double) amplitude for torsional energy
c period   - (integer) period of rotation (e.g.  2 for C=C, 3 for C-C)
c phase    - (double) phase factor for torsional energy
c
        integer itor1(maxtors),itor2(maxtors),itor3(maxtors)
        integer itor4(maxtors)
        integer period(maxtors)
        double precision ktors1(maxtors),ktors2(maxtors)
        double precision ktors3(maxtors),phase(maxtors)
c--------------------------------------------------------------------


c--------------------------------------------------------------------
c *** vectors of length nimp (number of improper torsions)
c             nimp =< maximp
c iimp1 iimp2 iimp3 iimp4 - pointers to particles in improper torsions
c kimp     - (double) force constant for all improper torsions
c impeq    - (double) equilibrium position for improper torsion
c
        integer iimp1(maximp),iimp2(maximp),iimp3(maximp),iimp4(maximp)
        double precision kimp(maximp)
        double precision impeq(maximp)
c--------------------------------------------------------------------


c--------------------------------------------------------------------
c Exclusion lists between 1-2 1-3 and  aspecial list for 1-4.
c excl is a pointer to exc2, i.e. exc2(excl(i)) is the last
c particles to be excluded from interaction with i
c Similarly spec1 is a pointer to spec2
c
        integer excl(0:maxpt),exc2(maxex)
        integer exc31(0:maxpt),exc32(maxex)
        integer spec1(0:maxpt),spec2(maxspec)
        integer spec31(0:maxpt),spec32(maxspec)
c--------------------------------------------------------------------


c--------------------------------------------------------------------
c *** integers define actual size of molecule, initially are set to 0
c totmon   - total number of monomers
c npt      - total number of particles
c nb       - total number of bonds
c nmb      - total numberof morse bonds
c nangl    - total number of angles
c ntors    - total number of torsion
c nimp     - total number of improper torsions
c lestyp   - total number of different types of LES monomers
```

```
c totex    - number of exclusions of particle interactions
c totspe   - number of special (1-4) interactions
c nexc     - number of exclusions of particle interactions
c nspec    - number of special (1-4) interactions
c NBULK    - number of "bulk" (molecular systems)
c nchg     - the number of charged particles in the molecule(s
c *** lesflag - if true, les particles are present
c
        integer totmon,npt,nb,nmb,nangl,ntors,nimp,lestyp,NBULK
        integer totex,totspe,nchg
        integer totex3,totspe3
        logical lesflag



c------------------------------------------------------------------
c Now gather everything into common blocks
c
        common /CONNLOG/lesflag,flagchr
        common /CONNINT/poimon,lesid,ptid,pbulk,poipt,ib1,ib2,iangl1,
    1        iangl2,iangl3,itor1,itor2,itor3,itor4,period,
    2        iimp1,iimp2,iimp3,iimp4,exc1,exc2,spec1,spec2,
    3        exc31,exc32,spec31,spec32,
    4        totmon,npt,nb,nangl,ntors,nimp,lestyp,NBULK,
    5        totex,totspe,totex3,totspe3,imb1,imb2,nmb,nchg
        common /CONNDBL/ptms,invms,ptchg,epsgm6,epsgm12,ptwei,
    1        kbond,req,kangl,angleq,ktors1,ktors2,ktors3,phase,
    2        kimp,impeq,rmeq,D,alpha,Arep,beta1,Brep
        common /CONNCHR/ptnm,moname,BULK
```

REFERENCES

1. U. Burkert and N.L. Allinger, "Molecular Mechanics", ACS, Washington D.C., 1982
2. S.J. Weiner, P.A. Kollman, D.A. Case, U.C. Singh, C.Ghio, G. Alagons, S. Profeta Jr. and P. Weiner, "A new force field for molecular mechanical simulation of nucleic acids and proteins", J. Amer. Chem. Soc. 106:765(1984)
3. B.R. Brooks, R.E. Bruccoleri, B.D. Olafson, D.J. States, S. Swaminathan and M. Karplus, "CHARMM: A program for macromolecular energy, minimization, and dynamics calculations", J. of Comput. Chem. 4:187(1983)
4. Groningen Molecular Simulation program manual, version GROM87, BIOMOS B.V. Groningen, The Netherland
5. R. Czerminski and R. Elber, "Self avoiding walk between two fixed points as a tool to calculate reaction paths in large molecular systems", Int.J.Quant.Chem. andthe Proc. of Sanibel Symposia 24:167(1990)
6. A. Roitberg and R. Elber, "Modeling side chains in peptides and proteins: Application of the Locally Enhanced Sampling (LES) and the simulated annealing methods to find minimum energy conformations", J. Chem. Phys. 95,9277(1991)
7. W.L. Jorgensen and J. Tirado-Rives, "The OPLS potential function for proteins. Energy minimizations for crystals of cyclic peptides and crambin", J. Amer. Chem. Soc. 110:1657(1988)
8. E.E. Nikitin, "Theory of elementary atomic and molecular processes in gases", (Clarendon, Oxford, 1974)
9. H. Li, R. Elber and J.E. Straub, "Molecular dynamics simulation of NO recombination to myoglobin mutants", J. Biol. Chem., in press
10. H.C. Anderson, "Rattle: A velocity version of the SHAKE algorithm for molecular dynamics simulations", J. Comput. Physics, 52:24(1983)

11. A. Ulitsky and R. Elber, "The equilibrium of the time dependent Hartree and the Locally Enhanced Sampling approximations: Formal properties, corrections and computational examples of rare gas clusters", J. Chem. Phys. 98:3380(1993)

12. R. Elber and M. Karplus, "Enhanced sampling in molecular dynamics: Use of the Time-Dependent Hartree approximation for a simulation of carbon monoxide diffusion through myoglobin", J. Amer. Chem. Soc., 112:9161(1990)

13. Chen Keaser, unpublished results

14. R. Czerminski and R. Elber, "Computational studies of ligand diffusion in globins: I. leghemoglobin", Proteins 10:70(1991)

15. Q.H. Gibson, R. Regan, R. Elber, J.S. Olson and T.E. Carver, "Distal pocket residues affect picosecond recombination in myoglobin: An experimental and molecular dynamics study of position 29 mutants", J. Biol. Chem. 267,22022(1992)

16. G. Verkhivker, R. Elber and W. Nowak, "Locally enhanced sampling in free energy calculations: Application of mean field approximation to accurate calculations of free energy differences", J. Chem. Phys. 97:7838(1992)

17. R. Czerminski and R. Elber, "Reaction path study of conformational transitions in flexible systems: Applications to peptides", J. Chem. Phys. 92:5580(1990)

18. J. Barker, "An algorithm for the location of transition states", J. Comput. Chem. 7:385(1986)

19. G. Verkhivker, R. Elber and Q.H. Gibson, "Microscopic modeling of ligand diffusion through the protein leghemoglobin: Computer simulations and experiments", J. Amer. Chem. Soc. 114:7866(1992)

20. R. Elber, "Calculations of the potential of mean force using molecular dynamics with linear constraints: An application to a conformational transition of a solvated dipeptide", J. Chem. Phys. 93:4312(1990)

21. R.H.J.M. Otten and L.P.P. van Ginneken, "The annealing algorithm", Kluwer Academic, Boston, 1989

22. W. Nowak, R. Czerminski and R. Elber, "Reaction path study of ligand diffusion in proteins: Application of the SPW method to calculate reaction coordinate for the motion of the CO through leghemoglobin", J. Amer. CHem. Soc. 113:5627(1991)

23. A. Roitberg and R. Elber, "The locally enhanced sampling for side chain modeling", a chapter in "Protein Structure Determination", Ed. K. Merz and S. Le Grand, Ed., Springer Verlag, in press.

FLEXIBILITY IN THE Fc REGION OF IMMUNOGLOBULIN G

R. Oomen[1], M.S. Chappel[2], and M. Klein[1,2]

[1]Connaught Center for Biotechnology Research
Willowdale, Ontario, Canada, M2R 3T4
[2]Department of Immunology, University of Toronto
Toronto, Ontario, Canada, M5S 1A8

INTRODUCTION

The immunoglobulin G (IgG) hinge region links the two Fab regions to the Fc region, allowing the Fab 'arms' to move relative to each other and to the Fc region. The length and sequence of the hinge varies between different Ig classes and subclasses, and these differences are implicated in differential binding to membrane-bound cellular receptors known as Fc receptors (FcR) (Burton et al., 1985; Burton, 1990). In particular, the human IgG binding site for the high affinity Fc receptor (FcγRI) has been identified as being on the C-terminal portion of the lower hinge, encoded by the C_H2 exon (Duncan et al. 1988, Chappel et al. 1991).

A Non-Hinge Residue Influences FcγRI Binding

Although the IgG2 subclass does not normally bind to FcγRI, it was found that by exchanging the lower hinge sequence in IgG2 (residues 233-235, Pro-Val-Ala) with that of IgG1 (Glu-Leu-Leu-Gly), the mutant IgG2 acquired FcγR binding activity that was 3-4 times greater than that of IgG1 (Chappel et al., 1991). The enhanced binding activity in this mutant implied that the Fc region of IgG2 contains determinants capable of enhancing FcγR binding that are distinct from the hinge region, and which are not present in IgG1. However, there are only five amino acid sequence differences between the C_H2 domains of IgG1 and IgG2; of these, the IgG2-specific residue Thr339 was shown by site-directed mutagenesis to enhance FcγRI binding (Chappel et al., 1993). Thr339 is located at the far end of the domain from the mutated hinge residues, a distance of about 28Å.

Using the crystal structures of Fc (Deisenhofer, 1981, Brookhaven Protein Data Bank (PDB) entries 1fc1 and 1fc2), we found that residue 339 is in the center of the contact region between the C_H2 and C_H3 domains, with the side chain directed into a pocket formed by residues Tyr373 and Pro375 of the C_H3 domain. The three residues

Statistical Mechanics, Protein Structure, and Protein Substrate Interactions
Edited by S. Doniach, Plenum Press, New York, 1994

193

are close together in space, and are well conserved in the IgG subclasses of many species (excepting mouse IgG2 subclasses), suggesting some structural or functional role.

Structural Features of the 'Fc Pivot'

The side chain of residue 339 is directed into the cleft formed by the junction of the C_H2 and C_H3 domains, packing against the side chains of Tyr373 and Pro374 of the C_H3 domain (Figure 1). This arrangement is similar to the flexible 'elbow' joint in the heavy chain of Fab regions, described as a molecular "ball and socket" joint by Lesk and Chothia (1988). Although differing in some details, the central elements of the Fab (V-D-J)$_H$:C_H1 elbow and the Fc C_H2:C_H3 pivot are the same. In each case, a threonyl side chain is directed into a pocket in the molecular surface resulting from the presence of a <u>cis</u>-proline. An aromatic ring from the residue immediately N-terminal to the <u>cis</u>-proline packs against the threonyl side chain. The aromatic residue is phenylalanine in the case of the Fab elbow joint, and tyrosine in the Fc C_H2:C_H3 pivot joint.

Distance comparisons between identical residues of the two Fc crystal structures suggest a combination of angular and rotational displacements of the C_H2 domains are possible relative to the C_H3 domains. However, residue 339 and other C_H2 residues near in space have small displacements, suggesting that residue 339 acts as a pivot, allowing rotary motion of the C_H2, as well as changes in the orientation of the axis of rotation.

The Fc joint differs from the Fab elbow (Lesk and Chothia, 1988) in that it has fewer component residues, suggesting that allowable movements about the Fc pivot may differ from those in the Fab elbow. In particular, movement of the two C_H2 domains relative to each other would be buffered by the interposed N-linked oligosaccharides. Movement of the C_H2 domains relative to the C_H3 dimer would be restricted by the covalent constraints of the linker connecting the C_H2 and C_H3 domains and the hinge disulfides immediately preceeding the lower hinge.

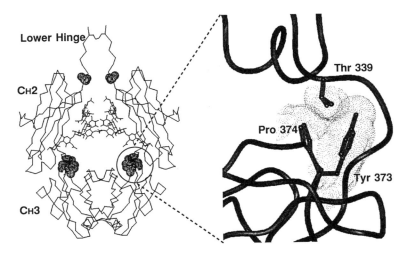

Figure 1. On the left, a Cα-tracing of the bilaterally symmetric Fc fragment is shown with the lower hinge region and the C_H2 and C_H3 domains labelled. N-linked oligosaccharides separate the two C_H2 polypeptide chains. Van der Waals surfaces are displayed on residue 235 in the lower hinge, which is implicated in Fcγ receptor binding, and on the residues 339, 373 and 374, which constitute the 'Fc pivot'. The close-up view of the pivot was obtained by rotating the structure ~90°. This view shows the position and packing of Thr339 relative to the molecular surfaces of the sidechains of Tyr373 and Pro374. These diagrams were derived from PDB file 1fc2.

Proposed Effect of Thr339

In the Thr339 mutant IgG2, the altered side chain likely causes a small change in packing at the C_H2:C_H3 interface, effectively causing the C_H2 domains to swivel about the Fc pivot. This small structural change would be amplified at the N-terminal end of the domain, where the hinge joins the C_H2 domain. Since the Fc has bilateral symmetry, this change would affect the relative orientations of both C_H2 domains, resulting in functionally significant changes in the exposure and/or affinity of the hinge FcγRI binding site.

REFERENCES

Burton, D.R., Jefferis, R., Partridge, L.J., Woof, J.M., 1985, Molecular recognition of antibody (IgG) by cellular Fc receptor (FcRI), *Mol. Immunol.* 25:1175.

Burton, D.R., 1990, The conformation of antibodies, *in*: "Fc Receptors and the Action of Antibodies", H. Metzger, ed., American Society for Microbiology, Washington.

Chappel, M.S., Isenman, D.E., Everett, M., Xu, Y.-Y., Dorrington, K.J., Klein, M.H., 1991, Identification of the Fcγ receptor class I binding site in human IgG through the use of recombinant IgG1/IgG2 hybrid and point-mutated antibodies, *Proc. Natl. Acad. Sci.* 88:9036.

Chappel, M.S., Isenman, D.E., Oomen, R., Xu, Y.-Y., Klein, M.H., 1993, Identification of an accessory FcγRI binding site within a genetically engineered human IgG antibody, *J. Biol. Chem.* submitted.

Deisenhofer, J., 1981, Crystallographic refinement and atomic models of a human Fc fragment and its complex with fragment B of protein A from *Staphylococcus aureus* at 2.9- and 2.8-Å resolution, *Biochem.* 20:2361.

Duncan, A.R., Woof, J.M., Partridge, L.J., Burton, D.R., Winter, G., 1988, Localization of the binding site for the human high-affinity Fc receptor on IgG, *Nature* 332:563.

Lesk, A.M. and Chothia, C., 1988, Elbow motion in the immunoglobulins involves a molecular ball-and-socket joint, *Nature* 335:188.

COMPUTER-SUPPORTED PROTEIN STRUCTURE DETERMINATION BY NMR

Peter Güntert

Institut für Molekularbiologie und Biophysik
Eidgenössische Technische Hochschule-Hönggerberg
CH-8093 Zürich, Switzerland

INTRODUCTION

The relationship between amino acid sequence, three-dimensional structure and biological function of proteins is one of the most intensely pursued areas of molecular biology and biochemistry. In this context, the three-dimensional (3D) structure has a pivotal role, as its knowledge is a prerequisite to understanding the physical, chemical, and biological properties of a protein (Schulz and Schirmer, 1979). Until 1984, structural information at atomic resolution could only be determined by X-ray diffraction techniques with protein single crystals (Blundell and Johnson, 1976). The introduction of nuclear magnetic resonance (NMR) spectroscopy as a technique for protein structure determination has made it possible to obtain structures at comparable resolution in solution (Wüthrich, 1986).

The utility of the NMR method for the study of molecular structures depends to a large part on the sensitive variation of the resonance frequency of a nucleus with the chemical structure, the conformation of the molecule, and the solvent environment (Ernst *et al.*, 1987). These so-called chemical shifts ensure the necessary spectral resolution, although they do not provide direct structural information. They arise because a given nucleus is shielded from the externally applied magnetic field to a different extent depending on its local environment. Three of the four most abundant elements in biological materials, *i.e.*, hydrogen, carbon and nitrogen, have naturally occurring isotopes with nuclear spin $I = 1/2$, and are therefore suitable for high-resolution NMR experiments in solution (Wüthrich, 1986). The proton (^1H) has the highest natural abundance (99.98%) and the highest sensitivity (due to its large gyromagnetic ratio) among these isotopes, and hence plays a central role in most NMR experiments with biopolymers. The low natural abundance of ^{13}C and ^{15}N (1.11% and 0.37% respectively) and low relative sensitivity of these nuclei normally require isotope enrichment. Today, this is routinely achieved by overexpression of proteins in isotope-labelled media, which is indispensable for structure determination with larger systems (Billeter *et al.*, 1993; Clore and Gronenborn, 1991). Structures of small proteins with molecular weight $\leq 10,000$ can be efficiently determined by ^1H NMR alone.

The procedure used for the determination of three-dimensional protein structures by NMR is outlined in Fig. 1. It includes preparation of the protein, NMR experiments, obtain-

Statistical Mechanics, Protein Structure, and Protein Substrate Interactions
Edited by S. Doniach, Plenum Press, New York, 1994

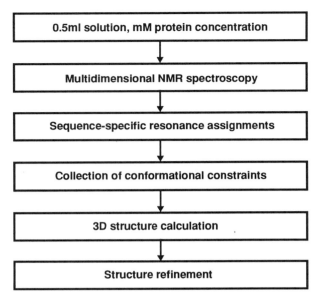

Figure 1. Outline of the procedure for NMR structure determination of proteins in solution.

ing assignments of the NMR lines to individual atoms in the polypeptide chain, calculation and refinement of the three-dimensional structure.

In practice, the data for the structure determination of a small protein are collected by performing homonuclear two-dimensional (2D) ^{1}H NMR experiments (Wüthrich, 1986). Typically, these spectra contain diagonal peaks (Fig. 2), with $\omega_1 = \omega_2$, which correspond to a conventional 1D spectrum. In addition there are a large number of off-diagonal cross peaks with $\omega_1 \neq \omega_2$. Each cross peak establishes a correlation between two protons. For example, in 2D nuclear Overhauser enhancement spectroscopy (NOESY; Anil-Kumar *et al.*, 1981) the cross peaks represent "trough-space" nuclear Overhauser effects (NOEs) and indicate that the two correlated protons are separated by a short distance (< 5 Å) in the three-dimensional structure. In 2D correlated spectroscopy (COSY; Rance *et al.*, 1983), the cross peaks represent "through-bond" scalar spin-spin couplings. For a protein, a COSY cross peak thus indicates that two protons are separated by at most three chemical bonds in the covalent structure. Using suitable NMR experiments with H_2O and/or D_2O solutions of the protein, one can obtain a complete or nearly complete set of sequence-specific resonance assignments (Wüthrich, 1986).

Once resonance assignments are available for all or nearly all polypeptide protons, one can start to collect conformational constraints for the determination of the 3D structure: NOEs which are converted into distance constraints, and spin-spin coupling constants which yield information on dihedral angles. The most important input for a structure calculation are upper bounds on ^{1}H-^{1}H distances derived from NOESY experiments (Fig. 3). The intensity of a NOESY cross peak is related to the distance r between the two correlated protons by

$$NOE \propto \langle r^{-6} \rangle f(\tau_c) \tag{1}$$

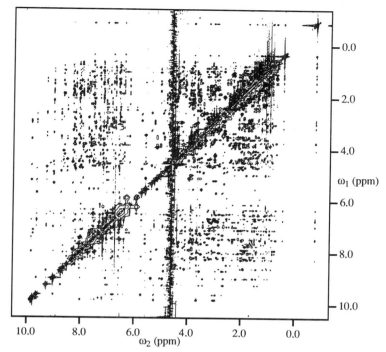

Figure 2. Two-dimensional ^{1}H NOESY spectrum of Toxin K in H_2O at pH 4.6 and 36 °C (protein concentration 10 mM, mixing time $\tau_m = 40$ ms; Berndt *et al.*, 1993).

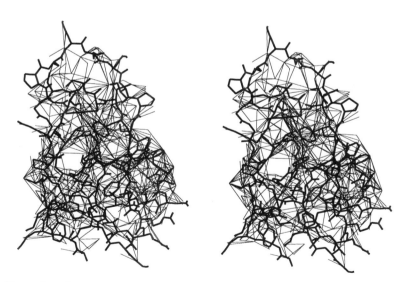

Figure 3. Stereo view of an all-heavy-atom representation of the protein Toxin K (bold lines). Each of the 809 upper distance constraints used in the structure calculation (Berndt *et al.*, 1993) is shown by a thin line connecting the two protons (or pseudo-atoms) involved in the constraint.

where the second term is a function of the correlation time, τ_c, which accounts for the influence of motional averaging processes on the observed NOE. $f(\tau_c)$ is governed by the combined effects of overall rotational tumbling and intramolecular mobility, and may vary for different locations in a protein molecule. Therefore only upper bounds on ^{1}H-^{1}H distances are usually derived from NOE measurements (Wüthrich, 1986). In a systematic analysis of the local conformation, measurements of vicinal spin-spin coupling constants $^3J_{HN\alpha}$ and $^3J_{\alpha\beta}$ may be combined with intraresidual and sequential NOEs to restrict the dihedral angles ϕ, ψ, and χ^1 and to obtain stereospecific assignments for β-methylene protons (Güntert et al., 1989).

STRUCTURE CALCULATION

The network of distance constraints between spatially proximate hydrogen atoms (Fig. 3) and angle constraints derived from the NMR experiments constitutes the input for the structure calculation. In the early stages of the development of the NMR method for protein structure determination it immediately followed from the nature of the conformational constraints that the techniques for structure determination from other experimental data, in particular X-ray diffractions, could not be adapted for the structural analysis of the NMR data, and hence new ways had to be developed. Initially, metric matrix distance geometry algorithms were used (Braun et al., 1981; Havel and Wüthrich, 1984). Subsequent work included a variable target function algorithm (Braun and Gō, 1985), interactive molecular modelling using computer graphics (Billeter et al., 1985), and restrained molecular dynamics (Brünger et al., 1986) in conjunction with model building (Kaptein et al., 1985) or distance geometry calculations (Clore et al., 1985). The following procedures are currently mostly employed: Embedding using metric matrix distance geometry, followed by simulated annealing using molecular dynamics (e.g., Driscoll et al., 1989; Lee et al., 1989) and structure calculation using the variable target function algorithm, supplemented by energy minimization (e.g., Berndt et al., 1992; Qian et al., 1989).

NMR structures of proteins in solution are commonly presented as a group of conformers, each of which has been calculated individually from the same experimental input data. In a high-quality structure determination each individual conformer has small residual violations of the conformational constraints, and the root mean square deviations (RMSD) among all conformers in the group are small (Braun, 1987; Brünger and Nilges, 1993). In this chapter we describe a new strategy for the use of the variable target function program DIANA (Güntert et al., 1991a) that enables efficient calculation of such groups of conformers.

Variable Target Function Algorithm

The basic idea of the variable target function algorithm (Braun and Gō, 1985) is to *gradually* fit an initially randomized starting structure to the conformational constraints collected with the use of NMR experiments, starting with intraresidual constraints only, and increasing the "target size" step-wise up to the length of the complete polypeptide chain. Advantages of the method are its conceptual simplicity and the fact that it works in dihedral angle space, so that the covalent geometry is preserved during the entire calculation.

The algorithm used by the program DIANA (Güntert et al., 1991a) is based on the minimization of a variable target function $T(\phi_1, ..., \phi_n)$, where the n degrees of freedom are the dihedral angles $\phi_1, ..., \phi_n$ about rotatable bonds of the polypeptide chain. During the calculation the bond lengths, bond angles and chiralities of the covalent structure are kept fixed at the ECEPP standard values (Momany et al., 1975; Némethy et al., 1983). The target function T, with $T \geq 0$, is defined such that $T = 0$ if and only if all experimental distance and dihedral angle constraints are fulfilled and all non-bonded atom pairs satisfy a check for the absence of steric overlap. $T(\phi_1, ..., \phi_n) \leq T(\phi'_1, ..., \phi'_n)$ if the conformation $(\phi_1, ..., \phi_n)$ satisfies the constraints better than the conformation $(\phi'_1, ..., \phi'_n)$. The problem to be solved is to find values $(\phi_1, ..., \phi_n)$ that yield low values of the target function.

To reduce the danger of becoming trapped in a local minimum with a function value much higher than the global minimum, the target function is varied during a structure calculation. At the outset only local constraints with respect to the polypeptide sequence are considered, and in subsequent rounds of calculations, constraints between atoms further apart with respect to the primary structure are included in a step-wise fashion (Fig. 4). Consequently, in the first stages of a structure calculation the local features of the conformation will be established, and the global fold of the protein will be obtained only towards the end of the calculation (Fig. 5).

Two different kinds of constraints are considered by the target function (Wüthrich, 1986; Braun, 1987): Upper and lower bounds $c_{\alpha\beta}$ on distances between two atoms α and β, and restraints on individual dihedral angles ϕ_a in the form of allowed intervals $[\phi_a^{min}, \phi_a^{max}]$ with $\phi_a^{min} < \phi_a^{max} < \phi_a^{min} + 2\pi$. Let $\Gamma_a = \pi - (\phi_a^{max} - \phi_a^{min})/2$ be the half-width of the forbidden interval of dihedral angle values, and Δ_a the size of the dihedral angle constraint violation. I_o, I_u and I_v denote the sets of atom pairs $\alpha\beta$, for which upper, lower or van der Waals distance bounds $u_{\alpha\beta}$, $l_{\alpha\beta}$ or $v_{\alpha\beta}$ exist, $I'_c \subseteq I_c$ ($c = u, l, v$) subsets thereof, I_a the set of restrained dihedral angles, w_u, w_l, w_v and w_a weighting factors for the different types of constraints, $d_{\alpha\beta} = d_{\alpha\beta}(\phi_1, ..., \phi_n)$ the distance between the atoms α und β, $\Theta_u(t) = \max(0, t)$ and $\Theta_l(t) = \Theta_v(t) = \min(0, t)$. Then the target function T is defined by (Güntert et al., 1991a):

$$T = \sum_{c = u, l, v} w_c \sum_{\alpha\beta \in I'_c} \left(\frac{\Theta_c(d_{\alpha\beta}^2 - c_{\alpha\beta}^2)}{2c_{\alpha\beta}} \right)^2 + w_w \sum_{a \in I_a} \left(1 - \frac{1}{2} \left(\frac{\Delta_a}{\Gamma_a} \right)^2 \right) \Delta_a^2. \qquad (2)$$

The subsets $I'_c \subseteq I_c$ ($c = u, l, v$) for which the program DIANA can calculate the target function consist of all distance constraints between residues whose sequence numbers differ by not more than a given *minimization level L* (Fig. 4). The target function is continuously differentiable over the entire conformation space, and is chosen such that the contribution of a single small violation δ_c is given by $w_c \delta_c^2$ for all types of constraints. Because only squared interatomic distances and no square roots have to be computed, the target function can be calculated rapidly.

A drawback of the variable target function algorithm is that for all but the most simple molecular topologies only a small percentage of the calculations converge with small residual constraint violations, which is a typical local minimum problem. Because of the low yield of acceptable conformers, calculations have typically been started with a large number of randomized starting conformers in order to obtain a group of good solutions, and sometimes a compromise had to be made between the requirements of small residual violations, the availability of approximately 10–20 "good" conformers to represent the solution conformation, and the available computing time (Kline et al., 1988). With the introduction of the highly optimized program DIANA, which significantly reduced the computation time needed for the calculation of a single conformer, a workable situation was achieved for α-proteins (Güntert et al., 1991b), but for β-proteins with more complex topology the situation remained unsatisfactory. With the use of redundant dihedral angle constraints (REDAC) described in this chapter, a greatly improved yield of converged conformers is obtained also for β-proteins.

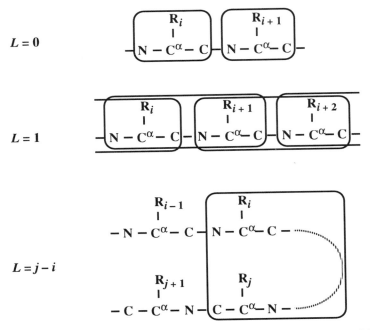

Figure 4. Illustration of the variable target function algorithm as implemented in the program DIANA (Güntert *et al.*, 1991). Boxes indicate examples of allowed ranges of active constraints at various minimization levels L.

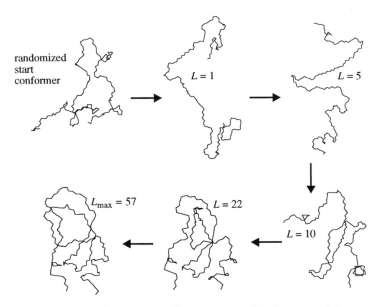

Figure 5. Randomized starting conformer, intermediate structures and final, converged structure of the protein Toxin K during variable target function minimization with the program DIANA. The backbone atoms N, C^{α}, and C' for all 57 residues of the random start conformation, and of the conformations at the end of the minimization levels $L = 1, 5, 10, 22,$ and $L_{max} = 57$ are shown (Berndt *et al.*, 1993).

Use of Redundant Dihedral Angle Constraints

In Fig. 6 the use of DIANA with redundant dihedral angle constraints (REDAC; Güntert and Wüthrich, 1991) is outlined and placed in perspective with the "direct" variable target function method as proposed originally by Braun and Gō (1985). In the direct approach, n start conformers with randomized dihedral angles are subjected to DIANA minimization against the experimental NMR constraints in step $B^{(0)}$. Experience has shown that the target function can be further reduced by repeating the DIANA refinement with all constraints and variable weights for the van der Waals constraints for a limited number k of well converged solutions ($m \leq k \leq n$) in step D. Among the resulting solutions, m conformers with the smallest final target function values are selected to represent the solution structure. In practice, n is adjusted so as to obtain $m = 10–20$ acceptable conformers.

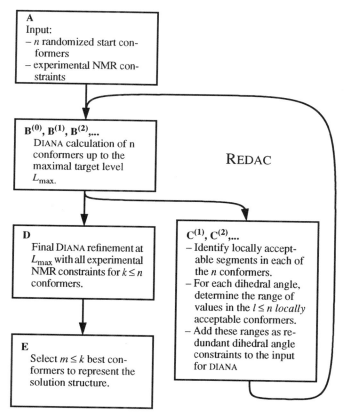

Figure 6. Flowchart outlining the course of a protein structure calculation with the program DIANA using either the "direct" way (A-B$^{(0)}$-D-E) or redundant dihedral angle constraints (REDAC; Güntert and Wüthrich, 1991) (A-B$^{(0)}$-[C$^{(1)}$-B$^{(1)}$-...]-D-E). Typically, the number of REDAC-cycles is 1 or 2.

To use REDAC, one or several cycles $C^{(i)}$–$B^{(i)}$ are added to the calculation, providing a partial feedback of structural information from all conformers that were calculated up to the maximal minimization level L_{max} in the step $B^{(i-1)}$. In the step $C^{(i)}$, a particular amino acid residue is considered to have an acceptably well defined conformation if the target function value due to constraint violations that involve atoms or dihedral angles of this residue is less than a predefined value, and if the same condition holds for the two sequentially neighbouring residues. Redundant dihedral angle constraints are generated and added to the input for the DIANA structure calculation in step $B^{(i)}$ for all those residues that were found to be acceptable in at least a predefined minimal number of conformers, typically 10, by taking the two extreme dihedral angle values in the group of acceptable conformers as upper and lower bounds.

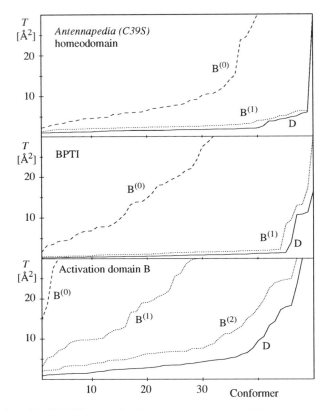

Figure 7. Distribution of the DIANA target function values, T, among the 50 conformers calculated using RE-DAC for each of the three proteins *Antennapedia(C39S)* homeodomain (HD), BPTI, and activation domain B (ADB). Along the horizontal axis the 50 conformers are ordered according to increasing target function value. For the HD and for BPTI the calculation consisted of the sequence of steps $A–B^{(0)}–C^{(1)}–B^{(1)}–D$; for ADB the sequence of steps was $A–B^{(0)}–C^{(1)}–B^{(1)}–C^{(2)}–B^{(2)}–D$. The letters identify the results at the end of the respective step in Fig. 6.

To compare the efficiency of DIANA calculations with and without use of REDAC, we calculated structures from high-quality experimental NMR data sets for the *Antennapedia (C39S)* homeodomain (HD; Güntert *et al.*, 1991b), the basic pancreatic trypsin inhibitor (BPTI; Berndt *et al.*, 1992), and the activation domain from porcine procarboxypeptidase B (ADB; Vendrell *et al.*, 1991). The HD is a typical α-protein, BPTI contains α and β secondary structure, and ADB has a more complex topology including a four-stranded β-sheet, two α-helices, and three loops that are only poorly determined by the NMR data (Vendrell *et al.*, 1991). For each protein we performed a structure calculation starting with $n = 50$ randomized conformations and using REDAC, and selected the $m = 20$ conformers with the smallest final target function values for further analysis (Güntert and Wüthrich, 1991). For the HD and BPTI more than 40 conformers with final target function values at L_{max} below 2.1 Å^2 and 1.3 Å^2, respectively, were obtained after one REDAC cycle. For ADB two REDAC cycles were needed to yield a group of 20 conformers with target function values below 2.9 Å^2 (Fig. 7)

For the HD and BPTI the DIANA calculations were repeated with the direct approach (Fig. 6), with the aim of producing a group of 20 conformers of equal quality, i.e., with final target function values in the same range as the 20 best conformers obtained from 50 starting conformers with the use of REDAC. For the HD this was achieved with $n = 400$ starting conformers, for BPTI with $n = 2000$ (Table 1). For the ADB it was found that calculations without use of REDAC produced only one acceptable converged conformer from 400 starting conformers, so that we would have had to compute of the order of 8000 conformers in order to obtain a comparable result. Table 1 shows that for the three proteins the effective overall CPU time was reduced through the use of REDAC by factors of 5.7, 29, and about 100, respectively. The improved efficiency when using REDAC is particularly pronounced for the β-proteins, i.e., those proteins where the direct approach gives the lowest yields of acceptable structures.

Table 1. Efficiency of DIANA calculations with and without use redundant dihedral angle constraints (REDAC).

	Antennapedia(C39S) homeodomain		BPTI		activation domain B	
	direct	REDAC	direct	REDAC	direct[1]	REDAC
Number of start conformers,[2] n	400	50	2000	50	≈8000	50
CPU time (h)[3]	3.8	0.66	17.7	0.61	≈140	1.48

[1]Estimated (see text).
[2]The numbers n listed under "direct" were chosen so as to obtain the same number of acceptable conformers as with $n = 50$ and use of REDAC.
[3]Measured on a Cray Y/MP using one processor.

For both the HD and BPTI the DIANA structure calculations with and without REDAC gave nearly identical values for the final target function, the different types of residual constraint violations, and the global pairwise RMSDs (McLachlan, 1979) among the 20 best conformers. In addition, for both proteins the global RMSDs between the two groups of conformers calculated with and without use of REDAC are for all practical purposes identical to the RMSDs calculated between different conformers within each group (Güntert and Wüthrich, 1991).

CONCLUSIONS

The empirically found higher yield of good conformers with the use of REDAC can be rationalized as follows: In many regions of a protein structure, in particular in β-strands, the local conformation is determined not only by the local conformational constraints derived from intraresidual, sequential and medium-range NOEs (Wüthrich, 1986), but also by longer-range constraints, e.g., interstrand distance constraints in β-sheets. Therefore, the local constraints alone may allow for multiple different local conformations at low target levels in a DIANA calculation, of which some may be incompatible with the longer-range constraints taken into account at higher minimization levels. Obviously, incorrect local conformations that satisfy the experimentally available local constraints are potential local minima, which could only be ruled out from the beginning if the information contained in the long-range constraints were already available at low levels of the minimization. The use of REDAC achieves this: information contained in the complete data set is translated into (by definition intraresidual) dihedral angle constraints. It further makes clear why the yield of good solutions with the direct strategy was in general higher for α-proteins than for β-proteins, since the conformation of an α-helix is particularly well-determined by sequential and medium-range constraints (Wüthrich, 1986).

In conclusion, the success of DIANA structure calculations using REDAC is primarily due to the feedback of useful structural information derived from conformers calculated up to the maximal level L_{max} into a subsequent round of structure calculations, which starts with local constraints only. In this way information gathered during the entire duration of the structure calculation is used in obtaining the final result, whereas most of this information is discarded in the direct approach. Groups of conformers of equal quality are obtained with and without use of REDAC, and the only significant effect of the use of REDAC is a large reduction of the overall computation time (Table 1). The use of REDAC should therefore become the standard strategy for protein structure calculations with the program DIANA and, more generally, with all implementations of the variable target function algorithm.

ACKNOWLEDGMENTS

I thank Dr. Martin Billeter and Prof. Kurt Wüthrich for helpful discussions. Financial support by the Schweizerischer Nationalfonds (project 31.32033.91) and the use of the Cray Y/MP computer of the ETH Zürich are gratefully acknowledged.

REFERENCES

Anil-Kumar, Wagner, G., Ernst, R. R. and Wüthrich, K., 1981, Buildup rates of the nuclear Overhauser effect measured by two-dimensional proton magnetic resonance spectroscopy: implications for studies of protein conformation, *J. Amer. Chem. Soc.* 103:3654.

Berndt, K. D., Güntert, P., and Wüthrich, K., 1993, Nuclear magnetic resonance solution structure of dendrotoxin K from the venom of *Dendroaspis polylepis polylepis*, *J. Mol. Biol.* 234, in press.

Berndt, K. D., Güntert, P., Orbons, L. P. M. and Wüthrich, K., 1992, Determination of a high-quality NMR solution structure of the bovine pancreatic trypsin inhibitor (BPTI) and comparison with three crystal structures, *J. Mol. Biol.* 227:757.

Billeter, M., Engeli, M. and Wüthrich, K., 1985, Interactive program for investigation of protein structures based on ^{1}H NMR experiments, *J. Mol. Graph.* 3:79.

Billeter, M., Qian, Y. Q., Otting, G., Müller, M., Gehring, W. and Wüthrich, K., 1993, Determination of the nuclear magnetic resonance solution structure of an *Antennapedia* homeodomain-DNA complex, *J. Mol. Biol.* 234, in press.

Blundell, T. L. and Johnson, L. N., 1976, "Protein Crystallography," Academic, New York.

Braun W., 1987, Distance geometry and related methods for protein structure determination from NMR data, *Q. Rev. Biophys.* 19:115.

Braun, W. and Gō, N., 1985, Calculation of protein conformations by proton-proton distance constraints. A new efficient algorithm, *J. Mol. Biol.* 186:611.

Braun, W., Bösch, C., Brown, L. R., Gō, N. and Wüthrich, K., 1981, Combined use of proton-proton overhauser enhancements and a distance geometry algorithm for determination of polypeptide conformations. application to micelle-bound glucagon, *Biochim. Biophys. Acta* 667:377.

Brünger, A. T. and Nilges, M., 1993, Computational challenges for macromolecular structure determination by X-ray crystallography and solution NMR-spectroscopy, *Q. Rev. Biophys.* 26:1.

Brünger, A. T., Clore, G. M., Gronenborn, A. M. and Karplus, M., 1986, Three-dimensional structure of proteins determined by molecular dynamics with interproton distance restraints: application to crambin, *Proc. Nat. Acad. Sci., U.S.A.* 83:3801.

Clore, G. M. and Gronenborn, A. M., 1991, Structures of larger proteins in solution: three- and four-dimensional heteronuclear NMR spectroscopy, *Science* 252:1390.

Clore, G. M., Gronenborn, A. M., Brünger, A. T. and Karplus, M., 1985, Solution conformation of a heptadecapeptide comprising the DNA binding helix F of the cyclic AMP receptor protein of Escherichia coli. Combined use of ^{1}H nuclear magnetic resonance and restrained molecular dynamics, *J. Mol. Biol.* 186:435.

Driscoll, P. C., Gronenborn, A. M., Beress, L. and Clore, G. M., 1989, Determination of the three-dimensional solution structure of the antihypertensive and antiviral protein BDS-I from sea anemone *Anemonia sulcata*: a study using nuclear magnetic resonance and hybrid distance geometry-dynamical simulated annealing, *Biochemistry* 28:2188.

Ernst, R. R., Bodenhausen, G. and Wokaun, A., 1987, "The principles of nuclear magnetic resonance in one and two dimensions," Clarendon, Oxford.

Güntert, P. and Wüthrich K., 1991, Improved efficiency of protein structure calculations from NMR data using the program DIANA with redundant dihedral angle constraints, *J. Biomol. NMR* 1:446.

Güntert, P., Braun, W., Billeter, M. and Wüthrich, K., 1989, Automated stereospecific ^{1}H NMR assignments and their impact on the precision of protein structure determinations in solution, *J. Amer. Chem. Soc.* 111: 3997.

Güntert, P., Braun, W. and Wüthrich, K., 1991a, Efficient computation of three-dimensional protein structures in solution from nuclear magnetic resonance data using the program DIANA and the supporting programs CALIBA, HABAS and GLOMSA, *J. Mol. Biol.* 217:517.

Güntert, P., Qian, Y. Q., Otting, G., Müller, M., Gehring, W. J. and Wüthrich K., 1991b, Structure determination of the *Antp(C39→S)* homeodomain from nuclear magnetic resonance data in solution using a novel strategy for the structure calculation with the programs DIANA, CALIBA, HABAS and GLOMSA, *J. Mol. Biol.* 217:531.

Havel, T. F. and Wüthrich, K., 1984, A distance geometry program for determining the structures of small proteins and other macromolecules from nuclear magnetic resonance measurements of intramolecular [1]H-^{1}H proximities in solution, *Bull. Math. Biol.* 46:673.

Kaptein, R., Zuiderweg, E. R. P., Scheek, R. M., Boelens, R. and van Gunsteren, W. F., 1985, A protein structure from nuclear magnetic resonance data. Lac repressor headpiece, *J. Mol. Biol.* 182:179.

Kline, A. D., Braun, W. and Wüthrich, K., 1988, Determination of the complete three-dimensional structure of the α-amylase inhibitor tendamistat in aqueous solution by nuclear magnetic resonance and distance geometry, *J. Mol. Biol.* 204:675.

Lee, M. S., Gippert, G. P., Soman, K. V., Case, D. A. and Wright, P. E., 1989, Three-dimensional solution structure of a single zinc finger DNA-binding domain, *Science* 245:635.

McLachlan, A. D., 1979, Gene duplication in the structural evolution of chymotrypsin, *J. Mol. Biol.* 128:49.

Momany, F. A., McGuire, R. F., Burgess, A. W. & Scheraga, H. A., 1975, Energy parameters in polypeptides. VII. Geometric parameters, partial atomic charges, nonbonded interactions, hydrogen bond interactions, and intrinsic torsional potentials for the naturally occuring amino acids, *J. Phys. Chem.* 79:2361.

Némethy, G., Pottle, M. S. & Scheraga, H. A., 1983, Energy parameters in polypeptides. 9. Updating of geometrical parameters, nonbonded interactions, and hydrogen bond interactions for the naturally occuring amino acids, *J. Phys. Chem.* 87:1883.

Qian, Y. Q., Billeter, M., Otting, G., Müller, M., Gehring, W. J. and Wüthrich, K., 1989, The structure of the *Antennapedia* homeodomain determined by NMR in solution: comparison with prokaryotic repressors, *Cell* 59:573.

Rance, M., Sørensen, O. W., Bodenhausen, G., Wagner, G., Ernst, R. R. and Wüthrich, K., 1983, Improved spectral resolution in COSY ^{1}H NMR spectra of proteins via double quantum filtering, *Biochem. Biophys. Res. Comm.* 117:479.

Schulz, G. E. and Schirmer, R. H., 1979. "Principles of Protein Structure," Springer, New York.

Vendrell, J., Billeter, M., Wider, G., Avilés, F. X. and Wüthrich, K., 1991, The NMR structure of the activation domain isolated from porcine procarboxypeptidase B. *EMBO J.* 10:11.

Wüthrich, K., 1986, "NMR of Proteins and Nucleic Acids," Wiley, New York.

ENUMERATION WITH CONSTRAINTS: A POSSIBLE APPROACH TO PROTEIN STRUCTURE RECONSTRUCTION FROM NMR DATA

Eugene I. Shakhnovich

Department of Chemistry
Harvard University
12 Oxford Street, Cambridge, Ma. 02138

We suggest a new algorithm which enumerates all and **only all** conformations of a protein which satisfy a given set of constraints presented as a set of known interatomic contacts.The algorithm is very effective and allowed to investigate the dependence of number of protein conformations on the number of constraints. This dependence is very dramatic so that introduction of few constraints decrease the number of conformations significantly making their enumeration and analysis feasible. The record calculation yielded 2134672156 conformations of Crambin on diamond lattice when only 6 interatomic contacts (3 originating from S-S bonds and three other) were given. These results suggest a new approach to determination of protein conformations using very limited NMR data about contacts in the native state.

INTRODUCTION

Prediction of protein conformation from their aminoacid sequences is a very important problem of molecular biophysics of proteins. However, such prediction a'priori, without using any information about the native structure is hardly possible now (see, however, Finkelstein and Reva,1991 and Thornton et al, 1991). Therefore many approaches have been developed to simulate folding of a protein. In some cases bias to the native state was introduced in the form of intrinsic propensities to be in "native secondary structure" (Skolnick and Kolinski, 1990) which provided weak but crucial energetic preference for the "native conformation" over all other conformations. Introduction of such preferences was shown (Skolnick and Kolinski, 1990) to stimulate relatively rapid and reliable folding to the native state. Other models with several types of biases to the native state were suggested and investigated by simulations and analytically (Taketomi et al, 1975, Go and Abe, 1981, Bryngelson and Wolynes, 1987, Shakhnovich and Gutin, 1989). These models represent simple examples of polymer-based associative memories. In the recent work (Friedrichs et al,1991) this method was used in attempts to create data-based approach to prediction of protein conformations by generalization from known protein structures; this approach was successful when

Statistical Mechanics, Protein Structure, and Protein Substrate Interactions
Edited by S. Doniach, Plenum Press, New York, 1994

209

the data-base contained proteins homologous to the one which structure was to be determined.

These approaches are typical attempts to reconstruct protein structure using some information about it. Very often this information is provided as set of interatomic distances obtained from NOESY experiments.

The usual procedure to reconstruct structure of a protein from this data includes (Braun and Go, 1985, Wutrich, 1986, Brunger at al ,1986). Required computational efforts are significant; they include many hours of simulations (see e.g. Michnick et al, 1991). Knowledge of about 1000 constraints is necessary to reconstruct structure of a protein of $\sim$ 100 aminoacids. This requirement makes corresponding NMR experiments costly and not possible for any protein.

We may ask now a question: is this amount of experimental data really necessary to reconstruct a protein structure or is it due to specific shortcomings of dynamics algorithms involved? How much information is really necessary to reconstruct a protein structure? How the experimentally derived constraints decrease the number of possible conformations?

In this study we will answer these questions with the aid of a simple analytical theory and exhaustive enumeration of protein conformations on a lattice. Exhaustive enumeration results suggest a new algorithm for structure prediction which requires much less experimental information and may be much more effective in predicting lattice represented protein structures than traditional dynamic methods.

THEORY

In this section we consider a mean-field theory of polymer globules and estimate how contacts decrease the number of conformations. The reader not interested in details may skip this section and go directly to the next one.

We estimate what fraction of all possible conformations of a globular polymer constitute conformations which possess a given contact between, say, monomers i and j. To this end we consider a globular homopolymer (as a basic example). Fraction of conformations which have a given contact (i,j) is given by the expression:

$$P_{ij} = \frac{\sum_{\{r_i\}} \Delta(r_i - r_j) e^{-\frac{U(r_i - r_j)}{kT}} \prod_i g(r_{i+1} - r_i)}{\sum_{\{r_i\}} e^{-\frac{U(r_i - r_j)}{kT}} \prod_i g(r_{i+1} - r_i)} \tag{1}$$

where Δ is a Kronecker delta which is 1 when there is a contact between monomers i and j and 0 otherwise. E.g. for a lattice model $\Delta(r_i - r_j) = 1$ when monomers i and j are lattice neighbors and 0 otherwise. This is introduced to single out only such conformations which have contact between monomers i and j. U is a potential of interaction between monomers of a chain; this potential must be attractive to provide chain compactisation. T is temperature and k is Boltzmann constant. Functions g describe chain connectivity; they define space positions available to monomer $i + 1$ provided that position of i-th monomer is fixed. The numerator in eq.(1) gives the total number of conformations which possess a given contact while denominator gives the total number of all compact conformations.

The principal simplification comes from the fact that when the polymer is in condensed (globular) state the mean-field approximation becomes applicable (Lifshitz et al 1978). To this end we may substitute interaction potential U by external potential $\phi(r)$ acting on each monomer; the profile of this potential must be a potential well to provide compactisation of the chain.

Substitution of interatomic potential $U(r_i - r_j)$ by external potential leads to the obvious modification of the eq.(1):

$$P_{ij} = \frac{\sum_{\{r_i\}} \Delta(r_i - r_j) e^{-\frac{\phi(r_i)}{kT}} \prod_i g(r_{i+1} - r_i)}{\sum_{\{r_i\}} e^{-\frac{\phi(r_i)}{kT}} \prod_i g(r_{i+1} - r_i)} \tag{2}$$

The equation (2) can be further simplified if we present it in the form:

$$P_{ij} = \frac{\sum_{\{r_i\}} \Delta(r_i - r_j) G_{i-1}(R_0, r_i) G_{j-i}(r_i, r_j) G_{N-j}(r_j, R_N) dr_i dr_j dR_N}{Z} \tag{3}$$

where $G_i(R_0, r_i)$ is a partition function (propagator) of the chain of i monomers in external field ϕ with first monomer positioned in point R_0 and the last monomer in r_i. Z is the partition function of the chain; it is given by the denominator of eq.(2). The simplification given by eq.(3) is due to the presentation of the chain in external potential which makes the problem Markovian (without self-interactions).

Propagators entering eq.(3) are calculated in the Appendix; this calculation is based on theory of globules (Lifshitz et al 1978).

Using eq.(10) of the Appendix we may determine now the quantity of interest - probability to find a contact (i, j) among all compact conformations. As most general consider the case when $i \gg 1$ and $N - j \gg 1$ (both monomers are far from chain ends). In this case substituting (10) from the Appendix into (3) and taking asymptotic eq.(7) of the Appendix for the "end" parts of the propagator and partition function in the denominator (G_i and G_{N-j}) we obtain finally:

$$P_{ij} = \begin{cases} \frac{v}{a^3 (j-i)^{3/2}} \Lambda^{j-i} & \text{if } (j - i) < g^* \\ const \sim \frac{1}{N} & \text{if } (j - i) > g^* \end{cases} \tag{4}$$

for maximally compact globule. Eq.(12) from Appendix was used to derive (4). Here g^* is determined as such part of the chain which random-coil size coincides with the size of the globule. (see Eq.(9) of the Appendix for more formal definition).

The most important implication of this result eq.(4) is that probability to find a contact between two monomers depends on their separation along the chain only when this separation is small ($|i - j| < g^*$). Monomers which are far apart from each other along the chain "forget" that they belong to the same polymer and effectively behave as teared links. This is due to the fact that proteins are in compact state and boundaries of the globule create finite correlation scale - after the chain "reflects" from the boundary.

This theoretical prediction can be compared with experimental data on known protein structures (Fig.1)

The statistics of interresidue contacts in real proteins shows that indeed as separation between monomers along the chain increases probability to find a contact between them becomes independent on their positions along the chain. This can be "visualized" as if after correlation scale g^* (for proteins it can be estimated from Fig.1 as $g^* \sim 5$) monomers "forget" that they belong to the same chain and behave as effectively disconnected ones. This is in a complete agreement with the main result of this section, Eq.(4).

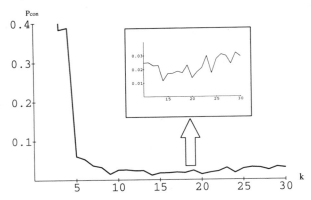

Figure 1 Probability of a contact between two monomers as a function of distance along the chain between them. Contact between two aminoacid residues was defined when distance R between their C_α atoms satisfied the condition $6A < R < 8A$ (Miyazawa and Jernigan, 1985). The insertion shows the region of distant contacts in more detail. The statistics was obtained using structures of 92 proteins from Protein Data Bank: 155c 1apb 1acx 1bds 1bp2 1cc5 1cpv 1crn 1cse 1est 1etu 1fb4 1fdx 1gcn 1gcr 1gp1 1hip 1hmg 1hmq 1hoe 1hvp 1ins 1lh4 1lz1 1mbd 1mlt 1ovo 1paz 1pcy 1pfk 1phh 1pp2 1ppt 1pyp 1rei 1rhd 1rn3 1sn3 1tim 1wrp 2abx 2alp 2aza 2b5c 2c2c 2cab 2ccy 2cdv 2cna 2cyp 2ebx 2gn5 2grs 2hhb 2lhb 2lzm 2mdh 2mt2 2pab 2pka 2rhv 2sbt 2sga 2sod 2ssi 2stv 2taa 2tbv 351c 3adk 3fxc 3fxn 3gap 3gpd 3icb 3pgk 3rp2 3wga 4adh 4ape 4cyt 4dfr 4lhd 4pti 4sbv 4tln 5api 5cpa 5rxn 8cat 9pap

This result shows also that even one specific long-range contact significantly decreases the number of conformations (by N - the number of sites on a lattice occupied by the polymer. Therefore this makes us hope that setting of relatively small fraction of contacts observed in the native structure will tremendously decrease the number of conformations. If this is so this will give the possibility to reconstruct the native structure of a protein using a very limited set of experimental data.

In order to see this for polymer models and for real proteins we develop an algorithm which enumerates exhaustively all conformations which satisfy a given set of distance constraints.

ALGORITHM OF ENUMERATION WITH CONSTRAINTS

The first-depth algorithm of enumeration of lattice self-avoiding walks was suggested as early as in 1947 by F.Orr (Orr, 1947) and was used by Domb and co-workers to investigate scaling properties of polymers with excluded volume (Domb, 1963). More recently this method was further developed and applied for enumeration of compact three-dimensional conformations of heteropolymers modeling protein globules (Shakhnovich and Gutin, 1990a,1990b) and for enumeration of lattice conformations of proteins within exactly given lattice shapes (Covel and Jernigan, 1990). The description of the algorithm is given, e.g. in (Shakhnovich and Gutin, 1990b). It is a tree-type search where chain grows and monomers of the growing chain are positioned in all possible directions allowed by lattice structure and chain connectivity. The important feature is the presence of "dead ends" which are the situations where chain cannot be continued any longer after some specific placement of a monomer since it encounters lattice boundaries and/or sites which are already occupied. Therefore the tree for search becomes relatively sparse and though the algorithm is still exponential (i.e. the number of conformations and required time grow exponentially as γ^N with the length of a chain N the exponent γ becomes smaller when additional constraints (e.g. compactness of a chain) is introduced.

The idea to incorporate distance constraints into enumeration algorithm is very simple. An ordered set of k constraints - list of pairs of monomers (L_i, M_i) with $i = 1, ...k$ and $M_i > L_i$ is introduced. These constraints have a meaning that, a pair of monomers L_i and M_i from the list are either nearest neighbors on the lattice or lie within given distance interval. At each step of the tree growth when new monomer is added it is checked whether it belongs to the list of constraints (as "senior" monomer M_i in a pair). If so, it is checked whether a monomer L_i which serves as a "partner" for newly added monomer M_i is or not within specified distance interval from M_i. If constraint is satisfied the chain grows further. If constraint is not satisfied this event is considered as a "dead end" for the tree-search algorithm and usual action is taken: a step back and attempt to grow the chain in new direction (see Shakhnovich and Gutin, 1990b). A block-scheme of this algorithm is shown on Fig.2.

Introduction of constraints has the effect that "tree" branches become cut more often than in unconstrained case; the "tree" becomes more sparse. This decreases the number of conformations in exponentially large number of times and makes it possible to enumerate all conformations (constrained, of course) of relatively large proteins. Below we will show explicitly how this algorithm works and how many constraints are necessary for exhaustive enumeration.

COMPACT CONFORMATIONS OF A 27-MER ON A $3 \times 3 \times 3$ LATTICE: THE TEST OF THE ALGORITHM

As a simplest and fundamental test of our approach we take a 27-monomer chain on a 3*3*3 fragment of cubic lattice. All compact self-avoiding conformations (CSA) of this polymer which are unrelated by symmetry have been enumerated (Shakhnovich and Gutin, 1990b); the total number of them is 103346 (Shakhnovich and Gutin, 1990a, 1990b, Shakhnovich et al 1991). Since the number of unconstrained conformations is known we may have exact estimate of the effect of constraints.

We start with the analysis of the effect of one constraint since for this we have an analytical estimate (4). We enumerate all conformations which have a given contact, say, between monomers L and M. It is worth noting that due to specifics of simple cubic lattice only contacts between monomers with different parity (e.g. L-even M-odd or L-odd, M-even) are possible. The result is shown on Fig.3. We see that the dependence of the number of conformations on the range of the contact $M - L$ is for only short range contacts with $M - L < 3$. At $M - L > 3$ the number of conformations becomes independent on the range of the contact. This is in full accord with our analytical estimate Eq.(4) and analogous experimental result for proteins (Fig.1). The threshold value $g^* = 3$ in this case. This effect of independence of number of conformations on the length of long-range contact ("flatness of configurational space" for the same model of 27-mer) was discussed in (Chan and Dill, 1991) but there it did not receive theoretical explanation.

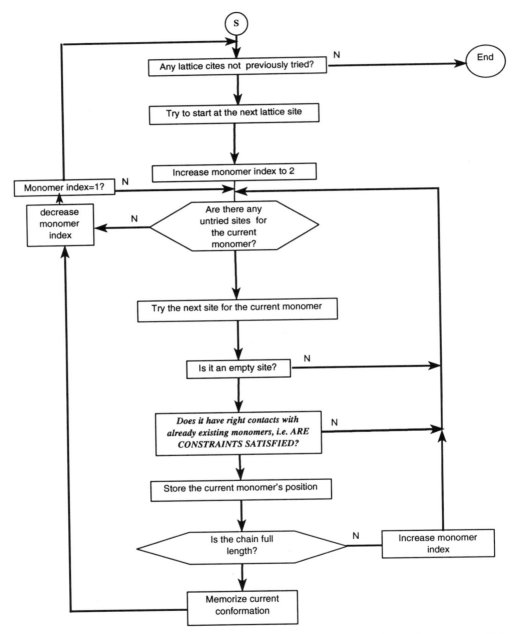

Figure 2 Flow chart of the first-depth enumeration with constraints. A stage at which constraints are checked is emphasized.

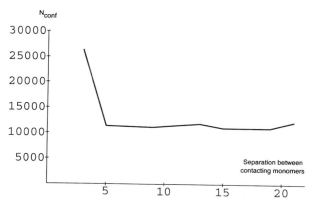

Figure 3 The number of conformations with one given contact as a function of distance along the chain between monomers participating in this contact in the model of 27-mer on 3*3*3 cubic lattice. The total number of CSA conformations is 103346.

We may ask now how does the number of conformations depend on the number of the "known" contacts which serve as constraints in the enumeration process. Using 27-mer as an example we can estimate the fraction of the total number of conformations which remain allowed when several constraints are applied. In order to do it we choose some arbitrary CSA conformation as a "target" native structure. The number of contacts in this structure as in any other compact self-avoiding conformation of 27-mer on the cubic lattice is 28. From this set we pick up randomly k contact pairs (L,M) of total 28 and enumerate all CSA conformations with this number of given pairs as constraints. The result of the calculation is shown in Fig.4. The most striking result of this calculation is that the dependence shown on Fig.4 is very steep suggesting that setting of only 5 contacts in the target structure reduces the constrained configurational space dramatically: five contacts (20% of total number) decrease the number of possible conformations by five orders of magnitude!

EXHAUSTIVE ENUMERATION OF LATTICE CONFORMATIONS OF PROTEINS

The fact that dependence of number of conformations on the number of constraints found for the model 27-mer is very steep makes us expect that introduction of few constraints will make it feasible to enumerate all lattice conformations of moderately sized proteins, thus at the expense of using constraints getting rid of very restrictive requirement that shape of a protein is given (Covel and Jernigan, 1989). We will be concerned with the question how does the number of conformations depend on the number of constraints and what is the minimal number of constraints which is necessary to set in order to make exhaustive enumerations feasible. As a model for this study we choose few proteins of different lengths which 3-dimensional structure is well-defined: Crambin (Hendrikson and Teeter, 1981), BPTI (Wlodawer et al, 1987) and Barnase (Maugen et al, 1982). The most detailed study was done for Crambin, however below we will discuss some results for all proteins mentioned above.

We consider here the simplest case of proteins fitted into diamond lattice with coordination number 4. Such fit provides an average 4.1 A RMS deviation from the native structure with 63% of all contacts reproduced correctly from the native conformation.

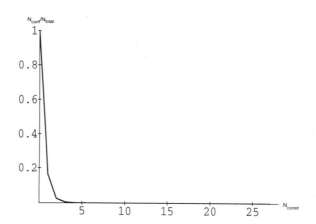

Figure 4 The dependence of ratio of the number of conformations satisfying a set of constraints to the total number of CSA conformations of 27-mer on the number of constraints. Maximal number of constraints equals the total number of nearest-neighbor contacts in CSA conformation which is 28 for a 27-mer.

First of all we investigate relative role of short- and long-range constraints in decreasing the number of conformations of a chain. In order to do so we did enumeration with fixed number of short-range contacts and variable number of long-range contacts (Fig.5) and fixed number of short-range contacts (Fig.6)

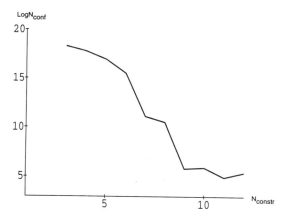

Figure 5 Logarithmic plot for the number of conformations of Crambin on diamond lattice vs number of long-range contacts when number of short-range contacts is 5. Short-range contacts were picked up randomly from the contact matrix and kept unchanged while long-range constraints were picked up randomly for each point of the plot. The total number of contacts in this model of Crambin is 104 from which 47 are short-ranged and 57 are long-ranged.

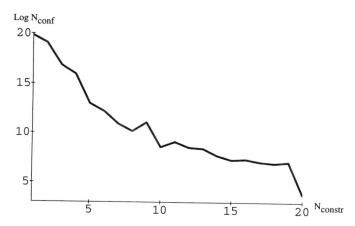

Figure 6 Logarithmic plot for the number of conformations of Crambin on diamond lattice vs number of short-range contacts when number of long-range contacts is 5. Long-range contacts were picked up randomly from the contact matrix and kept unchanged while short-range constraints were picked up randomly for each point of the plot. Total number of contacts of each type is given in Caption to Fig.5.(we define as a short-range a contact between monomers which are closer than 6 **monomers** apart).

The first inspection of Figs.5 and 6 reveals the most striking result of the present study: the number of conformations depend of the number of constraints in the most dramatic way making relatively simple number of them sufficient to allow exhaustive enumeration of conformations. The same conclusion can be drawn for the case of BPTI (Fig.7a) and Barnase (Fig.7b).

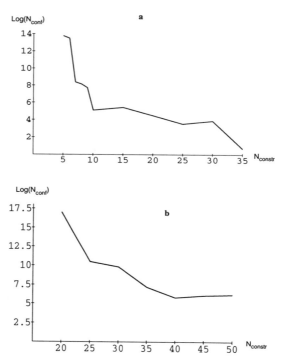

Figure 7 Logarithmic plot for the number of conformations of other proteins on diamond lattice vs total number of constraints.

(a):BPTI, 58 aminoacid residues and 146 contacts in the native diamond-fitted conformation.

(b):Barnase, 110 aminoacid residues and 247 contacts in the native diamond-fitted conformation.

Of course the actual number of conformations depends on which specific contacts were chosen and corresponding fluctuations are very high (see Fig.8).

We may see that element of luck is a necessary component of NOESY-based approach to the determination of structure: successful choice of constraints may lead to an order of magnitude decrease of the number of conformations compared with "average" case.

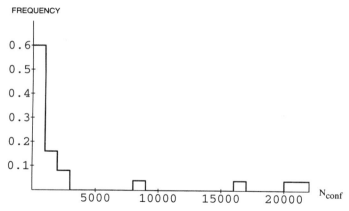

Figure 8 Histogram for the number of enumerated conformations of Crambin for different random choices of 12 short-range and 5 long-range contacts from the contact matrix. Total 25 calculations were done.

DISCUSSION

We suggested a simple and efficient method to reconstruct lattice structure of a protein from NOESY data. The principal idea of this approach is that only conformations which satisfy the constraints are enumerated. This simplifies the computational burden significantly and allows to enumerate all conformations with rather small number of constraints. The record calculation was obtained for Crambin where only 6 out of 104 constraints were taken and all corresponding conformations enumerated. Of these 6 3 were taken corresponding to S-S bonds and 3 other constraints were picked up randomly from contact matrix (two of them were short-ranged and one-long range). The number of enumerated conformations was 2134672156. This of course was only estimate of number of conformations without output of conformations itself. This number is impractical for analysis but knowledge of 20-50 % of contacts (which comprises for 100 aminoacid protein 40-80 constraints) reduces the number of conformations to few hundred and makes further analysis computationally straightforward.

Of course there remains a problem of choice of one - native structure from the remaining conformations. This can be done in principle using some version of potential of mean force (Sippl, 1990) to single out conformation with minimal energy. This, however, often can be avoided because larger number of constraints (in many cases 60-70 %) of total contact matrix leaves only one-native- conformation.

The described algorithm is computationally very effective. Enumeration which leaves few hundred of conformations takes less than a sec of CPU time on IBM RISC/6000 which was used in our calculations. Computation with 6 constraints took about 3hr of CPU time of the same machine.

The choice of optimal ratio of the number of short-range constraints to the number of long-range constraints is very important for the algorithm to be effective. Long-range constraints decrease the number of conformations more dramatically (compare Figs. 5 and 6); this follows also directly from eq.(4). However, short-range contacts are more effective in reducing time of computation. The reason is that before a long-range constraint will have its effect many conformations of a long part of the chain between monomers forming this constraints are to be enumerated. In contrary, short-range constraints have "immediate effect" and significantly increase speed of enumeration of structures.

Of course lattice representation does not give the final structure which is desirable at atomic level of resolution. This poses the problem of refinement from lattice representation to continuous space with lattice C_α conformation found in enumeration as initial one. This problem is discussed in literature and some approaches are already suggested (see, e.g. (Correa, 1990), (Levitt, 1992)). Simulated annealing or other dynamic approaches may also be quite useful for that. First encouraging results were obtained in the approach to placement of side-chains by simulated annealing Monte-Carlo procedure (Lee and Subbiah, 1991) and that makes us hope that this part of the problem can and will be solved.

In this work we used the simplest lattice - diamond which allowed to do lots of calculations in effective way. Of course this approach is not limited only to diamond lattice and lattices which better represent protein conformations can be used. We think however that the problem of optimal choice of a lattice should be approached from "minimalistic" point of view: the lattice with minimal coordination number which still allows "annealing" to the final complete structure is probably the best practical choice. Therefore future work will give an answer what is the best lattice representation for this combined lattice-off-lattice approach to prediction of protein structures.

Finally we would like to mention that the enumerative approach may be used successfully in a bunch of other problems: in determination of structure of small peptides for drug design purposes in reconstruction of loop conformation in homology modeling and other problems which involve global minimization of structures.

Acknowledgements This work was supported by Packard Fellowship.

APPENDIX

It was shown in (Lifshitz et al, 1978) that the chain propagator G allows bilinear expansion of the form:

$$G_N(R_0, R_N) = \sum_k \Lambda_k^N \Psi_k^+(R_0) \Psi_k(R_N) \tag{5}$$

Where Λ_k and Ψ_k are k-th eigenvalue and eigenfunction of the equation

$$\Lambda \Psi(x) = e^{-\frac{\phi(x)}{kT}} \int g(x-y) \Psi(y) dy \tag{6}$$

It is a well established result of the theory of homopolymer globules that in the globular state (in other words, when potential of interatomic interactions U is strong enough and attractive or when the potential well described by potential ϕ is deep enough the equation (6) has a discrete level (Λ_0, Ψ_0) separated by a finite gap from other levels (Λ_k, Ψ_k) which belong to continuous spectrum. This somewhat formal construction has a very simple physical interpretation. In order to see this we note that for $N \gg 1$ the term corresponding to the largest eigenvalue Λ_0 in the expression (5) dominates so that this equation has a simple form:

$$G_N(R_0, R_N) = \Lambda_0^N \Psi_0^+(R_0)\Psi_0(R_N) \qquad (7)$$

Factorization in the last equation to the term depending only on the coordinate of the first monomer R_0 and of the last monomer R_N means that distant along the chain parts of the globule become independent: they do not "feel" that they belong to the same chain. Free energy of such a chain equals to $-N \ln \Lambda$. This can be illustrated in a simple "blob" picture. Let the globule be maximally compact and have a size R (see Fig.1) so that $Nv \sim R^3$ where v is the excluded volume of a monomer and N is the number of monomers. Locally such a globule is indistinguishable from a melt and, in accord with Flory theorem (Flory, 1953, DeGennes, 1979) shorter pieces of a chain must have Gaussian (or other ideal) statistics. This ideal statistics of a shorter portion of a chain is violated however when this portion is large enough to feel boundaries of the globule. This happens when unperturbed (Gaussian) size of a portion of g^* monomers becomes equal to the size of a globule:

$$ag^{1/2} \sim R \sim v^{1/3}N^{1/3} \qquad (8)$$

from where we find:

$$g^* = \left(\frac{v}{a^3}\right)^{2/3} N^{2/3} \qquad (9)$$

from where we find that two regimes exist for the propagator G of an i-monomer fragment of a globule:

$$\begin{cases} \frac{1}{a^3 i^{3/2}} exp(-\frac{(R_i - R_0)^2}{2ia^2}) & \text{if } i < g^* \\ \Lambda_0^i \Psi_0(R_0)\Psi_0(R_i) & \text{if } i > g^* \end{cases} \qquad (10)$$

The first regime (for shorter chains) corresponds to the situation when a portion of the chain does not "feel" globule boundaries and behaves as in melt: it is Gaussian according to Flory theorem. Longer portions of the chain have essentially non-Gaussian statistics and their ends are effectively independent - they behave as if they do not belong to the same chain.

We can estimate now the value of Ψ_0 which is entered in Eq.(10). In order to do it we consider a long globular chain with one end fixed at the point R_0 and other end allowed to be elsewhere. In this case

$$P(R) = \frac{G_N(R_0, R)}{\int G_N(R_0, R')dR'} \qquad (11)$$

serves as a probability density to find an end monomer in point R. As the density distribution in a large globule is approximately uniform (Lifshitz et al, 1978) and therefore an end monomer can be with equal probability elsewhere we find that

$$\Psi_0 \sim V^{-1/2} \qquad (12)$$

where V is the volume occupied by the globule.

REFERENCES

Braun, W & Go, N. (1985) Calculation of protein conformations *J.Mol.Biol.* **186**, 611-626.

Bryngelson, J.B. & Wolynes, P.G. (1987) Spin glasses and the statistical mechanics of protein folding. *Proc. Nat. Acad. Sci.USA*, **89**, 7524-7528.

A.T. Brunger, G. M. Clore, A. M. Gronenborn and M. Karplus, (1986) *Proc. Natl. Acad.Sci.USA* **83**, 3801-3805.

Chan, H.S. & Dill, K.A.(1991) Polymer principles in protein structure and stability.*Annu Rev. Biophys. Biophys.Chem.* **20**,447-490.

Correa, P.E. (1990) Building of protein structure from α-carbon coordinates. *Proteins* **7**, 366-377.

Domb, C.J. (1963) Excluded volume effect for two- & three-dimensional lattice models.*J.Chem.Phys.* **38**, 2957.

Go, N. & Abe, H. (1981) Noninteracting local-structure model of folding & unfolding transition in globular proteins. II. Application to two-dimensional lattice proteins. *Biopolymers* **20**, 1013-1031.

Finkelstein, A.V. & Reva, B.V (1991) A search for the most stable folds of protein chains. **Nature** *351*, 497-499.

Flory, P. (1953) Principles of polymer chemistry. *Cornell Univ.Press*, Ithaca, N.Y

Friedrichs,M, Goldstein, R. & Wolynes P.G. (1991) Generalized protein structure recognition using associative memory Hamiltonians. *J.Mol.Biol.* **222**, 1013-1034.

de Gennes, P.G. (1979) Scaling concepts in polymer physics.*Cornell Univ.Press*, Ithaca, N.Y.

Hendrickson, W. & Teeter, M. Structure of the hydrophobic protein Crambin determined directly from the anomalous scattering of sulphur. *Nature* **290**, 107-113.

Lee, C. & Subbiah, S. (1991) Prediction of protein side-chain conformation by packing optimization. *J.Mol.Biol* **217**, 373-388.

Levitt, M. (1992) Accurate modeling of protein conformations by automatic segment matching.*J.Mol.Biol* **226**, 507-533.

Mauguen, Y., Hartley, R.,Dodson, G., Dodson, E., Bricogne,G., Chothia, C., &Jack,A. (1982) Molecular structure of a new family of ribonucleases. *Nature*, **297**, 162-164.

Michnick, S., Rosen, M., Karplus, M. Wandless, T. & Schreiber, S.(1991) Solution structure of FKBP, a rotamaze enzyme and receptor for FK506 and rapamycin. *Science* **252**, 836-839.

Myazawa, S. & Jernigan, R. (1985) Estimation of interresidue contact energies from crystal structures:quasichemical approximation. *Macromolecules*, **18** pp.534-552.

Lifshitz, I.M., Grosberg, A.Yu., & Khokhlov A.R. (1978) Some problems of the statistical physics of polymer chains with volume interactions. *Rev.Mod.Phys* **50** pp. 683-713.

Orr, W.J.C (1947) Statistical treatment of polymer solutions at infinite dilution. *Trans Faraday Soc.* **43** 12.

Shakhnovich,E.I. & Gutin, A.M. (1989) Protein folding as pattern recognition. *Studia Biophysica* **132**, 47-56.

Shakhnovich,E.I. & Gutin, A.M. (1990a) Implications of thermodynamics of protein folding for evolution of primary sequences. *Nature* **346**, 773-775.

Shakhnovich,E.I. & Gutin, A.M. (1990b) Enumeration of all compact conformations of copolymers with random sequence of links. *J.Chem.Phys.* **93**, 5967-5971.

Shakhnovich,E.I.,Farztdinov, G., Gutin, A.M. & Karplus, M (1991b) Protein folding bottlenecks: A lattice Monte-Carlo simulation. *Phys.Rev.Lett.* **67**, 1665-1668.

Sippl, M. (1990) Calculation of conformational ensembles from potentials of mean force.

An approach to the knowledge-based prediction of local structures in globular proteins. *J.Mol.Biol.* **213**, 859-883.

Skolnick, J. & Kolinski, A (1990) Simulation of the folding of a globular protein. *Science* **250**, 1121-1125.

Taketomi, H., Ueda, Y., Go N. (1975) Studies on protein folding, unfolding & fluctuations by computer simulations. I the effect of specific amino-acid sequence represented by specific inter-unit interactions. *Intl. Journ. pept. Prot.Res.* **7**, p.445-459.

Thornton, J. Flores, T., Jones, D. & Swindells, M (1991) Prediction progress at last. *Nature* **354**, 105-106.

Wodawer,A.,Nachman,J.,Gilli&,G.L.,Gallacher,W. & Woodward,C. (1987) Structure of form III Crystals of bovine pancreatic trypsin inhibitor *J.Mol.Biol.* **198** 469-481.

Wutrich,K *NMR of Proteins and Nuclear Acids* (Wiley, New York, 1986)

A DECOMPOSITION OF THE NOE INTENSITY MATRIX

Thérèse E. Malliavin and Marc A. Delsuc

Centre de Biochimie Structurale
Faculté de Pharmacie
15, av. Ch. Flahault
F-34 060 Montpellier, France

The protein structure determination by NMR is based on the estimation of the distances between the hydrogens, from the measurement of the nuclear Overhauser effect (nOe). Numerous methods were designed to calculate precise estimates of these distances from the nOe intensities (Borgias and James, 1990; Boelens et al., 1988; Madrid et al., 1991). But, as the exact calculation of the distances requires the knowledge of the entire nOe matrix (Olejniczak et al., 1986), the distance calculation methods are thus limited, not only by the experimental signal-to-noise ratio, but also by the superpositions on the experimental spectra.

Nevertheless, from a 3D NOESY-HMQC experiment, recorded on a ^{15}N-labelled protein (Frenkiel et al., 1990), accurate measurements of nOe intensities between amide hydrogens and between amide and non amide hydrogens can be obtained, and almost all overlapping intensities can be separated.

It was recently shown (Malliavin et al., 1992) that, if the distances between the amide hydrogens are accurately determined, the protein folding can be directly determined in the 3D space, without using any assignment information. We propose here an approximate expression of the nOe intensities between amide hydrogens from the distances between amide hydrogen and the sum of intensities between each amide hydrogens and other non amide hydrogens. The approximate intensities are simulated and compared to real simulated ones. Also, iterative distance calculations are performed using simulated data.

Statistical Mechanics, Protein Structure, and Protein Substrate Interactions
Edited by S. Doniach, Plenum Press, New York, 1994

The nOe intensity matrix can be expressed as a symmetric matrix $I(\tau_m)$, where the element (i,j) of $I(\tau_m)$ holds the intensity of the cross peak between the hydrogens i and j, for a mixing time of τ_m. The nOe intensity matrix $I(\tau_m)$ is bound to the matrix of relaxation rates R by the relation:

$$I(\tau_m) = \exp(-R\ \tau_m)\ I_0 \tag{1}$$

where I_0 is the matrix of the equilibrium magnetisations.

The non-diagonal relaxation matrix element R^{ij} is the dipolar relaxation rate between the spins i and j. The diagonal relaxation matrix element R^{ii} is the rate of magnetisation leakage from the spin i. The matrix R can be written as the sum of the matrices R_1 and R_2, where the elements of the matrices R_1 and R_2 are defined in this way:

The matrix R_1 is a "relaxation matrix" restrained to the amide hydrogen subset. As the R_1 matrix is limited to the amide hydrogens, all the elements of $I_1 = \exp(-R_1\ \tau_m)$ concerning non-amide hydrogens are null. The matrix R_2 is the complementary part of the matrix R.

It can be shown (Malliavin and Delsuc, 1993) that the intensity between the amide hydrogens i and j can be approximated by I^{ij}_*:

$$I^{ij}_* = \frac{I^{ij}_1}{2}\left(\exp\left(-R*^{ii}_2\ \tau_m\right) + \exp\left(-R*^{jj}_2\ \tau_m\right)\right) \tag{2}$$

$$\text{with: } R*^{ii}_2 = -\frac{2\alpha\ S^i_*}{\tau_m\ I^{ii}_1} \tag{3}$$

$$\text{where: } \alpha = \frac{J(2\omega) + J(\omega) + J(0)}{J(2\omega) - J(0)} \tag{4}$$

J is the spectral density function, ω is the spectrometer frequency.

The parameter

$$S^i = \sum_k I^{ik} \tag{5}$$

can be measured in a straightforward way by recording only the first plane ($t_1=0$) of the 3D NOESY-HMQC experiment. The parameter S^i_* can be obtained from S^i, by subtracting to it the NOESY intensities between the hydrogen i and the other amide hydrogens.

To test the quality of the approximation, we choose to apply it on the PDB structure (6PTI) of the BPTI (Wlodawer et al., 1987). All the calculation were done for a mixing time of 100 msec. The intensity calculation on all the hydrogens gives the intensities I^{ij} and the parameters S^i. The intensity calculation restrained on the only amide hydrogens gives intensities I_I^{ij}. The approximate intensities I_*^{ij} were then obtained from the quantities I_I^{ij} and S^i by the Eq. 2 and compared to the real intensities I^{ij}.

The percentage of relative error $\Delta I/I$ was 50%. But, the relative error $\Delta r/r$ on the distances can be evaluated as one sixth of the relative error on the intensity. The mean relative errors, which can be expected for the amide-amide hydrogen distances, are thus smaller than 10%. This value is in the range of accuracy, currently obtained in structure determination by NMR (Liu et al., 1992).

Iterative calculations (Borgias and James, 1990) of amide-amide distances were then ran from theoretical I^{ij} intensities and S^i parameters. The input data were the theoretical nOe intensities I^{ij} between amide hydrogens greater than 0.005, the parameters S^i for all amide hydrogens, and the initial inter proton distances between these hydrogens, obtained from a qualitative analysis of the nOe intensities. The initial inter proton distances were chosen (Ikura et al., 1992) into four ranges: 1.8 to 2.7, 1.8 to 3.3, and 1.8 to 5.0 Å., corresponding to the strong (>0.2), medium (between 0.2 and 0.05) and weak (<0.05) nOe intensities. The calculation was performed with three noise levels on the intensities: 0.0, 0.005 and 0.01, the mean value of non diagonal nOe intensity between amide hydrogens being 0.026. The noise levels are expressed in nOe intensity unit.

Table 1. Summary of the results obtained by the iterative distance calculation.

noise level	initial rms (Å)	final rms (Å)	number of final distances	$\langle \Delta \rangle$	mean final distance (Å)
0.0	1.42	0.17	34	0.98	2.84
0.005	1.46	0.39	33	0.99	2.86
0.01	1.48	0.36	30	0.97	2.71

The results were appreciated by calculating two parameters: the root-mean-square (rms) between final d_{ij}^f and real d_{ij}^r distances, and $\Delta = \langle d_{ij}^f/d_{ij}^r \rangle$. The results of this comparison are summarised in Table 1. First, the Δ values are always close to 1.0; the proposed method introduces thus little bias to the distance estimation. Then, in all cases, the distances between hydrogens closer than 3 Å were precisely obtained. As the noise level increases, the rms values worsen, but it should be noticed that the obtained values are of the order of the usual rms values obtained in inter proton distance calculation. Indeed, the first calculations using MARDIGRAS on the BPTI (Borgias and James, 1990), with a noise level of 0.003, an initial rms of 1.19 Å, and a mixing time of 100 msec, provides a final rms of 0.33 Å. The calculation run here with a mixing time of 100 msec, an initial rms of 1.46 Å, and a noise level of 0.005, gives a final rms of 0.39 Å, which is of the same order.

Nevertheless, the nOe correlations between the only amide hydrogens usually determine at most the protein secondary structures. The proposed method is thus not sufficient to determine the protein tertiary structure. In the 3D NOESY-HMQC experiments, correlations between amide and non amide hydrogens should be used to complete the structure determination in the way proposed here.

REFERENCES

Boelens, R., Koning, T.M.G. and Kaptein, R., Determination of biomolecular structures from proton-proton nOe's using a relaxation matrix approach, *J.Mol.Struct.*, 173:299 (1988).

Borgias, B.A. and James, T.L., MARDIGRAS-A procedure for matrix analysis of relaxation for discerning geometry of an aqueous structure, *J.Magn.Res.*, 79:493 (1990).

Frenkiel, T., Bauer, C., Carr, M. D., Birdsall, B. et Feeney, J., HMQC-NOEQY-HMQC, a three-dimensional NMR experiment which allows detection of nuclear Overhauser effects between protons with overlapping signals, *J. Magn. Res.*, 90:420 (1990).

Ikura M., Clore G.M., Gronenborn, A.M., Zhu, G., Klee C.B. and Bax, A., Solution structure of a calmoduline-target peptide complex by multidimensional NMR, *Science*, 256:632 (1992).

Liu Y., Zhao D., Altman R. and Jardetzky O., A systematic comparison of three structure determination methods from NMR data: dependance upon quality and quantity of data, *J. Biomolec. NMR* , 2:373 (1992).

Madrid, M., Llinas, E. and Llinàs, M., Model-independent refinement of interproton distances generated from ^{1}H NMR Overhauser intensities, *J.Magn.Res.*, 93:329 (1991).

Malliavin, T.-E. and Delsuc, M.-A., A decomposition of the NOE intensity matrix, *J. Chim. Phys.* , submitted.

Malliavin, T.-E., Rouh A., Delsuc, M.-A. and Lallemand, J.-Y., Approche directe de la détermination de structures moléculaires à partir de l'effet Overhauser nucléaire, *CRAS série Ii*, 315:653 (1992).

Olejniczak, E.T., Gampe, R.T. and Fesik, S.W., Accounting for spin diffusion in the analysis of 2D NOE data, *J.Magn.Res.*, 67:28 (1986).

Wlodawer, A., Nachman, J., Gilliland, G.L., Gallagher, W. and Woodward, C., Structure of form III crystals of bovine trypsin inhibitor, *J.Mol.Biol.*, 198:469 (1987).

STRUCTURE AND DYNAMICS OF MEMBRANE PROTEINS

Fritz Jähnig[1], Olle Edholm[2], and Jürgen Pleiss[1]

[1]Max Planck Institute for Biology
Corrensstrasse 38
D-72076 Tübingen
Germany
[2]Department of Theoretical Physics
Royal Institute of Technology
S-10044 Stockholm
Sweden

INTRODUCTION

The structure of proteins can not be calculated at present. Membrane proteins may represent an exception, because their structure may become accessible to calculation in the near future. The reason for that lies in the fact that membrane proteins can adopt only a few structures, basically an α-helix and a ß-barrel. Only in these two structures the hydrogen bonds among the backbone atoms can be saturated, in the α-helix between the helical turns, in the ß-barrel between the neighboring strands. For stable insertion into the membrane the side chains of the helix must be hydrophobic, while the side chains of the strands must be amphipathic, i.e. hydrophobic on the outer side and hydrophilic on the inner side since water will enter the interior of the barrel forming a pore. These basic structures may associate to form larger aggregates. In the case of several α-helices, the helices may then also be amphipathic providing a pathway for the transport of polar substrates. Altogether there are four classes of membrane proteins and all experimental data are consistent with this classification.

The important point for our purposes is that the basic elements of the structure of membrane proteins, hydrophobic and amphipathic α-helices and amphipathic ß-strands, can easily be predicted from the amino acid sequence by means of so-called hydrophobicity plots (Jähnig, 1989 and 1990). All what remains to be done then is to determine their arrangement in the membrane, their positions and orientations. This problem might be solved by molecular dynamics (MD) simulations. That at least was the idea when we started such an approach a couple of years ago.

Statistical Mechanics, Protein Structure, and Protein Substrate Interactions
Edited by S. Doniach, Plenum Press, New York, 1994

THE STRUCTURE OF BACTERIORHODOPSIN

The system we chose was bacteriorhodopsin, a light-driven proton pump found in the purple membrane of halobacteria and probably the best-investigated membrane protein. Its structure has been determined by Henderson et al. (1990), and consists of 7 membrane-spanning α-helices arranged in a kidney-shaped form with the inner 3 helices being perpendicular to the membrane plane and the outer 4 helices being slightly tilted. The parts of the protein outside the membrane were not resolved in the structure of Henderson et al.

We started from the amino acid sequence and worked out hydrophobicity and amphipathy plots to predict the 7 helices (Jähnig and Edholm, 1990). In the hydrophobicity plot, one can identify 5 regions extended over about 20 residues which are predicted as membrane-spanning helices. In the region of the C-terminus, a clear-cut prediction can not be made from the hydrophobicity but from the amphipathy plot leading to the prediction of 2 further membrane-spanning helices. Thus, one predicts 7 helices A-G from the sequence, and their positions agree well with the experimental result (Henderson et al., 1990).

Based on this prediction, we constructed 7 regular α-helices comprising the predicted membrane-spanning segments and joining loop regions, oriented them perpendicular to the membrane, and positioned them on a circle to begin with an arbitrary arrangement (Fig.1, left). Furthermore, the helices were rotated about their long axes so that their most hydrophobic sides pointed towards the lipids. The lipids were not treated explicitly, but modeled by a hydrophobic potential acting on the protein atoms (Edholm and Jähnig, 1988). Upon energy-minimization the loops between the helices closed, but apart from that the structure did not change considerably. In the ensuing MD run of 25 ps, more drastic changes occurred. Most significant is the movement of helix C towards the centre of the circle. If only the C-terminus of helix C is considered, one recognizes that the experimental kidney shape is approached. Thus, we did not completely reach the kidney shape, but the symmetry of the circle was broken in the correct way.

In the next step, we started from the experimental kidney shape by putting the predicted helices on the positions given by experiment (Henderson et al., 1990)(Fig.1, right). Furthermore, we chose the correct assignment as given by experiment of the predicted helices to the positions on the kidney shape. In the MD simulation, the experimental structure was closely approached. The inner 3 helices remained perpendicular to the membrane plane, while the outer 4 helices underwent a tilt. The tilt angles agree very well with the experimental values.

In a further step, we studied other assignments of the predicted helices to the positions on the kidney shape. In no other case, the helices tilted in the correct way, in most cases even the kidney shape was lost during the MD run. The energy was always higher than for the correct assignment. Hence, from our MD simulations we are able to predict the correct assignment.

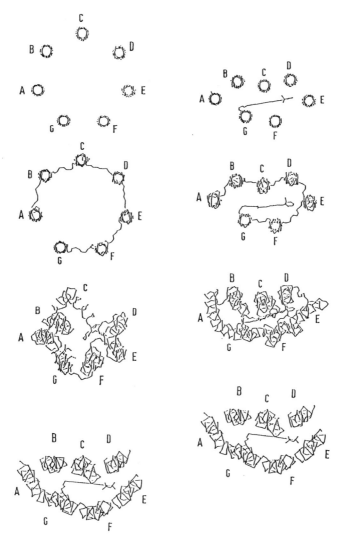

Figure 1. (*Left*) Projections into the membrane plane of (from top to bottom) the initial structure with the helices A-G arranged on a circle, the energy-minimized structure, the structure after 25 ps MD simulation, and the experimental structure of Henderson et al. (1990). (*Right*) Projections into the membrane plane of (from top to bottom) the initial structure with the helices A-G arranged on the experimental kidney shape with the correct assignment, the energy-minimized structure, the structure after 25 ps MD simulation, and the experimental structure.

The final question then is: How good is our MD structure with respect to atomic details? To answer it we calculated the root-mean-square (rms) deviation of all atoms between our structure and the experimental structure. The result is 3.8 Å. If one analyzes the contributions of the individual helices, one recognizes that helix D has the largest deviation. This is the helix which has the largest uncertainty in the experimental structure and, therefore, may actually be slightly wrong in the experimental structure. One may also analyze the contributions of the various degrees of freedom of the helices. Their positions contribute 20% to the total deviation, their tilt 14%, their orientation about the long axes 10%, and the internal degrees of freedom 56%. Hence, the internal degrees of freedom provide the largest contribution, the atomic details of our structure are wrong.

Summarizing our attempts to model the structure of bacteriorhodopsin we can say the following. We were not able to find the kidney shape by MD simulations, although the symmetry of a circular arrangement was broken in the correct way. This may be improved by using a Monte-Carlo technique which permits to sample the phase space more efficiently. Starting from the kidney shape, we got the correct tilt of the helices and could determine the correct assignment of the helices. On the level of atomic details, our structure is not correct. This may be improved by treating the protein environment in a more realistic manner such as simulating the crystalline lattice of the purple membrane with the trimer of bacteriorhodopsin and surrounding lipids plus water on both sides.

DYNAMICS OF BACTERIORHODOPSIN AND SINGLE HELICES

To study the internal dynamics of bacteriorhodopsin the MD simulation of our best structure was extended to 1 ns (Edholm and Jähnig, 1992). The rms fluctuations of the atomic positions did not yet level off but still increased. After 1 ns, they reached a value of 1.4 Å averaged over all atoms and 0.9 Å averaged over the C_α-atoms only. The variation of the rms fluctuations of the C_α-atoms along the sequence is shown in Fig.2. As expected, the loop regions exhibit stronger fluctuations than the helical regions. The rms fluctuations of our structure agree quite well with the fluctuations of the experimental structure obtained by restraining the MD structure to the experimental structure.

Of special interest are the rigid-body fluctuations of the helices, since they have been postulated in another case to be correlated with the functioning of a membrane protein (Jähnig and Dornmair, 1991). The tilt angle of helix A is plotted as a function of time in Fig.3. The helix oscillates about a mean tilt angle of 17° with an amplitude of about 2° and a period of about 10 ps. In the case of myoglobin, even larger helical fluctuations have been obtained (Elber and Karplus, 1987). It remains to be clarified if membrane proteins in general undergo smaller fluctuations than soluble proteins.

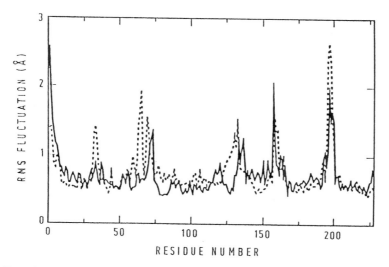

Figure 2. Rms fluctuations of the C_α-atoms in the MD structure (full line) and in the experimental structure (broken line) obtained by restraining the MD structure to the structure of Henderson et al. (1990). The data extend up to residue number 227, since the C-terminus has been omitted.

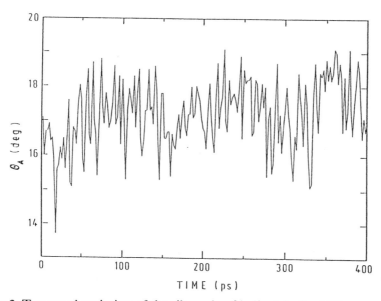

Figure 3. Temporal variation of the tilt angle of helix A in the MD structure.

The internal dynamics of a membrane-spanning helix has been studied for the simple system of an $(Ala)_{20}$-helix (Pleiss and Jähnig, 1991). A helix may exhibit two kinds of collective motions, stretch and bend vibrations. The periods of these two vibrations for the $(Ala)_{20}$-helix were obtained as 1.4 ps and 4.3 ps, respectively, the value for the stretch vibration being in good agreement with available experimental data (Peticolas, 1979). This behaviour of an α-helix can be modeled by an elastic rod with an elasticity modulus ten times smaller than that of steel.

To study the influence of a kink on the dynamics of an α-helix, a kink was introduced in the middle of the $(Ala)_{20}$-helix by altering the 3 backbone dihedrals φ_{10}, ψ_{10}, and φ_{11} appropriately. The dynamic behaviour of the helix was not altered very much, the periods of the stretch and bend vibrations are essentially the same as for a straight helix. The occurrence of these vibrations indicates that the kinked helix is in a local equilibrium. However, this equilibrium is not stable and the helix will undergo a conformational transition back to a straight helix. Such a transition could indeed be observed during our simulations (Pleiss and Jähnig, 1992). In Fig.4, the kink angle is plotted as a function of time. Over an extended period of time, it fluctuates about a value near 110° and exhibits bend oscillations with a period of about 5 ps. Suddenly at about 115 ps, the kink angle drops to essentially zero, i.e. the helix becomes straight. To improve the statistics on the time required for the transition to occur, several simulations with slightly different initial conditions were performed. The transition always occurred between 75 and 250 ps, with a mean of 150 ps.

An obvious question then is: What determines this reaction time, is it caused by an energy barrier to be overcome? From the time course of the energy, which is included in Fig.4, a high energy barrier is not obvious, but is may be hidden in the fluctuations of the energy. Therefore, an energy minimization was performed on 40 states distributed at equal time intervals across the transition, with the values of the 3 dihedrals φ_{10}, ψ_{10}, and φ_{11} kept fixed. The result is shown in Fig.5. There is still no high energy barrier, but the energy surface is rugged with deep valleys. If the system falls into one of the valleys, thermal energy is sufficient to leave it. The analogous behaviour is obtained for the entropy, more specifically for the local entropy calculated for each of the 40 states according to the procedure of Karplus and Kushick (1981). These properties of the energy and entropy in the transition region are the same in the equilibrium state of the kinked helix indicating that the transition region is not distinguished from an equilibrium state. Thus, the reaction time is not determined by a local energy or entropy barrier, but by the entropy to find a certain state in phase space from which the transition can proceed. It is like walking in a labyrinth and by chance find the exit. In our case, walking in phase space is associated with bend vibrations of the helical segments. Hence, the bend vibrations give rise to the conformational transition.

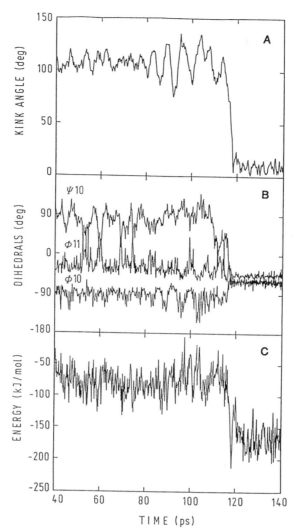

Figure 4. Temporal variation of the kink angle (A), the 3 dihedrals ϕ_{10}, Ψ_{10}, and ϕ_{11} (B), and the total potential energy (C) of an $(Ala)_{20}$-helix with an initial kink in the middle.

This picture is consonant with the hierarchical model of protein dynamics put forward by Frauenfelder and coworkers (Hong et al., 1990). It may be generalized to functionally important conformational transitions in membrane proteins. For example, a protein involved in transport of a substrate across the membrane may exist in two equilibrium states, one with the binding site for the substrate exposed to the outside and one with the binding site exposed to the inside, and these two states may differ in the orientation of some membrane-spanning helices. Transport would then be associated with a change in the orientation of these helices. From measured

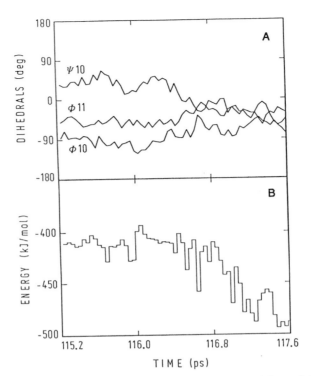

Figure 5. Temporal variation across the conformational transition of the 3 dihedrals ϕ_{10}, Ψ_{10}, and ϕ_{11} (A), and the total potential energy (B) obtained by keeping the 3 dihedrals fixed and relaxing the other degrees of freedom.

transport rates one concludes that such conformational transitions occur in the time range of 10-100 ms. Within the framework of the hierarchical model of protein dynamics, the changes in the orientation of the helices would be postulated to arise from wobbling fluctuations of the same helices. Such wobbling fluctuations have been observed in the range of 10-100 ns (Dornmair and Jähnig, 1989), but more experimental data, especially in the intermediate time range between ns and ms, are required to substantiate this model.

REFERENCES

Dornmair, K., and Jähnig, F., 1989, Internal dynamics of lactose permease, *Proc.Natl.Acad.Sci.USA* 86:9827.

Edholm, O., and Jähnig, F., 1988, The structure of a membrane-spanning polypeptide studied by molecular dynamics, *Biophys.Chem.* 30:279.

Edholm, O., and Jähnig, F., 1992, Molecular dynamics simulations of bacteriorhodopsin, *in*: "Membrane Proteins: Structures, Interactions and Models", A. Pullman et al., eds., Kluwer Press, Amsterdam.

Elber, R., and Karplus, M., 1987, Multiple conformational states of proteins: A molecular dynamics analysis of myoglobin, *Science* 235:318.

Henderson, R., Baldwin, J.M., Ceska, T.A., Zemlin, F., Beckmann, E., and Downing, K.H., Model for the structure of bacteriorhodopsin based on high-resolution electron cryo-microscopy, *J.Mol.Biol.* 213:899.

Hong, M.K., Braunstein, D., Cowen, B.R., Frauenfelder, H., Iben, I.E.T., Mourant, J.R., Ormos, P., Scholl, R., Schulte, A., Steinbach, P.J., Xie, A.-H., and Young, R.D., 1990, Conformational substates and motions in myoglobin, *Biophys.J.* 58:429.

Jähnig, F., 1989, Structure prediction for membrane proteins, *in*: "Prediction of Protein Structure and the Principles of Protein Conformation", G.D. Fasman, ed., Plenum, New York.

Jähnig, F., 1990, Structure predictions of membrane proteins are not that bad, *Trends Biochem.Sci.* 15:93.

Jähnig, F., and Dornmair, K., 1991, A possible correlation between fluctuations and function of a membrane protein, *in*: "Biologically Inspired Physics", L. Peliti, ed.,Plenum, New York.

Jähnig, F., and Edholm, O., 1992, Modeling of the structure of bacteriorhodopsin, *J.Mol.Biol.* 226:837.

Karplus, M., and Kushick, J.N., 1981, Method for estimating the configurational entropy of macromolecules, *Macromolecules* 14:325.

Peticolas, W.L., Mean-square amplitudes of the longitudinal vibrations of helical polymers, *Biopolymers* 18:747.

Pleiss, J., and Jähnig, F., 1991, Collective vibrations of an α-helix, *Biophys.J.* 59:795.

Pleiss, J., and Jähnig, F., 1992, Conformational transition of an α-helix studied by molecular dynamics, *Eur.Biophys.J.* 21:63.

STEPS TOWARDS PREDICTING THE STRUCTURE
OF MEMBRANE PROTEINS

P. Tuffery[2], C. Etchebest[1] and R. Lavery[1]

[1]Laboratoire de Biochimie Théorique
CNRS URA 77
Institut de Biologie Physico-Chimique
13 rue Pierre et Marie Curie
75005 Paris, France
[2]Unité de Recherches Biomathématiques et Biostatistiques
INSERM U 263
1, Place Jussieu
Tour 53, 1er Etage
75251 Paris, France

INTRODUCTION

Predicting protein structure using computer simulation is still very much in its infancy. Proteins are far from being easy systems to treat, firstly, because of their size, typically involving many thousands of atoms and, secondly, because of their complex structures which are stabilised by many weak interactions between sites often distant from one another in terms of the primary amino acid sequence. As a result of these features, current simulation techniques, even with the help of the fastest computers presently available, are far from being able to explain the folding pathways which lead from the denatured to the native state of typical globular proteins.

In the case of integral membrane proteins the situation is somewhat different due to the non-polar lipid environment in which such proteins are imbedded. Within the membrane, the absence of hydrogen bonding possibilities with the hydrocarbon chains of the lipids means that the protein must stabilise itself by internal hydrogen bonding. This can be achieved by the formation of regular structures such as α-helices or β-barrels which enable the NH and CO groups of the peptide backbone to bond with one another. In addition, β-barrels by themselves, or α-helices arranged in bundles, create structures with inside and outside surfaces enabling further stabilisation by the positioning of hydrophobic side chains on the outer surface in contact with the lipids. In contrast, within the cavities, hydrophilic side chains can be used to facilitate interactions with, or the transport of, polar or ionic species. Unfortunately, because of their natural environment, integral membrane proteins are

Statistical Mechanics, Protein Structure, and Protein Substrate Interactions
Edited by S. Doniach, Plenum Press, New York, 1994

extremely difficult to crystallise and in consequence only a few atomic resolution structures have been obtained using Xray crystallography. Other techniques such as high resolution electron microscopy, neutron diffraction and NMR spectroscopy have led to partial structural data on a number of membrane proteins. These results have provided examples of both α-helix bundles such as bacteriorhodopsin (Henderson et al, 1990) and the photosynthetic reaction centre (Deisenhofer et al, 1985) and β-barrel proteins such as the porins (Weiss and Schulz, 1992). Further structural information has come from studying the primary sequences of membrane proteins, in which the amphipathic profiles of membrane helices can be recognised rather clearly (Donnelly et al, 1993; Engelman et al, 1986; Popot and De Vitry, 1990; von Heijne, 1992). This data has provided evidence that the α-helix bundle is a very common structural motif for this class of proteins and bundles composed of widely different numbers of helices have been identified (Popot and de Vitry, 1990).

A further interesting feature of membrane proteins concerns the way in which they fold. A variety of experiments now support the view that such folding occurs in two stages which involve, firstly, the formation of individual transmembrane helices and, secondly, the packing of the helices to form the final bundle conformation (Borman and Engelman, 1992; Popot and Engelman, 1990). It should be added that the extra-membrane loops which link the transmembrane helices together do not appear to play a major role in determining the structure of the bundle. In several cases, it has been clearly demonstrated that cutting one or more of these loops does not prevent the formation of an active protein (reviewed in Popot et al, 1993).

MODELLING MEMBRANE PROTEINS

How do these special features of membrane proteins influence possible simulation strategies? First and foremost, in contrast to the situation with globular proteins, we know that at least the intra-membrane structure of membrane proteins is very likely to involve an α-helix bundle, moreover an analysis of the primary sequence of the protein in question enables us to identify the residues taking part in these secondary structures with reasonable accuracy. We can even go further in some cases since the amphipathic nature of the transmembrane helical segments often gives clues as to which face of the helices should be oriented towards the lipids. Secondly, the evidence supporting the two stage folding model for α-helix bundles implies that forming the bundle consists of packing pre-formed and structurally stable α-helices together, rather than a more complex combination of packing and secondary structure condensation.

Despite these simplifications, we are still confronted with a very complex problem. Even if we can assume that the backbones of the α-helices are rather rigid, forming the bundle requires locating their optimal relative positions. This, in turn, requires optimising the conformations of a large number of flexible side chains which are necessarily brought into contact at the helix-helix interfaces. The fact that the majority of the space inside folded proteins is occupied means that any given amino acid side will be in contact with a large number of neighbouring side chains. In consequence, changing the conformation of any side chain will imply crossing important energy barriers and, in general, side chain movements will be strongly coupled. This also implies that simple energy minimisation clearly cannot be used for optimising side chain packing since the procedure will rapidly be blocked by falling into a local energy minimum. Molecular dynamics will also have difficulties in solving such problems when the initial state chosen for the system is separated from the optimal conformation by important energy barriers. Although we have no complete solution to these problems, our recent research has suggested at least some elements of a strategy which may lead in the right direction. This work involves two main innovations, firstly, a new technique for finding the best conformations of a large number of strongly interacting amino acid side chains and, secondly, a means of simplifying the combinatorial search of rotational (and, partially, translational) positions within a helix bundle. These steps are described below and the first applications of this strategy to proteins of known conformation is then discussed.

One closing remark should be made in connection with early hopes that certain families of membrane proteins, notably those composed of 7 transmembrane helices, would all share more or less identical bundle conformations. This assumption was used as the basis for modelling G-protein coupled receptors starting from the structure of bacteriorhodopsin determined by high resolution electron microscopy (Henderson et al, 1990). Such an assumption was rather adventurous given that bacteriorhodopsin shows hardly 10% sequence homology with the G-protein coupled receptor family. A recent electron diffraction study of one member of this family, rhodopsin (Schertler et al, 1993), has indeed shown that the helices within its transmembrane bundle pack differently from those of bacteriorhodopsin. In consequence, predicting the structure of other α-helix bundle membrane proteins will certainly require the development of reliable methods for packing α-helices and for determining the optimal bundle conformation.

OPTIMISING SIDE CHAIN CONFORMATIONS

The first step we have taken to find an efficient way of optimising side chain conformations involves simplifying their conformation space. A preliminary remark is that, to a good first order approximation, we need only consider degrees of freedom corresponding to rotations around single bonds. Bond length changes and bond angle deformations clearly require higher energies than single bond rotations and will be of little importance at the degree of structural resolution which presently interests us. An analysis of proteins of known structure subsequently shows that each amino acid has a tendency to adopt a relatively limited number of preferential conformations defined by its dihedral angles and commonly termed "rotamers". This idea has already led a number of authors to propose finite sets of such conformations which can be used to replace the truly continuous conformational space of each side chain (see, for example, Janin et al 1978; McGregor et al, 1987; Piela et al, 1987; Ponder and Richards, 1987; Summers et al, 1987, Dunbrack and Karplus, 1993).

Since the quality of the rotamer set depends on the quantity and the quality of rapidly accumulating conformational data, we have repeated the work of earlier authors, using dynamic clustering techniques (Diday, 1972) to define a new rotamer set. The search was carried out using a total of 60 non-redundant proteins from the Protein Data Bank (Bernstein et al, 1977), all these structures being resolved to at least 2Å and having R factors of less than 0.25. Care was also taken to remove any steric contacts involving side chains by a simple energy minimisation of each side chain in the presence of its close neighbours. This procedure also allowed the hydrogens of methyl and hydroxyl groups, not present in the crystallographic data, to be appropriately positioned. The analysis of these proteins (Tuffery et al, 1991) led us to propose a total of 110 rotamers to represent the 20 common amino acids, the smallest number of rotamers per side chain being 1 (for glycine or alanine) and the largest number being 16 (for lysine). On average each side chain is represented by roughly 5 rotameric states. This value represents an important simplification since, on average, each side chain has 2 single bond rotations and if we assume that these bonds can generally adopt three stable conformations (g^+, g^-, t), we would deduce that each side chain would typically need a total of 3^2 or 9 rotamers. In fact many of these combinations are sterically prohibited, as the analysis of known protein structures shows.

Although using side chain rotamers clearly simplifies our search problem, one cautionary remark should be made. It is clear that rotamers only represent the average conformations within clusters of related conformational states. Consequently, a given side chain, in a given protein environment, will not, in general, exactly adopt a rotamer conformation. This simply means that while rotamers can certainly help in making a rapid initial search, the resulting conformations will have to be refined by some other technique which will allow deviations from the rotamer state chosen. However, in our opinion, this limitation does not invalidate the use of rotamers, as some authors have suggested (Eisenhaber and Argos, 1993).

We next turn to the problem of how to find an energetically optimal set of side chain rotamers for a given backbone structure. This problem requires a sophisticated algorithm and cannot be approached using a simple combinatorial search for more than a few residues (Ponder and Richards, 1987; Bruccoleri and Karplus, 1987). This is easily seen if we consider a protein with 150 residues. Assuming that each side chain is represented by an average of 5 rotamers, as we have found, then a total of roughly 10^{105} global conformations can be generated. It is clear that it is computationally impossible to search this number of states, however, appropriate optimisation algorithms can be used to obtain meaningful answers, in reasonable amounts of time, even for larger proteins. Our approach to this problem is a heuristic method termed SMD "Sparse Matrix Driven" (Tuffery et al 1991, 1993a) which takes advantage of the fact that many pairs of the side chains within a given protein have very weak interaction energies. This enables us to effectively decrease the size of the combinatorial problem of searching for ideal side chain rotamers. The aim of this approach is to limit exhaustive combinatorial searches to a set of residue clusters that contain most of the strong side chain-side chain interactions and are, in consequence, only weakly coupled to one another.

The first step in this approach involves pre-calculating the interaction energies between all the side chain rotamers for a fixed backbone conformation. The conformational energy calculations have been carried out using the "Flex" force field (Lavery et al, 1986a,b) which includes electrostatic, Lennard-Jones, bond torsion and angle deformation terms. An additional term takes into account the angular dependence of hydrogen bond interactions. The point charges used for the electrostatic term result from Hückel-Del Re calculations reparameterized to reproduce the SCF *ab initio* electrostatic properties of oligopeptides (Zakrzewska and Pullman, 1985). A dielectric constant of $\varepsilon=3$ was used in all calculations.

In order to calculate the energy of a protein with a given set of side chain rotamers, we divide the total conformational energy E into a sum over backbone-backbone, backbone-side chain and side chain-side chain interactions. If the backbone peptide groups are denoted by the indices a,b, the side chains by the indices i,j and the associated rotamers by k,l we may write the total energy as follows:

$$E = \Sigma_{a,b}\, E_{ab} + \Sigma_{b,i}\, E_{bi}{}^{k} + \Sigma_{i,j}\, E_{ij}{}^{kl}$$

Since the backbone conformation is fixed in our study the first summation of this expression is constant and can be ignored. The remaining two summations consist of terms which can be calculated and stored once and for all, for each possible rotamer of each side chain or pair of side chains. Once these terms are known the conformational energy of any set of rotamers can be calculated very rapidly by simply summing the appropriate terms. If, in addition, it is noted that the interaction energy between a given backbone element and all the rotamers of a given side chain always falls below a chosen absolute value (presently taken to be 1 kcal/mol) this interaction can be set to zero and ignored in future summations. The same cut-off can be employed between pairs of side chains for which all interactions over the rotamers k,l also fall below this chosen limit. It is finally remarked that the internal energy of side chains with no conformational freedom (e.g. Cys involved in disulphide bonds and any residues bound to prosthetic groups) is ignored as are the interaction energies between such side chains.

The SMD search algorithm consists, firstly, of filtering the side chain-backbone interactions in order to discard rotamers that lead to high energies, thus reducing the dimension of the overall combinatorial space. Next one constructs a Boolean matrix of side chain-side chain interactions. This is done by screening the interaction energies $E_{ij}{}^{kl}$ for each pair of side chains. An interaction "flag" is then set only if the maximum absolute value of the interactions exceeds a chosen threshold value (1 kcal/mol). The exact value of this threshold has little importance, however it improves the efficiency of the search by allowing us to ignore interactions which cannot lead to important energy gains for the overall conformation.

Using the Boolean matrix it is now possible to construct clusters of strongly interacting residues. This process is equivalent to reordering the matrix by interchanging rows and columns so that such groups of residues (which may be widely separated in terms of the primary protein sequence) are brought together to form submatrices centred on the leading diagonal. The maximum number of residues per cluster is controlled by the user. An exhaustive combinatorial search is then performed on the residues in each cluster. The combination of these single cluster solutions leads to local minima which can be extended to a global solution by an iterative averaging approach which takes into account extra-cluster interactions (that is, matrix elements far from the diagonal, which describe the coupling between clusters) by statistically weighting the intra-cluster rotamers. Averaging is carried out at two levels. The first level treats the interaction of an intra-cluster rotamer with the rotamers of each relevant extra-cluster residue represented by the median of the corresponding interaction energies. The second level averages these values for all the extra-cluster side chains having non-zero interactions with a given intra-cluster rotamer, the corresponding energies again being represented by the median value. After the first iteration all side chains have assigned rotamers and, consequently, the first level averaging described above is simply replaced by the corresponding interaction energy between the inter-cluster rotamer and the chosen extra-cluster rotamer.

The SMD search is stopped when no further improvement in the overall conformational energy is obtained between two successive steps. The method is found to converge in very few cycles (generally less than 5). Complementary refinement can be performed by repeating the steps listed above on alternative clusters that are formed using a Boolean matrix from which interactions contained within the previous clusters have been removed. This controls the validity of the previous solution, since a combinatorial search is now performed on the interactions that were previously used only in weighting.

In order to go beyond the limited resolution of the rotamer states, the conformations resulting from SMD are systematically refined using a fast procedure where a quasi-Newton minimisation is carried out for the torsion angles of each side chain in turn, in the presence of neighbouring side chains and backbone peptide units. This procedure allows methyl and hydroxyl groups rotations not described by the rotamer states to be correctly adjusted and also allows minor readjustments of the remaining side chain torsion angles. Conformational changes are however generally small compared to the differences between rotamer states. The speed of this refinement results from the fact that only one side chain is minimised at a time. The results obtained by applying this procedure to globular proteins of known conformation show that we are currently capable of positioning side chains to a mean RMS of 1.6Å, a value which further improves to 1.2Å for buried residues. It is worth adding that although the SMD method was originally developed for and tested on globular proteins, recent NMR results (Barsukov et al, 1992; Pervushin and Arseniev, 1992) on three of the helices from bacteriorhodopsin (A, G and F) are also very well reproduced by SMD searches. Overall, 50 out of a total of 57 measured $\chi 1$ values are correctly reproduced. In addition, for all the side chains where the $\chi 1$ value is correctly predicted the $\chi 2$ torsion is also correctly reproduced (Tuffery et al, 1993b). It should be added that in these tests and in the helix packing studies described below, the backbone geometries of the α-helices came from experimental data. We return to the question of predicting such geometries in the final section of this article.

OPTIMISING α-HELIX PACKING

The availability of a rapid and reliable technique for predicting side-chain conformations starting only from a knowledge of sequence and backbone conformation has allowed us to progress to defining a conformational search strategy for packing secondary structure elements within a protein. This involves scanning all possible relative orientations of the structural elements, optimising the side chain packing at each step. This approach assumes that packing will be conditioned mainly by optimal side chain interactions and that relatively little deformation of the peptide backbones will occur. This strategy is appropriate for studying transmembrane helical bundles because of two factors: Firstly, α-helices are strongly stabilised by their regular hydrogen bonding and their peptide dipole interactions. This limits possible deformations of the peptide backbone. In contrast, amino acid side-chains are generally very flexible suggesting that bundles can be built without major backbone deformation. Secondly, as discussed above, experimental results have shown that the cleavage of inter-helical loops which lie outside the membrane does not prevent the correct packing of the protein.

Even if we can solve the side chain problem rapidly, looking at all possible relative positions of the helices within a bundle (6 degree of freedom for each helix) represents a major computational problem. If we consider a 7-helix bundle and we limit ourselves to studying rotations around the axis of each helix using, say, 20° steps, a seven helix bundle will require testing 18^7 (roughly 600 million) different orientations. There are however ways to further simplify the problem.

The first step in this direction consists of temporarily dividing the helix bundle into smaller parts consisting of pairs or triplets of helices. How this sub-division is carried out will depend on the number and arrangement of helices within the bundle. For the moment, we will take the 7-helix bundle of bacteriorhodopsin as an example. The structure of this bundle is shown schematically in figure 1. The helices are named conventionally A through G. If we look for the helix pairs likely to interact most strongly, in addition to the 7 neighbouring pairs around the periphery of the bundle (AB, BC, CD, DE, EF, FG, GA), there are 4 pairs which are likely to come into contact across the core of the bundle (BG, CG, CF, FD). This makes a total of 11 principal pairs. Similarly an inspection of the form of the bacteriorhodopsin bundle suggests that it can be considered as a collection of 5 principal helix triplets (ABG, BCG, CFG, CDF, DEF). In bundles consisting of other numbers of helices and/or having different overall forms, these sub-divisions will obviously be different, but a similar approach can be adopted as long as the rough conformation of the bundle and the location of each helix within the bundle is known.

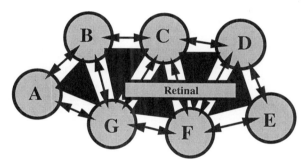

Figure 1. Schematic view of the 7 helices of bacteriorhodopsin viewed from the cytoplasmic face. The topology of the bundle leads to a decomposition of the bundle into 11 helix pairs (arrows), and 5 helix triplets (shaded triangles).

Although it is easy to make such a decomposition, is it related in any way to the energetics of packing within the bundle? The answer can be seen if we look at a sub-group of helices taken once more from bacteriorhodopsin. Table 1 shows the best side chains rotamers obtained for the helices ABCG. As well as performing SMD optimisation on the 4-helix group, we have carried out calculations on all the constituent helix pairs and triplets. The results show that in nearly all cases the optimal conformation of any given side chain within the bundle is actually determined either by the helix to which the side chain belongs or by the presence of only one other helix. Only 5 side chains out of a total of 105 do not obey this rule and 4 of these cases are in fact determined by the presence of only 2 other helices. Similar results were obtained in an earlier study of the 4-helix bundle of myohemerythrin (Tuffery and Lavery, 1992). These observations suggest that, to a good first approximation, the conformations of the side chains within a bundle may be determined by considering only pairs of interacting helices.

Table 1. Analysis of the side chain positioning for the helix bundle ABCG from bacteriorhodopsin. The backbone is defined as in the experimental structure (Henderson et al, 1990). For each of the four helices, the side chain conformations (rotamer number) are given in each of the possible helix sub-bundles.

A	1	2	1	1	1	2	1	2	1	1	1	2	4	1	2	1	1	2	4	2	2	1	2	1	3
AB	1	2	6	1	3	1	1	2	1	2	1	2	10	1	2	1	1	2	4	4	2	1	2	1	3
AC	1	6	1	1	1	2	1	2	1	1	1	2	4	1	2	1	1	2	4	2	2	1	2	1	3
AG	1	2	1	1	6	2	1	2	1	1	1	1	3	1	2	1	1	2	4	2	2	1	15	1	3
ABC	1	3	6	1	2	1	1	2	1	2	1	2	10	1	2	1	1	2	4	4	2	1	2	1	3
ABG	1	2	6	1	6	1	1	2	1	2	1	1	10	1	2	1	1	2	4	4	2	1	15	1	3
ACG	1	2	1	1	6	2	1	2	1	1	1	1	3	1	2	1	1	2	4	2	2	1	15	1	3
ABCG	1	2	6	1	6	1	1	2	1	2	1	1	10	1	2	1	1	2	4	4	2	1	15	1	3

Influence line: `BG B    G B        B   G B            B        G`

B	1	1	2	8	1	4	1	1	1	1	3	1	1	1	1	1	2	1	3	3	2	3	4	2	2
AB	1	1	5	8	1	1	1	1	1	1	3	1	1	1	1	1	2	1	3	2	2	3	5	2	2
BC	1	1	2	8	2	4	1	3	1	1	3	1	1	1	1	1	2	1	6	3	2	3	4	2	2
BG	1	1	2	8	1	4	1	1	1	1	3	1	1	1	1	1	2	1	3	4	2	3	4	2	2
ABC	1	1	5	8	2	1	1	3	1	1	3	1	1	1	1	1	2	1	6	2	2	3	5	2	2
ABG	1	1	5	8	1	1	1	1	1	1	3	1	1	1	1	1	2	1	3	2	2	3	5	2	2
BCG	1	1	2	8	1	4	1	3	1	1	3	1	1	1	1	1	2	1	6	3	2	3	4	2	2
ABCG	1	1	5	8	1	1	1	3	1	1	3	1	1	1	1	1	2	1	6	2	2	3	9	2	2

Influence line: `A   AG A  C                 C A      *`

C	5	3	1	1	1	3	1	1	8	4	1	1	1	2	2	1	1	1	2	2	2	2	1	2	1	2	1
AC	4	3	1	1	1	3	1	1	8	4	1	1	1	2	2	1	1	1	2	2	2	2	1	2	1	2	1
BC	5	3	1	1	1	3	1	1	8	4	1	2	1	2	4	3	1	1	5	2	2	2	2	2	1	2	1
CG	5	3	1	1	1	4	1	1	8	4	1	1	7	2	2	2	1	1	2	2	2	2	1	2	1	2	1
ABC	4	3	1	1	1	3	1	1	7	4	1	2	1	2	4	3	1	1	5	2	2	2	2	2	1	2	1
ACG	2	1	1	1	1	4	1	1	8	4	1	1	7	2	2	2	1	1	2	2	2	2	1	2	1	2	1
BCG	5	3	1	1	1	4	1	1	8	4	1	2	7	2	4	2	1	1	5	2	2	2	2	2	1	2	6
ABCG	2	1	1	1	1	4	1	1	8	4	1	2	7	2	4	2	1	1	5	2	2	2	2	2	1	2	6

Influence line: `# #         G      ?        B G    B G      B          B              #`

G	1	1	7	1	2	2	2	7	1	2	1	1	3	1	1	1	1	2	1	2	3	2	1	7
AG	1	1	3	3	4	2	2	4	1	2	1	1	3	1	1	1	1	2	1	2	3	2	2	7
BG	1	1	7	1	2	2	2	4	1	2	1	1	3	1	1	1	1	2	1	2	3	5	1	7
CG	1	1	3	1	2	2	4	7	1	2	1	1	3	1	1	1	1	2	1	2	3	2	1	7
ABG	1	1	3	3	4	2	2	4	1	2	1	3	3	1	1	1	1	2	1	2	3	5	2	7
ACG	1	1	3	3	4	2	4	4	1	2	1	1	3	1	1	1	1	2	1	2	3	2	2	7
BCG	1	1	3	1	2	2	4	4	1	2	1	1	3	1	1	1	1	2	1	2	3	5	1	7
ABCG	1	1	3	3	4	2	4	4	1	2	1	3	3	1	1	1	1	2	1	2	3	5	2	7

Influence line: `AC A A    C AB         #                     B A`

Note: Side chains conformations determined only by the helix to which they belong are indicated by a blank in the final line of each sub-table. Side chains influenced by one other helix are indicated by the appropriate letter and side chains influenced by two or three other helices are indicated by the symbols # and * respectively. ? indicates temporary changes in conformation for certain bundles that have no influence on the conformation of the corresponding side chain in the 4 helix system.

In the case of the 11 pair sub-division chosen for bacteriorhodopsin, this first level of approximation reduces the number of combinations to be considered from 18^7 to 11×18^2 or 3564. The energy of a given bundle conformation can then be approximated from the pair energies as,

$$E_{bundle} = \Sigma E_{pair} - \Sigma (n_s-1) E_{single}$$

where E_{pair} corresponds to the energy of each of the 11 pairs and E_{single} to the internal energy of the helices, which are considered a total of n_s times.

It should be noted that the full bundle energy obtained by the summation of pairs given above would only be exact if all the side-chain conformations within each constituent pair were identical to the conformations found within the full bundle. This cannot be exactly true since there are some side-chains whose optimal conformation within the bundle is only adopted in the simultaneous presence of two or more other helices. We therefore pass to a further level of approximation by breaking the helix bundle down into a combination of helix triplets. In this case, the energy of the full bundle becomes:

$$E_{bundle} = \Sigma E_{triplet} - \Sigma (n_p-1) E_{pair}$$

where n_p is the total number of times a pair appears within the set of triplets. Passing from pair summation to triplet summation will certainly improve our model of the full bundle, but also increases the cost of the computations. Higher levels of approximation involving helix quadruplets or larger clusters would probably further improve the results, but only at the cost of considerably increased computational effort. We have therefore chosen to stop at this level and then pass directly to calculations on the full helix bundle. The search strategy we have developed thus consists of three principal steps:

- Firstly, at the level of helix pairs, we make a full rotational scan of each helix in 20° steps. At each of the resulting 18^2 steps an SMD side chain optimisation is performed followed by the rapid minimisation procedure. The results are used to predict the best bundle conformations which are ordered in terms of their pair summation energies.

- Secondly, we select a large number of the best solutions obtained during the preceding step (typically of the order of several thousand, covering an energy range of several hundred kcal/mol) and analyse these solutions to determine which conformations of the pre-defined helix triplets they contain (generally many fewer than the theoretical number of 18^3 combinations). We then repeat the SMD side chain optimisation and the rapid minimisation procedure for all these triplet conformations and again estimate the corresponding bundle energies, now at the triplet summation level.

- We finally select a limited number of the best full bundle orientations obtained at the triplet level for further refinement. This refinement consists again of an SMD optimisation, now carried out for the complete bundle, followed by the rapid side chain optimisation procedure and then a quasi-Newton energy minimisation of all side chains combined with free helical rotations and axial displacements.

TESTS ON PROTEINS CONTAINING α-HELIX BUNDLES

The hierarchical approach we have developed to search for the optimal rotational packing of helices within helix bundles enables us to sample all the rotational states at the pair approximation level and several thousand states at the triplet level. Even the latter figure would be beyond computational possibilities if we had directly attacked the full helix bundle. Our approach thus enables focusing of the solutions towards the final bundle conformation. Several choices nevertheless had to be made during the procedure which must be verified. Preliminary results obtained on myohemerythrin (Tuffery and Lavery, 1993) and bacteriorhodopsin (Tuffery et al, 1993b) already allow us to comment on two of these points.

Firstly, it seems that 20° step size used for rotational scanning is sufficient since relaxation of the full bundle led individual helices to rotate by up to 15°. It may be added that, during full bundle refinement, helix displacements along the helical axis were limited to roughly 1Å and thus searching axial displacement would require a step size of the order of 1.5-2.0Å.

Secondly, it seems that the number of pseudo-optimal solutions kept at each stage of the search is reasonable. It is found that at the pair level, few helix orientations are completely rejected and also that the median deviation of the pair and triplet energies for corresponding conformations is much smaller than the energy cut-off used. It has also been found that the moderately stable solutions at the triplet level are mainly composed mixtures of the few best solutions or minor deviations from these (generally by only a single 20° step), deviations which can be removed during refinement of the full bundle.

Finally, how do our results compare with experimental data for the proteins tested? In the case of the 4-helix bundle of myohemerythrin, correct helix orientations were obtained if a simple correction was made for hydration effects not included in our force field. This consisted of using the amphipathic profiles of each helix to partially restrict the rotational orientations searched (A,D ±90°, C ±60°, B no restriction). This study also allowed us to show that the terminal residues of the helices seemed to have little influence on their optimal packing and also that it was possible to predict the correct lateral separation of helices at the pair approximation level. In the case of bacteriorhodopsin, good orientations were found for all 7 helices forming the bundle, with no more than a few degrees of rotation with respect to the best experimental model available, based on high resolution electron microscopy (Henderson et al, 1990). Our calculations also supported a suggestion by the authors of the experimental study that the D helix should be displaced roughly 2Å along its axis towards its N-terminus. If this same displacement is made in the experimental conformation, the Cα positions differ from those of our best model by an overall RMS of only 0.37Å. For the side-chains, the overall RMS is 2.06Å, with a value of 1.33Å for internal residues (for which the accessible area is less than 25% of the corresponding area computed for the residue within an Ala-X-Ala tripeptide in the α-helical conformation). It should however be recalled that the electron microscopy data used to generate the experimental structure was not detailed enough to place any but the largest aromatic side chains with any degree of certainty (Henderson et al, 1990). Several major changes in side chain conformation involving aromatic residues on the periphery of the bundle are seen (Trp 10, Trp 138 and Trp 189). These changes may be related to the lack of any explicit model for the lipids surrounding bacteriorhodopsin, but they seem to be of little importance in determining the optimal packing of the bundle. Table 2 shows the final bundle conformations of bacteriorhodopsin which led to negative packing energies (i.e. the total bundle energy less the energies of the 7 constituent helices optimised individually). These results show that the rotational positions of helices A, B, C, G are strongly determined by the bundle. In contrast, helices E and F and, to a lesser extent, D are less well determined. This distinction becomes even more marked if the calculations are performed in the absence of retinal (bound to Lys171 in the G helix) which, in consequence, appears to play an important role in fixing the positioning of the DEF helix triplet.

Table 2. Best solutions obtained for the rotational positioning of the helices within the bacteriorhodopsin bundle. Only the solutions presenting a stable packing are presented. Energies are in kcal/mol, axial rotations (r) in degrees and axial translations (t) in Å.

Set	Epck	Eint	rA	rB	rC	rD	rE	rF	rG	tA	tB	tC	tD	tE	tF	tG
	-160	-388	4	-1	-4	-13	4	-7	-1	0.5	0.7	0.1	-2.3	0.6	-0.1	0.2
	-123	-380	4	-1	-2	-16	14	-5	0	0.8	0.7	0.3	-2.5	-0.3	-0.1	0.2
	-121	-386	2	-3	-3	-5	-10	-11	-2	0.8	0.7	0.1	-1.7	0.8	-0.6	0.1
	-84	-385	1	0	-4	-5	-15	-7	-1	0.3	0.7	0.1	-1.8	0.8	-0.3	0.2
	-81	-376	0	0	-4	-2	-16	-8	-2	0.2	0.7	0.2	-1.7	0.8	-0.3	0.2
*	-79	-353	6	0	-1	17	2	-4	0	0.7	0.4	-0.4	-0.2	-0.2	-0.3	-0.1
	-76	-385	1	0	-9	-17	1	-6	0	0.4	0.8	0.4	-2.5	0.4	-0.1	0.3
	-74	-380	0	1	-8	-12	-1	-4	0	0.4	0.8	0.3	-2.5	0.2	0.1	0.4
*	-73	-342	4	3	0	21	2	-8	0	0.6	0.5	-0.5	-0.1	-0.4	-0.5	0.0
	-69	-358	2	6	-5	-13	7	-10	-1	0.2	0.6	0.2	-2.2	0.5	0.1	0.2
	-54	-354	1	-3	-6	-14	9	-3	-1	0.8	0.8	0.5	-2.4	0.4	-0.1	0.0
	-52	-360	2	6	-6	-14	-7	-11	-1	0.3	0.8	-0.1	-2.3	0.9	-0.4	0.3
*	-34	-320	6	1	3	17	6	-3	0	0.7	0.4	-0.1	-0.3	-0.4	-0.3	-0.1
	-24	-344	0	1	-4	0	-18	-126	1	0.5	1.0	-0.1	-2.6	-0.0	0.1	0.6
*	-20	-322	3	2	3	16	6	-3	0	0.2	0.5	-0.0	-0.4	-0.3	-0.2	0.0
	-18	-314	2	4	-3	24	99	-4	-1	0.3	0.8	-0.0	-1.9	0.2	-0.1	0.3
	-11	-338	5	1	-7	22	105	-120	4	0.4	0.5	-0.0	-2.5	0.2	0.6	0.4
	-2	-341	1	2	-9	-18	7	-1	-1	0.2	0.6	0.7	-2.3	0.4	0.0	0.2
*	-1	-341	4	2	0	18	111	-119	2	0.0	0.5	-0.6	-0.1	-0.3	0.4	0.0
	0	-332	2	-2	-14	-8	-22	-119	2	0.5	0.8	0.1	-3.8	0.1	1.1	0.7

Note: *Set* corresponds to the origin of the solution: * indicates that helix D was not displaced along its axis, while other conformations began with this helix displaced by 2.5Å. *Epck* is the difference between the energy of the bundle and the sum of the energies of the isolated helices. *Eint* is the sum of the helix interaction energies.

FUTURE DEVELOPMENTS

The approach we have developed to predict the rotational packing of helices within transmembrane bundles is based on the assumption that helices are pre-folded within the membrane before packing takes place and that inter-helix packing is principally determined by the helix side chains. We also assume, in line with available experimental data, that the loop regions of membrane protein do not play a crucial role in packing. The techniques we have developed for rapidly positioning amino acid side chains and for simplifying the combinatorial problems of bundle conformation have enabled us to treat very complex problems with only reasonable computational expenses. The results we have obtained for the 4-helix bundle of myohemerythrin and the 7-helix bundle of bacteriorhodopsin are in good agreement with experimental data. These results would appear to justify our hypotheses and suggest that this technique may now be applied to a variety of new problems.

The first such development we are considering is the study of site directed mutagenesis and, in particular, of locating which residues are most important for determining the helix bundle conformation. Secondly, the method may help in refining the structure of other membrane proteins, such as rhodopsin, for which only low resolution data is available (Schertler et al, 1993). In such cases, we know the axial positions of the helices and their peptide sequences, but we need to be able to determine both the detailed geometry of their backbones and the relative longitudinal displacements of the helices, in addition to performing the rotational search already discussed. Studies in this direction are currently underway. Two encouraging observations in this respect are that, firstly, judging from our present study of D helix displacement in bacteriorhodopsin, quite significant helix displacements cause only minor changes in rotational positioning and, secondly, based on our studies of myohemerythrin, residues at the ends of the helices composing the bundle do not appear to play a major role in determining optimal packing.

ACKNOWLEDGEMENT

The authors would like to thank Organibio and the French Ministère d'Education Supérieure et de la Recherche for their support of this work through the research program CM^{2}AO "Conception Macromoléculaire Assistée par Ordinateur". We also wish to thank Jean-Luc Popot for many helpful discussions.

REFERENCES

Barsukov, I.L., Nolde, D.E., Lomize, A.L. and Arseniev, A.S. (1992) Three-dimensional structure of proteolytic fragment 163-231 of bacterioopsin determined from nuclear magnetic resonance data in solution. *FEBS* 206, 665-672.

Bernstein, F.C., Koetzle, T.F., Williams, G.J.B., Meyer, E.F., Brice, M.D., Rodgers, J.R., Kennard, O., Shimanouchi, T. and Tasumi, M. (1977) The Protein Data Bank: A computer-based archival file for macromolecular structures. *J. Mol. Biol.* 112, 535-542.

Borman, B.J. and Engelman, D.M. (1992) Intramembrane helix-helix association in oligomerization and transmembrane signaling. *Annu. Rev. Biophys. Biomol. Struct.* 21, 223-242.

Bruccoleri, R.E. and Karplus, M. (1987) Prediction of the folding of short polypeptide segments b uniform conformational sampling. *Biopolymers* 26,137-168.

Deisenhofer, J., Epp, O., Huber, K. and Michel, H. (1985) Structure of the protein subunits in the photosynthetic reaction centre of Rhodopseudomonas viridis at 3Å resolution. *Nature* 318, 618-624.

Diday, E. (1972) "An Introduction to the Dynamic Cluster Method (DYC program)", Metra.

Donnelly, D., Overington, J.P., Ruffle, S.V., Nugent, J.H.A. and Blundell, T.L. (1993) Modeling α-helical transmembrane domains: the calculation and use of substitution tables for lipid-facing residues. *Prot. Sci.* 2, 55-70.

Dunbrack, R.L. and Karplus, M. (1993) Backbone-dependent rotamer library for proteins. *J. Mol. Biol.* 230, 543-574.

Engelman, D.M., Steitz, T.A. and Goldman, A. (1986) Identifying nonpolar transbilayer helices in amino acid sequences of membrane proteins. *Ann. Rev. Biophys. Chem.* 15, 321-353.

Henderson, R., Baldwin, J.M., Ceska, T.A., Zemlin, F., Beckmann, E. and Downing, K.H. (1990) Model for the structure of bacteriorhodpsin based on high-resolution electron cryo-microscopy. *J. Mol. Biol.* 213, 899-929.

Janin, J., Wodak, S., Levitt, M. and Maigret, B. (1978) Conformation of amino acid side-chains in proteins. *J. Mol. Biol.* 125, 357-3862.

Lavery, R., Parker, I. and Kendrick, J. (1986a) A general approach to the optimization of the conformation of ring molecules with an application to Valinomycin. *J. Biomol. Struct. Dyn.* 4, 443-461.

Lavery, R., Sklenar, H., Zakrzewska, K. and Pullman, B. (1986b) The flexibility of the nucleic acids: (II) The calculation of internal energy and applications to mononucleotide repeat DNA. *J. Biomol. Struct. Dyn.* 3, 989-1014.

Mc.Gregor, M.J., Islam, S.A. and Sternberg, M. J. E. (1987) Analysis of the relationship between side-chain conformation and secondary structure in globular proteins. *J. Mol. Biol.* 198, 295-310.

Pervushin, K.V. and Arseniev, A.S. (1992) Three-dimensional structure of (1-36) bacterioopsin in methanol-chloroform mixture and SDS micelles determined by 2D H-NMR spectroscopy. *FEBS. Lett* 308, 190-196.

Piela, L., Nemethy, G. and Scheraga, H. A. (1987) Conformational constraints of amino acid side chains in α-helices. *Biopolymers* 26,1273-1286.

Ponder, J.W. and Richards, F.M. (1987) Tertiary templates for proteins: Use of packing criteria in the enumeration of allowed sequences for different structural classes. *J. Mol. Biol.* 193, 775-791.

Popot, J.L. and De Vitry, C. (1990) On the microassembly of integral membrane proteins. *Annu. Rev. Biophys. Chem.* 19, 369-403.

Popot, J.L. and Engelman, D. (1990) Membrane protein folding and oligomerization: the two stage model. *Biochemistry* 29, 4031-4037.

Popot, J.L. (1993) Transmembrane α-helices of integral proteins as autonomous folding domains. *Curr. Opin. Struct. Biol.* in press.

Schertler, G.F.X., Villa, C. and Henderson, R. (1993) Projection structure of rhodopsin. *Nature* 362, 770-772.

Summers, N.L., Carlson, W.D. and Karplus, M. (1987) Analysis of side-chain orientations in homologous proteins. *J. Mol. Biol.* 196,175-198.

Tuffery, P., Etchebest, C., Hazout, S. and Lavery, R. (1991) A new approach to the rapid determination of protein side chain conformations. *J. Biomol. Struct. Dyn.* 8, 1267-1289.

Tuffery, P., Etchebest, C., Hazout, S. and Lavery, R. (1993a) A critical comparision of search algorithms applied to the optimisation of protein side-chain conformations. *J. Comp. Chem.* 14, 790-798.

Tuffery, P., Etchebest, C., Popot, J.L. and Lavery, R. (1993b) Prediction of the positioning of the 7 transmembrane α-helices of bacteriorhodopsin: A molecular simulation study. *J. Mol. Biol.* submitted.

Tuffery, P. and Lavery, R. (1993) Packing and recognition of protein structural elements: A new approach applied to the 4-helix bundle of the Myohemerythrin. *Proteins* 15, 413-425.

von Heijne, G. (1992) Membrane protein structure prediction. Hydrophobicity analysis and the positive-inside rule. *J. Mol. Biol.* 225, 487-494.

Weiss, M.S. and Schulz, G.E. (1992) Structure of porin refined at 1.8Å resolution. *J. Mol. Biol.* 227, 493-509.

Zakrzewska, K. and Pullman, A. (1985) Optimized monopole expansions for the representation of the electrostatic properties of polypeptides and proteins. *J. Comp. Chem.* 6, 265-273.

PROTEIN FOLD FAMILIES AND STRUCTURAL MOTIFS

C.A. Orengo[1], T.P. Flores[1,2], W.R. Taylor[2] and J.M. Thornton[1]

[1]Biomolecular Structure and Modelling Unit
Department of Biochemistry and Molecular Biology
University College
Gower Street
London WC1E 6BT

[2]Laboratory of Mathematical Biology
National Institute for Medical Research
The Ridgeway
Mill Hill
London NW7 1AA

INTRODUCTION

Identifying Unique Fold Families

Over the last twenty years the number of known protein structures has risen exponentially. However, there are still at least tenfold more proteins whose sequences are known but whose structures have not yet been determined. For these proteins, information about sequence/structure relationships are used to predict a probable structure. This can include residue preferences for specific secondary structure conformations or residue contact potentials. It is critically important, though, that any statistically based study, to derive such information, should use a non-degenerate dataset. That is, a set which contains a single representative from each of the protein fold families.

Proteins can adopt the same fold by divergent or convergent evolution. In the former, there is a common ancestor, structures are homologous and often retain some sequence identity and functional equivalence. In the latter, proteins are analogous and there is no sequence similarity and the fold may have converged due to inherent structural restraints or functional pressure on an active site. As well as improving structure prediction, detailed knowledge of these fold families and their relationships should aid our understanding of evolutionary mechanisms.

In 1981 Richardson reviewed the known protein structures and classified them according to different classes and types of protein folds. Since then, individual families have

Statistical Mechanics, Protein Structure, and Protein Substrate Interactions
Edited by S. Doniach, Plenum Press, New York, 1994

253

been studied but the complete set of folds has not been re-examined or updated. However, the rapid increase in known structures makes it essential to regularly review and describe the set of unique folds. Although, there are more than 2000 protein structures determined, nearly two thirds of these are practically identical (>95% similar by sequence). In addition, some families are very overrepresented; there are nearly 200 mutant lyzozyme structures and over 100 highly similar globin structures. Furthermore, proteins having 30% or more of their sequences in common are known to adopt very similar folds and methods which group proteins on the basis of sequence similarity suggest that there are 200 or fewer unique folds. As more than 50 new structures are being added to the databank each month, it is important to develop consistent and automatic methods for regularly assessing structural similarity.

Early approaches at fold classification were based on sequence comparisons alone. Several groups (Chothia & Lesk (1986, 1987), Hubbard & Blundell (1987), Flores et al) have shown that when at least 50% of the sequences are common, the structures are highly conserved, with the majority of residues adopting the same fold and any small deviations occurring within the loops. Sander and Schneider (1991) derived a relationship between sequence identity and structural similarity, which showed that structures with much lower sequence identities also adopt similar folds. This enabled them to generate an extended structure databank in which proteins whose sequences match those of known structures by 30% or more, were assigned the same fold. Similarly, using sequence comparison approaches to assign homologous folds, Hobohm et al. (1992) proposed two automatic methods to derive non-homologous datasets.

However, it is becoming increasingly apparent that structures with very low sequence identities (<10%) can adopt the same fold (Orengo et al (1993), Orengo et al (1993), Murzin (1993), Holm and Sander (1993)). At these levels of similarity, sequence alignment techniques become unreliable and only structural comparison can reveal remote relationships. This is generally more time-consuming than sequence comparison and therefore need only be applied once highly related proteins (>30% sequence identity) have been identified. Several groups now include structural information when defining non-homologous sets. Boberg et al. (1992) use the percentages of residues in alpha and beta conformations to better identify analogous proteins. However, since this is only a superficial measure of structural similarity a final manual inspection of the set is also necessary. Pascarella and Argos (1992) have compiled a database of structural families, based on information from the literature. Structural comparisons were performed using rigid body techniques. Unfortunately, as the method does not automatically check for similarities, the sets are not easily updated. More recently. Holm and Sander have performed extensive comparisons between all protein families in order to check for structural relatedness. However, their method often combines families containing common motifs, so that it is harder to identify a set of unique folds.

We have developed a procedure which automatically orders the complete structure databank into families according to sequence and structure relatedness. Detailed analyses of global similarities between families are performed, using a flexible structural alignment program. For each release of new structures, families can be rapidly updated and representatives chosen. Rationalising the databank in this way will make it easier to select non-degenerate datasets. It should also enable structural characteristics and sequence preferences to be studied for individual families. This will become increasingly important as the volume of data grows.

Alpha plus Beta Fold Families

As part of this study, we have re-examined the structural relationships between $\alpha+\beta$ folds. Amongst the four major classes of protein structure described by Levitt and Chothia

(1976), these folds are the most diverse. They can be described as having both α-helix and β-strand secondary segments generally segregated along the chain and the folds observed vary from predominantly α-helix domain with small β-sheet regions to mainly β-proteins which include a few α-helices. Diversity results from the broad range of structural motifs available and the different connectivities between them. Structural comparisons between representatives from the α+β fold families showed extensive similarities between a large group of folds. This group constitutes nearly one third of the α+β class. The structures within it are all 2-layer α-β sandwiches, in which one layer consists of β-strands, the other α-helices. A split βαβ motif recurs frequently amongst these folds.

Concensus Templates

Individual folds within multidomain proteins are often hard to recognise. However, it is essential to identify these, to completely classify the databank into fold families. A single domain may constitute 25% or less of the total structure, so there can be a large amount of noise generated during any structural comparison. This makes it harder to match corresponding domains. Also, since evolutionary constraints on single and multidomain proteins may differ, considerable divergence may have occured. We have developed a method for generating a consensus structure for a given fold family. This encapsulates the most conserved characteristics of the fold and therefore makes it easier to find an equivalent region within a larger multidomain structure.

METHODS

The analysis uses release 60 of the protein structure databank (April 1992, Bernstein *et al.* (1977)). Only proteins resolved to better than 3Å are considered if they contain at least 45 residues. An outline of the method is shown in Figure 1 (see also Orengo *et al.* 1993). Pairwise sequence comparisons were first performed between all well-resolved proteins in the brookhaven databank. This gave a sequence homology matrix for establishing families of related proteins (at 35% identity). Best representatives of these families are then structurally compared and a structure relatedness matrix generated. This allows homologous and analagous proteins to be found even at low sequence identity.

Sequence Comparisons

Pairwise sequence comparisons are performed using the Needleman and Wunsch algorithm (1977), scoring 1 for identities and using a length independent gap penalty of 4. For small proteins (less than 80 residues) the percentage identity measured is adjusted using the equation of Sander and Schneider (1991). Percent identities are calculated using the length of the smaller protein, and significance checked by ensuring the length of the larger is comparable (no more than 30% larger).

A two dimensional homology matrix, SEQMAT, is produced from the pairwise sequence identities and then single linkage cluster analysis (Krzanowski (1990)) used to find families of proteins, related at 35% identity (**35Seq family**). A representative structure, having the best resolution and R-factor, is selected from each 35Seq family.

Structure Comparisons

As homologous proteins can have much lower sequence identities than 35% (e.g. down to 10% in the globin family) and still retain the same fold, structural comparisons are performed between all the 35Seq representatives. These are generated by the method of Taylor and Orengo (SSAP) (1989a,b) which is based on the Needleman and Wunsch algorithm, but compares structural environments rather than sequence identities and uses dynamic programming at two levels. For a given residue, the structural environment is defined as the set of vectors from the Cβ atom to the Cβ atoms of all other residues in the same structure. Structural environments are compared between pairs of residues in **lower** level matrices, whose dimensions correspond to the lengths of the proteins being aligned. Optimal pathways through these matrices are determined using dynamic programming and summed in an **upper** level matrix. Finally, a concensus path is obtained from the upper level matrix, again using dynamic programming. This identifies pairs of equivalent residues which can be used to generate a superposition of the structures, by the method of Rippmann and Taylor (1991).

In this analysis, a fast version of SSAP first compares secondary structures to determine whether the proteins are related, in which case, residues having similar structural characteristics (accessibility, torsional angles) are compared (Orengo et al. (1992)). The score is normalised on the size of the smaller protein and is in the range of 0 to 100, where 100 is for identical proteins. When assessing significance, the number of equivalent residues should correspond to at least 60% of the larger protein to ensure that domains are matched rather than motifs.

SSAP scores above 80 indicate highly similar structures. Single linkage cluster analysis was used to link proteins aligning with a score of 80 (**80Str family**), together with the other members of their 35Seq families, where appropriate. Scores between 70 and 80 suggest a more diverse relationship, generally with fewer equivalent pairs and more variation in the loops and orientations of secondary structures. For example, most of the virus jelly roll proteins give an alignment score in this range. Families of analogous structures were therefore generated using a cutoff on the SSAP score of 70 (**70Str family**).

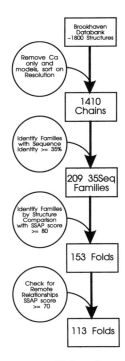

Figure 1. Flowchart showing the steps used to identify homologous and analagous fold families in the protein structure databank.

Structures containing related domains also generally align in the 70 - 80 range but can be identified by comparing protein sizes and checking the number of equivalent pairs. However it is often difficult to establish boundaries and therefore multidomain proteins were placed in a single group, for this analysis. It is hoped that further examination using concensus fold templates and local alignment techniques (Orengo and Taylor (1993), will improve the classification of these structures.

Checking and Classifying the Non-Homologous and Non-analagous Sets

Single linkage cluster analysis of the structure homology matrix gives clusters of similar folds which can be used as a rough guide to fold groups or types. No cutoff on the SSAP score is used, and the resulting trees are used to indicate similarities between more weakly related folds (i.e SSAP scores below 70).

As a final check, structures are examined on a Silicon Graphics workstation and with the aid of schematic representations such as MOLSCRIPT drawings (Kraulis, (1991)). Topology diagrams showing secondary structure connectivity, produced by the program TOPS (Flores *et al.* (1993)) are also used. These can be generated according to different constraints; neighbour contacts, hydrogen bond patterns, crossover connections, which can be adjusted for a given protein class.

Concensus Structure Templates

For a given fold family or fold group, a concensus template can be generated from a multiple structural alignment of a set of non-homologous proteins taken from the family. The method used is based on the algorithm MULTAL (Taylor, 1991) developed for multiple sequences. All pairwise comparisons are performed using SSAP, to give a homology matrix for the set. Pairs of structures are then realigned in an order dictated by the matrix. At each stage, a concensus structure or template is obtained by calculating average structural environments for aligned residue pairs.

Information on the variability of the environments is also stored and used to weight further comparisons. After generating each average structure, the SSAP homology matrix is recalculated to determine the next pair of structures or templates to combine. The final concensus template contains average vectors between all the equivalenced positions and information on the conservation of those positions over the whole family. When searching for a given fold in a multidomain protein, only the most conserved positions in the template need be compared, which reduces the amount of noise generated and increases the probability of a match.

RESULTS

There are **1792** protein chains in the April 1992 release of the databank. Removing model and alpha-carbon only structures, this reduces to **1410** of which 430 are identical (> 98% sequence identity). A sequence homology matrix was generated in 51 CPU hours on a Solbourne 506, 22 MIPS machine. Applying single linkage cluster analysis to this matrix, **209** (35Seq) fold families were identified using a 35% identity cutoff. Table 1, column 1 lists the number of sequences in each family, and column 2 gives the representative with best resolution.

All 209 members of this set were subsequently compared structurally, requiring 3 CPU days on a Solbourne 506, and a structure homology matrix generated. Single linkage clustering identified proteins matching with an SSAP score of 80 and above, and these were combined into 80Str families, together with their associated sequences. The non-homologous set was thereby reduced to **153** folds (table 1, column 3). Most of the proteins matching at this stage have the same or related functions.

Table 1. Lists of family representatives. The first column shows **35Seq** families, whilst the second and third give nonhomologous (**80Str**) and non-analagous (**70Str**) fold families.

35Seq Rep	80Str Rep	70Str Rep	Title	Res Å	Length
ALPHA: GLOBIN					
13 1mbc	1mbc*	1mbc+	Myoglobin (*sperm whale*)	1.5	153
5 1mba			Myoglobin (*Sea Hare*)	1.6	146
4 1ecd			Erythrocruorin (*Chironomous Thummi Thummi*)	1.4	136
52 1thbA			Hemoglobin (*Human*)	1.5	141
14 2lh3			Leghemoglobin (*Yellow Lupin*)	2.0	153
4 2sdhA			Hemoglobin (*Clam*)	2.4	146
2 1ithA			Hemoglobin (*Innkeeper Worm*)	2.5	141
1 2lhb	↓		Hemoglobin (*Sea Lamprey*)	2.0	149
2 1colA	1colA	↓	Colicin A (*Escherichia Coli*)	2.4	197
ALPHA: ORTHOGONAL					
2 1lmbA	1lmbA*	1lmbA*	Lambda Repressor Operator Complex (*Bacteriophage Lambda*)	1.8	87
1 1r69	↓		434 Repressor (*Phage 434*)	2.0	63
4 1utg	1utg	1utg+	Uteroglobin (*Rabbit*)	1.3	70
5 1fiaA	1fiaA		Factor for Inversion Stimulation (*Escherichia Coli*)	2.0	79
3 2wrpR	2wrpR*		DNA Binding Regulatory Protein (*Escherichia Coli*)	1.6	104
4 1hddC	↓	↓	Engrailed Homeodomain Complex (*Fruit Fly*)	2.8	57
2 3sdpA	3sdpA	3sdpA	Iron superoxide Dismutase (*Pseudomonas Ovalis*)	2.1	186
ALPHA: EFHAND					
16 4cpv	4cpv*	4cpv+	Calcium Binding Parvalbumin (*Carp*)	1.5	108
2 2scpA	↓		Sarcoplasmic Calcium Binding Protein (*Sandworm*)	2.0	174
3 4icb	4icb	↓	Bovine Calbindin D9K (*Bovine*)	1.6	76
ALPHA: UP/DOWN					
2 256bA	256bA*	256bA*	Cytochrome b562 (*Escherichia Coli*)	1.4	106
2 2ccyA			Cytochrome C' (*Rhodospirillum Molischianum*)	1.7	127
17 2hmzA			Hemerythrin (*Sipunculid worm*)	1.7	114
1 1le2			Apolipoprotein E2 (*Human*)	3.0	144
1 1ropA	↓	↓	Rop: Col E1 repressor (*Escherichia Coli*)	1.7	56
1 2tmvP	2tmvP	2tmvP	Tobacco Mosaic Virus (*Tobacco Mosaic Virus*)	2.9	154
2 1gmfA	1gmfA	1gmfA	Granulocyte-Macrophage Colony-Stimulating Factor (*Human*)	2.4	119
ALPHA: COMPLEX UP/DOWN					
2 1prcL	1prcL*	1prcL*	Photosynthetic Reaction Centre (*Rhodopseudomonas Viridis*)	2.3	273
2 1prcM	↓	↓	Photosynthetic Reaction Centre (*Rhodopseudomonas Viridis*)	2.3	323
ALPHA: METAL RICH					
13 1ycc	1ycc*	1ycc+	Cytochrome C (reduced) (*Bakers yeast*)	1.2	108
2 451c			Cytochrome C551 (*Pseudomonas Aeruginosa*)	1.6	82
1 1cc5	↓		Cytochrome C5 (*Azobacter Vinelandii*)	2.5	83
1 1c5a	1c5a	↓	Des-Arg-74 Complement c5a (*Pig*)	nmr	65
1 1cy3	1cy3	1cy3	Cytochrome c3 (*Desulfovibro Desulfuricans*)	2.5	118
1 1prcC	1prcC	1prcC	Photosynthetic Reaction Centre (*Rhodopseudomonas Viridis*)	2.3	332

Table 1 (Continued).

35Seq Rep	80Str Rep	70Str Rep	Title	Res Å	Length
BETA: ORTHOGONAL BARREL					
3 1ifc	1ifc	1ifc+	Intestinal fatty Acid Binding Protein (*Rat*)	1.2	131
1 1rbp	1rbp*	|	Retinol Binding Protein (*Human*)	2.0	174
4 1bbpA	↓	↓	Bilin Binding Protein (*Cabbage Butterfly*)	2.0	173
BETA: SUPER BARREL					
1 2por	2por	2por	Porin (*Rhodobacter Capsulatus*)	1.8	301
BETA: GREEK KEY					
68 4ptp	4ptp*	4ptp+	Beta Trypsin (*Bovine*)	1.3	223
1 1sgt	|	|	Trypsin (*Streptomyces Griseus*)	1.7	223
7 2sga	|	|	Proteinase A (*Streptomyces Griseus*)	1.5	181
19 2alp	↓	|	Alpha-Lytic Protease (*Lysobacter Enzymogenes*)	1.7	198
1 1snv	1snv	↓	Sindbis Virus capsid Protein (*Sindbis Virus*)	3.0	151
38 2rhe	2rhe*	2rhe+	Immunoglobulin (*Human*)	1.6	114
1 1cd8	|	|	CD8 (*Human*)	2.6	114
2 2cd4	↓	|	CD4 (*Human*)	2.4	176
29 2fb4H	2fb4H*	|	Immunoglobulin FAB (*Human*)	1.9	229
2 3hlaB	↓	↓	Human Class I Histocompatability Antigen (*Human*)	2.6	99
4 1hoe	1hoe	1hoe+	Amylase Inhibitor (*Streptomyces Tendae*)	2.0	74
1 1acx	1acx	|	Actinoxanthin (*Actinomyces Globisporus*)	2.0	107
6 1cobA	1cobA	↓	Superoxide Dismutase (*Bovine*)	2.0	151
2 1paz	1paz*	1paz+	Pseudoazurin (*Alcaligenes Faecalis*)	1.5	120
7 1pcy	↓	|	Plastocyanin (*Poplar*)	1.6	99
3 2azaA	2azaA	↓	Azurin (*Alcaligene Denitrificans*)	1.8	129
2 2pabA	2pabA	2pabA	Prealbumin (*Human*)	1.8	114
2 1gcr	1gcr	1gcr	Gamma Crystallin (*Calf*)	1.6	174
BETA: JELLY ROLLS					
1 2stv	2stv	2stv+	Satellite Tobacco Necrosis Virus (*Tobacco Necrosis Virus*)	2.5	184
1 1bmv1	1bmv1	|	Bean Pod Mottle Virus (*Bountiful Bean*)	3.0	185
1 1bmv2	1bmv2	|	Bean Pod Mottle Virus (*Bountiful Bean*)	3.0	374
15 2plv1	2plv1	|	Poliovirus (*Human*)	2.9	287
3 1tnfA	1tnfA	|	Tumour Necrosis Factor (*Human*)	2.6	152
1 2mev1	2mev1	|	Mengo Encephalomyocarditis Virus Coat Protein (*Monkey*)	3.0	268
1 2mev2	2mev2*	|	Mengo Encephalomyocarditis Virus Coat Protein (*Monkey*)	3.0	249
15 2plv2	↓	|	Poliovirus (*Human*)	2.9	268
1 2mev3	2mev3*	|	Mengo Encephalomyocarditis Virus Coat Protein (*Monkey*)	3.0	231
15 2plv3	|	|	Poliovirus (*Human*)	2.9	235
3 4sbvA	4sbvA*	|	Southern Bean Mosaic Virus (*Southern Bean Mosaic Virus*)	2.8	199
3 2tbvA	↓	↓	Tomato Bushy Stunt Virus (*Tomato Bushy Stunt Virus*)	2.9	283
2 2cna	2cna	2cna	Concanavalin A (*Jack Bean*)	2.0	237
2 2ltnA	2ltnA	2ltnA	Pea Lectin (*Garden Pea*)	1.7	181

Table 1 (Continued).

35Seq Rep	80Str Rep	70Str Rep	Title	Res Å	Length
BETA: COMPLEX SANDWICH					
23 2er7E	2er7E*	2er7E+	Endothia Aspartic Protease (*Chestnut Blight Fungus*)	1.6	330
6 1psg	|	|	Pepsinogen (*Porcine*)	1.6	365
20 5hvpA	↓	|	HIV-1 Protease (*NY5 Strain of Human Immunodeficiency Virus I*)	2.0	99
6 2rspA	2rspA	↓	Rous Sarcoma Virus Protease (*Rous Sarcoma Virus*)	2.0	115
1 1f3g	1f3g	1f3g	Phosphocarrier III (*Escherichia Coli*)	2.1	150
1 1hcc	1hcc	1hcc	16th Complement Control Protein (*Human*)	nmr	59
BETA: TREFOIL					
2 3fgf	3fgf*	3fgf*	Basic Fibroblast Growth Factor (*Human*)	1.6	124
9 8i1b	↓	↓	Interleukin 1-Beta (*Murine*)	2.4	146
BETA: PROPELLOR					
2 1nsbA	1nsbA	1nsbA	Neuraminidase Sialidase (*Influenza Virus*)	2.2	390
BETA: METAL RICH					
6 5rxn	5rxn	5rxn	Rubredoxin (*Clostridium Pasteurianum*)	1.2	54
2 2hipA	2hipA	2hipA	High Potential Iron Sulfur Protein (*Ectothiorhodospira Halophila*)	2.5	71
BETA: DISULPHIDE RICH					
2 1pi2	1pi2	1pi2	Bowman Birk Proteinase Inhibitor (*Tracy Soybean*)	2.5	61
1 1atx	1atx*	1atx*	Sea Anemone Toxin (*Sea Anemone*)	nmr	46
1 1sh1	|	|	Neurotoxin (*Sea Anemone*)	nmr	48
1 2sh1	↓	↓	Neurotoxin (*Sea Anemone*)	nmr	48
13 3ebx	3ebx	3ebx	Erabutoxin (*Sea Snake*)	1.4	62
4 1epg	1epg	1epg	Epidermal Growth Factor (*Mouse*)	nmr	53
1 4tgf	4tgf	4tgf	Transforming Growth Factor (*Human*)	nmr	50
ALPHA/BETA: TIM BARREL					
16 2timA	2timA	2timA+	Triosephosphate Isomerase (*Trypanosoma Brucei*)	1.8	249
1 1gox	1gox	|	Glycolate Oxidase (*Spinacia Oleracia*)	2.0	350
1 1ald	1ald	|	Aldolase (*Human*)	2.0	363
2 1fcbA	↓	|	Flavocytochrome B2 (*Yeast*)	2.4	494
1 1pii	1pii	|	Anthranilate Isomerase (*Escherichia Coli*)	2.0	452
1 1wsyA	1wsyA	|	Tryptophan Synthase (*Salmonella Typhimurium*)	2.5	248
21 6xia	6xia	|	Xylose Isomerase (*Streptomyces Albus*)	1.6	387
8 5rubA	5rubA	↓	Rubisco (*Rhodospirillum Rubrum*)	1.7	436
1 2taaA	2taaA	2taaA	Taka-Amylase A (*Aspergillus Oryzae*)	3.0	478
5 4enl	4enl	4enl	Enolase (*Bakers Yeast*)	1.9	436

Table 1 (Continued).

35Seq Rep	80Str Rep	70Str Rep	Title	Res Å	Length
ALPHA/BETA: DOUBLY WOUND					
2 5p21	5p21*	5p21+	Ras P21 Protein (*Human*)	1.4	166
1 1etu	↓		Elongation Factor Tu (*Escherichia Coli B*)	2.9	177
2 4fxn	4fxn*		Flavodoxin (*Clostridium MP*)	1.8	138
5 2fx2	↓		Flavodoxin (*Desulfovibrio Vulgaris*)	1.9	147
1 2fcr	2fcr		Flavodoxin (*Chondrus Crispus*)	1.8	173
2 3chy	3chy		Chey Protein (*Escherichia Coli B*)	1.7	128
7 5cpa	5cpa		Carboxypeptidase A (*Bovine*)	1.5	307
3 2trxA	2trxA*		Thioredoxin (*Escherichia Coli*)	1.7	108
2 3trx			Thioredoxin (*Human*)	nmr	105
2 1ego			Glutaredoxin (*Escherichia Coli*)	nmr	85
2 1gp1A	↓		Glutathione Peroxidase (*Bovine*)	2.0	183
20 1cseE	1cseE	↓	Subtilisin Carlsberg (*Bacillus Subtilis*)	1.2	274
9 4dfrA	4dfrA*	4dfrA*	Dihydrofolate Reductase (*Escherichia Coli B*)	1.7	159
13 8dfr			Dihydrofolate Reductase (*Chicken*)	1.7	186
1 3dfr	↓	↓	Dihydrofolate Reductase (*Lactobacillus Casei*)	1.7	162
5 3adk	3adk	3adk+	Adenylate Kinase (*Porcine*)	2.1	194
1 1gky	1gky	↓	Guanylate Kinase (*Bakers yeast*)	2.0	186
1 3pgm	3pgm	3pgm	Phoshphoglycerate Mutase (*Dried Bakers Yeast*)	2.8	230
ALPHA/BETA: TWO DOUBLY WOUND DOMAINS (MULTIDOMAIN)					
1 1rhd	1rhd	1rhd	Rhodanese (*Bovine*)	2.5	293
7 4pfk	4pfk	4pfk	Phosphofructokinase (*Bacillus Stearothermophilus*)	2.4	319
1 3pgk	3pgk	3pgk	Phosphoglycerate Kinase (*Bakers Yeast*)	2.5	415
2 2yhx	2yhx	2yhx	Yeast Hexokinase B (*Bakers Yeast*)	2.1	457
2 2gbp	2gbp*	2gbp+	Glucose Binding Protein (*Escherichia Coli*)	1.9	309
7 8abp	↓		Arabinose Binding Protein (*Escherichia Coli*)	1.5	305
2 2liv	2liv	↓	Leucine/Isoleucine/Valine Binding Protein (*Escherichia Coli*)	2.4	344
ALPHA/BETA: ONE DOUBLY WOUND (MULTIDOMAIN)					
4 3grs	3grs*	3grs*	Glutathione Reductase (*Human*)	1.5	461
1 1trb	↓	↓	Thioredoxin Reductase (*Escherichia Coli*)	2.0	315
3 8catA	8catA	8catA	Catalase (*Bovine*)	2.5	498
9 6ldh	6ldh*	6ldh*	Lactate Dehydrogenase (*Dogfish*)	2.0	329
2 4mdhA	↓	↓	Malate Dehydrogenase (*Porcine*)	2.5	333
1 1ipd	1ipd*	1ipd*	Isopropylmalate Dehydrogenase (*Thermus Thermophilus*)	2.2	345
7 4icd	↓	↓	Phosphorylated Isocitrate Dehydrogenase (*Escherichia Coli*)	2.5	414
1 1pgd	1pgd	1pgd	6-Phosphogluconate Dehydrogenase (*Sheep*)	2.5	469
5 8adh	8adh	8adh	Alcohol Dehydrogenase (*Horse*)	2.4	374
16 1gd1O	1gd1O	1gd1O	Glyceraldehyde Phosphate Dehydrogenase (*Bacillus Stearothermophilus*)	1.8	334
9 7aatA	7aatA	7aatA	Aspartate Aminotransferase (*Chicken*)	1.9	401
4 2ts1	2ts1	2ts1	Tyrosyl-Transfer RNA Synthetase (*Bacillus Stearothermophilus*)	2.3	317
2 1phh	1phh	1phh	P-Hydroxybenzoate Hydroxylase (*Pseudomonas Fluorenscens*)	2.3	394

Table 1 (Continued).

35Seq Rep	80Str Rep	70Str Rep	Title	Res Å	Length
ALPHA + BETA: ALPHA-BETA SANDWICHES					
1 1rnh	1rnh*	1rnh*	Selenomethionyl Ribonuclease (*Escherichia Coli*)	2.0	145
2 1hrhA	↓	↓	Ribonuclease H Domain of HIV Reverse Transcriptase (*Human Immunodeficiency Virus Type I*)	2.4	125
4 1il8A	1il8A*	1il8A*	Interleukin 8 (*Human*)	nmr	71
4 2sicI	2sicI	2sicI	Subtilisin Inhibitor (*Bacillus Amyloliquefaciens*)	1.8	107
8 1cseI	1cseI*	1cseI*	Subtilisin Inhibitor (*Bacillus Subtilis*)	1.2	63
2 2ci2I	↓	↓	Chymotrypsin Inhibitor (*Barley*)	2.0	65
1 1sn3	1sn3	1sn3	Scorpion Neurotoxin (*Scorpion*)	1.8	65
5 2ovo	2ovo*	2ovo+	Ovomucoid Third Domain (*Silver Pheasant*)	1.5	56
3 1tgsI	↓		Trypsinogen Inhibitor (*Bovine*)	1.8	56
1 1crn	1crn*	1crn+	Crambin (*Abyssinian Cabbage*)	1.5	46
1 1ctf	↓		L7/L12 50 S Ribosomal Protein (*Escherichia Coli*)	1.7	68
1 1aps	1aps		Acetylphosphatase (*Horse*)	nmr	98
2 2hpr	2hpr		Histidine-Containing Phosphocarrier Protein (*Bacillus Subtilis*)	2.0	87
1 1pba	1pba		Procarboxypeptidase (Activation Domain) (*Porcine Pancreas*)	nmr	81
1 1fxd	1fxd*		Ferredoxin II (*Desulfovibrio Gigas*)	1.7	58
2 2fxb	|		Ferredoxin (*Bacillus Thermoproteolyticus*)	2.3	81
1 1fdx	|	↓	Ferredoxin (*Peptococcus Aerogenes*)	2.0	54
6 3rubS	3rubS	3rubS	Ribulose 1,5-Bisphosphate Carboxylase/Oxygenase (*Nicotiana Tabacum*)	2.0	123
19 8atcB	8atcB	8atcB	Aspartate Carbamoyl Transferase (*Escherichia Coli*)	2.5	146
ALPHA + BETA: ALPHA-BETA ROLLS					
27 7rsa	7rsa	7rsa	Ribonuclease A (*Bovine*)	1.3	124
10 9rnt	9rnt	9rnt+	Ribonuclease T1 (*Aspergillus Oryzae*)	1.5	104
4 2sarA	2sarA*	|	Ribonuclease SA (*Streptomyces Aurofaciens*)	1.8	96
1 1rnbA	↓	↓	Barnase (*Bacillus Amyloliquefaciens*)	1.9	109
2 1msbA	1msbA	1msbA	Mannose Binding Protein (*Rat*)	2.3	115
1 1fkf	1fkf	1fkf	FK506 Binding Protein (*Human*)	1.7	107
5 1bovA	1bovA	1bovA	Verotoxin (*Escherichia Coli*)	2.2	69
4 1snc	↓	↓	Staphylococcal Nuclease (*Staphylococcus Aureus*)	1.6	135
1 1ubq	1ubq*	1ubq*	Ubiquitin (*Human*)	1.8	76
3 1fxiA	|	|	Ferredoxin I (*Blue Green Algae*)	2.2	96
2 2gb1	↓	↓	Protein G (*Streptomyces Griseus*)	nmr	56
ALPHA + BETA: SMALL MISCELLANEOUS					
19 5pti	5pti	5pti	Trypsin Inhibitor (*Bovine*)	1.0	58
3 1tpkA	1tpkA	1tpkA	Kringle 2 (*Human*)	2.4	88
ALPHA + BETA: LARGE/COMPLEX					
8 9wgaA	9wgaA	9wgaA	Wheat germ Agglutinin (*Wheat*)	1.8	171
6 2tscA	2tscA	2tscA	Thymidilate Synthase (*Escherichia Coli*)	2.0	264
4 4cla	4cla	4cla	Chloramphenicol Acetyltransferase (*Escherichia Coli*)	2.0	213
4 1rveA	1rveA	1rveA	Eco Rv Endonuclease (*Escherichia Coli*)	2.5	244
1 1pyp	1pyp	1pyp	Inorganic Pyrophosphatase (*Bakers Yeast*)	3.0	280
ALPHA + BETA: METAL RICH					
1 2cdv	2cdv	2cdv	Cytochrome c3 (*Desulfovibrio Vulgaris*)	1.8	107
1 3b5c	3b5c	3b5c	Cytochrome b5 (*Bovine*)	1.5	85

Table 1 (Continued).

35Seq Rep	80Str Rep	70Str Rep	Title	Res Å	Length
MULTI DOMAIN					
10 9pap	9pap	9pap	Papain (*Papaya*)	1.6	212
1 3blm	3blm	3blm	Beta Lactamase (*Staphylococcus Aureus*)	2.0	257
2 3hlaA	3hlaA	3hlaA	Human Class I Histocompatibility Antigen (*Human*)	2.6	229
11 2cpp	2cpp	2cpp	Cytochrome P450Cam (*Pseudomonas Putida*)	1.6	405
19 8atcA	8atcA	8atcA	Aspartate Carbamoyl Transferase (*Escherichia Coli*)	2.5	310
12 1csc	1csc	1csc	Citrate Synthase (*Chicken Heart Muscle*)	1.7	429
1 1ace	1ace	1ace	Acetyl Cholinesterase (*Electric Ray*)	2.8	526
1 1cox	1cox	1cox	Cholesterol Oxidase (*Brevibacterium Sterolicum*)	1.8	502
1 1cpkE	1cpkE	1cpkE	C-Amp-Dependent Protein Kinase (*Recombinant Mouse*)	2.7	336
6 1fbpA	1fbpA	1fbpA	Fructose 1,6-Bisphosphatase (*Pig*)	2.5	316
2 1fnr	1fnr	1fnr	Ferredoxin (*Spinach*)	2.2	296
2 1gstA	1gstA	1gstA	Isoenzyme 3-3 of Glutathione S-Transferase (*Rat*)	2.2	217
1 1gly	1gly	1gly	Guanylate Kinase (*Aspergillus Awamori*)	2.2	470
1 1lap	1lap	1lap	Leucine Aminopeptidase (*Bovine*)	2.7	481
4 1ovaA	1ovaA*	1ovaA*	Ovalbumin (*Hen*)	1.9	385
3 7apiA	↓	↓	Modified Alpha-1 Antitrypsin (*Human*)	3.0	339
2 1vsgA	1vsgA	1vsgA	Variant Surface Glycoprotein (*Trypanosoma Brucei*)	2.9	362
3 1lfi	1lfi	1lfi	Lactoferrin (*Human*)	2.1	688
1 1wsyB	1wsyB	1wsyB	Tryptophan Synthase (*Salmonella*)	2.5	385
5 2cyp	2cyp	2cyp	Cytochrome C Peroxidase (*Bakers Yeast*)	1.7	293
12 2glsA	2glsA	2glsA	Glutamine Synthetase (*Salmonella Typhimurium*)	3.5	468
2 2pmgA	2pmgA	2pmgA	Phosphoglucomutase (*Rabbit*)	2.7	561
1 3bcl	3bcl	3bcl	Bacteriochlorophyll A (*Prosthecochloris Aestuarii*)	1.9	344
4 8acn	8acn	8acn	Aconitase (*Bovine*)	2.0	753
1 2reb	2reb	2reb	RecA protein (*Escherichia Coli*)	2.3	303
11 6tmnE	6tmnE	6tmnE	Thermolysin (*Bacillus Thermoproteolyticus*)	1.6	316
33 1hgeA	1hgeA	1hgeA	Hemagglutinin (*Influenza Virus*)	2.6	328
33 1hgeB	1hgeB	1hgeB	Hemagglutinin (*Influenza Virus*)	2.6	175
6 2ca2	2ca2	2ca2	Carbonic Anhydrase (*Human*)	1.9	256
6 3gapA	3gapA	3gapA	Gene regulatory Protein (*Escherichia Coli*)	2.5	208
2 1prcH	1prcH	1prcH	Photosynthetic Reaction Centre (*Rhodopseudomonas Viridis*)	2.3	258
92 3lzm	3lzm	3lzm	Lysozyme (*Bacteriophage T4*)	1.7	164
26 1lz1	1lz1	↓	Lysozyme (*Human*)	1.5	130
12 4bp2	4bp2	4bp2	Prophospholipase A2 (*Bovine*)	1.6	115

Families of analagous structures were generated in a similar way, but using a cutoff of 70 on the SSAP score. For each, the representative with best resolution is shown in table 1, column 4, flagged by a +. This gave a non-analagous set of **113** folds. At this degree of similarity, matching proteins can have different functions despite having related folds. For example, the TIM barrels match at this level but are known to exhibit diverse catalytic functions. It can be seen from figure 2 that although the number of structures determined has grown exponentially in the last 20 years, the number of new folds has increased at a much slower rate.

TOPS representations for all the single domain proteins in the non-analagous set are shown in figure 3. It can be seen from these, that there are still topological similarities between many of the folds, even though they give SSAP scores below 70 when compared. For example many of the alternating α/β proteins shown in figure 3 contain common motifs which constitute a large proportion of the fold. Single linkage cluster analysis of the structure SSAP homology matrix gave three conspicuous clusters containing mainly-α, mainly-β and alternating α/β folds respectively. The non-alternating α+β structures were distributed amonst these clusters.

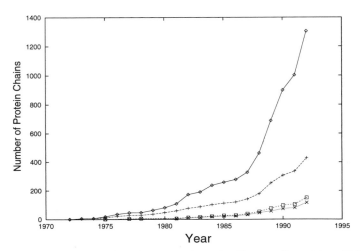

Figure 2. Annual increases in the number of structures determined by crystallographic and NMR techniques. The top line (◇) gives the total number of structures. The line below (+) shows the increase in numbers of non-identical structures (<98% sequence identity). the lower two lines are the numbers of non-homologous(□)(<35% sequence identity and >80 SSAP score) and non-analagous (x) (<35%identity and >70 SSAP score).

Alpha

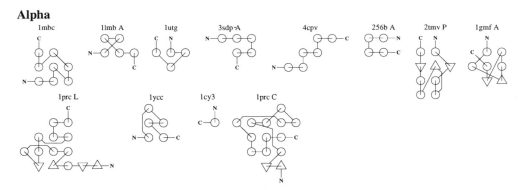

Figure 3. Schematic TOPS representation for the mainly-α protein folds.

Beta

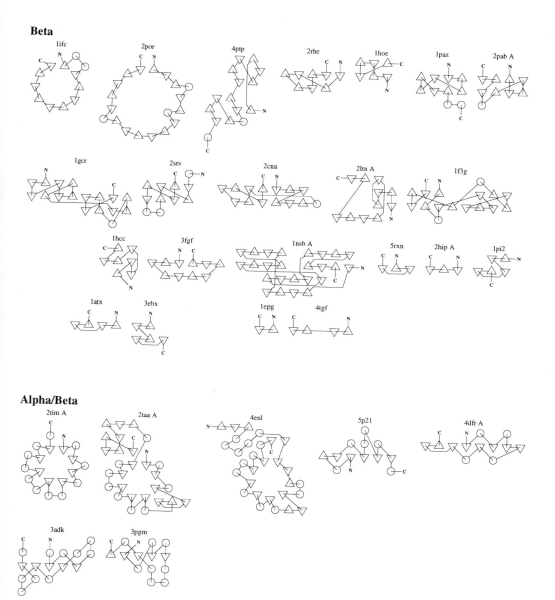

Alpha/Beta

Figure 3. (Continued)

Alpha + Beta

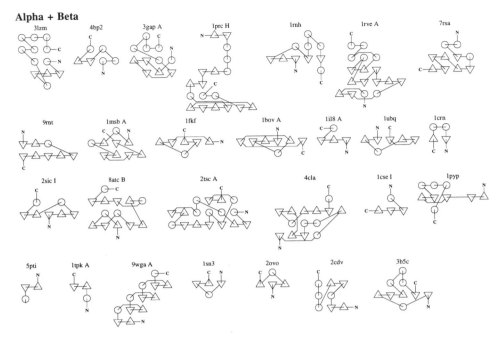

Figure 3. Schematic TOPS diagrams for single domain proteins in the non-analagous data set. α-Helices are represented by circles and β-strands by triangles. Lines drawn over a symbol are connections to the top of the secondary structure; otherwise connection is to the base. Multidomain structures are not shown.

Trees generated for each class are shown in figures 4-7, from which groups of folds were suggested. Figure 4 shows the globin fold group and up/down folds for the all-α proteins. The jelly roll folds and greek key fold groups are shown for the all-β class in figure 5. Whilst figure 6 shows the TIM barrel folds and different doubly wound α/β fold groups for the alternating α/β proteins. The α-β sandwich and α-β roll groups found in the α+β class are given in figure 7. Fold groups suggested by these trees were subsequently checked and refined by reference to TOPS and Molscript drawings and using information from the literature.

In table 1, the set of non-analagous structures has been classified according to the four protein groups as defined by Levitt and Chothia (1976), mainly-α, mainly-β, alternating α/β, and non-alternating α+β. Fold groups were assigned using the structure relatedness trees. Known groups were given common names from the literature and for new groups suitable names chosen. Large multidomain proteins which only weakly match single domain proteins are placed in a single group.

Fold Groups

Table 2 list the different fold groups identified within each protein class and the number of analagous folds for each. In the the mainly-α folds, there are 6 groups, of which the globin family (1mbc) are the most closely related in the dataset. Although identities fall as low as 10%, the SSAP scores are all above 80. Conformation is also well conserved in the four helical bundle or up/down family (256bA), despite differences in function and low sequence identity. The β-class is more diverse. There are 9 fold groups amongst which there are 23 non-analogous folds (Table 2). The Greek key group currently has the largest number of structures (195) including the trypsins, the immunoglobulins and the plastocyanins.

Table 2. Fold groups identified in the set of single domain non-analagous folds.

Class	Fold Groups	No. of Non-Analogous Folds	No. of Chains
ALPHA	Globin	1	97
	Orthogonal	3	23
	EF Hand	1	21
	Up/Down	3	26
	Complex Up/Down	2	4
	Metal Rich	3	19
	TOTAL	**13**	**190**
BETA	Orthogonal Barrel	1	8
	Superbarrel	1	1
	Greek Key	6	195
	Jelly Roll	3	66
	Complex Sandwich	3	57
	Trefoil	1	11
	Propellor	1	2
	Metal Rich	2	8
	Disulphide Rich	5	22
	TOTAL	**23**	**370**
ALPHA/BETA	TIM Barrels	3	56
	Doubly Wound	4	69
	(in multidomain proteins)	15	86
	TOTAL	**22**	**211**
APHA+BETA	Alpha-Beta Sandwich	9	64
	Alpha-Beta Roll	6	60
	Small Miscellaneous	2	22
	Large complex	5	23
	Metal Rich	2	2
	TOTAL	**24**	**171**

There are only two fold groups in the α/β class, the singly wound TIM barrels and the doubly wound folds. In the TIM barrels, the topology is fairly consistent though large insertions are possible. By contrast, the doubly wound folds are more diverse (see figure 3). Structure is only weakly conserved within the TIM barrel family due mainly to considerable variation in the dimensions and arrangements of the helices. Within the doubly wound folds, it is better conserved even though the average sequence identity is low. Many structures in this group bind a nucleotide in the cleft formed by the chain reversal.

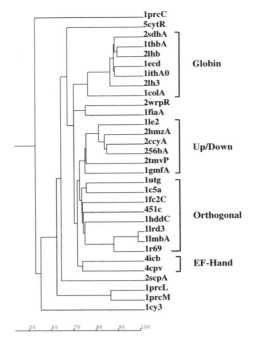

Figure 4. Structure relatedness trees for the mainly-α class of proteins, drawn from the SSAP homology matrix.

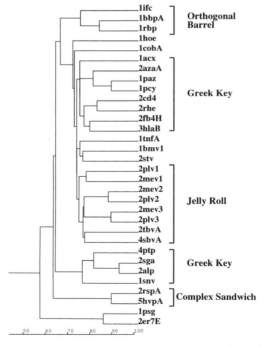

Figure 5. Structure relatedness trees for the mainly-β class of proteins, drawn from the SSAP homology matrix.

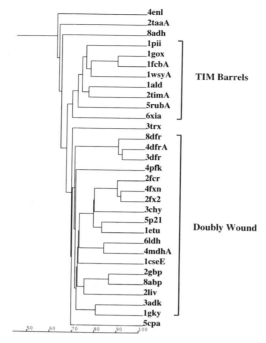

Figure 6. Structure relatedness trees for the alternating α/β class of proteins, drawn from the SSAP homology matrix.

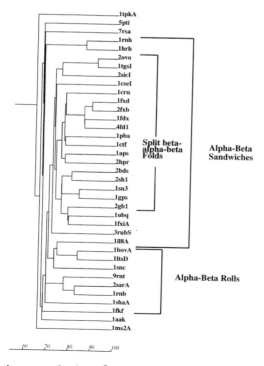

Figure 7. Structure relatedness trees for the α+β class of proteins, drawn from the SSAP homology matrix.

Alpha plus Beta Folds

The α+β class contains a very diverse set of structures having 5 fold groups amongst which there are 24 non-analogous folds (Table 2). Folds in this group have been reviewed in more detail elsewhere (Orengo and Thornton (1993)). Two major groups can be identified from the TOPS diagrams (figure 3) and the structure relatedness trees (figure 7). The first group are α-β sandwiches which consist of a layer of α-helices packed against an antiparallel β-sheet and the second are α-β rolls. These contain one or more α-helices cradled in a β-sheet which rolls over to form a partial barrel. The type of fold adopted is partly determined by the ratio of α-helices to β-strands along the sequence.

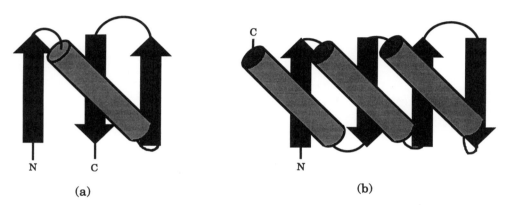

(a)   (b)

Figure 8. Schematic representation of a single (a) and double (b) split βαβ motif. α-helices are drawn as grey cylinders and β-strands as black arrows. The α-helix can also adopt the opposite diagonal orientation, in which case long loops are observed connecting the helix to the edge strands.

Split βαβ Motif Within the α-β sandwiches, there are extensive similarities between a large subgroup of proteins all of which contain a split βαβ as a recurring motif. This motif involves two parallel strands connected by an α-helix (see figure 8). The connection is right handed in all cases observed and the two strands are separated by a third antiparallel strand. The loops connecting the first and second strands to the helix tend to be long and two alternative diagonal orientations of the helix with respect to the sheet are observed with equal frequency. These split βαβ sandwiches account for nearly 30% of the α+β class.

Figure 9 and 10 show TOPS and MOLSCRIPT drawings of the 35Seq representatives of the split βαβ based folds. Most match three or more members of the group with significant SSAP scores above 70. However, the extensive structural relatedness is not accompanied by much functional similarity. Instead a wide variety of functions and active sites is observed (see figure 11). This suggests that the fold can adapt a number of surface regions for different biological functions. Furthermore, the frequency with which the split βαβ folds occur may suggest that it is an energetically favoured fold.

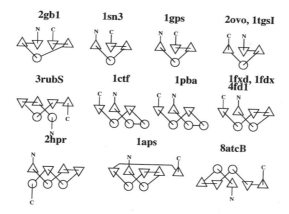

Figure 9. Schematic topology (TOPS) diagrams for the split βαβ sandwiches. Helices are represented by circles and strands by triangles. Lines drawn over a symbol are connections to the top of the secondary structure; otherwise connection is to the base.

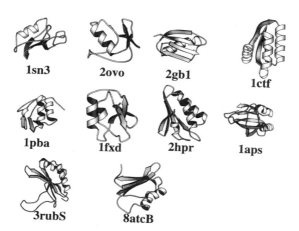

Figure 10. MOLSCRIPT diagrams for the split βαβ sandwich folds. Beta strands are represented by arrows and helices by ribbons.

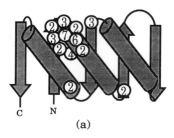 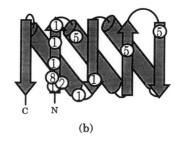

(a)   (b)

Figure 11. Location of functional sites in the split βαβ sandwich folds. (a) front view (b) back view. 1 - 1sn3, 2 - 2gbl, 3 - 2ovo, 4 - 1fxd, 5 - 3rubS, 6 - 2hpr, 7 - laps, 8 - 8atcB.

Concensus Templates for Doubly Wound Folds

A template was generated for a set of doubly wound alternating α/β folds. This was used to test the ability of concensus structures to identify folds within large multidomain proteins. Single representatives were selected from ten doubly wound 35Seq families. Each selected structure aligned against 5 or more others with SSAP scores above 70 (see figure 12). No pairs of structures had more than 40% sequence similarity to prevent bias towards a particular family. A multiple structure alignment was performed, requiring 10CPU minutes on a Solbourne 506, 22 MIPS machine. Figure 13 shows that the core β-strands (a,b,d,e) and α-helices(1,2) of this fold are well conserved accross the group, whilst edge secondary structures are more variable and align poorly.

```
Conc    85   84   82   72   83   84   84   83   82   83
61dh         78   77   68   66   76   73   77   72   71
8adh              78   73   65   72   73   77   69   68
1gd1                   74   66   67   68   76   70   69
1grcA                       70   73   74   77   69   70
2fcr                             87   89   78   71   70
4fxn                                  89   77   74   74
2fx2                                       79   77   77
3chy                                            79   80
5p21                                                 81
1etu
```

Figure 12. SSAP structure homology matrix derived from pairwise alignments of ten doubly wound α/β folds. The top entry shows the SSAP scores obtained by aligning the concensus structure against each member of the set.

A concensus structure was generated from the most highly conserved positions and this was used to search for a similar fold in the multidomain protein, malate dehydrogenase (4MDH). The template included the central four β-strands and two α-helices. A match was obtained with the N-terminal domain, giving a SSAP score of 81 over 70 equivalent residue pairs. Figure 14 shows the superposition of the concensus structure on the equivalent doubly wound domain in 4MDH. The SSAP scores given by matching the template against all the structures in the set are better overall than for matching any individual against the rest of the set (see figure 12). Therefore, it may be hoped, that generation of templates will improve recognition of individual folds within many of the multidomain proteins and allow these structures to be properly classified.

272

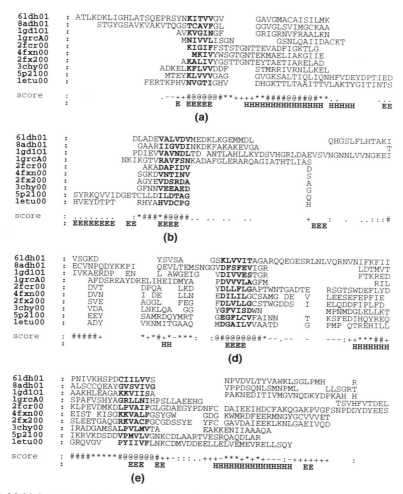

```
61dh01   : ATLKDKLIGHLATSQEPRSYNKITVVGV      GAVGMACAISILMK
8adh01   : STGYGSAVKVAKVTQGSTCAVFGL         GGVGLSVIMGCKAA
1gd101   :               AVKVGINGF          GRIGRNVFRAALKN
1grcA0   :              MNIVVLISGN          GSNLQAIIDACKT
2fcr00   :             KIGIFFSTSTGNTTEVADFIGKTLG
4fxn00   :             MKIVYWSGTGNTEKMAELIAKGIIE
2fx200   :            AKALIVYGSTTGNTEYTAETIARELAD
3chy00   : ADKELKFLVVDDF             STMRRIVRNLLKEL
5p2100   : MTEYKLVVVGAG              GVGKSALTIQLIQNHFVDEYDPTIED
1etu00   : FERTKPHVNVGTIGHV          DHGKTTLTAAITTVLAKTYGITINTS

score    : .--+#@@@@@#**++++**####@@##@#**..          ...
         :       E EEEEE       HHHHHHHHHHHHHHH HHHHH          EE
```
(a)

```
61dh01   :              DLADEVALVDVMEDKLKGEMMDL         QHGSLFLHTAKI
8adh01   :              GAARIIGVDINKDKFAKAKEVGA              T
1gd101   :              PDIEVVAVNDLTD ANTLAHLLKYDSVHGRLDAEVSVNGNNLVVNGKEI
1grcA0   :              NKIKGTVRAVFSNKADAFGLERARQAGIATHTLIAS
2fcr00   :              AKADAPIDV                            D
4fxn00   :              SGKDVNTINV                           S
2fx200   :              AGYEVDSRDA                           A
3chy00   :              GFNNVEEAED                           G
5p2100   : SYRKQVVIDGETCLLDILDTAG                            Q
1etu00   : HVEYDTPT     RHYAHVDCPG                            H

score    : ........     :*####*#@@##.. ... ..        +   :   .  ..::::#
         : EEEEEEEE  EE    EEEE                           EEE
```
(b)

```
61dh01   : VSGKD           YSVSA    GSKLVVITAGARQQEGESRLNLVQRNVNIFKFII
8adh01   : ECVNPQDYKKPI     QEVLTEMSNGGVDFSFEVIGR             LDTMVT
1gd101   : IVKAERDP   EN    L AWGEIG  VDIVVESTGR              FTKRED
1grcA0   : AFDSREAYDRELIHEIDMYA      PDVVVLAGFM               RIL
2fcr00   : DVT       DPQA    LKD    YDLLFLGAPTWNTGADTE     RSGTSWDEFLYD
4fxn00   : DVN       I DE    LLN    EDILILGCSAMG DE     V  LEESEFEPFIE
2fx200   : SVE       AGGL    FEG    FDLVLLGCSTWGDDS      I  ELQDDFIPLFD
3chy00   : VDA       LNKLQA  GG     YGFVISDWN              MPNMDGLELLKT
5p2100   : EEY       SAMRDQYMRT     GEGFLCVFAINN          T KSFEDIHQYREQ
1etu00   : ADY       VKNMITGAAQ     MDGAILVVAATD          G PMP QTREHILL

score    : #####+          *+*#+*-***:  :@#@@@@@@#*--.--    -  ---:++***##+
         :                    HH          EEEE                   HHHHHHH
```
(d)

```
61dh01   : PNIVKHSPDCIILVVS          NPVDVLTYVAWKLSGLPMH       R
8adh01   : ALSCCQEAYGVSVIVG          VPPDSQNLSMNPML        LLSGRT
1gd101   : AAKHLEAGAKKVIISA          PAKNEDITIVMGVNQDKYDPKAH   H
1grcA0   : SPAFVSHYAGRLLNIHPSLLAEEHG
2fcr00   : KLPEVDMKDLPVAIFGLGDAEGYPDNFC DAIEEIHDCFAKQGAKPVGFSNPDDYDYEES
4fxn00   : EIST KISGKKVALFGSYGW      GDG KWMRDFEERMNGYGCVVVET
2fx200   : SLEETGAQGRKVACFGCGDSSYE   YFC GAVDAIEEKLKNLGAEIVQD
3chy00   : IRADGAMSALPVLMVTA         EAKKENIIAAAQA
5p2100   : IKRVKDSDDVPMVLVGNKCDLAARTVESRQAQDLAR
1etu00   : GRQVGV   PYIIVFLNKCDMVDDEELLELVEMEVRELLSQY

score    : ####****#@@@@@@#++-::::...+++-**+*+*+---:-++++++       :
         :           EEE  EE    HHHHHHHHHHHHH  EE
```
(e)

Figure 13. Multiple structural alignment of ten non-homologous doubly-wound α/β proteins. Conserved β-strand regions are shown bold. Secondary structure assignments by the program DSSP (Kabsch and Sander (1983)) are shown in the bottom line, E for β-strand and H for α-helix. Symbols in the line above reflect the degree of conservation of the residue structural environments and are in the order (.:-+*£@).

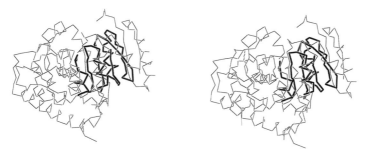

Figure 14. Stereo view of the superposition of a concensus structure for a set of doubly-wound α/β folds (shown as bold) on malate dehydrogenase.

DISCUSSION

Sequence and structure comparisons have been used to find related folds in the protein structure databank. A set of 1410, well determined protein chains, was thereby reduced to a set of 113 analagous fold families. An essential feature of the method is the structural comparison. As the average sequence identity between proteins matched in this step, is less than 18%, these relationships could not have been detected by sequence based methods. Sequence and structure homology matrices have been generated, which can be easily updated with each release of the databank. New structures would first be compared by sequence against all the 35Seq family representatives. Any not matching at 35% identity will be compared structurally against the set of non-homologous folds (80Str representatives) to check for structural similarity. Apart from reducing the degeneracy of the dataset by an order of magnitude, the classification of structures into fold families should enable a better understanding of evolutionary mechanisms within different classes and fold groups.

There is most diversity within the mainly-β and $\alpha+\beta$ classes. Both contain a large number of folds, many of which are small, irregular, with few secondary structures, and often stabilised by disulphide bridges. The α/β class has the smallest number of fold motifs (2) and folds (7). This may be a consequence of the geometric constraints and stability of the righthanded adjacent $\beta\alpha\beta$ motif. As the structures diverge, the basic fold remains the same but may extend by adding strands to the sheet or extra secondary structures in the loops. Within the $\alpha+\beta$ class of structures, two major fold groups could be identified, the α-β sandwiches and the α-β rolls. One of these the α-β sandwiches, showed extensive structural similarities due to a recurring non-adjacent split $\beta\alpha\beta$ motif.

Our analysis highlighted global similarities between folds and ensured that at least 60% of the larger fold was matched. This prevented folds containing common motifs being united in an analagous family. However, relationships involving motifs were found within all classes and suggested a continuum of folds available to proteins. For example within the β-class several Greek key folds aligned against jelly roll folds and other smaller β-proteins were occurring as motifs within Greek key folds. Some doubly wound α/β proteins aligned against parts of TIM barrel proteins. In other words, as the size increases, the repertoire of folds available appears in many cases to consist of extensions to existing motifs.

The large multidomain proteins are the hardest to classify. Many consist of one or more folds from different protein classes. Using a consensus template derived from a set of non-homologous doubly wound folds we were able to correctly identify an equivalent region in a large multidomain protein. It may be hoped, therefore, that generation of templates for different fold families and groups may help in identifying folds in multidomain proteins and improve the classification of these structures. Furthermore, concensus templates should also aid fold recognition by optimal sequence threading (Jones *et al.*) and so improve prediction of structure from sequence.

ACKNOWLEDGEMENTS

We are indebted to David Jones for useful suggestions and discussions. Christine Orengo acknowledges financial support under the EC bridge initiative and Tom Flores from the SERC.

REFERENCES

F.C. Bernstein, T.F. Koetzle, G. J. D. Williams, E.F. Meyer, M.D. Brice, J. R. Rodgers, O. Kennard, T. Shimanochi, and M. Tasumi, The protein databank: a computer-based archival file for macromolecular structures, *J. Mol. Biol.* 112:535 (1977).

J. Boberg, T. Salakoski and M. Vihinen, Selection of a representative set of structures from the Brookhaven protein databank, *PROTEINS: Structure, Function and Genetics*, 14: 265 (1992).

C. Chothia and A. M. Lesk, The relationship between the divergence of sequence and structure in proteins, *The EMBO Journal*, 5:823 (1986).

C. Chothia and A. M Lesk, The evolution of protein structures, *Coldspring Harbour Symposia in Quantitative Biology*, LII: 399 (1987).

T. P. Flores, C. A. Orengo, D. S. Moss and J. M. Thornton, *Protein Science*, Conservation of conformational characteristics in structurally similar protein pairs (submitted) (1993).

T. P. Flores, D. S. Moss and J. M. Thornton, An algorithm for automatically generating protein topology cartoons, *Protein Engineering*, (submitted) (1993).

U. Hobohm, M. Scharf, R. Schneider and C. Sander, Selection of representative datasets, *Protein Science*, 1:409 (1992).

L. Holm and C. Sander, A databank of protein structure families with common folding motifs, *J. Mol. Biol.* 225:93 (1992).

T. J. P. Hubbard and T. L. Blundell, Comparison of solvent-inaccessible cores of homologous proteins. Definitions useful for protein modeling, *Protein Engineering.*, 1:159 (1987).

D. T. Jones, W. R. Taylor and J. M. Thornton, A new approach to protein fold recognition, *Nature*, 358:86 (1992).

W. Kabsch and C. Sander, Dictionary of protein secondary structure - pattern recognition of hydrogen-bonded and geometrical features, *Biopolymers*, 22:2577 (1983).

P. J. Kraulis, MOLSCRIPT - A program to produce both detailed and schematic plots of protein strutcures, *J. Appl. Cryst.*, 24:946 (1991).

W. J. Krzanowski, *'Principles of Multivariate Analysis', Oxford Statistical Science*, Series **3**, Oxford University Press (1990).

M. Levitt and C. Chothia, Structural patterns in globular proteins, *Nature*, 261:552 (1976).

A. G. Murzin and C. Chothia, *Curr. Opin. in Struct. Biol.*, Protein architecture: new superfamilies, 2:895 (1992).

S. B. Needleman and C. D. Wunsch, A general method applicable to the search for similarities in the amino acid sequences of two proteins, *J. Mol. Biol.*, 48:443 (1970).

C. A. Orengo and W. R. Taylor, A rapid method for protein structure alignment *J. Theor. Biol.* 147:517 (1990).

C. A. Orengo, N. Brown and W. R. Taylor, Fast structure alignment for protein databank searching, *PROTEINS: Structure, Function and Genetics*, 14:139 (1992).

C. A. Orengo and W. R. Taylor, A local alignment method for protein structure motifs, *J. Mol. Biol.* (In Press) (1993).

C. A. Orengo, T. P Flores, W. R. Taylor and J. M. Thornton, Identification and Classification of Protein Fold Families, Protein Engineering, 6:485 (1993).

C. A. Orengo and J. M. Thornton, Structure (submitted) (1993).

S. Pascarella and P. Argos, A data-bank merging related protein structures and sequences, *Protein Engineering*, 2:121 (1992).

J. S. Richardson, The anatomy and taxonomy of protein structure, *Advances In Protein Science*, 34:167 (1981).

F. Rippmann and W. R. Taylor, Visualisation of structural similarity in proteins, *J. Mol. Graph.* 9:169 (1991).

C. Sander and R. Schneider, Database of homology derived protein strutcures and the structural meaning of sequence alignment, *PROTEINS: Structure, Function and Genetics*, 9:56 (1991).

W. R. Taylor, Multiple sequence alignment by a pairwise algorithm, Comput. Appl. Biosci., 3:81 (1987).

W. R. Taylor and C. A. Orengo, Protein structure alignment, *J. Mol. Biol.*, 208:1 (1989a).

W. R. Taylor and C. A. Orengo, A holistic approach to protein structure alignment, *Protein Engineering*, 2:505 (1989b).

W. R. Taylor, T. P. Flores and C. A. Orengo, Multiple Protein structure Alignment, *CABIOS* (submitted) 1993.

DATA BASED MODELING OF PROTEINS

L. Holm, B. Rost, C. Sander, R. Schneider, and G. Vriend

Protein Design Group, EMBL
Meyerhofstrasse 1
69117 Heidelberg, Germany

INTRODUCTION

Knowledge of the three-dimensional structure is a prerequisite for the rational design of site-directed mutations in a protein and can be of great importance for the design of drugs. X-ray crystallography and NMR spectroscopy are the only ways to obtain such detailed structural information. Unfortunately, these techniques involve elaborate technical procedures and many proteins fail to crystallize at all [Giege 89] and/or cannot be obtained or dissolved in large enough quantities for NMR measurements. The size of the protein is also a limiting factor for NMR. In the absence of experimental data, model-building on the basis of the known three dimensional structure of a homologous protein is at present the only reliable method to obtain structural information [Swindells et al 91]. Comparisons of the tertiary structures of homologous proteins have shown that three-dimensional structures have been better conserved in evolution than protein primary structures [Bajai et al 87; Bashford et al 87; Vriend & Sander 91; Holm et al 92], indicating the feasibility of model-building by homology.

In the absence of a structure with sufficient homology to the protein of interest, secondary structure prediction can provide important structural information.

All structure prediction techniques depend in one way or another on experimental data. This is most easily seen for model building by homology, but also secondary structure prediction programs are trained on proteins with a known three dimensional structure. A description of structure prediction techniques would be incomplete without recognition of the fact that this is only possible thanks to the many crystallographers who deposit data in publicly accessible databases.

Here, several aspects of 3D model building by homology will be described. It will become clear that sequence alignments are often the limiting step for modeling, and an evaluation of the required homology between the sequence and the structure that will serve as the modeling template is very important.

Differences between three-dimensional structures increase with decreasing sequence identity and accordingly the accuracy of models built by homology decreases [Chothia & Lesk 86; Sander & Schneider 91]. The errors in a model built on the basis of a structure with 90% sequence identity may be as low as the errors in crystallographically

Statistical Mechanics, Protein Structure, and Protein Substrate Interactions
Edited by S. Doniach, Plenum Press, New York, 1994

determined structures, except for a few individual side chains [Chothia & Lesk 86; Blundell et al 87]. If as a test case a known structure is built from another known structure, then in case of 50% sequence identity the RMS error in the modeled coordinates can be as large as 1.5 Å, with considerably larger local errors. If the sequence identity is only around 25% the alignment is the main bottleneck for model building by homology, and large errors are often observed. With less than 25% sequence identity the homology often remains undetected. Figure 1 shows the key limiting factors in modeling as a function of sequence identity.

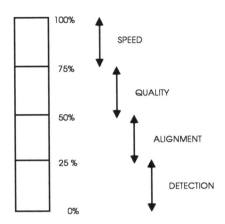

Figure 1. The main limiting steps for model building by homology as function of the percentage sequence identity between the structure and the model.

The deviations between a model built by homology and the 'real' X-ray structure vary throughout the molecule, the largest deviations occurring in loops at the protein surface. In secondary structure elements and in the hydrophobic core models normally are more accurate.

At present most model building by homology protocols start from the assumption that except for the insertions and deletions (indels) the backbone of the model is identical to the backbone of the structure. In practice, however, domain motions and 'bending' of parts of molecules with respect to each other is often seen. Even in case of significant bending short range interactions will not differ very much and the model will be perfectly adequate for rational protein engineering, etc. However, the prediction of differences in the backbone between structures that are homologous in sequence still requires much research.

In recent years automatic model-building by homology has become a routine technique that is implemented in most molecular graphics software packages. Currently four main lines of research are pursued:
• In case of high homology: Improvement of the way side chains are positioned in the modelled structure, and improvement of the database techniques needed to do this fast and accurately.
• In case of intermediate homology: Improvement of the way loops are inserted, and development of techniques to check the quality of models.
• In case of low homology: Improvement of the sequence alignment that is the basis of the model-building procedure, and evaluation of the significance of alignments.
• In case of no detectable homology: Development of secondary structure prediction and ab initio 3D modeling techniques.

HIGH HOMOLOGY: PLACING NEW SIDE CHAINS IN THE STRUCTURE

In case of very high sequence homology between the structure and the model the main problem lies in the replacement of just a few side chains. Often this can be done by hand [Read et al 89], but if done in an automatic fashion two main strategies can be followed.

• Residue sidechains can be 'grown' during molecular dynamics runs [e.g. van Gunsteren & Mark 92; Goodfellow & Williams 92; Karplus & Petsko 90; McCammon 91; Cornell et al 91; Berendsen 91 and references therein]. This is a very time consuming operation, which is normally only used if the difference between the structure and the model is only one residue. It has the advantage that one can get an impression of the energetics of the residue exchange, which might influence the decision about making a mutation or not.

• Alternatively one tries to find examples of residues in a similar environment in the database of known structures. The problems here are how to define similar environments, and how to extract the information quickly from the database of known structures. In practice, similar environments for residues have most often been defined as having the same secondary structure (α-helix, β-strand or loop).

Until recently side chains were normally placed using standard rotamer libraries [McGregor et al 87; Ponder et al 87; Schrauber et al 93], and a variety of procedures, ranging from manual adaptation [Read et al 1989] to complicated schemes involving energy calculations [Summers et al 89] or Monte Carlo procedures combined with energy calculations [Holm & Sander 1992; Eisenmenger et al 93], have been used to get rid of Van der Waals clashes. The latest developments are position specific rotamers. Dunbrack *et al.* [Dunbrack et al 93] showed that the preferred rotamer is to a large extend determined by the local backbone.

Position Specific Rotamers Are the Best Tool to Put New Side Chains in a Protein Structure

The conformation of an amino acid sidechain is indeed determined much more by the local backbone and much less by the three-dimensional environment than was previously assumed [Dunbrack et al 93; Vriend & Sander 93]. Therefore, position specific rotamer searches are an important tool in modeling. In fact, it is our experience that using position specific construction of side chains gives the best model building by homology results (for all-atom RMS deviations) obtainable to date when one known protein structure is modeled from another. Position specific rotamer distributions are obtained by searching the database for stretches of residues that fulfill two criteria:

• The stretch has a local backbone conformation similar to that of the stretch around the position in the actual structure for which the rotamers are required;

• The stretch has the same residue type at the middle position.

Additional constraints can be placed on the accessibility of the middle residue and on the types of residues next to the central one (e.g., hydrophobicity, bulkiness, etc.). Typically these stretches are 5 or 7 residues long and the RMS difference between superposed alpha carbon atoms is kept at maximally 0.5 Å. Levitt recently described such a method [Levitt 92].

Figure 2 shows the position specific rotamers for histidine, phenylalanine and tyrosine as the second residue of the N- and C-terminal ends of a short helix. Only one rotamer is allowed at the C-terminal end whereas the rotamer distribution is residue dependent at the N-terminal end of the helix. It is difficult to completely explain these effects, but figure 2 should suffice to stress the importance of position specific rotamers for homology modeling.

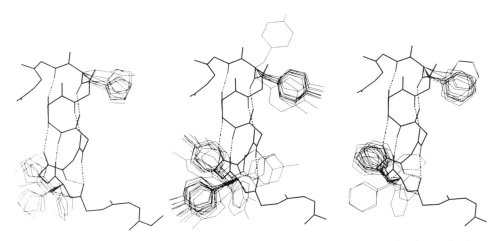

Figure 2. Position specific rotamer distributions for His, Tyr and Phe at the second and second to last position in a short alpha helix.

Database Retrieval Systems Are Needed To Extract Protein Structure Information from the Databases

It is obvious that database systems that allow for fast, easy and flexible retrieval of specific information are crucial for model building procedures. Several general [Bryant 89; Vriend 90; Islam & Sternberg; 89; Gray et al 90; Huysmans et al 91] and single purpose [Sander & Schneider 91; Lesk et al 89] data storage and retrieval systems have been developed to extract information about protein sequences and structures from databases. Some of them hardly (re)organize data, but merely combine a database of three-dimensional protein structures with a set of algorithms for pattern recognition, data analysis and graphics. In general, these systems provide very flexible tools, but this flexibility is paid for by a rather low speed when the algorithms are applied to large amounts of data. PKB [Bryant 89] and to some extent the parameter correlation method [Vriend 90] are good examples of such systems. They are well suited for prototyping queries or searches in small subsets of the database, but less suitable for use if fast data extraction is required.

If retrieval times must be reduced to a minimum, one resorts to systems that pre-process and reorganize data to speed up the process of extracting information. Two important classes of such systems are object-oriented database systems (OODBS) and relational database systems (RDBS) [Parsaye et al 89]. An OODBS can easily search for many related objects, but the organization of the data makes it slow at doing sequential scans [Gray et al 90]. P/FDM [Gray et al 90] is a good example of an OODBS. Its high level query language Daplex is very concise and approaches the power of a programming language for complex queries.

In a protein RDBS many structural properties such as accessibility, torsion angles and secondary structure are stored in tables and queries are performed by logical combination of these tables. BIPED [Islam & Sternberg] and SESAM [Huysmans et al 91] are examples of such systems. SESAM does not fit the relational model exactly as it also provides some algorithms on top of the RDBS to allow for otherwise impossible or prohibitively slow queries. Advantages of a generalized RDBS [Parsaye et al 89] are the generally high speed of searches and the intuitive way in which queries are constructed.

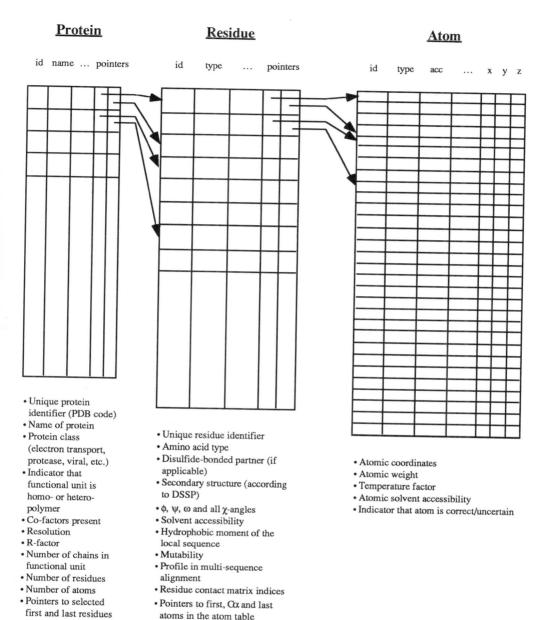

Protein

id name ... pointers

- Unique protein
  identifier (PDB code)
- Name of protein
- Protein class
  (electron transport,
  protease, viral, etc.)
- Indicator that
  functional unit is
  homo- or hetero-
  polymer
- Co-factors present
- Resolution
- R-factor
- Number of chains in
  functional unit
- Number of residues
- Number of atoms
- Pointers to selected
  first and last residues
  in the residue table

Residue

id type ... pointers

- Unique residue identifier
- Amino acid type
- Disulfide-bonded partner (if
  applicable)
- Secondary structure (according
  to DSSP)
- ϕ, ψ, ω and all χ-angles
- Solvent accessibility
- Hydrophobic moment of the
  local sequence
- Mutability
- Profile in multi-sequence
  alignment
- Residue contact matrix indices
- Pointers to first, Cα and last
  atoms in the atom table

Atom

id type acc ... x y z

- Atomic coordinates
- Atomic weight
- Temperature factor
- Atomic solvent accessibility
- Indicator that atom is correct/uncertain

Figure 3. The hierarchical organisation of the SCAN3D database tables. Under each of the three tables the properties stored in its columns are listed. "id" stands for identifier, "acc" for accessibility. The "pointers" establish the links from the highest level (Protein) via the Residue level down to the atom level.

The SCAN3D Database System Can Quickly Extract the Data Needed for Modeling from the Databases

When one wants to use a standard RDBS to aid with model building by homology one major problem is encountered: Entries in the same database table are assumed to be unrelated; or in other words, the database does not know which residues sit next to each other in the sequence. So, a query like 'buried - accessible - buried - accessible' to find surface β-strands is inherently beyond the capabilities of a standard relational system [Islam & Sternberg 89].

We have specifically designed the SCAN3D database system [Vriend et al 93] to bypass this problem. SCAN3D exploits the sorted character of protein structures in that it stores the residues or their characteristics in the database tables in the sequential order in which they occur in the protein. This allows to easily search for stretches of consecutive residues with specified characteristics. An example is shown in figure 4.

The Brookhaven Protein Databank (PDB) [Bernstein et al 77] contains atomic coordinates and some related information for more than 1000 macromolecular structures in plain text files. SCAN3D uses a representative set of slightly more than 200 protein structures [Hobohm et al 92]. This is done to avoid bias towards a small number of abundantly present protein families. For example the June 1992 version of the PDB contains more than 100 lysozyme, 30 hemoglobin and 10 rhinovirus (with 3 related chains each) structures.

The PDB files with crystal structures often contain errors, such as residues with wrong chirality [Morris et al 92], the absence of atoms and residues, atoms with incorrect names, histidines with six membered rings, etc. The structure input modules of the WHAT IF program [Vriend 90] adequately deal with these problems by issuing warnings and optionally applying corrections. This is done to avoid that residues or fragments with errors later end up in a model built by homology. Hereafter, accessibility for solvent, secondary structure, mutability, torsion angles, position in the chain and other properties are derived from PDB files in an automated fashion and stored in the database tables. The hierarchical, three-level organization of the SCAN3D data tables is described in figure 3.

Profile Matching Allows for Very Fast Extraction of Rotamer Information from the Database of Known Structures

Many queries that concern single residue properties can be handled efficiently by a general protein RDBS. Queries such as "Find all prolines with a positive φ-angle" require two table lookups and an **AND** operation. Responses to such queries are generally very fast. However, the analysis of protein structures and sequences most often leads to queries that concern characteristics of neighboring residues or residues that are a certain sequence distance apart. Examples are "Find all prolines that do not have another proline within 27 residues in the sequence," or "Find all prolines that have an atomic contact between their side-chain and a cysteine Sγ that is part of a disulfide bond." Such queries are much more difficult, or sometimes even impossible, to perform when using a strictly relational system. For example, to obtain information about the sequence spanning 15 residues on either side of the one being inspected, one would need 30 additional columns (as in BIPED [Islam & Sternberg 89]) or 30 join operations to the residue table to pick up these values (as in SESAM [Huysmans et al 91]). Another 30 join operations would be needed to retrieve the solvent accessibility of those residues, etc. Clearly this is not the most efficient approach.

SCAN3D solves this key problem of relational approaches by comparing a profile of the desired length and with the desired residue properties with all stretches of that length in the entire database. The answer to a query is a group of hits (GOH), i.e., a list of residue stretches that satisfy the constraints imposed by the profile on the individual residues.

SCAN3D's GOHs can be combined with logical operators such as **AND**, **OR**, and **NOT**. **AND** should be read as "Is present in both GOHs" and **OR** as "Is present in at least one GOH." Internally these operations are performed using a standard sort-merge-like algorithm. Because the data in each GOH are always sorted this algorithm works much faster than a normal RDBS join operation. This feature of comparing more than one residue at a time gives SCAN3D a great deal of its power.

Figure 4 shows an example of a search for sequences that satisfy certain constraints. The stretches to be found should have the following sequence profile: G-aromatic-any-G-small-G, where "any" stands for any of the twenty residues and G for glycine. The leftmost columns in this figure represent the GOH generated by this query and the protein sequence table that is being inspected. The stretch of amino acids searched for is 6 amino acids long, and thus the profile box is dimensioned 6 * 20. Assuming that the profile width is less than the length of the shortest sequence in the table, N * (M–P+1) profile matches have to be performed, where N is the number of proteins in the database, M the average length of each protein, and P the profile width (6 in this example). Two of the hexapeptide stretches that meet the constraints are indicated in figure 4. The top profile box matches the corresponding amino acids in the sequence table perfectly, and consequently a pointer to this stretch of amino acids is stored in the GOH. The second profile box shows one mismatch (the glycine at position 4 in the profile is a serine in the sequence table). A pointer to this stretch of amino acids would therefore only be stored in the GOH if the user defined number of allowed mismatches is one or more.

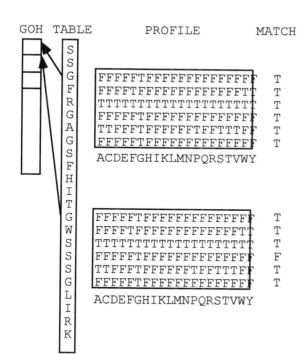

Figure 4. Example of a sequence profile search. Two of the matches between the data table and the sequence profile are shown. The second profile match would only enter the group of hits (GOH) if the allowed number of mismatches is set to 1 or more. The amino acid string at the bottom is given for easy reference only.

The same principle can be used for secondary structure searches. Here the table stores a code for each residue that represents helix, strand, turn or coil. The profile now does not contain 6 * 20 true/false flags (a 6-residue stretch with 20 possible amino acids at each position), but only 6 * 4 (a 6-residue stretch with 4 possible states for each residue).

Local folds are very important for modeling and, therefore, SCAN3D allows the user to search for stretches of amino acids that have a desired backbone and/or side chain conformation. The fragment databases required to perform such searches have been described by Jones and Thirup [Jones & Thirup 86]. In SCAN3D this is implemented by storing for every residue the C_α-C_α distance to the next N-1 residues adjacent in the sequence, where N is the maximal length allowed for a stretch. Results of these searches can be combined with the results of other queries, and this opens a new range of possibilities. It is exactly this type of search that provides the tool to arrive at the best possible models built by homology.

INTERMEDIATE HOMOLOGY 1: BUILDING LOOPS IF THERE IS AN INSERTION IN THE MODEL

In case of 75% or more sequence identity between the structure and the model one seldomly encounters insertions or deletions, and when they are encountered, they normally are short.

One of the two major problems in model building with intermediate homology is the insertion of loops. If an insertion in the sequence occurs relative to the structure, there is no template to model on, and other techniques have to be applied. The main techniques used to model loops are:

• Searching databases for loops with endpoints that match the residues in the structure between which the loop has to be inserted. This technique is based on the same distance geometry loop searching algorithm as described above for the retrieval of position specific rotamers from the structure database [Jones & Thirup 86]. Many molecular graphics and modeling programs use this technique (e.g., WHAT IF [Vriend 90], QUANTA ,Hydra [Hubbard 86], FRODO [Jones 78], Insight [Dayringer et al 86], O [Jones et al 91], BRAGI [Schomburg et al 88]).

• Ab initio building can be done in many ways, but a few points are common to all ab initio techniques: the fact that endpoints have to match is a very strong constraint. Further, Van der Waals clashes have to be prevented, and all rules about proper stereochemistry and energetics should be obeyed. Many articles have been published about these topics, [e.g. Moult et al 86; Bruccoleri et al 87; Fine et al 86; Havel et al 91; Sippl et al 92; Simon et al 91; Ripoll et al 90].

INTERMEDIATE HOMOLOGY 2: VERIFICATION OF THE QUALITY OF THE MODEL(S)

All models built by homology will have errors. Sidechains can be placed incorrectly, or whole loops can be misplaced. As with most errors, they become less of a problem when they can get localized. For example, in the study of a protease it is probably not important that a loop far away from the active site is modeled incorrectly. The most important step in the process of model building by homology is undoubtedly the verification of the model, and the estimation of the likelihood and magnitude of errors.

There are two principally different ways to estimate errors in a structure.
• Determination of the energy of the structure using energy minimization and molecular dynamics programs.
• Determination of normality indices that indicate how well a given characteristic of the model resembles the same characteristic in real structures.

The key aspect is the development of criteria with sufficient discriminatory power to distinguish a good model from a bad one. An example is provided by deliberately misfolded proteins in which the sequence of a protein known to have an all-helical 3D structure is placed into a known structure of a completely different type, an antiparallel β-barrel, and *vice versa*. For the evaluation of the quality of these clearly incorrect hypothetical structures, intramolecular energy, calculated in vacuum using standard empirical potentials, is not a sensitive criterion [Novotny et al 84, 88]. The free energy difference between the folded and unfolded states would be an ideal criterion, but present theories are not capable of calculating free energy differences to sufficient accuracy.

Faced with the lack of an accurate theory of protein folding, empirical observations of regularities gleaned from the database of solved structures can be very useful. Thermodynamic arguments suggest that the hydrophobic effect is a major driving force of protein folding [Kauzmann 59; Sharp 91]. A variety of statistical criteria, which measure the preferential distribution of hydrophobic side chains in the interior of proteins, have been used successfully to discriminate between deliberately misfolded and native structures [Baumann et al 89; Bryant & Amzel 87; Novotny et al 88; Hendlich 90]. Solvent-modified potential energy functions have been generated, for example, by omitting the attractive part of the Lennard-Jones potential of exposed nonpolar carbon atoms [Novotny et al 88], effectively giving them a preference for the interior, or by introducing an effective residue pair potential calibrated on hydrophobicity that is attractive for pairs of hydrophobic residues [Levitt 76]. Empirical solvation free energy functions have been derived using atomic solvent accessible surface areas [Eisenberg & McLachlan 86; Ooi et al 87], or the volume of the hydration shell [Kang et al 88], and various hydrophobicity scales or observed frequencies of 'buried' residues [Janin 79]. For example, the atomic transfer free energy parameters of Eisenberg and McLachlan (1986) for five atom types, based on accessible surface areas and calibrated on $DG_{octanol/water}$ [Fauchere & Pliska 83], were capable of discriminating between correct and deliberately misfolded conformations [Novotny et al 88; Eisenberg & McLachlan 86; Chiche et al 90]. Recently, Vila *et al.* tested surface area based models with several empirical scales derived from model compounds, the best of which showed the desired discrimination among the native and a set of near-native conformations of pancreatic trypsin inhibitor [Vila et al 91].

There is continuous progress in molecular dynamics and energy minimization. The main topics are force field improvement [e.g. Cornell et al 91; Weiner et al 84; Gerber 92; Gibson & Scheraga 91; Kollman & Dill 91], incorporation of effective solvation terms [Stouten et al 93], ab initio modeling of small proteins [Sippl et al 92; Simon et al 91; Ripoll et al 90; Vajda & Delisis 90; Andersen & Hermans 88], the incorporation of real data in Xray refinement [Brunger et al 87] and NMR structure determination [Vlieg et al 91]. Much work will still have to be put into the force fields used for energy calculations before any but trivial errors can be detected this way.

Normality indices for structures, however, have already proven their power in structure verification. Many characteristics of protein structures lend themselves for normality analysis. Most of them are directly or indirectly based on the analysis of contacts, either inter residue contacts, or contacts with water. Some published examples are:
• General checks for the normality of torsion angles, bond angles and bond lengths etc. [Morris et al 92; Vriend 90, Holm & Sander 91] are good checks for the quality of experimentally determined structures, but are less suitable for the evaluation of models because the better model building programs simply do not make this kind of errors.
• Inside outside distributions of polar and apolar residues can be used to detect completely misfolded models [Baumann et al 89].
• Solvation potentials can detect local errors as well as complete misfolds [Holm & Sander 92].
• Residue transfer energy is another way of looking at inside/outside distributions [Eisenberg & McLachlan 86].
• Packing rules have been implemented for structure evaluation [Gregoret & Cohen 90].

• Atomic contacts that are not abundant in the protein structure database are good indicators of local model building problems [Vriend & Sander 93].

Atomic contacts are observed because they are energetically favored. Real structures cannot tolerate too many unfavorable interactions. Thus for a model to be correct only a few infrequently observed atomic contacts are allowed. We made a detailed analysis of atom atom contacts [Vriend et al 93]. From this we derived a method for normality analysis based on the comparison of 3D contacts in the model with the same contacts in a database of 'real' proteins. This method heavily depends on the non-random character of the spatial distribution of atomic contacts.

Singh *et al.* determined directional preference patterns for residue-residue contacts [Singh et al 90]. They found that many types of contacts show a strong asymmetric distribution. Figure 5 shows some examples of observed contact distributions extracted from the protein structure database using the WHAT IF relational database module SCAN3D [Vriend et al 93].

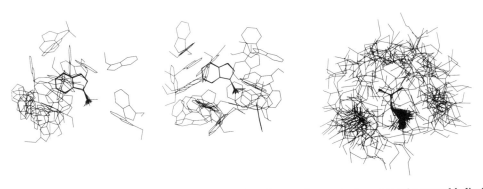

Figure 5. Asymmetric contact patterns. Left: Distribution of contacting tryptophans around a central helical tryptophan. Middle: same, but central tryptophan in a strand. Right Glutamic acid side chains around a central, helical arginine.

WHAT IF also holds a module that compares the local contact patterns with the average contact patterns for similar residue-residue contacts such as shown in figure 5. This method has been described extensively elsewhere [Vriend & Sander 93] It can be summarized as follows: If a residue-residue contact has the same contact patterns and the same spatial orientation as a contact that occurs often in the database then a high score is given. If a contact in the modeled molecule seems rather unique, either from a point of view of which residues make the contact, or from a point of view of directionality of the contact, a low score is given. This 'quality control' of local packing has proven to be a powerful tool for the detection of abnormal structures.

LOW HOMOLOGY 1: DATABASE SIZE AND THE SIGNIFICANCE OF ALIGNMENTS

Many aspects of modeling are limited by the size of the database in which one performs searches or from which one derives structural preferences.

As described above, the transfer of structural information to a potentially homologous protein is straightforward if the sequence similarity is high, but the assessment of the structural significance of sequence similarity can be difficult when sequence similarity is low or restricted to a short region. That indicates one of the two key problems (the other problem being refined measures of sequence similarity): the shorter the length of the alignment, the higher the level of similarity required for structural significance.

To solve this problem, we need to calibrate the length dependence of structural significance of sequence similarity. Empirically, this can be done by deriving from the database of known structures a quantitative description of the relationship between sequence similarity, structural similarity and alignment length. The resulting definition of a length-dependent homology threshold can provide the basis for reliably deducing the likely structure of globular proteins down to the size of domains and fragments. Previously, Chothia and Lesk (1986) have quantified the relation between the similarity in sequence and three-dimensional structure for the cores of entire globular proteins.

Structural Significance of Sequence Alignments

We performed an empirical determination of homology thresholds by studying thousands of sequence alignments within the PDB database. Each protein from a selected set of high and low resolution protein structures is compared with all others from the set. We use the dynamic sequence alignment algorithm [Smith and Waterman 81] as implemented in MaxHom [C.S. & R.S., unpublished]. For details see [Sander & Schneider 91].

For each alignment, sequence similarity, structural similarity and alignment length are noted. Sequence similarity as percentage identity of amino acids; structural similarity of two aligned proteins fragments either as the rms difference of equivalent C_α positions in 3D space after optimal superposition (tertiary structure similarity) [Kabsch 78] or as the percentage identity of secondary structure symbols according to DSSP (secondary structure similarity) [Kabsch & Sander 83] and alignment length as the number of amino acids. Note that while it is important to use a more sophisticated measure of sequence similarity in producing the alignments, the simpler measure of percentage identity is useful when comparing widely different alignment methods.

We determined a safe threshold for each alignment length such that any alignment with a similarity value above this threshold represents structural homology. The results are summarized in figure 6.

One of the remarkable features of this analysis [Sander & Schneider 91] is the saturation behavior of structure similarity with increasing sequence similarity. For large sequence similarity, secondary structure similarity is well within 30 percentage points of perfect (100%) and tertiary structure similarity within 2.5 Å of perfect (0.0 Å). However, perfect sequence similarity does not always imply perfect structural agreement: a protein crystal structure may vary, typically in loop regions or in domain orientation, as a result of different crystal packing, different substrate or cofactor interaction or complex formation.

The Significance Threshold for an Alignment Varies with Its Length

With a clear definition of structural homology we are now able to define a cutoff in sequence similarity above which homology of structure can be inferred. For each alignment length, the cutoff is determined by inspection of the Structure-Similarity/Sequence-Similarity scatter plot as that value of sequence similarity above which almost all alignments are structurally homologous, i.e., fall within the plasticity margin of perfect structural identity. The resulting homology cutoff (fig. 6) is a strongly varying function of alignment length up to a length of about 70-80 residues. For example, for alignment length 30, sequence similarity has to be at least 43% to infer structural homology. For very long alignment lengths, 25% sequence identity is sufficient. Note that below these values of sequence similarity structural homology cannot be asserted nor excluded - the region of weaker sequence similarity is a *don't know* region.

The threshold for structural homology can be used to improve the evaluation of matches in sequence database searches. Although the length dependence of statistical significance of sequence matches is well known mathematically [Smith & Waterman, 85], the most popular database alignment search programs (e.g. FASTA, Wordsearch [Pearson & Lipman 88; Devereux et al 84]) sort the best hits on total similarity, without reference to length. We suggest that a homology threshold curve like the one presented here can be used to order the database matches by the extent to which their score exceeds the threshold, in appropriate units.

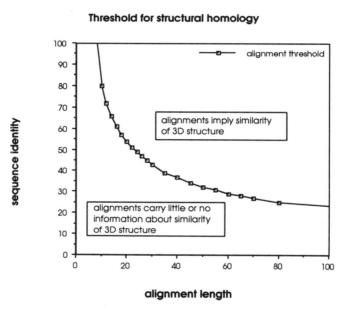

Figure 6. Homology threshold for structurally reliable alignments as a function of alignment length. The homology threshold (curved line) divides the graph into a region of safe structural homology where essentially all fragment pairs are observed to have good structural similarity and a region of homology unknown or unlikely where some fragment pairs are structurally similar but most are not.

Multiple Sequence Alignments of all Known Sequences Against all Known Structures Are Available

For all protein structure files (PDB files [Bernstein et al 77]) a multiple sequence alignment using the above criteria has been made. The results, stored in so-called HSSP files are available from the EMBL file server (see appendix A,B). This significantly increased database with its structural family alignments, sequence profiles, and sequence variability can be used:

288

• To study the evolution of protein sequence and structure. For example, the correlation between residue side chain contacts and sequence variation.

• To derive statistically more reliable preference parameters or sequence patterns for structure prediction [Maxfield & Scheraga 79; Rooman & Wodak 88; Gibrat et al 87].

• To extract weighted sequence profiles for database searches. For example, sequence positions along the profile can be given a weight corresponding to the degree of conservation, such that strongly varying positions are effectively ignored in a profile sequence comparison [Gribskov et al 87; Bashford et al 87; Smith & Smith 90; Staden 88].

• To define the core region of a structural family for model building by homology, even when only one structure is known. Strongly varying positions are considered not be part of the invariant core.

• To derive structure-dependent similarity tables for amino acid types (or tuples of types) for use in aligning sequences to proteins of known 3D structure and for use in planning point mutations based on known or predicted protein structures.

LOW HOMOLOGY 2: USING THE STRUCTURE TO IMPROVE THE ALIGNMENT

If the percentage sequence identity between the structure and the sequence to be modeled is well below 50% then the quality of the sequence alignment is the limiting factor for the quality of the overall model. Normal sequence alignment programs only use sequence information. However, in case of model building by homology more information is available in the form of one structure. Use of this information in principle must lead to a better alignment.

Several authors have described techniques to align two proteins for cases for which the structure of one of the proteins is known [Taylor 86; Novotny et al 88; Overington et al 92; Crippen 90; Sippl et al 92; Luthy et al 91; Bowie et al 91; Luthy et al 92; Jones et al 92; Scharf 89; Thornton et al 91; Ouzounis et al 93; Maiorov et al 92; Godzik et al 92; Stultz et al 93]. Rather than using the sequence of the known structure as input to the alignment procedure, a structure dependent sequence profile is set up, and the sequence of the unknown structure is aligned with this profile. The goal of these new methods is the detection of structural homologies in the 'twilight zone', e.g., the structural similarity between the two domains of rhodanese (sequence identity below 15%). However, none of the sequence-structure alignment methods has been proven to consistently and generally detect remote structural homologies with a high score and with the correct alignment. Nevertheless, several interesting examples have been reported, e.g., the relation between actin and heat shock protein hsp70 [Bowie et al 91] or between phycocyanin and myoglobin [Jones et al 92].

It would be interesting to perform a comprehensive test of these different methods using a large number of control examples, jack-knifing (removal of homologues of test proteins from the parameter database), predetermined gap penalties (rather than adjusting gap penalties to fit the known examples), and reporting not only the score of best hits, but also the quality of the resulting alignments. Work along these lines is continuing in several groups. Exhaustive examples of sequence-remote homologues to known structures can be taken from the database of structurally aligned protein families [Holm et al 92] available through anonymous file transfer protocol as described in appendix B. A flow diagram of the structure - sequence alignment method described by Ouzounis et al (1993) is given in figure 7.

Improvement of sequence-structure alignments for protein structure prediction in the twilight zone may come from a more adequate definition of structural states, from use of homologous sequence information, e.g. in the form of conservation weights or multiple sequence alignments, from mixing sequence and structure information in correct proportions, from improved optimization methods in combining different interaction terms, from genuine 2D alignment optimization that takes into account the change of contact partner with alignment, and from variation in backbone geometry and general plasticity of homologous structures. The problem is an urgent one, as genome projects will soon flood us with many protein sequences lacking functional or structural information. When developed and used properly, computational sequence analysis and structure prediction algorithms can save countless hours in the laboratory by deriving probable 3D structures.

Figure 7 is a scheme of the basic procedure of structure-sequence alignment (a) From three-dimensional structure to a two-dimensional residue-residue contact map. Starting from the three-dimensional structure, the strength of residue-residue and residue-solvent contacts in the known 3D structure ("Contacts") is calculated. Contact strength is indicated by gray levels; contacts with solvent are in right-most column ("W"). (b) From contact map to contact interface profile. For each residue, i.e., for each row in the contact map, contacts are summed over all contacting residues in each interface type (five types in this figure). The resulting set of five interface strengths is called the contact interface vector and describes the structural environment of a residue. The array of interface vectors, one for each residue, is a simplified representation of protein structure, called the interface contact profile ("Interfaces"). This description is one-dimensional in the sense that each interface vector describes the local structural environment, independently of the type of the contacting residues. Preferences for a residue type in each of the interface states ("Preferences") are derived from the interface profiles of all known structures in the database, removing from the database the protein to be aligned. (c) From contact interface profile to fitness profile. In preparation of alignment, the fitness of each of the twenty

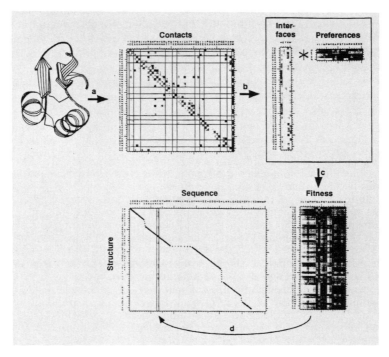

Figure 7. Scheme of the basic procedure of structure-sequence alignment: (a) From three-dimensional structure to a two-dimensional residue-residue contact map. (b) From contact map to contact interface profile. (c) From contact interface profile to fitness profile. (d) From fitness profile to structure-sequence alignment.

amino acids at each structural position is evaluated by simply summing over the preference of that residue type for each interface, weighted with the strength of the interface at the structural position. The resulting table ("Fitness" profile) represents the fitness of each of the 20 amino acid residue types for this structural position. Mathematically, the fitness profile $f(R,j)$ is simply the matrix product of the interface profile $c(j,I)$ with the preference table $p(R,I)$. Such fitness profiles can also be derived from other types of structural preferences, e.g., those for secondary structure, as well as from multiple sequence alignments (Gribskov et al. 1987, 1990). (d) From fitness profile to structure-sequence alignment. The fitness profile is just the right form of input to an alignment problem, here that of aligning some amino acid sequence to the given structure. The local similarity value at a given sequence position is simply the appropriate column copied from the fitness table, representing the fitness of that residue type for each position in the structure. A dynamic programming algorithm then finds the optimal trace, such that the sum of fitness values along the trace is optimal. The result is an alignment of the sequence with the given structure. The alignment is the basis of an explicit three-dimensional model, and the total fitness value quantifies how well the sequence fits into this structure.

NO DETECTABLE HOMOLOGY: PREDICTION OF PROTEIN SECONDARY STRUCTURE BY USE OF SEQUENCE PROFILES AND NEURAL NETWORKS

If building by homology of a three dimensional model is not feasible because no structure with significant sequence homology can be found one would like to be able to get at least an impression about the secondary structure. Because the sequence of a protein determines the structure, it must in principle be possible to reconstruct the secondary and tertiary structure from the sequence information alone. In practice however, secondary structure prediction methods have long been unable to produce higher precision than 55-64% [Chou & Fasman 74; Garnier et al 78; Gibrat et al 87; Levin & Garnier 88; Levin et al 86; Lim 74; Ptitsyn & Finkelstein 83; Zvelebil; et al 87]. Keeping in mind that a program that always predicts all residues to be helical will score 35%, one realizes that 55% correct (that means 45% wrong!) is not enough to draw conclusions about the structure.

We have recently achieved a substantial increase in both the accuracy and quality of secondary structure predictions, using a neural network algorithm [Rost & Sander 92]. The main improvements come from the use of multiple sequence alignments (better overall accuracy), from 'balanced training' (better prediction of β-strands) and from 'structure context training' (better prediction of helix and strand lengths). The best method, cross-validated on seven different test sets purged of sequence similarity to learning sets, achieves a three-state (α-helix, β-strand and loop) prediction accuracy of 71.4%, significantly better than any previous method. The improved distribution and accuracy of helices and strands makes the predictions well suitable for use in practice as a first estimate of structural type of newly sequenced proteins.

How accurate are predictions likely to be in practice? As a tough check, the network system was trained on 130 chains (25% cutoff in pairwise sequence identity) and then tested on 26 proteins whose structures became available after the network architecture had been finalized. None of these had more than 25% sequence identity with any of the training proteins. In this final set, 72% of the observed helical and 62% of strand residues were predicted correctly in a three state prediction. The overall accuracy for this set of completely new proteins was 71.6%. Table 1 shows some recent results of three multiple sequence alignment based secondary structure prediction methods for SH3. For comparison the results of several well known 'classical' methods are also listed.

Secondary structure predictions using the currently best version of the PHD profile network are available via electronic mail (see appendix B).

Table 1. Secondary structure predictions of SH3.

```
      ....,....1....,....2....,....3....,....4....,....5....,....6
AA    |TGKELVLALYDYQEKSPREVTMKKGDILTLLNSTNKDWWKVEVNDRQGFVPAAYVKKLD|
Obs   |   EEEE    B       B  B      EEEEE      EEEEE   EEEEEEGGGEEEE

PHD   |    EEEEEEE           EE    EEEEE        EEE       EEE    EEEE
ETH   |    EEEEEEE        EEEEE  HHHHHHHH      EEEE       EEE    EEE

C+F   |   HHHHHH          HHHHH  EEEEEE        HHHHH    EEEEEHHHHHHHHH
ROB   |  HHHHHHHHHHH    HHHHHHHHHHHHEEEEE           EEE    EEEHHHHHHHH
S83   |    EEEEEEE          EEEE  HHHHH        EEEEEE         HHHHHHH
GOR   |  HHHHHHHH          HHHHH  HHEEHH       HEEE H   H H   HHHHHHH
ALB   |    EEEEEE                  EEEE        EEEEE
MUS   |    EEEEE      EEEEE  EEEEEE  EEEEEEEE
```

Comparison of the secondary structure of the SH3 domain, determined experimentally and predicted on the basis of multiple sequence alignments by three different methods. AA = amino acid sequence; EXP = crystal structure [Musacchio et al 92]; PHD = profile neural network prediction [Rost & Sander 92; Rost & Sander 93]; ETH = *de novo* prediction of conformation [Benner & Gerloff 92]; C+F = statistical prediction by Chou and Fasman [Chou & Fasman 74]; ROB = statistical prediction also known as GORI [Garnier et al. 78]; S83 = Segment 83, unpublished prediction method by Kabsch and Sander; GOR = GORIII prediction method [Levin et al 86]; ALB = prediction based on physico-chemical principles by Ptitsyn and Finkelstein [Ptitsyn & Finkelstein 83]; MUS = partial prediction [Musacchio et al 92]. E = β-strand (extended), H = α-helix, G = 3_{10}-helix, B = isolated β-bridge.

REFERENCES

Giege, R., Mikol, V., Trends Biotech., 7 (1989) 277-282.

Swindells, M.B., Thornton, J.M., Curr.Op.Struct.Biol., 1 (1991) 219-223.

Bajai, M., Blundell, T.L., Ann.Rev.Biophys.Bioeng., 13 (1984) 453-492.

Bashford, D., Chothia, C., Lesk, A.M., J.Mol.Biol., 196 (1987) 199-216.

Vriend, G., Sander, C., PROTEINS, 11 (1991) 52-58.

Holm, L., Ouzounis, C., Sander, S., Tuparev, G., Vriend, G., Prot.Sci., 1 (1992) 1691-1698.

Chothia, C., Lesk, A.M., EMBO J., 5 (1986) 823-836.

Sander, C., Schneider, R., PROTEINS, 9 (1991) 56-68.

Blundell, T.L., Sibanda, B.L., Sternberg, M.J.E., Thornton, J.M., Nature, 326 (1987) 347-352.

Read, L.S., Thornton, J.M., PROTEINS, 5 (1989) 170-182.

Gunsteren, W.F. van, Mark, A.E., Eur.J.Biochem., 204 (1992) 947-961.

Goodfellow, J.M., Williams, M.A., Cur.Op.Struct.Biol., 2 (1992) 211-216.

Karplus, M., Petsko, G.A., Nature, 347 (1990) 631-639.

McCammon, J.A., Cur.Op.Struct.Biol., 1 (1991) 196-200.

Cornell, W.D., Howard, A.E., Kollman, P., Cur.Op.Struct.Biol., 1 (1991) 201-212.

Berendsen, H.J.C., Cur. Op. Struct. Biol., 1 (1991) 191-195.

Mc Gregor, M.J., Islam, S.A., Sternberg, M.J.E., J.Mol.Biol., 198 (1987) 295-310.

Ponder, J.W, Richards, F.M., J.Mol.Biol., 193 (1987) 775-791.

Schrauber, H., Eisenhaber, F., Argos, O., J.Mol.Biol., 230 (1993) 592-612.

Dunbrack, R.L., Karplus, M., J.Mol.Biol., 230 (1993) 543-574.

Summers, N.L., Karplus, M., J.Mol.Biol., 210 (1989) 785-811.

Eisenmenger, F., Argos, O., Abagyan, R., J.Mol.Biol. Submitted.

Vriend, G., Sander, C., In preparation.

Levitt, M., J.Mol.Biol., 226 (1992) 507-533.

Bryant, S.H., PROTEINS, 5 (1989) 233-247.

Vriend, G., Prot.Engin., 4 (1990) 221–223.

Islam, S.A., Sternberg, M.J.E., Prot.Engin., 2 (1989) 431–442.

Gray, P.M.D., Paton, N.W., Kemp, G.J.L., Fothergill, J.E., Prot.Engin., 3 (1990) 235–243.

Huysmans, M., Richelle, J., Wodak, S.J., PROTEINS, 11 (1991) 59–76.

Lesk, A.M., Boswell, D.R., Lesk, V.I., Lesk, V.E., Bairoch, A., Protein Seq. Data Anal., 2 (1989) 295–308.

Parsaye, K., Chignell, M., Khoshafian, S., Wong, H., (1989). "Intelligent databases. Object-oriented, deductive hypermedia technologies." Wiley & Sons, New York.

Bernstein, F.C., Koetzle, T.F., Williams, G.J.B., Meyer, E.F., Brice, M.D., Rodgers, J.R., Kennard, O., Shimanouchi, T., Tasumi, M., J.Mol.Biol., 112 (1977) 535-542.

Hobohm, U., Scharf, M., Schneider, R., Sander, C., Prot.Sci., 1 (1992) 409-417.

Vriend, G., J.Mol.Graph., 8 (1990) 52–56.

Jones, T.A., Thirup, S., EMBO J., 5 (1986) 819-822.

Schomburg, D., Reichelt, J., J.Mol.Graph., 6 (1988) 161-165.

Dayringer, H.E., Tramontano, A., Fletterick, R.J., J.Mol.Graph., 4 (1986) 82-87.

Jones, T.A., J.Appl.Cryst., 11 (1978) 268-272.

Hubbard, R.E., In: Computer Graphics and Molecular modeling. Edt. Fletterick, R.j., Zoller, M., Cold Spring Harbor, (1986) 9-12.

Jones, T.A., Zou, J.Y., Cowan, S.W., Kjelgaard, M., Acta Cryst. A, 47 (1991) 110-119.

Moult, J., James, M.N.G., PROTEINS, 1 (1986) 146-163.

Bruccoleri, R.E., Karplus, M., Biopolymers, 26 (1987) 137-168.

Fine, R.M., Wang, P.S., Shenkin, D.L., Yarmush,#### Levinthal, C., PROTEINS, 1 (1986) 342-362.

Havel, T.F., Snow, M.E., J.Mol.Biol., 217 (1991) 1-7.

Sippl, M.J., Hendlich, M., Lackner, P., Prot.Sci., 1 (1992) 625-640.

Simon, I., Glasser, L., Scheraga, H.A., PNAS, 88 (1991) 3661-3665.

Ripoll, D.R., Scheraga, H.A., Biopolymers, 30 (1990) 165-176.

Weiner, S.J., Kollman, P.A., Case, D.A., Singh, U.C., Ghio, C., Alagona, G., Profeta, S., Weiner, P., J.Am.Chem.Soc., 106 (1984) 765-784.

Gerber, P., Biopolymers, 32 (1992) 1003-1017.

Gibson, K.D., Scheraga, H.A., J.Bio.Mol.Struct.Dyn., 8 (1991) 1109-1111.

Kollman, P.A., Dill, K.A., J.Bio.Mol.Struct.Dyn., 8 (1991) 1103-1107.

Vajda, S., Delisi, C., Biopolymers, 29 (1990) 1755 1772.

Andersen, A.G., Hermans, J., PROTEINS, 3 (1988) 262-265.

Brunger, A.T., Kutiyan, J., Karplus, M., Science, 235 (1987) 458-462.

Vlieg, J. de, Gunsteren, W.F. van, Meth. in Enzym., 202 (1991) 268-300.

Stouten, P.F.W., Frommel, C., Nakamura, H., Sander, C., Mol.Sim. 10 (1993) 97-120.

Ouzounis, C., Sander, C., Scharf, M., Schneider, R., J.Mol.Biol. 232(1993) 805-825.

Holm, L., Sander, C., J.Mol.Biol., 218 (1991) 183-194.

Novotny, J., Bruccoleri, R., Karplus, M., J.Mol.Biol., 177 (1984) 787-818.

Novotny, J., Rashin, A.A., Bruccoleri, R.E., PROTEINS 4 (1988) 19-30.

Kauzmann, W., Adv. Protein Chem., 14 (1959) 1-63.

Sharp, K.A., Cur.Op.Struct.Biol., 1 (1991) 171-174.

Baumann, G., Frömmel , C., Sander, C., Prot.Eng., 2 (1989) 329-334.

Bryant, S.H., Amzel, L.M., Int.J.Pept.Prot.Res., 29 (1987) 46-52.

Hendlich, M., Lackner, P., Weitckus, S., Floeckner, H., Froschauer, R., Gottsbacher, K., Casari, G., Sippl, M.J., J.Mol.Biol., 216 (1990) 167-180.

Levitt, M. J., J.Mol.Biol., 104 (1976) 59-107.

Eisenberg, D., McLachlan, A.D., Nature, 319 (1986) 199-203.

Ooi, T., Oobatake, M., Nemethy, G., Scheraga, H.A., PNAS, 84 (1987) 3086-3090.

Janin, J., Nature, 277 (1979) 491-492.

Kang, Y.K., Gibson, K.D., Nemethy, G., Scheraga, H.A., J.Phys.Chem., 92 (1988) 4739-4742.

Fauchere, J.-L., Pliska, V., Eur.J.Med.Chem.-Chim.ther., 18 (1983) 369-375.

Chiche, L., Gregoret, L.M., Cohen, F.E., Kollman, P.A., PNAS, 87 (1990) 3240-3243.

Vila, J., Williams, R.L., Vasquez, M., Scheraga, H.A., PROTEINS, 10 (1991) 199-218.

Morris, A.L., MacArthur, M.W., Hutchinson, E.G., Thornton, J.M., PROTEINS, 12 (1992) 345-364.

Holm, L., Sander, C., J.Mol.Biol., 225 (1992) 93-105.

Gregoret, L.M., Cohen, F.E., J.Mol.Biol., 211 (1990) 959-974.

Vriend, G., Sander, C., J.Appl.Cryst., 26 (1993) 47-60.

Singh, J., Thornton, J.M., J.Mol.Biol., 211 (1990) 595-615.

Smith, T.F., Waterman, M.S., J.Mol.Biol., 147 (1981) 195-197.

Kabsch, W., Acta Cryst., A34 (1978) 827-828.

Kabsch, W., Sander, C., Biopolymers, 22 (1983) 2577-2637.

Pearson, W.R., Lipman, D.J., PNAS, 85 (1988) 2444-2448.

Devereux., J., Haeberli, P., Smithies, O., NAR, 12 (1984) 387-395.

Maxfield, F.R., Scheraga, H.A., Biochemistry, 18 (1979) 697-704.

Rooman, M., Wodak, S,J., Nature, 335 (1988) 45-49.

Gibrat, J.-F., Garnier, J., Robson, B., J.Mol.Biol., 198 (1987) 425-443.

Gribskov, M., McLachlan, M., Eisenberg, D., PNAS. 84 (1987) 4355-4358.

Smith, R.F., Smith, T., PNAS., 87 (1990) 118-122.

Staden, R., CABIOS, 4 (1988) 53-60.

Taylor, W.T., J.Mol.Biol., 188 (1986) 233-258.

Overington, J., Donnelly, D., Johnson, M.S., Sali, A., Blundell, T.L., Prot.Sci., 1 (1992) 216-226.

Crippen, G.M., Biochemistry, 30 (1990) 4232-4237.

Sippl, M.J., Weitckus, S., PROTEINS, 13 (1992) 258-271.

Lüthy, R., McLachlan, A.D., Eisenberg, D., PROTEINS, 10 (1991) 229-239.

Bowie, J.B., Lüthy, R., Eisenberg, D., Science, 253 (1991) 164-170.

Lüthy, R., Bowie, J.B., Eisenberg, D., Nature, 356 (1992) 83-85.

Jones, D.T., Taylor, W.R., Thornton, J.M., Nature, 358 (1992) 86-89.

Scharf, M., Thesis, Univ. Heidelberg (1989).

Thornton, J.M., Flores, T.P., Jones, D.J., Swindells, M.B., Nature, 354 (1991) 105-106.

Stultz, C.M., White, J.V., Smith, T.F., Prot.Sci., 2 (1993) 305-314.

Maiorov, V.N., Crippen, G.M., J.Mol.Biol., 227 (1992) 876-888.

Godzik, A., Kolinski, A., Skolnick, J., J.Mol.Biol., 227 (1992) 227-238.

Chou, P. Y., Fasman, G. D., Biochem., 13 (1974) 211-215.

Garnier, J., Osguthorpe, D. J., Robson, B., J.Mol.Biol., 120 (1978) 97-120.

Gibrat, J.-F., Garnier, J., Robson, B., J.Mol.Biol., 198 (1987) 425-443.

Levin, J. M., Garnier, J., BBA, 955 (1988) 283-295.

Levin, J. M., Robson, B., Garnier, J., FEBS Lt., 205 (1986) 303-308.

Lim, V. I., J.Mol.Biol., 88 (1974) 857-872.

Ptitsyn, O. B., Finkelstein, A. V., Biopolymers, 22 (1983) 15-25.

Zvelebil, M. J., Barton, G. J., Taylor, W. R., Sternberg, M. J. E., J.Mol.Biol., 195 (1987) 957-961.

Benner, S.A., Cohen, M.A., Gerloff, D., Nature, 359 (1992) 781.

Musacchio, A., Noble, M., Pauptit, R., Wierenga, R., Saraste, M., Nature, 359 (1992a) 851-855.

Musacchio, A., Gibson, T., Lehto, V. P., FEBS Lt., 307 (1992b) 55-61.

Rost, B., Sander, C., Nature, 360 (1992) 540.

Rost, B., Sander, C., J.Mol.Biol, 232(1993) 584-599.

APPENDIX A. SAMPLE HSSP FILE

```
HSSP      HOMOLOGY DERIVED SECONDARY STRUCTURE OF PROTEINS , VERSION 1.0 1991
PDBID     1ppt
DATE      file generated on 31-Aug-92
SEQBASE   RELEASE 22.0 OF EMBL/SWISS-PROT WITH 25044 SEQUENCES
PARAMETER SMIN: -0.5 SMAX: 1.0
PARAMETER gap-open: 3.0 gap-elongation: 0.1
PARAMETER conservation weights
PARAMETER no insertions/deletions in secondary structure allowed
PARAMETER alignments sorted according to: DISTANCE
THRESHOLD according to t(L)=(290.15 * L ** -0.562) + 5
REFERENCE Sander C., Schneider R. : Database of homology-derived protein structures. Proteins, Proteins, 9:56-68 (1991).
CONTACT   e-mail (INTERNET) Schneider@EMBL-Heidelberg.DE or Sander@EMBL-Heidelberg.DE / phone +49-6221-387361 / fax
   +49-6221-387306
AVAILABLE Free academic use. Commercial users must apply for license.
HEADER    PANCREATIC HORMONE
COMPND    AVIAN PANCREATIC POLYPEPTIDE
SOURCE    TURKEY (MELEAGRIS GALLOPAVO) PANCREAS
AUTHOR    T.L.BLUNDELL,J.E.PITTS,I.J.TICKLE,S.P.WOOD
SEQLENGTH 36
NCHAIN    1 chain(s) in 1ppt.DSSP data set
NALIGN    17
NOTATION : ID: EMBL/SWISSPROT identifier of the aligned (homologous) protein
NOTATION : STRID: if the 3-D structure of the aligned protein is known, then STRID is the Protein Data Bank identifier as taken
NOTATION : from the database reference or DR-line of the EMBL/SWISSPROT entry
NOTATION : %IDE: percentage of residue identity of the alignment
NOTATION : %SIM (%WSIM): (weighted) similarity of the alignment
NOTATION : IFIR/ILAS: first and last residue of the alignment in the test sequence
NOTATION : JFIR/JLAS: first and last residue of the alignment in the alignend protein
NOTATION : LALI: length of the alignment excluding insertions and deletions
NOTATION : NGAP: number of insertions and deletions in the alignment
NOTATION : LGAP: total length of all insertions and deletions
NOTATION : LSEQ2: length of the entire sequence of the aligned protein
NOTATION : ACCNUM: SwissProt accession number
NOTATION : PROTEIN: one-line description of aligned protein
```

NOTATION : SeqNo,PDBNo,AA,STRUCTURE,BP1,BP2,ACC: sequential and PDB residue numbers, amino acid (lower case = Cys),
    secondary
NOTATION : structure, bridge partners, solvent exposure as in DSSP (Kabsch and Sander, Biopolymers 22, 2577-2637(1983)
NOTATION : VAR: sequence variability on a scale of 0-100 as derived from the NALIGN alignments
NOTATION : pair of lower case characters (AvaK) in the alignend sequence bracket a point of insertion in this sequence
NOTATION : dots (....) in the alignend sequence indicate points of deletion in this sequence
NOTATION : SEQUENCE PROFILE: relative frequency of an amino acid type at each position. Asx and Glx are in their
NOTATION : acid/amide form in proportion to their database frequencies
NOTATION : NOCC: number of sequences spanning this position (including the test sequence)
NOTATION : NDEL: number of sequences with a deletion in the test protein at this position
NOTATION : NINS: number of sequences with an insertion in the test protein at this position
NOTATION : ENTROPY: entropy measure of sequence variability at this position
NOTATION : RELENT: relative entropy, i.e. entropy normalized to the range 0-100
NOTATION : WEIGHT: conservation weight

```
## PROTEINS : EMBL/SWISSPROT identifier and alignment statistics
 NR.    ID        STRID  %IDE %WSIM IFIR ILAS JFIR JLAS LALI NGAP LGAP LSEQ2 ACCNUM    PROTEIN
   1 : paho_chick 1PPT   1.00 1.00    1   36    1   36   36    0    0    36 P01306    PANCREATIC HORMONE.
   2 : paho_strca        0.94 0.97    1   36    1   36   36    0    0    36 P11967    PANCREATIC HORMONE.
   3 : paho_allmi        0.80 0.85    2   36    2   36   35    0    0    36 P06305    PANCREATIC HORMONE.
   4 : paho_ansan        0.78 0.76    1   36    1   36   36    0    0    36 P06304    PANCREATIC HORMONE.
   5 : neuy_sheep        0.60 0.75    2   36    2   36   35    0    0    36 P14765    NEUROPEPTIDE Y (NPY).
   6 : neuy_pig          0.57 0.73    2   36    2   36   35    0    0    36 P01304    NEUROPEPTIDE Y (NPY).
   7 : pyy_human         0.54 0.70    2   36    2   36   35    0    0    36 P10082    PEPTIDE YY (PYY).
   8 : neuy_rat          0.54 0.72    2   36   31   65   35    0    0    98 P07808    NEUROPEPTIDE Y PRECURSOR (NPY).
   9 : neuy_human        0.54 0.72    2   36   30   64   35    0    0    97 P01303    NEUROPEPTIDE Y PRECURSOR.
  10 : neuy_rabit        0.54 0.72    2   36    2   36   35    0    0    36 P09640    NEUROPEPTIDE Y (NPY).
  11 : pyy_rat           0.54 0.71    2   36   30   64   35    0    0    98 P10631    PEPTIDE YY PRECURSOR (PYY).
  12 : pyy_pig           0.54 0.71    2   36    2   36   35    0    0    36 P01305    PEPTIDE YY (PYY).
  13 : pp_lepsp          0.49 0.68    2   36    2   36   35    0    0    36 P09473    PANCREATIC POLYPEPTIDE (PP)
                                                                                     (NEUROPEPTIDE
  14 : pp_oncki          0.49 0.68    2   36    2   36   35    0    0    36 P09474    PANCREATIC POLYPEPTIDE (PP).
  15 : paho_canfa        0.46 0.63    2   36   31   65   35    0    0    93 P01299    PANCREATIC HORMONE PRECURSOR.
  16 : paho_pig          0.46 0.63    2   36    2   36   35    0    0    36 P01300    PANCREATIC HORMONE.
  17 : paho_didma        0.46 0.63    2   36    2   36   35    0    0    36 P18107    PANCREATIC HORMONE.
```

```
## ALIGNMENTS    1 -   17
SeqNo PDBNo AA STRUCTURE BP1 BP2  ACC NOCC VAR   ....:....1....:....2....:....3....:....4....:....5....:....6....:....7
   1    1   G             0   0  101    4    0   GG G
   2    2   P     -       0   0   60   18    0   PPPPPPPPPPPPPPPPP
   3    3   S     -       0   0  106   18   45   SALSSSISSSAAPPLLQ
   4    4   Q     -       0   0  139   18   32   QQQQKKKKKKKKKEEE
   5    5   P     -       0   0   26   18    0   PPPPPPPPPPPPPPPPP
   6    6   T     -       0   0  126   18   46   TTKTDDEDDDEEEEVVV
   7    7   Y     -       0   0  121   18   47   YYYYNNANNNAANNYYY
   8    8   P     -       0   0   55   18    0   PPPPPPPPPPPPPPPPP
   9    9   G     >       0   0   27   18    0   GGGGGGGGGGGGGGGGG
  10   10   D   T 3   S+  0   0  128   18   15   DDDNDEEEEEEEEDDD
  11   11   D   T 3   S+  0   0  165   18    4   DDGDDDDDDDDDDDDDD
  12   12   A   S <   S-  0   0   17   18    0   AAAAAAAAAAAAAAAAA
  13   13   P     >>      0   0   73   18   27   PPPPPSPPPSSPPTTT
  14   14   V   H 3>  S+  0   0  108   18   38   VVVVAAPAAAPPPPPPP
  15   15   E   H 3>  S+  0   0  111   18    0   EEEEEEEEEEEEEEEE
  16   16   D   H <>  S+  0   0   58   18   18   DDDDDDEDDDEEEQQQ
  17   17   L   H   X S+  0   0   62   18    4   LLLLLLMMMLLLMMM
  18   18   I   H   X S+  0   0   91   18   42   IVIRAANAAASSAAAA
  19   19   R   H   X S+  0   0  142   18   30   RRQFRRRRRRRKKQQK
  20   20   F   H   X S+  0   0   39   18    2   FFFYYYYYYYYYYYYY
  21   21   Y   H   X S+  0   0  143   18   26   YYYYYYYYYYYYYYYAAA
  22   22   D   H   X S+  0   0   79   18   39   DDDDSSASSSAASTAAA
  23   23   N   H   X S+  0   0   97   18   41   NNDNAASAAASSAAEEE
  24   24   L   H   X S+  0   0   58   18    0   LLLLLLLLLLLLLLLL
  25   25   Q   H   X S+  0   0   90   18   23   QQQQRRRRRRRRRRR
  26   26   Q   H   X S+  0   0  112   18   27   QQQQHHHHHHHHHRRR
  27   27   Y   H   X S+  0   0   56   18    0   YYYYYYYYYYYYYYYY
  28   28   L   H   X S+  0   0   90   18   24   LLLRIILIIILLIIII
  29   29   N   H   <>S+  0   0   33   18   10   NNNLNNNNNNNNNNNN
  30   30   V   H ><5S+   0   0   21   18   30   VVVNLLLLLLLLMMR
  31   31   V   H 3<5S+   0   0   85   18   16   VVVVIIVIIIVVIILLL
  32   32   T   T 3<5S-   0   0   91   18    0   TTTFTTTTTTTTTTT
  33   33   R   T < 5S-   0   0  215   18    0   RRRRRRRRRRRRRRRRR
  34   34   H     < +     0   0  101   18   31   HHPHQQQQQQQQQPPP
  35   35   R             0   0  180   18    0   RRRRRRRRRRRRRRRR
  36   36   Y             0   0  224   18    1   YYFYYYYYYYYYYYYYY
```

```
## SEQUENCE PROFILE AND ENTROPY
SeqNo PDBNo  V  L  I  M  F  W  Y  G  A  P  S  T  C  H  R  K  Q  E  N  D NOCC NDEL NINS ENTROPY RELENT WEIGHT
   1    1    0  0  0  0  0  0  0 100  0  0  0  0  0  0  0  0  0  0  0  0    4    0    0   0.000      0   1.03
   2    2    0  0  0  0  0  0  0   0  0 100 0  0  0  0  0  0  0  0  0  0   18    0    0   0.000      0   1.27
   3    3    0  0 17  0  0  0  0   0 17 11 44  0  0  0  0  0  6  0  0  0   18    0    0   1.523     53   0.75
   4    4    0  0  0  0  0  0  0   0  0  0  0  0  0  0  0 56 28 17  0  0   18    0    0   0.981     34   0.74
   5    5    0  0  0  0  0  0  0   0  0 100 0  0  0  0  0  0  0  0  0  0   18    0    0   0.000      0   1.25
   6    6   17  0  0  0  0  0  0   0  0  0  0 22  0  0  0  6  0 28  0 28   18    0    0   1.505     52   0.60
   7    7    0  0  0  0  0  0 44   0 17  0  0  0  0  0  0  0  0  0 39  0   18    0    0   1.026     36   0.60
   8    8    0  0  0  0  0  0  0   0  0 100 0  0  0  0  0  0  0  0  0  0   18    0    0   0.000      0   1.17
   9    9    0  0  0  0  0  0  0 100  0  0  0  0  0  0  0  0  0  0  0  0   18    0    0   0.000      0   1.28
  10   10    0  0  0  0  0  0  0   0  0  0  0  0  0  0  0  0  0 50  6 44   18    0    0   0.868     30   1.19
  11   11    0  0  0  0  0  0  0   6  0  0  0  0  0  0  0  0  0  0  0 94   18    0    0   0.215      7   1.26
  12   12    0  0  0  0  0  0  0   0100  0  0  0  0  0  0  0  0  0  0  0   18    0    0   0.000      0   1.28
  13   13    0  0  0  0  0  0  0   0  0 67 17 17  0  0  0  0  0  0  0  0   18    0    0   0.868     30   0.98
  14   14   28  0  0  0  0  0  0   0 28 44  0  0  0  0  0  0  0  0  0  0   18    0    0   1.072     37   0.60
  15   15    0  0  0  0  0  0  0   0  0  0  0  0  0  0  0  0  0100  0  0   18    0    0   0.000      0   1.31
  16   16    0  0  0  0  0  0  0   0  0  0  0  0  0  0  0  0 17 28  0 56   18    0    0   0.981     34   1.00
  17   17    0 67  0 33  0  0  0   0  0  0  0  0  0  0  0  0  0  0  0  0   18    0    0   0.637     22   1.13
  18   18    6  0 17  0  0  0  0   0 56  0  0 11  0  0  0  0  0  0  6  0   18    0    0   1.351     47   0.60
  19   19    0  0  0  0  0  6  0   0  0  0  0  0  0  0 61 17 17  0  0  0   18    0    0   1.059     37   0.55
  20   20    0  0  0  0 22  0 78   0  0  0  0  0  0  0  0  0  0  0  0  0   18    0    0   0.530     18   1.44
  21   21    0  0  0  0  0  0 83   0 17  0  0  0  0  0  0  0  0  0  0  0   18    0    0   0.451     16   1.00
  22   22    0  0  0  0  0  0  0   0 33  0 33  6  0  0  0  0  0  0 17 22   18    0    0   1.249     43   0.66
  23   23    0  0  0  0  0  0  0   0 39  0 17  0  0  0  0  0  0  6  0  0   18    0    0   1.459     50   0.61
  24   24    0100  0  0  0  0  0   0  0  0  0  0  0  0  0  0  0  0  0  0   18    0    0   0.000      0   1.14
  25   25    0  0  0  0  0  0  0   0  0  0  0  0  0  0 72  0 28  0  0  0   18    0    0   0.591     20   0.87
  26   26    0  0  0  0  0  0  0   0  0  0  0  0  0 56 17  0 28  0  0  0   18    0    0   0.981     34   0.80
  27   27    0  0  0  0  0  0100   0  0  0  0  0  0  0  0  0  0  0  0  0   18    0    0   0.000      0   1.12
  28   28    0 39 56  0  0  0  0   0  0  0  0  0  0  0  0  0  0  0  0  0   18    0    0   0.854     30   0.80
  29   29    0  6  0  0  0  0  0   0  0  0  0  0  0  0  0  0  0  0 94  0   18    0    0   0.215      7   1.08
  30   30   22 56  0 11  0  0  0   0  0  0  0  0  0  0  0  6  0  0  0  0   18    0    0   1.226     42   0.82
  31   31   44 17 39  0  0  0  0   0  0  0  0  0  0  0  0  0  0  0  0  0   18    0    0   1.026     36   1.01
  32   32    0  0  0  0  0  6  0   0  0  0  0 94  0  0  0  0  0  0  0  0   18    0    0   0.215      7   1.13
  33   33    0  0  0  0  0  0  0   0  0  0  0  0  0  0100  0  0  0  0  0   18    0    0   0.000      0   1.22
  34   34    0  0  0  0  0  0  0   0  0 22  0  0  0 22  0  0  0  0  0 56   18    0    0   0.995     34   0.84
  35   35    0  0  0  0  0  0  0   0  0  0  0  0  0  0100  0  0  0  0  0   18    0    0   0.000      0   1.14
  36   36    0  0  0  0  0  6  0  94  0  0  0  0  0  0  0  0  0  0  0  0   18    0    0   0.215      7   1.33
```

APPENDIX B. SERVICES AVAILABLE VIA THE EMBL FILE SERVER

Derived protein databases available from the Protein Design Group, EMBL, Heidelberg

DSSP, HSSP, FSSP and PDB_SELECT are derived databases. They are based on data from the Protein Data Bank of 3D structures and the Swissprot database of 1D protein sequences. The data and related test software can be obtained fia anonymous file transfer protocol or from an electronic mail file server.

DSSP	secondary structures extracted from 3D coordinates
HSSP	sequences homologous to known 3D structures
FSSP	structurally aligned families of proteins
PDB_SELECT	representative selection of PDB datasets

unix> ftp ftp.embl-heidelberg.de

Protein_Extras databases

/pub/databases/protein_extras/dssp
/pub/databases/protein_extras/hssp
/pub/databases/protein_extras/fssp
/pub/databases/protein_extras/pdb_select

Protein_Extras software

/pub/software/unix/dssp
/pub/software/unix/protquiz

mail to> netserv@embl-heidelberg.de

Protein_Extras databases

send a message containing, for example:

send proteindata:1ppt.dssp
send proteindata:1ppt.hssp
send proteindata:pdb_select.185

netserv replies by be sending the datasets requested

This service is a result of work by Uwe Hobohm, Liisa Holm, Reinhard Schneider, Michael Scharf, Chris Sander, Georg Tuparev, Gert Vriend (EMBL Heidelberg) and Wolfgang Kabsch (MPIMF, Heidelberg).

Secondary structure prediction server from the Protein Design group, EMBL, Heidelberg

mail to>
predictprotein@embl-heidelberg.de

Protein secondary structure prediction from sequence
send a message containing:

- the name and species of your protein in lines that start with a *
- an amino acid sequence in one letter code

predictprotein replies by sending the secondary structure prediction according to the PHD method. PHD = profile network prediction from Heidelberg by Burkhard Rost and Chris Sander, rated at 70.8% sustained accuracy. In practice, the predictions give a good first hypothesis of the structural properties of any newly sequenced water-soluble protein and may be an aid in the planning of point mutation experiments and in the prediction of tertiary structure. The neural network was trained on water-soluble proteins and should thus not be used for the prediction of membrane proteins.

Commercial users must apply for a license.

APPLICATIONS OF

KNOWLEDGE BASED MEAN FIELDS

IN THE DETERMINATION OF PROTEIN STRUCTURES

Manfred J. Sippl, Markus Jaritz, Manfred Hendlich,
Maria Ortner, and Peter Lackner

Center of Applied Molecular Engineering
University of Salzburg
Salzburger TechnoZ
Jakob Haringer Str.1
A-5020 Salzburg, Austria

1   INTRODUCTION

A major goal in protein structure theory is the development of physical models for protein solvent systems in terms of energy functions or force fields. The successful construction of a reasonable energy model would have an enormous impact on protein science and molecular biology paving the way for a vast number of industrial and scientific applications like the design and production of proteins with improved activity, increased thermostability or desired specificity. In the wake of the genome project the sequences of genes and their protein products are deciphered with an ever increasing rate but in spite of significant progress in experimental structure determination the ratio of known structures to known sequences is still decreasing (Bowie et al. 1991).

The ultimate goal is, of course, the calculation of native protein structures solely from the information contained in amino acid sequences. But there are other less demanding problems which have to be solved in advance. One such problem concerns the analysis of protein folds. Suppose we have the three dimensional structure of a given amino acid sequence. The structure may have been determined by X-ray analysis or nuclear magnetic resonance, it may be the result of modeling and prediction or it could be a deliberately misfolded structure - just designed to fool specialists in protein structure theory.

Statistical Mechanics, Protein Structure, and Protein Substrate Interactions
Edited by S. Doniach, Plenum Press, New York, 1994

Confronted with this structure and endowed with present day protein structure theory are we able to determine whether the structure is correct or incorrect and can we detect its faulty parts? The problem is not only of theoretical interest. Recently several experimentally determined structures have been shown to contain errors (Brändén and Jones 1990, Janin 1990). In other words experimental data are in some cases ambiguous and may be interpreted in alternative ways. In X-ray analysis problems may arise, for example, when the resolution is low or when the electron density is weak due to disorder in the crystal. Nuclear magnetic resonance is confronted with similar problems when the set of distance constraints derived from experiment is incomplete or contains incompatible distances.

Errors in experimentally determined structures reveal that there are no reliable theoretical techniques at hand which can be used to detect faulty structures or to judge the quality of a given protein fold. However, it is clear that as long as we are unable to detect errors in protein folds efforts to predict the structure of a protein from its amino acid sequence are doomed to failure.

During the last two years the situation has changed considerably and several exciting new approaches are currently pursued by several groups. The common theme is the search for a reasonable model of protein solvent systems based on the information contained in the data base of known structures (for reviews see Bowie and Eisenberg 1993, Wodak and Rooman 1993, Fetrow and Bryant 1993, Sippl 1993a, Jones and Thornton 1993, Unger and Sussman 1993, Godzik et al. 1993) The result of these knowledge based approaches is either a set of statistical parameters which describe the preferences of amino acids to exist in a particular structural context or a knowledge based force field composed of mean force potentials derived from the data base. The exciting point is that these approaches have already produced techniques which are applicable to problems in protein structure. Currently there are at least two methods available which can reveal misfolded parts of a structure. The method reported by Eisenberg and coworkers (Lüthy et al. 1992) is based on profiles derived from surface accessibilities and preferences for secondary structures. A second method (Sippl 1993b) uses a knowledge based force field to analyse the energy distribution in protein folds. A particular attractive feature of the latter method is its applicability to low resolution structures since the minimum requirement for an energy calculation is the C^α-trace of the protein.

A second impetus comes from improved strategies in force field testing. Although several force fields for macromolecular modeling have been developed over the last two decades little information is available on their predictive power and failures to calculate the native structures of proteins are often attributed to the multiple minimum problem. There is however, a simple concept which can be used to challenge the predictive value of force fields or parameter sets. A reasonable model of protein-solvent systems must be able to recognize the native fold of a given amino acid sequence among a set of decoys. If this condition is violated then the model will necessarily fail in an attempt to predict native folds and it cannot be used with any confidence in the analysis of structures, e.g. to uncover incorrect parts.

In the following exposition we outline the basic principles of test strategies which can be used to judge the predictive power of models for protein solvent systems (Sippl 1993ab, Sippl and Jaritz 1993). In addition these techniques are important tools in the design and refinement of force fields and parameter sets. Then we present an overview of the predictive power of a knowledge based force field derived from a data base of known structures. Finally we show how the force field can be applied to reveal misfolded parts of protein folds.

2 STRATEGY FOR THE ASSESSMENT OF THE PREDICTIVE POWER OF FORCE FIELDS

A most important role of energy in protein structure prediction is that it can be used as a guiding principle. If the equivalence between global minima of energy functions and native structures holds, then, at least in theory, native folds of proteins can be identified and calculated by minimizing the energy as a function of the conformational variables. If the condition is violated the search for the native fold necessarily goes astray.

Unfortunately, the 'global minimum equals native state' property is hard to verify for a given energy model. A rigorous proof requires the generation of all possible conformations of a single amino acid chain, the calculation of the associated conformational energies and the comparison of these energies to the energy of the native fold. The number of possible folds of small proteins is in the order of 10^{100}. Generation of these structures is far beyond the capabilities of present day computer technology.

On the other hand, if we find structures of lower energy as compared to the native fold, the equivalence of global minimum and native fold is falsified. This is the principle we adopt. We try to disqualify the knowledge based mean field by pretending that we have a large number of possible candidates for the native fold, but we do not know the correct one. We then ask whether the force field will be able to identify the native fold. If it is confused by one or several folds, then the correspondence between global minimum and native state cannot hold. If the force field succeeds, we still have no rigorous proof of this correspondence, but at least we get some estimate of the predictive power of the force field.

Novotny et al. (1984,1988) have used this principle in their investigation of the predictive power of semi-empirical force fields in its most basic form. They used a native fold and one grossly misfolded structure and found that the force field had difficulties in distinguish the correct from the incorrect fold. The intention should be to make the correct identification of folds as hard as possible. The set of decoys therefore, should contain as many reasonable folds as possible.

In our earlier studies (Hendlich et al. 1990) a set of alternative conformations was prepared from the data base of known structures and the ability of the knowledge based mean field to recognize the native fold was investigated for a large fraction of proteins in the Brookhaven data base (Bernstein et al. 1977). By shifting a sequence of length l through a given fold of length h, $h - l + 1$ conformations are obtained. It is clear that decoys can be derived from a fold only if $h \geq l$. Hence, for short sequences l, a large set of decoys is obtained from the data base, but for longer sequences the yield is small. Consequently, the result that the force field is able to recognize the native fold of a large protein is insignificant and this prevented the investigation of the performance of the force field when applied to the larger proteins in the data base.

A quality check of force fields should cover all proteins whose structures are available. Besides the insignificance of the results obtained for large proteins an additional drawback is that results obtained for proteins of different sequence length are incomparable, since the number of decoys obtained from the data base depends on the sequence length of the protein under study. This prevents the definition of a meaningful performance measure over the whole test set.

When the structures in the current data base are combined to a single polyprotein these problems can be avoided (Sippl and Jaritz 1993). The total length L of the polyprotein is then in the order of $50,000$ residues and for any sequence l we have $l \ll L$ so that for practical purposes the number of decoys $L - l + 1$ is independent of the particular length of the test protein. For any protein we have approximately the same number of decoys and the test is comparably hard for all proteins in the data base. Individual results can be compared and the average predictive power of the force field over the whole data base can be estimated.

In the construction of the polyprotein it is important to model the joining regions between successive protein modules carefully. If linker regions have unusual properties polyprotein fragments containing such regions will be easily recognized as unfavourable folds. The basic requirements are that linker regions have good stereochemistry and that there are no steric overlaps between linker regions and protein modules, or between adjacent protein modules along the polyprotein. A detailed account of polyprotein construction is found elsewhere (Sippl & Jaritz 1993).

The polyprotein is a source of possible conformations for a given amino acid sequence. Each fragment encountered along the polyprotein is combined with the sequence and the energy is evaluated. The energy range of a protein sequence depends on its sequence length so that the results obtained for distinct proteins are not immediately comparable. But the individual energy ranges can be transformed to z-scores:

$$z = \frac{(E - \bar{E})}{\sigma} \tag{1}$$

where $\bar{E}$ is the average energy of a sequence when combined with all fragments along the polyprotein and σ is the corresponding standard deviation.

To evaluate the predictive power of a force field the sequences of a set of proteins of known native structure are sent through the polyprotein. For a particular protein i we are interested in the following points: (1) Is there any conformation (i.e. a fragment in the polyprotein) of lower energy than the energy of the native fold E_i - in this case the force field is unable to recognize the native fold of protein i, and (2) if the native fold is recognized, how significant is the result? The significance or strength of recognition is measured by the score z_i of protein i obtained from

$$z_i = \frac{(E_i - \bar{E}_i)}{\sigma_i} \tag{2}$$

Here E_i is the energy of the native fold of sequence i and $\bar{E}_i$ and σ_i are the average energy and standard deviation of sequence i derived from the polyprotein. Finally the predictive power of the force field can be estimated by the average score of native folds in the test set:

$$\bar{z} = \frac{1}{N} \sum_{i=1}^{N} z_i \tag{3}$$

We emphasize that in the construction of the polyprotein all side chains have been removed so that it represents a polyalanine chain. Therefore, it is not possible to deduce the original amino acid sequence of the individual proteins from the polyprotein. The combination of amino acid sequences with polyprotein fragments is straightforward as long as the potentials used to calculate energies depend on the backbone atoms only (including C^β).

In the discussion of the results presented below it is necessary to understand the role of the polyprotein. In the construction of the polyprotein successive protein modules are joinded by linkers. The linkers are obtained from the data base of known structures. In constructing the polyprotein the primary goal is not to increase the number of globular folds which can be used as decoys, but to have a constant supply of possible folds for all chain lengths which are typical for globular proteins. This enables the definition of average scores for native folds. However, it is important that linker containing fragments do not produce energies distinct from those of fragments which are derived from individual protein modules. To achieve this goal the linker regions have to be designed carefully (i.e. no overlaps, good stereochemistry).

One argument against linker containing fragments would be that they are not compact nor globular and therefore easy to distinguish (even in terms of energy) from fragments which are derived solely from protein modules. This is not the case. Fragments derived from protein modules are in many cases also not compact nor globular and, on average, the energies of linker containing and linker free fragments are indistinguishable.

Nevertheless, by using the polyprotein as a source of possible conformations we gain a substantial number of globular and compact folds as compared to the original test reported by Hendlich et al. (1990). Consider ten structures, say, of sequence lengths between 150 and 160 residues. In the original method the number of folds which can be derived from this set, depending on the sequence length of the test protein, is quite distinct. For a sequence of 150 residues we obtain a substantial number of globular decoys (a structure of N residues yields $N - l + 1$ decoys for a sequence of length l), but for sequences of 160 residues we have only one structure. When a polyprotein is constructed from these ten folds then for all sequences (of the order of 150 to 160 residues) we have a constant supply of compact globular folds since the terminal residues of the longer sequences are allowed to protude into the linker regions. But, as already pointed out, a necessary prerequisite is that the connections between protein modules and linkers are carefully designed, so that the total energy is not artificially raised by the few residues protruding into the linkers.

3   PREDICTIVE POWER OF SEVERAL VARIANTS OF KNOWLEDGE BASED FORCE FIELDS

We present the results for three different force fields. The first is defined in terms of C^β mean force pair interactions. The second is based on a simple model for surface interactions and the third is a combination of pair interactions and surface terms.

The mean force pair interactions are of the form

$$\Delta E_r^{abk} = -kT \ln \frac{f_r^{abk}}{f_r^k} \tag{4}$$

as defined by Sippl (1990), where a and b correspond to the two amino acids which participate in the interaction, k is the separation of these residues along the sequence and r is the spatial distance between the C^β atoms of a and b. The required probability densities f are compiled from a data base of 192 known structures. The pair interactions are represented by several variants of potentials, depending on the separation k along the sequence. In the short range, $k = 1, 2, 3, 4, 5,$ and 6, individual potentials are compiled for each value of k. But for medium, $7 \le k \le 9$, and large separations, $10 \le k$, the pair interactions are condensed to a single type of potential. Figure 1 shows several examples of long range mean force pair potentials of hydrophobic amino acids and Figure 2 of charged amino acids.

The model for protein-surface solvent interactions is based on a simple neighborhood calculation. The center of a sphere of constant radius ($12\mathring{A}$ in the present study) is placed at the C^β atom of a particular residue and the number s of residues within this sphere is calculated. The mean force solvent interaction for amino acid a is then defined as (Sippl 1993a)

$$\Delta O_s^a = -kT \ln \frac{f_s^a}{f_s}. \tag{5}$$

The pair interaction energy $P(S, C)$ of sequence S in conformation C is obtained from a summation over all interacting pairs whose spatial distance is less than $20\mathring{A}$. Similarily the surface energy $O(S, C)$ is obtained as the sum over all amino acids. The combined energy is defined by

$$E(S, C) = P(S, C) + wO(S, C) \tag{6}$$

where the weight $w = \sigma_p / \sigma_o$ is calculated from the standard deviations σ_p and σ_o of energies derived from the polyprotein, so that the two energy contributions have comparable weight.

We are interested in the following points:

1. What is the predictive power of the pair interaction energy?
2. What is the predictive power of the surface energy?
3. Is the predictive power raised if pair interactions and surface energies are combined?
4. In how many cases is the force field unable to recognize the native fold for the three types of force fields?

A set of 192 proteins was used to compile the mean force pair and surface potentials. The same structures were used to construct the polyprotein and they also comprise the test set. In all cases the protein subjected to the test calculation was excluded from the potentials.

Figure 3 presents the results obtained for pair interactions. Scores of native folds are plotted as a function of sequence length. Diamonds ($\Diamond$) correspond to proteins whose native fold is recognized by the force field, and ($+$) represents proteins whose native fold is not the most favorable in terms of the energy calculated. There are 18 cases, mostly small proteins, where the combination of a protein sequence with a non native fold is more favorable (Table 1).

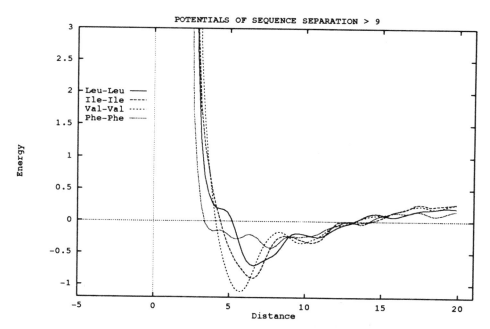

Figure 1. Several examples for mean force pair potentials of hydrophobic amino acid pairs. The potentials shown are long range potentials where $k > 9$. The energy scale is in units of $1/kT$.

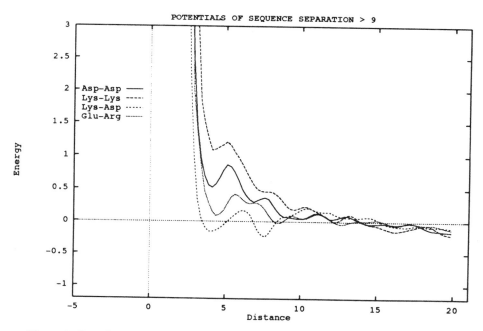

Figure 2. Several examples for mean force pair potentials of charged amino acid pairs. The potentials shown are long range potentials where $k > 9$. The energy scale is in units of $1/kT$.

The strength of recognition of native folds increases with sequence length. There may be several reasons responsible for this increase. For example, the number of (favourable) interactions increases with protein size. In other words large proteins contain more information and the recognition signal therefore becomes stronger. Another reason may be that the relative size of the accessible conformation space represented by the fragments of the polyprotein is small for the larger proteins.

Figure 3 (right) assembles the results for miscellaneous groups of structures. As a control several deliberately misfolded proteins were constructed by combining sequences with folds in an arbitrary manner. As expected these folds are not recognized as the most favourable for the associated sequence and the scores of these misfolded proteins ($\times$) are scattered around zero (i.e they have energies close to the average energy calculated from the polyprotein) and some are even strongly positive pointing to completely incompatible sequence structure combinations (i.e. worse than average).

The three squares in Figure 3 represent folds which have been determined by experiment, but which have subsequently been shown to contain errors (from left to right: gene 5 DNA-binding protein, 2gn5; small subunit of rubisco; thaumatin 1thi). The scores of these folds are rather insignificant and in all these cases there are other folds derived from the polyprotein having more favourable energy. The membrane proteins ($\Diamond$) and virus coat proteins (+) form distinct classes. Nevertheless, in terms of pair interaction energies these folds produce scores comparable to soluble globular proteins.

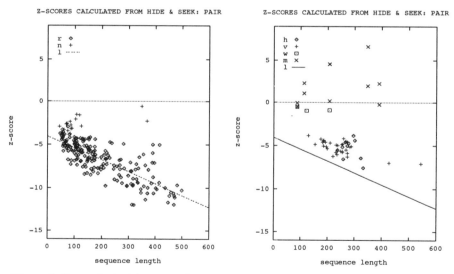

Figure 3. Scores obtained from mean force pair interactions. Left: Scores of soluble globular proteins. Proteins whose folds are recognized are marked by $\Diamond$, folds not identified are marked by +. Right: Scores for membrane proteins ($\Diamond$), virus coat proteins (+), folds known to be incorrect ($\Box$) and deliberately misfolded proteins ($\times$). The regression line l is calculated from the recognized proteins shown in the left figure.

Table 1. Proteins whose reported fold is not recognized by pair interactions

Protein	length	rank	z_n †	z_{opt} ‡	Protein	length	rank	z_n †	z_{opt} ‡
1pte	348	12116	-0.62	-3.83	1hip	85	152	-2.78	-4.36
2cdv	107	2640	-1.56	-4.26	3fgf	124	69	-2.90	-3.91
1cy3	118	2191	-1.63	-4.02	1fc2-c	43	78	-2.92	-4.24
1mda-f	103	578	-2.17	-3.92	1ten	89	8	-3.16	-3.86
1cpb	82	126	-2.55	-4.41	1zaa-c	85	18	-3.21	-3.56
2hip-a	71	172	-2.57	-3.72	1cdt-a	60	7	-3.32	-3.93
2ctx	71	5	-2.61	-2.67	1hoe	74	6	-3.45	-3.72
1mda-c	368	204	-2.63	-4.05	351c	82	6	-3.76	-4.16
1trc-b	68	242	-2.69	-6.43	1cho-i	53	3	-3.82	-4.11

†, z-score of native fold

‡, z-score of fold of lowest energy derived from the polyprotein

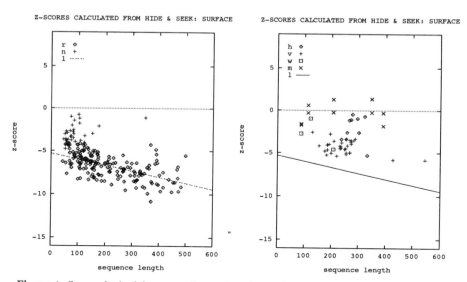

Figure 4. Scores obtained from mean force solvent interactions. See Figure 3 for the symbols used.

Table 2. Proteins whose reported fold is not recognized by surface interactions

Protein	length	rank	z_n†	z_{opt}†	Protein	length	rank	z_n†	z_{opt}†
1tro-g	101	10219	-0.75	-4.07	2ctx	71	21	-3.36	-4.29
3fis-a	73	7493	-0.98	-4.59	1grd-b	81	3	-3.51	-3.58
1pte	348	5939	-1.12	-4.30	hmbm-i	54	3	-3.62	-3.93
2wrp-r	104	5497	-1.19	-4.38	1crn	46	52	-3.66	-5.09
1zaa-c	85	3405	-1.43	-3.61	3ebx	62	14	-3.67	-4.85
1cmb-a	104	1852	-1.77	-3.91	1hoe	74	10	-3.76	-4.09
1ccd	77	1302	-2.01	-3.81	1hdd-c	57	6	-3.78	-4.15
2hmg-b	175	800	-2.10	-4.20	hcys-i	108	5	-3.81	-3.96
1cpb	82	425	-2.54	-4.47	1tgs-i	56	21	-3.84	-4.71
2kai-a	80	136	-2.75	-4.10	5pti	58	5	-3.95	-4.17
1utg	70	338	-2.67	-4.82	1tpk-a	88	3	-3.96	-4.16
rop2-a	56	344	-2.71	-4.27	rub-c	122	3	-4.04	-4.46
1cdt-a	60	99	-2.89	-4.49	1aap-a	56	18	-4.13	-5.11
2kai-b	152	98	-2.98	-4.70	1pgx	70	4	-4.28	-4.74
1cc5	83	70	-3.02	-4.34	5rxn	54	19	-4.33	-5.40
4sgb-i	51	121	-3.03	-4.46	1sn3	65	3	-4.40	-4.48
8rub-s	123	17	-3.26	-4.16	1trc-b	68	3	-4.43	-6.69
2abx-a	74	9	-3.29	-3.66	1rdg	52	8	-4.79	-5.66

†, see Table 1

The results for the surface energy are shown in Figure 4. There are a number of small proteins where the force field fails to recognize the native fold (Table 2). For deliberately misfolded and virus coat proteins the results are similar to those for the pair interaction but there are several differences with respect to other classes of folds. One of the known incorrect folds (1thi) yields a good score for the surface energy. The largest difference is found for integral membrane proteins. Their scores are close to zero reflecting the obvious fact that the solvent interactions derived from soluble globular proteins are not applicable to integral membrane proteins, which interact with a hydrophobic environment. It is noteworthy however, that the situation is different for pair interactions, where the potentials derived from soluble globular proteins provide a seemingly adequate model for the intramolecular interactions in membrane proteins.

When pair interactions and surface interactions are combined, then there is a strong increase in recognition success (Figure 5). There are only a few proteins whose folds are not recognized (Table 3). Results for integral membrane proteins are intermediate.

Table 4 summarizes recognition success for the three types of force fields using two test sets of protein structures. The first set consists of 192 proteins which were used to compile the mean force potentials (but in the tests the protein under study was removed). The second set is obtained from the first set by addition of 23 (mostly low resolution) structures.

Some of the proteins which are not identified are peculiar. 1trc-b is a low resolution structure (3.6Å) of a fragment of calmodulin (residues 78-148). The most favourable structure for this sequence has a score of -9.12 corresponding to a fragment of 5tnc a calmodulin which is part of the polyprotein. Another example is 1zaa-c composed of three zinc-finger domains. The zinc-ions play a dominant role in the stabilization of this structure. Obviously, in this case the energy model lacks important contributions due to the neglect of zinc-protein interactions. For the low resolution structures of D-alanyl carboxypeptidase (1pte; 2.8Å) and carboxipeptidase B (1cpb; 2.8Å) only the C^α coordinates are available.

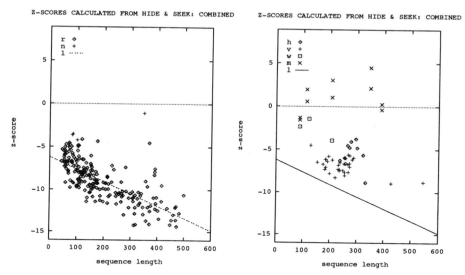

Figure 5. Scores obtained from the combined mean force pair and solvent interactions. See Figure 3 for the various symbols used.

Table 3. Proteins whose reported fold is not recognized by combined pair and surface interactions

Code	length	rank	z_n †	z_{opt} †	protein
1pte	348	6179	−1.09	−4.26	D-alanyl carboxypeptidase
1zaa-c	85	25	−3.53	−4.70	zinc finger (3 domains)
1cpb	82	16	−3.65	−5.21	carboxypeptidase B
1tro-g	101	6	−4.26	−5.09	TRP-repressor in complex with operator
3fis-a	73	3	−4.62	−4.81	Fis protein
1trc-b	68	4	−5.02	−9.12	calmodulin

†, See Table 1

Table 4. Summary of recognition success

Potential	mean field data base (192)		enlarged data base (215)	
	average score	% recognized	average score	% recognized
pair	−6.72	94	−6.43	92
surface	−6.48	90	−6.16	83
combined	−9.25	100	−8.86	97

The average predictive power obtained for the first set is -6.72 for pair interactions and -6.48 for the solvent terms. When combined the predictive power rises to -9.25. This indicates that the information contained in the two terms is largely complementary.

In summary with the exception of integral membrane proteins and a few low resolution structures of soluble proteins the combined force field is able to recognize the native folds of practically all proteins in the current data base. The folds of membrane bound proteins are recognized by pair interactions but not by solvent interactions.

4 SEQUENCE RECOGNITION

In the experiment described, the force field was tested for its ability to recognize the native fold belonging to a particular sequence. Another interesting point is whether the force field is able to recognize the sequence belonging to a particular fold. This problem, known as the inverse folding problem (Bowie et al. 1991) or structure recognizes sequence protocol (Wodak and Rooman 1993), is somewhat different from the 'canonical' folding problem investigated in the previous section.

There is a subtle difference between the folding problem and its inverse as pointed out by Sippl (1993a). Suppose that we have an energy function which provides a reasonable model for protein solvent systems. In this case the folding postulate (Anfinsen 1973) guarantees that the native structure of a particular protein sequence corresponds to the global minimum of this energy function. However, there is no such principle in inverse folding.

There is no gurantee that a native sequence structure pair is found when, starting from a native fold and its corresponding amino acid sequence, the energy is minimized by changing the sequence leaving the structure unchanged. In other words, sequences of lower energy may be found but the starting structure may not correspond to the native fold of the minimized sequence. Subsequent minimization of conformational energy may reveal structures of still lower energy. It is therefore an interesting question whether or not the force field is able to recognize sequences belonging to a known native fold in an experiment similar to the fold recognition study described in previous sections.

The implementation of such a test is straightforward. In the inverse problem we use the sequence of the polyprotein (which is obtained by joining the sequences of the proteins which were used in the construction of the polyprotein). We then take a fold, represented by its backbone, so that the original sequence cannot be deduced from the scaffold, and combine it with all fragments along the polysequence. The force field successfully recognizes the native sequence-structure pair if the combination of native conformation and its associated sequence has lowest energy among all alternatives. The strength of recognition is again expressed in terms of a z-score, where the average energy and standard deviation are now calculated from the energy distribution obtained from the combination of a single fold with all sequences derived from the polyprotein.

The results obtained from this study for the combined force field (Figure 6) are quite similar to those obtained for the 'canonical' problem (Figure 5). Again the deliberately misfolded proteins and folds known to be incorrect have insignificant scores. The same result is obtained for the integral membrane proteins. Sequences of soluble globular proteins which are not recognized are confined to small proteins (Table 5).

The predictive power of the force field for the inverse problem is smaller, but the differences to the canonical problem are rather marginal (compare Tables 4 and 6). From the results it seems that there is little difference between the canonical folding problem and its inverse.

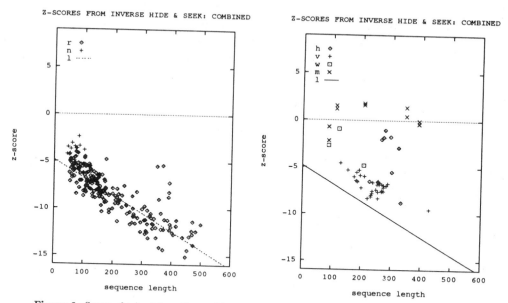

Figure 6. Scores obtained from the combined mean force pair and solvent interactions for the inverse problem. See Figure 3 for the various symbols used.

Table 5. Proteins whose reported fold is not recognized by combined pair and surface interactions in the inverse problem.

Code	length	rank	z_n†	z_{opt}†	protein
1cpb	82	350	-2.34	-3.89	carboxypeptidase B
1cdt-a	60	55	-3.05	-4.18	cytotoxin
1zaa-c	85	9	-3.34	-4.11	zinc finger (3 domains)
1utg	70	22	-3.42	-3.92	uteroglobin
1crn	46	13	-3.46	-3.97	crambin
1ccd	77	16	-3.58	-4.06	Clara cell 17 kD protein (3Å)
3fis-a	73	11	-3.66	-4.71	FIS factor for inversion stimulation
rop2-a	56	12	-3.67	-4.47	repressor of primer
1tro-g	101	10	-3.78	-4.53	TRP-repressor in complex with DNA
1cho-i	53	3	-4.19	-4.50	protease inhibitor
1fc2-c	43	3	-4.23	-4.38	protein fragment
2abx-a	74	2	-4.50	-4.60	neurotoxin
1trc-b	68	2	-4.95	-5.29	calmodulin (3.6Å)
2hhb-b	146	2	-6.56	-7.35	hemoglobin β (first position 1mbd)

†, See Table 1

Table 6. Summary of recognition success for the inverse problem

Potential	mean field data base (192)		enlarged data base (215)	
	average score	% recognized	average score	% recognized
pair	-7.36	95	-7.10	93
surface	-5.61	82	-5.31	75
combined	-8.58	97	-8.20	94

The polyprotein contains only a vanishingly small fraction of the 20^l possible sequences of length l and the chance to find a sequence of lower energy is very small. The situation is different when the energy of a protein is minimized by mutating amino acid residues (keeping the conformation rigid). This directed search converges rapidly to a sequence which is completely different from the starting sequence. However, when the sequence structure pair obtained from minimization is subsequently subjected to the 'canonical' test, in general an insignificant score is obtained, i.e. there are other structures which have more favourable energy when combined with the minimized sequence (manuscript in preparation).

In summary it is easy to demonstrate, that by minimizing in sequence space (structure constant) we can generate sequence structure pairs of lower energy as compared to the starting sequence-structure pair (i.e. a protein whose structure was solved by experiment). However, sequences obtained in this way are likely to adopt a different fold, as can be demonstrated by subjecting the minimized sequence to the canonical test.

The piece still missing in our puzzle is that we are able to proof that native folds correspond to global minima of knowledge based mean fields in conformation space. From the results obtained in the canonical test, we get some idea on the predictive power of the knowledge based mean field. But recognition of native folds is only a necessary condition for the correspondence of global minima and native structures. It is not sufficient. Hence, the next step should be energy minimization in conformation space. This is technically far more demanding than minimization in sequence space and, provided that no other conformation of lower energy is found, is equivalent to the solution of the folding problem.

The canonical test is a powerful tool for protein structure analysis. If a model for the native fold of a given amino acid sequence produces an insignificant score it is likely that the structure has some problems. This is demonstrated by the scores obtained for deliberately misfolded proteins and structures known to be incorrect. We will now analyse the distribution of energies in protein folds in terms of the mean field in more detail (Sippl 1993b).

5   ENERGY ANALYSIS OF PROTEIN FOLDS

The confromational energy is calculated as a sum of pairwise interactions and surface terms. This energy can be decomposed into the contributions from the different residues. The energy of a particular residue in a protein is

$$E_i = \sum_j e_{ij} + o_i \qquad (7)$$

Here e_{ij} is the pair interaction energy of residues i and j and o_i is the interaction energy of residue i with the solvent. Since in the present study we use C^{β} based potentials only

$$e_{ij} = \Delta E_r^{a(i),b(j),k} \tag{8}$$

and

$$o_i = O_s^{a(i)} \tag{9}$$

where $a(i)$ and $b(j)$ denote the amino acids at sequence positions i and j respectively.

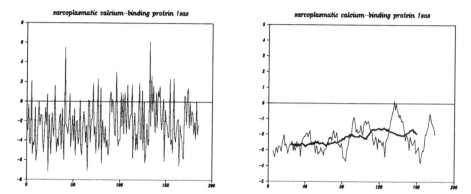

Figure 7. Energy graph for 1sas (sarcoplasmatic calcium binding protein). The graph on the left displays the interaction energies of the individual residues along the chain (no averaging). On the right the graph is smoothed using a window size of 10 (thin line) and 50 residues (bold line), respectively.

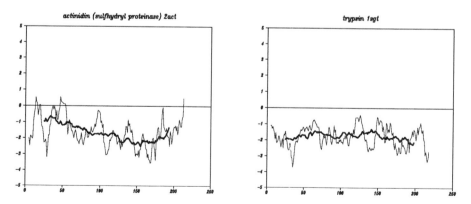

Figure 8. Energy graphs for actinidin (2act) and trypsin (1sgt. The graphs correspond to averages over 10 (thin line) and 50 residues (bold line), respectively.

When E_i is plotted as a function of sequence position i we obtain an energy graph displaying the energy distribution in a protein fold. Since energy graphs fluctuate strongly some averaging over a window of neighboring residues is required. Small windows reveal fine details of the energy distribution in protein folds whereas larger windows yield an overall view.

Figure 7 shows energy graphs calculated from the structure of sarcoplasmatic calcium binding protein 1sas using several window sizes. The graph for window size 1 (no averaging) is rather chaotic. The energies fluctuate strongly from one amino acid position to the next. A window size of 10 residues suffices to damp the fluctuations but strained regions (high energies) are still discernible. When a window size of 50 residues is used the resulting graph is rather flat.

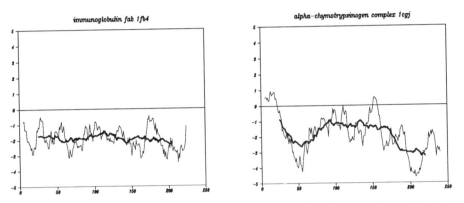

Figure 9. Energy graphs for immunoglobulin FAB (1fb4) and α-chymotrypsinogen (1cgj). The graphs correspond to averages over 10 (thin line) and 50 residues (bold line), respectively.

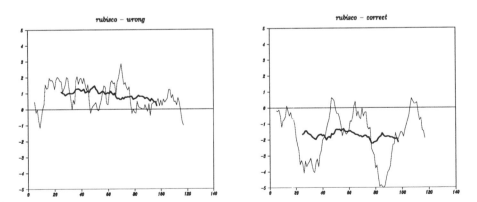

Figure 10. Energy graphs for an incorrect model of rubisco (left) and the subsequently corrected structure (right). Window sizes for averaging are again 10 (thin line) and 50 (bold line), respectively.

Several graphs calculated from native folds of soluble globular proteins are shown in Figures 8 and 9. Graphs averaged over a window of 10 residues show occasional peaks of positive energies. Energy graphs calculated from a 50 residue window are rather uniform staying in the region of negative energy. The main feature of native folds is the rather uniform distribution of favourable energies throughout the molecule.

The situation is quite different for incorrect folds. As an example Figure 10 compares the energy graphs of an incorrect model of rubisco and its correct structure (both structures resulted from X-ray analysis). The graph of the incorrect model has overall positive energies. In contrast the correct structure has an energy distribution typical for native protein folds.

Another erroneous structure is 1thi, thaumatin. The graph in Figure 11 displays positive energies in the N-terminal two thirds of the structure. At the C-terminus the graph drops to negative energies. The structure of 1thi was determined to a resolution of 2.8Å (DeVos et al. 1985).. Recently the structure was refined to 1.65Å resolution (Ogata et al. 1992). During refinement several frame shift errors were detected. These errors are located in the N-terminal region of the protein corresponding to the high energy regions in Figure 11.

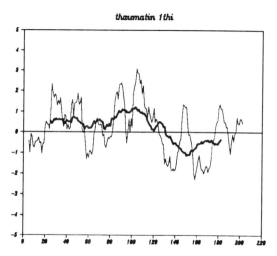

thaumatin 1thi

Figure 11. Energy graph of thaumatin 1thi. The high energy regions in the N-terminal two thirds of the graph correspond to frameshift errors in the low resolution model.

6   CONCLUSION

We have demonstrated that knowledge based force fields have considerable predictive power. They can be used to recognize incorrect folds and to detect faulty parts in protein structures. The results were obtained from a most simple description of protein solvent systems where the pair interactions and solvent terms were calculated from C^β atoms only (a force field based on C^α-atoms performes comparably well). A more sophisticated force field which incorporates more atoms and interactions should have still more predictive power.

Energy graphs are a most effective tool in protein structure research. The possiblity to analyse energy distributions in protein folds based on C^α or C^β interactions only is an enormous advantage, both in the interpretation of experimental data as well as in the calculation of protein folds. The technique allows the judgement and correction of early chain tracings through electron densities even at low resolution, with practically no effort. Calculation and display of energy graphs is completed within a second on typical workstations, so that during the modeling process changes in energy distributions can be monitored in real time.

Native folds have characteristic energy distributions, like those shown in Figures 7 to 10. Energy graphs therefore, open interesting new ways to approach the protein folding problem. The goal in ab initio calculations, i.e. in calculations based solely on the amino acid sequence and the force field employed, but with no additional information which could possibly derived from experiment, is to calculate the native fold by minimizing the energy as a function of the conformational variables. In native folds the energy, on a scale averaged over 50 residues, is well balanced throughout the molecule. An obvious strategy uses interactive modeling supported by energy graphs and energy minimization, where in each step of the modeling process, the energy distribution, reveals strained or faulty parts of the structure. One of the main goals in our current work is to establish interfaces to programs used for structure calculations in X-ray analysis and nuclear magnetic resonance spectroscopy and the combination of knowledge based mean fields with energy minimization and molecular dynamics.

Acknowledgements

The coordinates for the correct and incorrect structures of rubisco were obtained from David Eisenberg. This work was supported by the Austrian Fonds zur Förderung der wissenschaftlichen Forschung, grant number 08361-CHE and by the Jubiläumsfonds der Österreichischen Nationalbank, grant number 4356.

References

Anfinsen,C.B. (1973). Principles that govern the folding of protein chains *Science.* **181**, 223-230

Bernstein,F.C., Koetzle,T.F., Williams,G.J.B., Meyer,E.F.,Jr., Brice,M.D., Rodgers,J.R., Kennard,O., Shimanouchi,T. & Tasumi,M. (1977). The protein data bank: A computer based archival file macromolecular structures. *J.Mol.Biol.* **112**, 535-542

Bowie,J.U., Lüthy,R.& Eisenberg.D. (1991). A method to identify protein sequences that fold into a known three-dimensional structure. *Science* **253**, 164-170

Bowie,J.U. & Eisenberg,D. (1993). Inverted protein structure prediction. *Curr.Opin.-Struct.Biol.* **3**, 437-444

Bränden,C.-I., Jones,T.A. (1990). Between objectivity and subjectivity. Nature (London) **343**, 687-689

DeVos,A.M., Hatada,M., Van der Wel,H., Krabbendam,H., Peerdeman,A.F., Kim,S.-H. (1985). Three dimensional structure of Thaumatin I, an intensely sweet protein. *Proc.Natl.Acad.Sci.,USA.* **82**, 1406-1409

Fetrow,J.S. & Bryant,S.H. (1993). New programs for tertiary structure prediction. *Biotechnology.* **11**, 479-484

Godzik,A., Kolinski,A. & Skolnik,J. (1993). De novo and inverse folding predictions of protein structure and dynamics. *J.Comput.-Aided Mol.Design.* **7**, 397-438

Hendlich,M., Lackner,P., Weitckus,S., Floeckner,H., Froschauer,R., Gottsbacher,K., Casari,G. & Sippl,M.J. (1990). Identification of native protein folds amongst a large number of incorrect models. *J.Mol.Biol.* **216**, 167-180

Janin,J. (1990). Errors in three dimensions. Biochimie, **72**,705-709

Jones,D.T. & Thornton J. (1993). Protein fold recognition. *J.Comput.-Aided Mol.Design,* **7**, 439-456

Lüthy,R., Bowie,J.U., Eisenberg,D. (1992). Assessment of protein models with three-dimensional profiles. *Nature (London),* **356**, 83-85

Novotny,J., Brucceroli,R.E. & Karplus,M. (1984). An analysis of incorrectly folded protein models. Implications for structure predictions. *J.Mol.Biol.* **177**, 787-818

Novotny,J., Rashin,A.A., & Bruccoleri,R.E. (1988). Criteria that discriminate between native proteins and incorrectly folded models. *Proteins* **4**, 19-30

Ogata,C.M., Gordon,P.F., de Vos,A.M. & Kim,S.H. (1992). Crystal structure of a sweet tasting protein Thaumatin I, at 1.65 A resolution. *J.Mol.Biol.* **228**, 893-908

Sippl,M.J. (1990). Calculation of conformational ensembles from potentials of mean force. An approach to the knowledge-based prediction of local structures in globular proteins. *J.Mol.Biol.* **213**, 859-883

Sippl,M.J. (1993a). Boltzmann's principle, knowledge based mean fields and protein folding. *J.Comput.-Aided Mol.Design,* **7**, 473-501

Sippl,M.J. (1993b). Recognition of errors in three dimensional structures of proteins. *Proteins,* in press

Sippl,M.J., Jaritz,M. (1993). Predicitive power of mean force pair potentials. submitted

Unger,R. & Sussman,J.L. (1993). The importance of short structural motifs in protein structure analysis. *J.Comput.-Aided Mol.Design.* **7**, 457-472

Wodak,S.J. & Rooman,M.J. (1993). Generating and testing protein folds. *Curr.Opp.-Struct.Biol.,* **3**, 247-259

A NEW APPROACH TO PROTEIN FOLDING CALCULATIONS

R.L. Jernigan and K.-L. Ting

Laboratory of Mathematical Biology
Division of Cancer Biology, Diagnosis and Centers
National Cancer Institute
National Institutes of Health
Bethesda, MD 20892 USA

INTRODUCTION

Previously we demonstrated the feasibility of generating large number of protein conformations within the shape outlined by the known structure and subsequently evaluating them to demonstrate that the residue-residue potential functions can be usefully applied to assess them(Covell and Jernigan, 1990). In this paper we are going to outline some ways to improve the efficiencies of such generations and evaluations.

Restricting conformations to a set of n points for placing the n residues is a highly effective way to reduce the numbers of conformations. For example, consider a chain of 8 bonds in 2 dimensions, with three choices at each added bond. If the chain were unrestricted, with no volume exclusion, then there would be 3^7 forms or 2187 conformations. However, if multiple occupancy is forbidden and the molecule is constrained to the square 3x3 in size, then there are only a few forms possible, as we will enumerate further.

Other ways to achieve even further reductions are to maximize certain types of interactions. For instance, restricting cases to those with the maximum number of non-bonded contacts within the interior is highly effective. Consider a 6x6 2-dimensional square with 35 chemical bonds. Without any restrictions there would be 3^{34} or 1.6×10^{16} conformations. Restricting it to the square alone, with volume exclusion, as above, yields about 57,000 conformations. However, maximizing the number of intra-layer non-bonded contacts reduces the number of forms to only 5.

There are also many restrictions on the possible shapes themselves (Jernigan, 1991). For instance, because proteins are linear chain molecules, only two ends are permitted. If there are more than two, it is not an allowed shape. The shape

       X X X
       X X X
         X   for example is impossible because it has three ends.

Another type of impossible shape is one in which too many ends would be required to pass through a single point

         X
       X X X
         X X

Statistical Mechanics, Protein Structure, and Protein Substrate Interactions
Edited by S. Doniach, Plenum Press, New York, 1994

317

In this example, the top end and the left end both must connect into the same point, which in turn must be connected to the other points.

A third type of impossible case arises because of the lattice characteristics, a sort of odd/even effect for the square lattice. The chain ends can be located with respect to one another at impossible positions. For example, the following shape cannot accommodate a single chain.

It is an odd/even characteristic of the square lattice that such a set of lattice points cannot be occupied completely by a single chain. Try to trace a chain on these points yourself to see that it is impossible to occupy each point once and only once with a linear chain.

Residue-Residue Potential Functions

There is extensive experience in the field of polymers for treating interaction energies among groups of atoms as parameters to be evaluated from experimental data on small analog molecules or even by fitting data on large macromolecules (Flory, 1969; as an example, see Bohacek et al., 1991). Such approaches were utilized in developing the energy parameters for diverse sets of polymer repeat units with rotational isomeric state models. Part of the rationale for treating these interaction terms as adjustable parameters is that they can change, depending on the experimental conditions. For example studies by Mizushima (1954) demonstrate changes in the relative likeliness of isomeric states for different solvents and other differing conditions. These approaches validate the approach of adjusting potential functions based on some limited molecular data. In the case of proteins, it is possible to use the crystal structures directly to develop relative interaction preferences based strictly upon the observed frequencies of occurrences of residue-residue contact pairs. The direct approach taken in extracting such functions from the crystal structures is shown in Fig. 1a, and the results from Miyazawa and Jernigan (1985) are given in Fig. 1b. Previously Tanaka and Scheraga (1976) had performed a similar averaging of structural data. Subsequently, several others (including Gregoret and Cohen, 1990 and Sippl, 1990) have made similar analyses. Our residue-residue interaction energies are derived strictly from crystal structures and resemble a pair-wise hydrophobicity index. Conceptually it would be possible to obtain a different set of potentials based on NMR solution derived structures alone. Comparisons between the two should be informative about the effects of the crystal compared to the solution environment. The most important features of the structures can be obtained only if averaging of the data is performed, to remove errors in the structures as well as to remove individual aberrant cases. Use of all of the structural data is less likely to yield ultimate success than using results from judicious averaging. We have also looked at these interactions in more general ways (Bahar and Jernigan, 1993a) as a uniform homogeneous term, the same for all interactions, and have also probed individual positions to see where changes in interactions would most easily cause changes in conformation (Bahar and Jernigan, 1993b).

These potential functions have numerous uses to quickly assess conformational quality. They can be used to select sets of good conformations from large sets of conformations (Covell and Jernigan, 1990), and to derive amino acid substitution matrices for use in sequence comparisons (Miyazawa and Jernigan, 1993a) Another use of the potentials is for evaluating effects of amino acid substitutions on stability (Miyazawa and Jernigan, 1993b). In a similar way, hydrophobicities can be used to locate peptide binding sites on protein surfaces (Jernigan and Covell, 1991a; Young et al., 1993).

There are many questions to ask about these potentials. For instance, how variable are they from one situation, or from one protein to another?

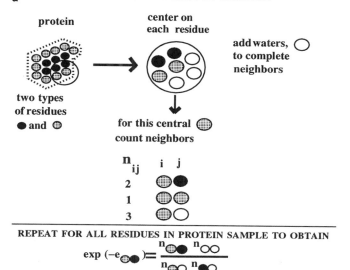

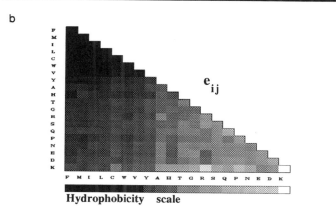

Figure 1. a) Top - Schema showing how residue-residue contact potentials are derived. A sphere is centered on each residue and the types of residues of the non-bonded neighbors are counted. Two types of residues are shown as filled and shaded circles; groups of waters are given as open circles. The counting is performed over all proteins in the sample of crystal structures to obtain effective interaction energies for each type of residue pair. b) Bottom. Results from Miyazawa and Jernigan (1985). Most frequently occurring pairs are shown in black which correspond to the hydrophobic pairs. Shading is by gray scale, with the least frequently occurring polar pairs shown in the lightest shades. Hydrophobic-polar pairs are of intermediate frequency. The hydrophobicity scale given at the bottom was derived in a similar way from the same structural data.

How Constant Are Environments of a Residue Type?

In an attempt to investigate this question, we undertook the following data collection: For each of 709 crystal subunits, we have calculated the average interaction energy per interacting pair. In Fig. 2 is shown the distribution of these values for the set of proteins for each type of residue.

$$e_i{}^* \text{(each protein)} \ = \ \sum_j n_{ij} e_{ij} \ / \ \sum_j n_{ij}$$

where n_{ij} is the number of i-type, j-type pairs within spheres of 6.5 A radius for one protein. The nearest sequence neighbors have been excluded in the counts.

The values do appear to reflect the hydrophobicities; the polar residue types have weaker total interactions and usually lie in a band between a value of -1.5 and -2.5 for $e_i{}^*$. The non-polar types have lower values, typically between -4 and -5. Several residues fall principally between these two ranges, notably alanine, cysteine, histidine, proline and tyrosine. It is interesting to consider whether or not the smaller variability of the polar residue types arises principally because of the smaller number of neighbors, under the premise that larger numbers of neighbors with similar individual variability would result in greater overall variability.

The fascinating part of this result is that, even over the wide range of proteins that are included in the sample, there are clearly separable regions of total hydrophobicity of the environments for all occurrences of a residue type. In addition, the smaller variability of hydrophobic environment for the polar residues indicates that they would be particularly good residues for specifying their relative location in the protein globule.

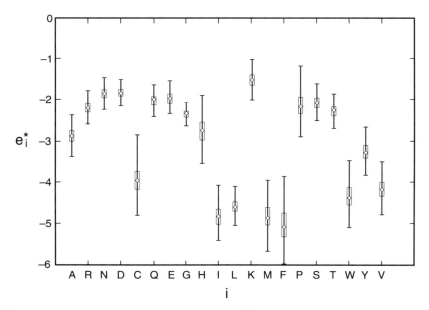

Figure 2. Variability of environmental interaction energies. These were calculated from the e_{ij} values in Fig. 1 by summing over all neighboring residues within 6.5A in 709 protein subunit crystal structures. Means are given as points; half of the data are enclosed in boxes. The outer limits are at the ends of a bar. See text for further details.

SCHEME FOR GENERATING FOLDED FORMS

There are several points to take advantage of here. First is the significant efficiency to be obtained when conformations are generated for a fixed shape (Covell and Jernigan, 1990; Jernigan and Covell, 1991b). Specifically, if n points corresponding to n residues are to be placed in a unique conformation upon n lattice points, the possible number of conformations is enormously reduced as described in the introduction. One possible scheme would be to generate shapes for a given composition. Shape is defined as the adjacent set of lattice points that have no specified occupancy. For example, a shape could be

```
    X X
  X X X
  X X X X
```

Then generate all conformations for this given shape.

One example is

```
          1 - 2
           •   |
        5 - 4 - 3
        |    •  •
        6 - 7 - 8 - 9
```

While generating the conformations, limit all residue environments to the observed ranges for hydrophobic energies of interaction. Calculate this on the fly, by seeing whether or not the remaining residues can be placed in a bound calculation, for the remaining residues, to obtain an energy within the accepted range for a residue of this type. Such a calculation would not be done until a residue has a certain substantial fraction of its neighbor positions occupied. In the above example, at one stage in the conformation generation, we would have

```
          1 - 2
           •   |
        X  4 - 3
        X  X  X  X
```

For such a 2 dimensional square lattice, the point occupied by unit 4 has a maximum of two non-bonded interactions, and one of them has been formed with unit 1. To find the limit of what is achievable with the remainder of the chain, without any constraints of conformation, one would consider the residue type of unit 4 and its interaction energy with type 1 together with 4•{5 or 6 or 7 or 8 or 9} to see if some combination of these interactions would fall within the customary range of energy of interaction of 4 type residues. Basically, such a test would determine whether or not each residue has been placed within an environment that is either too polar or too hydrophobic to be acceptable, and that this chain generation should be terminated.

If polar residues for example could be restricted to the exterior of the shape, then this restriction will be more advantageous in three dimensions than in two.

It would seem that independent folding domains could be defined on the basis of the polar composition of a segment of the sequence. For each size chain, there would be an optimal polar fraction, that could be specified within some range, for globular structures.

There are improvements in the efficiency of conformation generation, if both sequence and conformation are considered simultaneously. In the following section sample conformations are generated for a 3x3 and 4x4 square.

Conformation Generation Example

For a 3x3 square shape

```
.  .  .
.  .  .
.  .  .
```

Let us assume that this shape also specifies a composition with one hydrophobic, i.e. buried, residue x to be placed at the central position. The sequence composition then will be

(O_8X).

Let us simultaneously consider both sequence and conformation.

Consider an interior sequence placement of the x residue. There are two possible conformations (tiles) for the oxo sequence triplet.

```
o-x  and  o-x-o
|
o
```

There is only one way to place the first of these in the shape (ignoring symmetric placements),

```
      .  .  .
      o-x  .
      |
      .  o  .
```

We can see that the lower left corner must be a chain end connected to either of the neighboring o's. The remainder of the chain up to the other end can be generated only in one way to yield

```
o-o-o              o-o-o
|   |              |
o-x o     or       o-x o        These are mirror images.
  | |              | | |        Forms achieved by
o-o o              o o-o        rotation are not given here
                                explicitly.
```

The straight triplet can be extended in only two ways

```
o-o-o              o-o-o
|                      |
o-x-o     or       o-x-o        which are mirror images
  |                    |        of one another.
o-o-o              o-o-o
```

If the hydrophobic residue is at a chain end then there are two sequence and conformational triplets

```
                         o
                         |
           x-o           x-o
           |
           o
```

And only two conformations for the entire chain are possible

```
o-o-o              o-o-o
|                  |   |
o  x-o    or       o  x-o              which are mirror images.
|   |              |
o-o-o              o-o-o
```

The actual sequences can be collected from these six conformations, and there are only the following five sequences, to be considered in both a forward and a backward direction, in the 3x3 square:

$$O_2XO_6 \quad O_6XO_2 \quad O_4XO_4 \quad XO_8 \quad O_8X$$

Note that the odd/even effect of the square lattice is reflected artificially in these sequences: The hydrophobic residue can only be placed with an even number of O-type sequence neighbors. The efficiency of this approach of combining the sequence with a small conformational tile is significantly greater than other approaches that we pursued earlier (Jernigan, 1991; Covell and Jernigan, 1990), because specification of the residue type on the tile results in many fewer ways to place the individual tile.

Next we consider the 4x4 planar square

```
.  .  .  .
.  .  .  .
.  .  .  .
.  .  .  .
```

The appropriate composition will be (O_{12},X_4). There is a total of 16 quadruplet sequences

```
 1.  XXXX   2.  XXXO   3. OXXX   4. XXOX
 5.  XOXX   6.  XXOO   7. OXXO   8. OOXX
 9.  XOXO  10.  XOOX  11. OXOX  12. XOOO
13.  OXOO  14.  OOXO  15. OOOX  16. OOOO
```

We follow the same approach as for the 3x3 case, except that mirror images have been omitted. The only quartet conformation for sequence 1 is

```
x-x
|
x-x
```
which leads to

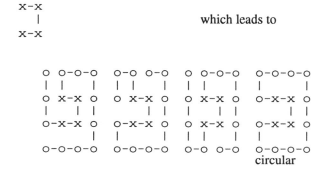

circular

This last case is circular, and its several individual chain conformations correspond to forms in each of which a different single bond is broken to specify the location of the chain ends.

If one end of the quartet is a chain end, then additional forms are obtained

```
o-o-o-o   o-o o-o   o o-o-o
      |   | | |     | | |   |
o-x-x o   o x-x o   o x-x o
|     | |  |  | | |  |  | |
o x-x o   o x-x o   o x-x o
|         | | |     | |     |
o-o-o-o   o-o-o-o   o-o-o-o
```

Sequences 2 and 3 can form two conformational tiles

```
x-x    and    x-x
  |              |
x-o              x
                 |
                 o
```
which give

```
o-o-o-o   o-o-o-o   o-o-o-o   o-o-o-o   o-o-o-o
|         |     |   |             |  |  |     |
o-x-x o   o-x-x o   o-x-x o   o-x-x o   o x-x o
    |         |     |  | |   |  | |  |  | |
o-x x-o   o-x x-o   o-x x o   o x x o   o-x x o
| |   |   |         | | |   | | | |      | |
o o-o-o   o-o-o-o   o-o o-o   o-o o-o   o-o o-o
```

Sequences 4 and 5 are impossible to place because of the square lattice connections.

Sequences 6 and 8 correspond only to the conformations

```
x-x-o    and  x-o     which give the following additional
    |         | |      conformations
    o         x o
```

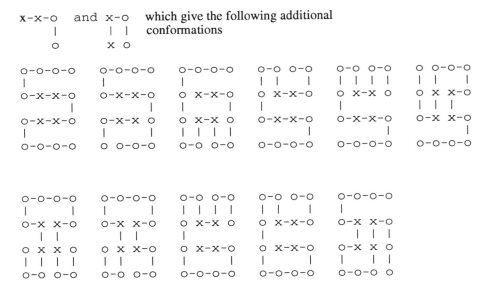

In addition, for sequence 7 the two forms

```
o-x-x-o   and   o          give no new conformations.
                |
          o-x-x
```

However, the tile form

```
x-x
| |
o o
```

yields the following additional forms

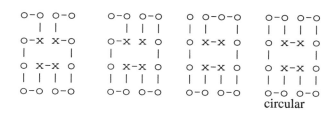

```
o-o o-o      o-o o-o      o o-o-o      o-o o-o
 |   |        | | |       | |   |      | | | |
o-x x-o      o-x x o      o x-x o      o x-x o
|     |       |   |       |     |      |     |
o x-x o      o x-x o      o x-x o      o x-x o
| | | |       | | | |     | | | |      | | | |
o-o o-o      o-o o-o      o-o o-o      o-o o-o
                                       circular
```

Sequences 9 and 11 are impossible to fit on the square lattice.
Sequence 10 has only one possible tile conformer

```
x x
| |
o-o
```

which leads to no new overall conformations.

Sequences 12 and 15 give no additional conformations. This sequence xooo requires consideration of many fewer cases, since the x can be treated as a chain terminus because any sequences with xx have already been treated and sequences with oxo will be treated subsequently.

Sequences 13 and 14 give the following two tiles

```
o-x       o-x          which lead to no new conformations.
 |         |
o-o       o-o
```

Sequence 16 can exist in the following two conformers

```
                              o
                              |
o-o-o-o            o-o-o
```

However, all conformations will have at least two sequential xx's, and so these have already been generated. One advantage of using these sequence considerations in the conformation generations is that anytime a conformation is being generated that has a sequence that has already been treated, its generation can be stopped because all of its forms have previously been generated.

Alternative Schemes

Consider separately the segments between polar residues. As a first approximation, the polar residues ought to be placed on the exterior. Although this would be difficult in two dimensions unless these sub-chains were perfectly aligned, such as in an anti-parallel beta sheet, in three dimensions this is not such a problem. Each such sub-chain has a specific length that serves to define the range of possible placements of the polar residues. The combinations of self-consistent placements would yield acceptable folds.

DISCUSSION

In the present paper we have presented several schemes to improve the conformation generation of folded forms. The established efficiency gained from restricting the conformations to a single shape can be improved upon if tiles with specified sequence are placed upon the shape. The number of combinations, in comparison with tiles of unspecified residue types (Jernigan, 1991), is greatly reduced. Such tiles could be formed quite effectively from segments of secondary structure.

Restrictions of a residue type to certain exposed or buried positions in the shape can be extremely effective in reducing the number of combinations. The investigation of the variability of environment of different residue types confirms that the polar residues manifest the smallest variability in environment. Further efficiencies can be achieved if these polar residues were restricted to the surface of the shape. In a similar way, tiles could be formed for the segments between sequential polar residues. Because of the requirement to place the polar residues on the exterior, these could then be placed in a limited number of ways upon the shape.

REFERENCES

Bahar, I. and Jernigan, R.L., 1993a, Stabilization of intermediate density states in globular proteins by homogeneous intramolecular attractive interactions. Submitted.

Bahar, I. and Jernigan, R.L., 1993b, Cooperative structural reorganizations induced by non-homogeneous intramolecular interactions in compact globular proteins. Submitted.

Bohacek, R.S., Strauss, U.P. and Jernigan, R.L., 1991, Configurational statistics of methyl vinyl ether-maleic anhydride copolymer: selection of important atomic interactions and conformations. *Macromolecules* 1991; 24:731.

Covell, D.G. and Jernigan, R.L., 1990, Conformations of folded proteins in restricted spaces. *Biochemistry* 29: 3287.

Flory, P. J., 1969, "Statistical Mechanics of Chain Molecules", Hanser Publ., Munich.

Gregoret, L.M. and Cohen, F.E., 1990, Novel Method for the Rapid Evaluation of Packing in Protein Structures. *J. Mol. Biol.* 211: 959.

Jernigan, R.L. and Covell, D.G., 1991a, Coarse graining conformations: a peptide binding example, *in:* "Theoretical Biochemistry and Molecular Biophysics Vol. 2: Proteins", Ed. D. L. Beveridge and R. Lavery, 69-76, Adenine Press, Schnectady, NY.

Jernigan, R.L. and Covell, D.G., 1991b, Compact protein conformations, *in:* "Proteins: Structure, Dynamics and Design", ESCOM Science Publishers B.V., Leiden, pp 346-351.

Jernigan, R.L., 1991, Generating general shapes and conformations with regular lattices, for compact proteins, *in :*"Structure and Function: Proceedings of Seventh Conversation in Biomolecular Stereodynamics", Eds. R.H. Sarma & M. H. Sarma, Vol. 2, Adenine Press, Schnectady, NY, pp 169-182.

Miyazawa, S., and Jernigan, R.L., 1985, Estimation of effective inter-residue contact energies from protein crystal structures: quasi-chemical approximation. *Macromolecules* 18: 534.

Miyazawa, S. and Jernigan, R.L., 1993a, A new substitution matrix for protein sequence searches based on contact frequencies in protein structures. *Prot. Eng.* 6: 267.

Miyazawa, S. and Jernigan, R.L., 1993b, Estimation of protein stability changes for single site substitution. Submitted.

Mizushima, S.-I., 1954, "Structure of Molecules and Internal Rotation" , Academic Press, N.Y.

Sippl, M. J., 1990, Calculation of conformational assemblies from potentials of mean force. An approach to the knowledge-based prediction of local structure in globular proteins. *J. Mol. Biol.* 213: 859.

Tanaka, S. and Scheraga, H.A, 1976, Medium- and long-range interaction parameters between amino acids for predicting three-dimensional structures of proteins. *Macromolecules* 9: 945.

Young, L., Covell, D. and Jernigan, R.L., 1993, Locating peptide binding sites on proteins. Submitted.

KNOWLEDGE BASED POTENTIALS FOR PREDICTING THE

THREE-DIMENSIONAL CONFORMATION OF PROTEINS

M.J. Rooman, J.-P.A. Kocher, R. Wintjens and S.J. Wodak

U.C.M.B., Université Libre de Bruxelles
CP 160, av. Héger, 1050 Bruxelles, Belgium

Several potentials are derived from a data set of 75 non−homologous known protein structures by computing statistical relations between amino acid sequence and different descriptions of the protein conformation. These potentials formulate in different ways backbone dihedral angle preferences, pair−wise distance dependent interactions between amino acid residues and solvation effects based on accessible surface areas. Parameters affecting the characteristics and the performance of the potentials are critically assessed by monitoring the recognition of the native fold in a strict screening test, where each sequence in the data set is threaded through a repertoire of motifs, generated from all corresponding structures. For a detailed description of this analysis, the reader is referred to ref. 1; recent reviews on the subject can be found in refs. 2−4.

Mean force potentials

Assuming a Boltzmann distribution, probabilities that sequence patterns are associated with structural states are computed from known structures and translated into relative free energies; the reference state corresponds to non−specific amino acid sequences adopting on the average the observed proportion of structural states. The partition function is not evaluated, as only energies of different conformations of a given sequence are compared. Several potentials, based on different structure descriptions, are derived.

Backbone torsion potentials. They describe local interactions along the chain, and are based on 7 (ϕ, ψ, ω) Ramachandran domains. Two types of potentials are considered. In the residue−to−torsion potential, the probability that a residue at position i along the sequence has its torsion angles in a given (ϕ, ψ, ω) domain is computed as the product of the probabilities of all individual residues and residue pairs in the sequence interval [i−1,i+1] to be associated with that domain at position i; the torsion−to−residue potential considers probabilities of pairs of torsion angle domains in [i−1,i+1] to be associated with a given amino acid type at i. The former potential assumes that the values of the torsion angles of neighboring residues along the sequence are non−correlated,

Statistical Mechanics, Protein Structure, and Protein Substrate Interactions
Edited by S. Doniach, Plenum Press, New York, 1994

while the latter takes these correlations into account, at the expense of neglecting correlations between amino acid types.

Distance dependent residue–residue interaction potentials. They measure the propensities of pairs of amino acids to be at a given spatial distance, according to their separation along the sequence. Potentials with different characteristics are obtained when distances are considered between Cα's, Cβ's, between average side chain centroids Cμ, or between the surfaces of the residue spheres Sμ, centered at the Cμ's. The first of these potentials reflects the proximities between backbone atoms, the second is sensitive to the direction of the side chains, the third takes into account the average side chain conformation, and in the last one the influence of residue size is eliminated.

Hydrophobicity potential. It evaluates the propensities of amino acid types to expose a certain fraction of their surface area to the solvent. Accessible surface areas are computed from the simplified model where each residue is represented by a sphere, using the fast analytic algorithm of Wodak & Janin[5].

Sequence–structure screening procedure

To test potentials and compare their efficiency, each sequence from the data set is threaded onto all structures, in search for compatible sequence–structure combinations. All shifts between sequence and structure are considered, but gaps in the sequence are not allowed, to avoid assuming that the partition function remains constant when the sequence is modified.

Performance of the individual potentials

The performance of the potentials tested by the screening procedure are summarized in Table I. Backbone torsion potentials appear to work surprisingly well, considering that they include only local interactions along the chain, known to be

Table I. Column 2 defines the abbreviations used in Table II. The last 3 columns summarize the recognition performance of the potentials, estimated by different measures : the number N of sequences whose lowest energy structure corresponds to the native one, and the relative energy preference *Pref* and energy gap *Gap* averaged over all proteins of the data set. *Pref* is defined as the difference between the average energy of all threadings and the energy of the native threading, divided by the corresponding standard deviation, and *Gap* is defined as the energy difference between the best ranking non–native structure and the native one, divided by the standard deviation.

Individual potentials				
Type of potential		$N_{recognized}/74$	*<Pref>*	*<Gap>*
residue–to–torsion	$T_{res-tor}$	64	7.3	3.9
torsion–to–residue	$T_{tor-res}$	68	7.3	3.9
Cα – Cα		40	4.7	1.8
Cβ – Cβ		60	6.2	2.9
Cμ – Cμ	**RR**	73	8.2	5.0
Sμ – Sμ		69	8.0	4.6
hydrophobicity	**H**	58	6.1	2.8

incapable, on their own, of determining protein folds. This leads us to question the ability of procedures that screen a limited repertoire of structures to act as a stringent test for the potentials. We also see that the torsion−to−residue potential performs somewhat better than the more classical residue−to−torsion potential. This could be attributed to the fact that the former includes correlations between structural states along the polypeptide chain, while the latter does not. More plausibly however, since there are less structural states (7) than amino acids (20), pairs of structural states yield more reliable statistics than amino acid pairs. This stresses that it may sometimes be better to consider less specific sequence or structure motifs, which occur more often in the data set.

The screening test shows quite clearly that the $C\alpha-C\alpha$, $C\beta-C\beta$, $C\mu-C\mu$ and $S\mu-S\mu$ potentials do not perform equally well: the least effective of all is the $C\alpha-C\alpha$ potential, the performance improves significantly with the $C\beta-C\beta$ and $S\mu-S\mu$ potentials and by far the best performance (73 recognized proteins out of 74) is obtained with the $C\mu-C\mu$ potential. This emphasizes the need of taking into account both side chain position and size when evaluating residue−residue interactions.

With the exception of the $C\alpha-C\alpha$ potential, the hydrophobicity potential has the poorest recognition performance. Reasons for failing to recognize certain proteins can be found in their atypical characteristics, such as higher than average hydrophobic surface. Interestingly, the integral membrane subunits L and M of the photosynthetic reactions center 1PRC have by far the worst ranking of all the proteins tested. On the contrary, for subunit H, which is inside the cell except for a trans−membrane helix, only 1 structure ranks better, and subunit C, which is outside the membrane, is well recognized.

Performance of combined potentials

All possible combinations of the individual potentials have been tested. Table II. summarizes the results for the most representative combinations. Combining potentials increases their performance, as witnessed from the increase in *Pref* and *Gap* values. The best performance is obtained by combining the torsion−to−residue and residue−to−torsion potentials, with the hydrophobicity and the $C\mu-C\mu$ potentials. The only unrecognized protein is then the short melittin monomer, which lacks a core.

Table II. See Table I for the definition of the abbreviations used and for details.

Combined potentials			
Type of potential	$N_{recognized}/74$	*<Pref>*	*<Gap>*
$T_{tor-res}+T_{res-tor}$	68	7.5	4.0
$T_{tor-res}+RR$	73	9.6	6.4
$T_{tor-res}+RR+H$	73	9.9	6.7
$T_{res-tor}+T_{tor-res}+RR+H$	73	10.0	6.7

REFERENCES

1. J.−P.A. Kocher, M.J. Rooman and S.J. Wodak, Factors influencing the ability of knowledge based potentials to identify native sequence−structure matches. submitted (1993).
2. S.J. Wodak and M.J. Rooman, Generating and testing protein folds. *Curr. Op. Struc. Biol.* **3**, 247−259 (1993).

3. J.S. Fetrow and S.J. Bryant, New programs for protein tertiary structure prediction. *Bio−Technology* **11**, 479−484 (1993).

4. M.J. Sippl, Boltzmann's principle, knowledge−based mean fields and protein folding. An approach to the computational determination of protein structure. *Comp. Mol. Design* **7** (1993).

5. S.J. Wodak and J. Janin, Analytical approximation to the accessible surface area of proteins. *Proc. Natl. Acad. Sci. USA* **77**, 1736−1740 (1980).

PROTEIN-PROTEIN RECOGNITION:
AN ANALYSIS BY DOCKING SIMULATION

Joël Janin and Jacqueline Cherfils

Laboratoire de Biologie Structurale, UMR 9920
CNRS-Université Paris-Sud, Bât. 433, 91405-Orsay, France

Introduction

In this paper, we make use of a docking algorithm developped by Wodak & Janin[1-3] and simulate protein-protein recognition by reconstituting complexes from component molecules. When applied to protease-inhibitor or antigen-antibody complexes of known X-ray structure, the procedure efficiently retrieves native modes of association. However, it also selects a number of non-native modes with structural and physical-chemical features that would be expected only from native complexes[4]. We find that these 'false positives' cannot be discriminated from the correct solution on simple criteria such as the number of H-bonds, buried polar groups or cavities at the interface, or more elaborate ones like conformational energy[5]. To predict a complex of unknown structure, additional information must be available. As an example, we model the complex between the hemagglutinin antigen from the flu virus and the Fab fragment of a monoclonal antibody raised against this antigen[6], taking into account the known location of mutations which affect interaction of the antigen with the antibody.

Docking proteins as rigid-bodies

Docking algorithms proceed by bringing two molecules into contact and giving a score to the contact[5] (Fig.1). In doing so, the molecules are kept rigid and have only six degrees of freedom, which in our implementation[1,2] are five rotations and a translation. Two polar angles fix the location of the binding region on molecule 1, two others, that on molecule 2. The last two parameters are a spin angle and a distance ρ. For each set of the five angles, the value of ρ that brings the the two molecules into Van der Waals contact can be computed analytically to fix the relative position of the molecules. After computing a score, the angles are changed and the procedure repeated.

Statistical Mechanics, Protein Structure, and Protein Substrate Interactions
Edited by S. Doniach, Plenum Press, New York, 1994

331

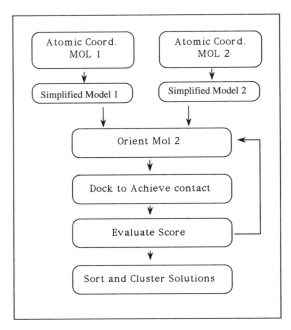

Figure 1. Flow-chart of a docking algorithm.

The score we use is essentially a measure of the interface area B between the two molecules. B is the sum of the Lee & Richards[7] solvent accessible areas of molecules 1 and 2 less that of the complex. Exact algorithms to compute accessible surface areas being too slow for our purpose, we use an analytical approximation[8], which is very fast and yields values B* which are about 10% smaller than B. At this stage, the two protein molecules are described by Levitt's 'simplified protein' model, where each amino acid is a single sphere[9]. This makes calculations faster by almost two orders of magnitude, and, more important, it by-passes the problem of side chains conformation, which can be left to later stages of the analysis.

A repulsive potential must be added to prevent molecules from interpenetrating. We use the repulsive part of the 4-8 potential derived by Levitt for the spheres simulating amino acid side chains[9]. Calling its value E*, we use as a scoring function the pseudo-energy:

$$E_B = E^* - \gamma B^*$$

The constant γ relates accessible or buried surface areas to the free energy of the system. It is the microscopic equivalent to the surface tension at the protein-water interface. We found empirically that $\gamma = 50$ cal.mol^{-1}.Å^{-2} suited our purpose. This value happens to coincide with that derived by Sharp et al.[10] from the free energy of transfer of organic molecules. Though the agreement is probably fortuitous, it confirms that our scoring function describes mostly the hydrophobic contribution to association.

Sampling rigid-body parameter space

The systematic sampling of all orientations and positions requires a large number of steps: of the order of 10^8 if all patches of the surface of one protein molecule are tested against all patches of the second molecule with an angular sampling of 6 degrees. On a workstation and with protein molecules that have 200 residues or more, this takes of the order of weeks. So, either we restrict the search to parts of the molecular surface that are known to be important for association, or we use Monte Carlo simulated annealing[4]. The Monte-Carlo search proceeds along a random path through the five angular parameter space, taking downwards steps in E_B. Upwards steps are also allowed, but with a probability that depends on the energy change ΔE_B and on a temperature parameter T as in Boltzmann's formula. To do annealing, T is progressively lowered until the system freezes in an energy minimum. Whereas this should be the global minimum if enough steps were performed, we find it more useful to explore all deep minima by iterating the whole procedure, which can be done in a matter of hours, not weeks.

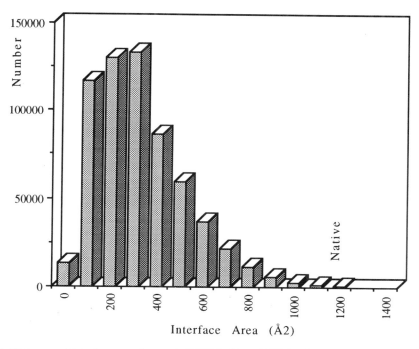

Figure 2. Histogram of interface areas in trypsin-BPTI dockings. BPTI in all orientations is docked onto the active site of trypsin with the Wodak-Janin algorithm. Values of B* achieved after about 6.10^5 dockings are plotted as a histogram. The best 'native' orientation has B*=1250 Å^2 and is 5 degrees off that of the crystal structure. About 100 other BPTI orientations yield equivalent or larger interfaces. Note that B*, calculated with the analytical approximation, is about 10% less than the actual interface area.

Fig. 2 shows results of a Monte-Carlo search performed while docking the trypsin inhibitor BPTI onto the active site of trypsin. The histogram of B* indicates that most dockings achieve much smaller interfaces than those observed in real structures[11], where B is within 20% of 1600 Å^2. The native trypsin-BPTI complex, the X-ray structure of which is known[12], has an interface area of 1400 Å^2 equivalent to B*=1250 Å^2 when calculated with the analytical approximation. After docking BPTI onto trypsin in its correct orientation, we obtain a reconstituted complex that is nearly identical to the X-ray structure, has the same value of B* and is very close to a local minimum of E_B. Nevertheless, the reconstituted complex is not at the global minimum of E_B and, while it stands at the very tail of the histogram, about 0.15% of the orientations tested achieve comparable or slightly scores. After clustering together those that are closely similar and involve the same surface regions of both components, we are left with about ten solutions: the correct one and nine 'false positives'.

Analysis of candidate complexes

Our docking studies[4] of other protease-inhibitor and Fab-lysozyme complexes of known X-ray structure confirm that the correct solution systematically appears within the top 1% of E_B or B* histograms. However, it always comes together with non-native solutions that have equally high scores. Does that lack of discrimination only reflect the crudeness of the model which ignore all atomic details, and of a score that does not even include electrostatic interactions?

To answer that question, we pick one representative of each cluster and submit it to conventional energy minimization with all atoms present. Atoms at the interface are allowed to move freely, others are constrained to stay near their starting position. The results are striking. Whereas the initial structure has a poorly packed interface with many bad contacts and essentially no H-bond, it nearly always relaxes to a low energy after a few hundred cycles of conjugate gradient[4]. The side chains, which were featureless spheres during docking, now reorient to form H-bonds and fill cavities. In about half of the energy-minimized complexes, 8 to 15 H-bonds appear at the interface, often including salt bridges. This is the same number as in real complexes[11]. The correct solution cannot therefore be picked up on the basis of the number of polar interactions. Indeed, algorithms have been designed[13] that pair polar groups on two protein surfaces and form H-bonds or salt bridges. The results show the problem to be largely undetermined, especially for polar groups carried by the long mobile side chains of acidic and basic residues.

As we stop minimization well before convergence, we cannot evaluate a proper energy of interaction. It would in any case be highly inaccurate and even misleading unless solvent is included. Rather than deriving an energy, we examined our 'false positives' in detail, and found some that could be rejected as having several unpaired buried polar groups or too many cavities at the interface. However, these were a minority and we were left with several solutions that looked just as good as the correct one. Other authors[14-16] performed similar studies with different docking algorithms. They checked electrostatic energies and Eisenberg-MacLachlan solvation energies[17] to find that they perform no better in discriminating the native complexes from the false positives.

Modelling an antigen-antibody complex

Having checked our docking algorithm against complexes of known structure, we now try and model one that is yet unknown. This has been done before, but not with automatic procedures. The trypsin-BPTI complex was modelled twenty years ago[18] with complete success on the basis of available chemical information before it was done crystallographically. There have been few serious attempts to repeat this performance. Recently, the complex between the periplasmic *E. coli* maltose binding protein (MBP) and its membrane receptor, was modelled by separately docking two short MBP peptides onto the receptor[19]. The peptides found locations which were compatible with the structure of MBP, fully determining its position. The result is striking, and experiments will show whether it is correct or accidental.

In our case, the two proteins are the hemagglutinin of the flu virus and an antibody raised against it. The hemagglutinin (HA) is the major antigen of the virus and its X-ray structure is known[20]. It is a trimer with a fibrous tail and a globular head that carries most of the antigenic sites. The X-ray structure of the Fab part of the monoclonal antibody HC19 has been determined in our laboratory[21], and that of the complex is under way. As the antibody inactivates the virus, escape mutants can be selected that grow in its presence. All known escape mutants are substituted in the same three residues of the globular head[6]. These residues are part of two loops that are close in space and approximately define the epitope recognized by HC19. Other mutations have been located all around this region of HA. Those which do not have a major effect on HC19 binding mark the bounds of the epitope.

We performed systematic docking of the Fv fragment (the two variable domains) of HC19 onto the globular head of HA. Each component has about 220 residues. Thanks to the known locations of the combining site on the antibody and of the escape mutation sites on the antigen, the search was restricted to about 30% of the total angular space on HA, and 10% on the Fv fragment. Of some 3.10^6 dockings that were performed, less than 0.4% achieved B* values larger than 1000 Å^2. They were clustered into 315 candidate solutions and checked against the information provided by the escape mutations. Complexes in which one of the three escape residues remained accessible to solvent were discarded, and also those where fewer than five of the six hypervariable loops of the HC19 combining site were in contact with HA. These two conditions reduced the number of candidates from 315 to 18, which were subjected to conformational energy minimization. In a last step, we checked whether the neutral mutation sites were at the interface and, if they were, rejected candidate complexes where known substitutions at these sites would create an excessive volume overlap or leave buried polar groups unpaired.

There were only five candidate complexes that fitted all our criteria and were compatible with mutant information. As in our earlier studies, they come out of the minimization step with low energies, reasonably well-packed interfaces and 7 to 10 H-bonds. Therefore, we still do not have a unique answer. Nevertheless, one of the five stands out as generally better from several points of views. Whether it represents the correct complex or not should be known in a near future when the crystal structure of the complex is completed.

Conclusion

Docking algorithms provide an efficient approach to the simulation of protein-protein recognition within the rigid-body approximation. Though there are clear-cut exceptions to this approximation, state-of-the-art procedures can handle many interesting problems such as enzyme-inhibitor or antigen-antibody recognition, and find the correct solution. Other applications can also be considered. Janin & Wodak[2] analyzed the quaternary structure change in hemoglobin by dimer-dimer docking, drawing a reaction pathway for the R to T transition from one allosteric state to another. The pathway was recently extended to the third allosteric state Y discovered in a mutant hemoglobin[22]. Returning now to specific recognition in protein-protein complexes, a remarkable result of the docking studies is that several alternative solutions fit accepted physical-chemical criteria. They form a large, well-packed interface excluding water and forming 8 to 15 polar interactions, yet they are incorrect. Other criteria must therefore be sought. One may choose to rely on biochemical or genetic data as in the hemagglutinin-antibody complex mentionned above, or hope that specially designed empirical potentials will show more discriminating power.

Acknowledgements

We acknowledge long-standing collaborations with Pr. S. Wodak (Brussels) and Dr. C. Chothia (Cambridge). The modelling study of the HC19-hemagglutinin complex has been performed with T. Bizebard and M. Knossow in this laboratory. They provided atomic coordinates and their expertise on this system.

References

1. Wodak S. & Janin J. (1978) Computer analysis of protein-protein interactions.
   J. Mol. Biol. 124:323-342
2. Janin J. & Wodak S. (1985) A reaction pathway for the quaternary structure change in hemoglobin.
   Biopolymers 24:509-552
3. Wodak S., de Crombrugghe M. & Janin J. (1987) Computer studies of interactions between macromolecules. Prog. Biophys. Molec. Biol. 49:29-63

4. Cherfils J., Duquerroy S. & Janin J. (1991) Protein-protein recognition analyzed by docking simulation. Proteins 11:271-280

5. Cherfils J. & Janin J. (1993) Protein docking algorithms: simulating molecular recognition. Curr. Op. Struct. Biol. 3:265-269

6. Cherfils J., Bizebard T., Knossow M. & Janin J. (1993) Rigid-body docking with mutant constraints of influenza hemagglutinin with antibody HC19 (submitted)

7. Lee B.K., Richards F.M. (1971) The interpretation of protein structures. Estimation of static accessibility. J. Mol. Biol. 55:379-400

8. Wodak S. & Janin J. (1980) Analytical approximation to the accessible surface area of proteins. Proc. Natl. Acad. Sc. USA 77:1736-1740

9. Levitt M. (1976) A simplified representation of protein conformation for rapid simulation of protein folding. J. Mol. Biol. 104:59-107

10. Sharp K.A., Nicholls A., Fine R.F. & Honig B. (1991) Reconciling the magnitude of the microscopic and macroscopic hydrophobic effects. Science 252:106-109

11. Janin J. & Chothia C. (1990) The structure of protein-protein recognition sites. J. Biol. Chem. 265:16027-16030

12. Huber R., Kukla D., Bode W., Schwager P., Bartel K., Deisenhofer J. & Steigemann W. (1974). Structure of the complex formed by bovine trypsin and bovine pancreatic trypsin inhibitor. Crystallographic refinement at 1.9 Å resolution. J. Mol. Biol. 89:73-101

13. Kasinos N., Lilley G.A., Subbarao N. & Haneef I. (1992) A robust and efficient automated docking algorithm for molecular recognition. Protein Engng 5:69-75

14. Shoichet B.K. & Kuntz, I.D (1991) Protein docking and complementarity. J. Mol. Biol.2261:327-246

15. Jiang F. & Kim S.H. (1991) "Soft docking" : Matching of molecular surface cubes. J. Mol. Biol. 219:79-102

16. Walls P.H. & Sternberg M.J.E. (1992). An algorithm to model protein-protein recognition based on surface complementarity: application to antibody-antigen docking. J. Mol. Biol. 228:277-297

17. Eisenberg D. & McLachlan A.D. (1986) Solvation energy in protein folding and binding. Nature 319:199-203

18. Blow D.M., Wright C.A.,Kukla D., Rühlmann A.,Steigemann W. & Huber R. (1972) A model for association of bovine pancreatic trypsin inhibitor with chymotrypsin and trypsin. J. Mol. Biol. 69:137-144

19. Stoddard B.L. & Koshland Jr D.E. (1992) Prediction of the structure of a receptor-protein complex using a binary docking method. Nature 358:774-776

20. Wilson I.A., Skehel J.J. &Wiley D.C. (1971) Nature 289:366-373

21. Bizebard, T., Daniels R., Kahn R., Golinelli-Pimpaneau B., Skehel J.J. & Knossow M. (1993) Refined 3-dimensional structure of the Fab fragment of a murine IgG1, λ antibody. Acta Cryst. D. (submitted)

22. J. Janin & S.J. Wodak (1993) The quaternary structure of Carbon-monoxy Hemoglobin Ypsilanti. Proteins 15:1-4

MOLECULAR DYNAMICS SIMULATION OF AN ANTIGEN-ANTIBODY COMPLEX: HYDRATION STRUCTURE AND DISSOCIATION DYNAMICS

Jean Durup, Fabienne Alary, and Yves-Henri Sanejouand

Laboratoire de Physique Quantique[*], I.R.S.A.M.C.
118 route de Narbonne, 31062 Toulouse, France

INTRODUCTION

This is the overview of an ongoing molecular dynamics study of an antigen-antibody complex, which includes the dynamics of the complex in the crystal state and in solution, the study of its hydration structure, the dynamics and thermodynamics of dissociation of the complex into its components, together with the computation of the effect of a point mutation on the equilibrium constant. Methods and problems will be emphasized rather than results. Most of the work described is to be published shortly, and proper references to papers submitted or to be submitted will be given.

The system under study is the hen egg-white lysozyme (antigen) - Fab D1.3 (antibody) complex, surroundered by explicit water molecules and counter-ions. This complex was crystallized by Mariuzza et al. (1983) and analyzed by X-ray diffraction by Fischmann et al. (1991), in Institut Pasteur, Paris. We used the atomic coordinates provided by these authors as a starting point for our simulations, and their B factors as a reference test for our fluctuation dynamics.

---

[*] Université Paul Sabatier, and Unité Associée 505 of the C.N.R.S.

Statistical Mechanics, Protein Structure, and Protein Substrate Interactions
Edited by S. Doniach, Plenum Press, New York, 1994

339

MODELS DEVELOPED FOR THE COMPLEX IN THE CRYSTAL STATE AND IN SOLUTION

The model system we used for the complex *in the crystal state* (Alary et al., 1993b) was the asymmetric unit cell of the crystal network, belonging to group $P2_1$. Ions identical to those used as a buffer by the crystallographers, namely potassium and dihydrogenophosphate, were used as counter-ions to neutralize the charged residues exposed to the solvent, as well as to simulate the ionic strength of the buffer. Water molecules (including those observed by crystallographers) were placed at selected locations around the charged atoms, and randomly spread water molecules were added to fill the free space in the cell.

The unit cell model thus included 5348 protein atoms, 26 potassium and 44 dihydrogenophosphate ions, and 1864 water molecules.

The force field, the dynamics algorithm, and the various sorting codes needed were from release 20 of the CHARMM package (Brooks et al., 1983) with slight modifications: inclusion of small negative charges on the aromatic carbon atoms of the phenylalanine and tyrosine residues (compensated by a positive charge on the β carbon), and description of the water molecules by an altered TIP3P model using different Lennard-Jones parameters for water-solute and water-water pairs (Alary et al., 1993a). Otherwise standard CHARMM-19 parameters were used. We applied a 7.5-Å cutoff radius to non-bonding interactions, we used a switching function for the van der Waals potential and 'shifted electrostatics' for the Coulomb interaction, and we handled the CH, CH_2 and CH_3 groups as 'extended atoms' (Brooks et al., 1983). With CHARMM parameters, no special function for hydrogen bonding is required.

We took into account the interactions of the system with the 26 neighboring cells, which were derived from the reference cell by the appropriate translations and rotations; out of these cells 8 have direct protein-protein contacts with the complex of the reference unit.

For modelling the complex *in solution* (Alary et al., 1993a) we decided to avoid any periodic or boundary condition in order to obtain an as realistic as possible description of the hydration structure. We therefore introduced the 'egg' model, whose principles are the following (see fig. 1).

(i) A convex surface is designed which closely embeds the protein complex and has the same symmetry as the inertia ellipsoid of this complex; the content of this surface will be designated as zone 1.

(ii) Outer surfaces are generated by adding constant seg-
ments Δr_1, Δr_2, Δr_3 to the radii vectors from the origin to
each point of the first surface; these surfaces define shells
which we denote as zones 2, 3, 4; the Δr_i's are adjusted in
such a way that these outer zones all have the same volume.

(iii) The free space in zones 1 and 2 is filled with random water molecules, after the required counter-ions and specific water molecules have been located in the same way as it
was described for the crystal model; however some of these
random water molecules are replaced with ions of both signs
in accordance with the desired ionic strength. Zones 1 and 2
together constitute the system under study.

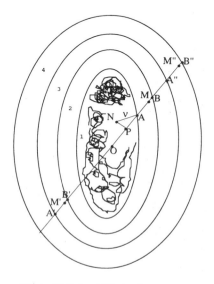

AB=Δr_1, B'A'=Δr_2, A"B"=Δr_3, OM=r, OM'=r', AM=Δr,

OA=ρ ; PA=$\hat{r}$ is the unit vector along r

Figure 1. The 'egg' model.

(iv) The rôle of zones 3 and 4 (which might be prolounged
by additional zones, but the tests showed that this was not
necessary) is to provide a simulated random environment of
solvent molecules, without requiring that the location and
orientation of these molecules be handled as free parameters
in the dynamics. For this purpose, zones 3 and 4 are filled
with *virtual* solvent molecules derived from those of zone 2
by specific operations: in terms of a point transformation,
along any straight line passing by the origin any Δr_1 vector
(which spans zone 2) linearly projects onto a Δr_2 vector
spanning zone 3, and located on the opposite half-axis and

therefore on the other side of the model, but pointing towards the same direction; the latter choice ensures that those water molecules of zone 2 located closest to the protein complex have their images in the outermost part of zone 3. Finally, zone 4 is derived from zone 3 in the same way as zone 3 from zone 2. It may be noticed that, since zones 2, 3 and 4 have an identical volume, the water density in the virtual zones is equal to that in zone 2; however the volume metric of zones 3 and 4 is distorted.

(v) While all energy and force terms pertaining to zones 1 and 2 are handled normally, the interaction energy between these two zones, on one hand, and image zones 3 and 4, on the other hand, has to be multiplied by 1/2. No interaction within zones 3 and 4 is taken into account. The choice of the factor 1/2 to handle the way the interaction between two non-identical subsystems S_1 and S_2 has to be shared, when only system S_1 is handled whereas system S_2 automatically follows system S_1's motion, is a delicate point which required both careful inspection and numerical tests of the dynamics.

Briefly, the fact that a force acting on an atom A of S_1 gives rise to an acceleration both of A and of its image A' in S_2 means a doubling of the effective mass of A, which is equivalent to dividing the force by 2 without changing the mass. This line of thought however leads to difficulties when considering the forces directly acting on A'. Therefore it is more convenient to handle directly the expression of the potential energy.

Evidently if the contents of S_1 and S_2 are the same, the potential energy of the total system is, with obvious notations:

$$U = U_{11} + U_{12} + U_{22} \approx 2 U_{11} + U_{12} . \qquad (1)$$

Now if only the dynamics of S_1 is considered U has to be shared into two equal parts, and thus the potential energy of S_1 is:

$$U_1 = \frac{1}{2} U = U_{11} + \frac{1}{2} U_{12} , \qquad (2)$$

which yields the factor 1/2 by which to multiply the interaction energy between S_1 and S_2.

This example applies e.g. to zones 2 and 3 of our model.

If now S_1 is different from S_2, as is the case for our zones 1 and 3, there is no simple prescription. The factor 1/2 then simply arises from a continuity requirement between zones 1 and 2.

(vi) From the choices taken in item (v) the expression of the forces entering Newton equations follows unequivocally, in such a way as to keep the whole system mathematically hamiltonian. The algebra for the interaction between zones 1 or 2 and zone 3 is as follows.

Fig. 1 illustrates the relevant notations. Let M be an atom in zone 2 and M' its image in zone 3. Vector $\overrightarrow{AB}$, of length Δr_1, is linearly projected onto vector $\overrightarrow{A'B'}$, of length Δr_2. Then (see caption in fig. 1)

$$r = \rho + \Delta r , \qquad (3)$$

$$r' = \rho + \Delta r_1 + \left(1 - \frac{\Delta r}{\Delta r_1}\right) \Delta r_2 . \qquad (4)$$

Hence the OM-to-OM' transformation factor is given by

$$\lambda = -\frac{\overrightarrow{OM'}}{\overrightarrow{OM}} = \frac{r'}{r} = \frac{(\Delta r_1 + \Delta r_2)(\rho + \Delta r_1) - \Delta r_2 \, r}{\Delta r_1 \, r} . \qquad (5)$$

Now let $\mathbf{r}_i$ be the radius vector of an atom M_i of zone 1 or 2 and $\mathbf{r}_j$ that of an atom M_j of zone 2 whose image $M_j{'}$ in zone 3 interacts with M_i. The potential energy term due to this interaction is, with obvious notations:

$$V_i = \frac{1}{2} \sum_j V(r_{ij}) , \qquad (6)$$

where $V(r_{ij})$ includes van der Waals and electrostatic interactions between M_i and $M_j{'}$, 1/2 is the factor earlier discussed in item (v), and

$$r_{ij} = |\mathbf{r}_i + \lambda_j \, \mathbf{r}_j| . \qquad (7)$$

λ_j is defined by equation (5), using r_j and ρ_j (defined like ρ in fig. 1) on the right-hand side.

For the whole system to keep hamiltonian, to each force component on atom M_i due to $M_j{'}$, namely

$$\mathbf{f}_{ij} = \nabla_{r_i} V(r_{ij}) = \frac{\partial V_i}{\partial r_{ij}} \nabla_{r_i} r_{ij} = \frac{1}{2r_{ij}} \frac{\partial V_i}{\partial r_{ij}} \nabla_{r_i} r_{ij}^2 , \qquad (8)$$

there will correspond a force component on real atom M_j,

$$\mathbf{f}_{ji} = \nabla_{r_j} V(r_{ij}) = \frac{\partial V_i}{\partial r_{ij}} \nabla_{r_j} r_{ij} = \frac{1}{2r_{ij}} \frac{\partial V_i}{\partial r_{ij}} \nabla_{r_j} r_{ij}^2 . \qquad (9)$$

From equation (7) it turns out that $\nabla_{r_j} r_{ij}^2$ is related to $\nabla_{r_i} r_{ij}^2$ by the following relation:

$$\nabla_{r_j} r_{ij}^2 = \lambda_j \nabla_{r_i} r_{ij}^2 + \left(\mathbf{r}_j . \nabla_{r_i} r_{ij}^2\right) \left(\frac{\partial \lambda_j}{\partial r_j} \overset{\wedge}{\mathbf{r}_j} + \frac{\partial \lambda_j}{\partial \rho_j} \nabla_{r_j} \rho_j\right) \qquad (10)$$

343

$$\mathbf{f}_{ji} = \lambda_j \, \mathbf{f}_{ij} + (\mathbf{r}_j \cdot \mathbf{f}_{ij}) \left(\frac{\partial \lambda_j}{\partial r_j} \stackrel{\wedge}{\mathbf{r}_j} + \frac{\partial \lambda_j}{\partial \rho_j} \nabla_{r_j} \rho_j \right), \qquad (11)$$

where $\stackrel{\wedge}{\mathbf{r}_j}$ stands for the unit vector along $\mathbf{r}_j$.

Equation (11) shows that the additional force $\mathbf{f}_{ji}$ on atom M_j when its image $M_j{}'$ interacts with a real atom M_i is easily derived from the regular force $\mathbf{f}_{ij}$ acting on M_i and from quantities $\mathbf{r}_j$, λ_j, etc. which concern only atom M_j and therefore have to be computed only once, providing in the algorithm the loop over M_j is the outer one and the loop over M_i is the inner one.

We shall now compute the terms in the bracket of the right-hand side of equation (11).

$$\frac{\partial \lambda_j}{\partial r_j} \stackrel{\wedge}{\mathbf{r}_j} = - \frac{(\Delta r_1 + \Delta r_2)(\rho_j + \Delta r_1)}{\Delta r_1 \, r_j^2} \stackrel{\wedge}{\mathbf{r}_j} = - \frac{\stackrel{\wedge}{\mathbf{r}_j}}{r_j} \left(\lambda_j + \frac{\Delta r_2}{\Delta r_1} \right), \qquad (12)$$

and

$$\frac{\partial \lambda_j}{\partial \rho_j} = \frac{\Delta r_1 + \Delta r_2}{\Delta r_1 \, r_j}. \qquad (13)$$

Finally, the term

$$\nabla_{r_j} \rho_j = \frac{\rho_j}{r_j} \left(\stackrel{\wedge}{\mathbf{r}_j} - \frac{\stackrel{\wedge}{v_j}}{\stackrel{\wedge}{v_j} \cdot \stackrel{\wedge}{\mathbf{r}_j}} \right), \qquad (14)$$

where $\stackrel{\wedge}{v_j}$ is the unit vector of the normal to the inner surface at point A (see fig. 1), describes the angular dependence of ρ_j on $\mathbf{r}_j$. It would be zero if the inner surface defining zone 1 were a sphere. The vector in brackets in equation (14) is vector $\overrightarrow{PN}$ in fig. 1. Due to the convexity of the inner surface this vector is small compared with unity, and as a whole the second term in the bracket in equation (11) is far smaller than the first term. We verified that we could neglect it without observing a noticeable effect in the dynamics, but as explained above its introduction would not have significantly increased the computing time.

(vii) It follows from item (iv) that whenever a solvent molecule diffuses from zone 2 to zone 3 its image moves from zone 3 to zone 2, on the other side of the model. From time to time, when the various interaction lists are updated, those molecules which moved from zone 2 to zone 3 are simply exchanged with their images. Whenever a solvent molecule diffuses from zone 2 to zone 1 or conversely, the loss or acquisition of its image (located in the outermost part of zone 3) becomes effective only when updating the lists, and care may

344

be taken that no discontinuity occurs in the dynamics. Thus as a whole the model introduces no bias favouring or disfavouring stochastic diffusion between neighboring zones.

In our model the lengths of the three axes of zone 1 were 113, 59, and 53 Å (respectively 154, 67, and 63 before truncation of the ellipsoid). The number of water molecules at the starting point was 9154.

THE HYDRATION SHELL OF THE PROTEIN COMPLEX

The protocols we used for energy minimization, equilibration at 300 K, and fluctuation dynamics of the lysozyme-Fab D1.3 in the crystal state (Alary et al., 1993b) and in solution (Alary et al., 1993a) are described in detail in these papers and do not differ fundamentally from conventional ones.

In both simulations we observed a drift of water molecules towards the protein complex. It was much less apparent in the crystal model probably because of the tight packing of the crystal, which essentially contains no bulk aqueous phase (75% of the water molecules in the crystal are located less than 3 Å away from the van der Waals envelope of a protein molecule of the same unit cell or of a neighboring one). In contrast, in the 'egg' model of the complex in solution, where some water molecules are as far as 16 Å from any protein atom, the water drift from zone 2 to zone 1 was quite obvious. We performed many tests for checking that this was not an artefact of the model: freezing the protein motion did not alter this drift; artificially suppressing all full or partial charges of protein atoms reduced the water drift but did not suppress it. In contrast, the same model but used with a pure water content did not present any drift.

Due to this water drift a number of cavities (in the sense of connex sets of nodes, from a 1-Å 3-dimensional mesh, exterior to the van der Waals spheres of all protein or solvent atoms) appeared in the parts of both zones 1 and 2 farthest away from the protein, so that after a total 67-ps simulation time of the complex dynamics in solution we decided to fill the cavities with water and continuously add further water molecules, as described elsewhere (Alary et al., 1993a).

The analysis we performed bears on the state of the system at 93 ps total simulation time. It is described in detail in the just above quoted paper. The most important results are the following.

A very well characterized first hydration shell appears

with a maximum density of water at 1.5 Å from the van der Waals surface of the protein, and a pronounced minimum at 2.5 Å. These data are from the statistics of the number of water oxygen atoms around arbitrary points of the free space as a function of the distance of these points to the closest protein or counter-ion atom. The densities at the maximum and minimum correspond to 1.5 and 0.9 g/cm^3, respectively. Using the distributions of distances of O and H water atoms to the protein van der Waals surface yields a value of 3.0 ± 0.2 Å for the width of this first hydration layer, in agreement with the experimental value of 3.2 Å obtained by Bourret and Parello (1984) for the width of the hydrodynamically equivalent layer around lysozyme, from pulsed Kerr-dielectric relaxation measurements. Water density oscillations around various solutes have been reported by many authors using molecular dynamics or Monte Carlo simulations. A density maximum at *ca* 2.3 Å from the hydrogen atoms of the protein surface in carbonmonoxymyoglobin crystals was observed by neutron diffraction by Cheng and Schoenborn (1990). The point we emphasized (Alary et al., 1993a) from our simulations is that the *integrated* water density around the protein surface largely exceeds pure water density, since we observed a net migration of water molecules from zone 2 to zone 1 whereas the whole free space initially was filled with water at unit density. This illustrates the fact that the local standard chemical potential of water near the protein is smaller than in pure water, and therefore has to be compensated by a higher density to maintain thermodynamic equilibrium with the bulk phase.

Considering in more detail the structure of the hydration layer of our protein complex, we noticed that the pair correlation functions of water O and H atoms were as expected in pure water, whereas the distance distributions and the orientations of water molecule of the first hydration layer with respect to the closest protein (or counter-ion) atom were very dependent on the charge of the latter (Alary et al., 1993a). Performing separate statistics on water molecules close to positively or negatively charged, or nonpolar, parts of the protein surface yielded results remarkably similar to those reported by many authors studying various hydrophilic or hydrophobic solutes, including results presented at this Workshop by Michael E. Paulaitis on methane in water. The same holds for the second hydration layer, which extends continuously towards bulk behavior, with only its first few Angströms showing slight context-dependent anisotropies reminiscent of those of the first hydration layer.

The effect of the hydration layer on the protein dynamics

also was very striking (Alary et al., 1993a). The root-mean-square fluctuations of the protein backbone atoms appeared significantly smaller in the simulation of the complex in solution than in the crystal simulation, an effect we interpreted as due to the higher density of water near the protein in the former than in the latter model, leading to a stronger damping of protein motion by the water environment. We argued that this should be true as well in the *actual* mobility of a protein in solution *vs* in the crystal state as in our models, since the crystallization process requires water pumping away from the protein molecules by agents like polyethylene glycol.

DISSOCIATION DYNAMICS AND THERMODYNAMICS

Dissociation path

After completing the above-described equilibration and fluctuation dynamics of the lysozyme-Fab D1.3 complex in solution, we performed the computation of a conformational path for the dissociation of this complex into its components. For this purpose we performed a *directed separation* of the complex, using a method inspired with substantial modifications from the 'directed dynamics' determination of a reaction path connecting two known conformations of a protein (Ech-Cherif El-Kettani and Durup, 1992).

The method we developed is as follows. The protein complex was under a molecular dynamics simulation at 300 K, in solution, using the above-described 'egg' model. At some moment a very small velocity component was added to all lysozyme atoms in the direction joining the centers of mass of the complex partners, and an opposite velocity component (scaled to the relative masses) was added to the Fab atoms. This additional velocity corresponded in kinetic energy units to adding from 0.5 to 20 K to the initial 300-K temperature. This directed excess kinetic energy of course is not a "temperature", inasmuch as it does not spread into all degrees of freedom. Due to the elasticity both of the antigen-antibody binding and of the water mattress, the centers-of-mass distance Z expanded in about 400 fs by an amount

$$\Delta Z \approx 0.44 \times (\Delta T)^{1/2} , \qquad (15)$$

where ΔT is the above-mentioned instantaneous 'temperature' increase, in Kelvin, and thereafter the system retracted but not back to the original centers-of-mass distance. The process then was iterated. All along this directed separation,

water was added each 50 fs in order to fill the cavities created by the forced separation, with the same procedure as
earlier used for compensating the water drift towards the
protein complex (see preceding section). It turned out that
in the first step of this procedure a ΔT of 0.5 K was sufficient, since the complex started at equilibrium and therefore
the restoring force was essentially zero. Later, a ΔT up to
20 K was needed, probably around the inflexion point of the
potential energy profile along the separation coordinate. At
most steps, 10 K appeared to be a convenient value. Allowing
the solvent to relax alone (the protein atom coordinates being held fixed) from the maximum expansion point during 2 ps
and thereafter releasing the whole system did not affect
significantly its retractation, which if it were confirmed by
more prolounged tests would indicate that solvent compression
is not the main cause of elasticity in this procedure.

The first results of this protocol showed the progressive
exposure to solvent of the hypervariable loops of the antibody and of the epitopes of the antigen. A detailed analysis of
the way the side chains of these contact areas separate from
each other and form new nonbonding interactions either by
sliding with respect to each other or by forming hydrogen
bonds with interstitial water is in progress.

Dissociation thermodynamics

Two separate studies of the dissociation free energy of
the egg-white lysozyme-Fab D1.3 complex have been undertaken.
The first one uses the thermodynamic perturbation method
along with a thermodynamical cycle for computing the difference in dissociation free energies arising from the Gln 121
Asn mutation in lysozyme (which is present in both Japanese
quail and pheasant lysozyme, as compared to hen lysozyme),
whereas the second one uses a thermodynamic integration method aiming at a direct determination of the dissociation free
energy.

In the first of these studies (Alary et al., 1993b) the
required difference in dissociation free energies is computed
according to a thermodynamical cycle (Wong and McCammon,
1986) as the difference between the *mutation free energies* of
the complex and of the free lysozyme, both surrounded with
solvent molecules. The system under study is a sphere of 15-Å
radius surrounding the binding site of Gln 121 to the Fab
D1.3 light and heavy chains. Free energy variations are computed by the thermodynamic perturbation method with a coupling parameter λ varying by steps of 0.1 from 0 to 1 and conversely, the systems with Gln 121 corresponding to $\lambda = 0$ and
those with Asn 121 to $\lambda = 1$. As usually observed a slight hy-

steresis occurs when comparing the reverse free energy variations. This cycle then was iterated several times both for
the complex and for lysozyme alone. The values retained were
the lowest free energies attained along the iteration for
each of the four corners of the thermodynamical cycle.

In the second study, a protocol was set up for a direct
computation of the free energy variation along the dissociation path of the hen egg-white lysozyme - Fab D1.3 complex,
obtained by the above-described *directed separation* method.
It is easily shown on the basis of thermodynamic perturbation
theory (McQuarrie, 1976) that the derivative of the free energy A with respect to a given coordinate Z is given by the
equation:

$$\frac{\partial A}{\partial Z} = < \frac{\partial U}{\partial Z} >_Z , \tag{16}$$

where the average is taken, at a fixed value of coordinate Z,
over the canonical ensemble at a given temperature T, and
where U is the potential energy of the system. In our implementation Z is the projection of the vector joining the centers of mass of the two partners on its initial direction,
which practically coincides with the principal inertial axis
of the complex. The canonical ensemble is simulated by a molecular dynamics run which in principle scans the microcano-
nical ensemble; actually, temperature drifts due to algorith-
mic imperfections are compensated by a periodic scaling of
the velocities to maintain a fixed temperature. A test of
this protocol in a 26.1-ps run on the complex in solution,
before application of the directed separation algorithm, indicated that it tends to converge within $ca \pm 0.1$ kcal.mol^{-1}.
$\overset{o-1}{A}$.

CONCLUSIONS

Molecular dynamics studies of antigen-antibody recognition using extensive models including explicit solvent molecules and trying to simulate the grand-canonical ensemble,
although very demanding in computer time, appear to provide
valuable results which compare reasonably well with the available experimental data. Further analyses of the outcomes of
this study of the hen egg-white lysozyme - Fab D1.3 complex
are under way: a comparison between the structures of the
complex in the crystal state and in solution, a detailed description of the antigen-antibody binding, an analysis of the
conformational changes along the dissociation path, and
finally a study of the diffusion of the various classes of water molecules of the hydration shell.

REFERENCES

Alary, F., Durup, J., and Sanejouand, Y.-H., 1993a, "Molecular dynamics study of the hydration structure of an antigen-antibody complex", submitted to *J. Phys. Chem.*

Alary, F., Sanejouand, Y.-H., and Durup, J., 1993b, "A molecular dynamics and free-energy perturbation study of the specific recognition of hen egg-white lysozyme by monoclonal antibody Fab D1.3", submitted to *Proteins*.

Bourret, D., and Parello, J., 1984, "Hydration of lysozyme in aqueous solution as studied by self-diffusion NMR measurements and by Kerr-dielectric relaxation", *J. Physique* 45:C7-255.

Brooks, B.R., Bruccoleri, R.E., Olafson, B.D., States, D.J., Swaminathan, S., and Karplus, M., 1983, "CHARMM: a program for macromolecular energy, minimization, and dynamics calculations", *J. Comput. Chem.* 4:187.

Cheng, X., and Schoenborn, B.P., 1990, "Hydration in protein crystals. A neutron diffraction analysis of carbonmonoxymyoglobin", *Acta Cryst. B* 46:195.

Ech-Cherif El-Kettani, M.A., and Durup, J., 1992, "Theoretical determination of conformational paths in citrate synthase", *Biopolymers* 32:561.

Fischmann, T.O., Bentley, G.A., Bhat, T.N., Boulot, G., Mariuzza, R.A., Phillips, S.E.V., Tello, D., and Poljak, R.J., 1991, "Crystallographic refinement of the three-dimensional structure of the Fab D1.3-lysozyme complex at 2.5 Å resolution", *J.Biol. Chem.* 266:12915.

McQuarrie, D.A., 1976, "Statistical Mechanics", Harper & Row, New York, section 14-1.

Mariuzza, R.A., Jankovic, D.Lj., Boulot, G., Amit, A.G., Saludjian, P., Le Guern, A., Mazié, J.C., and Poljak, R.J., 1983, "Preliminary crystallographic study of the complex between the Fab fragment of a monoclonal anti-lysozyme antibody and its antigen", *J. Mol. Biol.* 170: 1055.

Wong, C.F., and McCammon, J.A., 1986, "Dynamics and design of enzymes and inhibitors", *J. Amer. Chem. Soc.* 108:3830.

MULTIPLE CONFORMATIONS OF CYSTATIN, MUNG BEAN INHIBITOR, AND SERPINS

Richard A. Engh, Robert Huber, and Wolfram Bode

Max Planck Institut für Biochemie
82152 Martinsried bei München
Federal Republic of Germany

INTRODUCTION

The complexity of the physical interactions which govern macromolecules, along with natural selection processes, has driven the evolution of the subtly balanced biochemical systems seen in biology, and as part of this, protein behaviour ranging from surprising to mysterious. The balances of the complex enzymatic mixtures found *in vivo* enable sensitive and rapid triggering of biochemical response mechanisms, for example, while maintaining the stability and adaptability required for life. This combination of stability and subtlety is found among protein structures as well: a few amino acid types with unlimited variety in sequences; a few types of secondary structure but many globular folds; common folding scaffolds (predictable?) with hypervariability at functional segments.

X-ray crystallography has provided the greatest information about protein structure, including its diversity, but has often been accused of promoting the prejudice that proteins, when folded, function immutably. Although the charge is unjustified--many examples of functional mobility have been studied through crystallography--an ordered crystal is a prerequisite for diffraction, and the structural information from single crystal X-ray studies must be understood in the appropriate context, with awareness of possible effects of the crystalline environment and crystallization conditions, possible ambiguities--even different philosophies--in structural refinement, the factors contributing to thermal parameters, and so on.

In this article we describe structural studies on three systems, all involving proteinase inhibitors, which display a variety of subtle behaviours. A segment of the cysteine proteinase inhibitor cystatin is partially disordered in a crystal environment, but most likely adopts a helical conformation not seen in the solution structure by NMR. The double-headed Bowman-Birk inhibitor possesses a uniquely symmetric fold with water molecules playing a central role in main chain hydrogen bonding interactions, while its ternary complex with two trypsin molecules is uniquely disordered in the crystal. Finally, the serpins, so far not crystallised in an inhibitory complex or conformation, show unique mobile folding behaviour, as surmised by a combination of experimental techniques. These examples demonstrate some of the complicating processes seen with proteins, including specific

Statistical Mechanics, Protein Structure, and Protein Substrate Interactions
Edited by S. Doniach, Plenum Press, New York, 1994

structural effects of water in protein folding, peptide segments with a propensity to adopt more than one specific conformation, structural changes and protein association, hypervariability at functional regions, and forces in protein crystallisation.

CONFORMATIONAL HETEROGENEITY OF CYSTATIN

The cystatins form a superfamily of tightly and reversibly binding inhibitors of cysteine proteinases, including the human cathepsins B, H, L, and S, as well as other enzymes of the papain superfamily (Barrett *et al.*, 1986a; Turk and Bode, 1991). It is currently subdivided into three families (nomenclature established by Barrett, *et al.*, 1986b): Family 1 (the stefins), family 2 (the cystatins), and family 3 (the kininogens), although new sequences and ambiguities in sequence alignments suggest classification of one or more new families (Rawlings and Barrett, 1990; Turk and Bode, 1991; Murzin, 1993). Chicken cystatin belongs to family 2, which is distinguished by a conserved sequence containing two disulphide bridges.

The X-ray crystal structure of chicken cystatin (Bode *et al.*, 1988) at a resolution of 2.0 Å provided the basis for modelling the mechanism of cystatin inhibition (Bode *et al.*, 1989; Machleidt *et al.*, 1989; Bode *et al.*, 1990). The conserved segments previously suggested to be involved in binding (Turk *et al.*, 1985; Abrahamson *et al.*, 1987) were located in adjacent turn structures along the wedge edge, providing a molecular surface complementary to the cysteine proteinase active site cleft. The subsequently determined crystal structure of stefin B, a family 1 inhibitor, in an inhibitory complex with carboxymethylated papain (Stubbs *et al.*, 1990) confirmed the general model of inhibition.

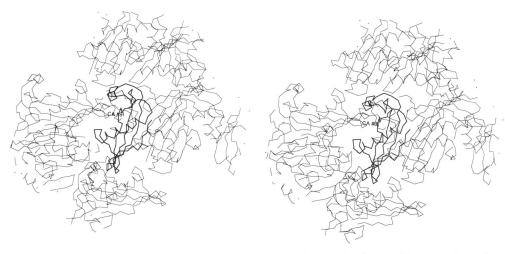

Figure 1. Stereo figure of the crystal packing of cystatin. Helix 2, located at the top of the molecule, nestles into a partially hydrophobic pocket formed by crystal neighbours. Below the helix, the turns of the binding wedge (including the N-terminal Gly9) pack against the central monomer.

Overall Structure

The overall fold of the inhibitor consists of a five stranded, twisted β-pleated sheet partially wrapped around a straight five turn α-helix (helix 1, in a 'hot dog' fold, see Figure 1). The first eight residues are disordered. In the crystal structure, the segment characteristic of family 2 cystatins (approximately residues 70-90) is found as a flexible loop (71-76), an α-helical segment (helix 2, 77-85), and a disordered region (86-89). However, the NMR structure of recombinant cystatin shows this region to adopt a completely different fold (Dieckmann *et al.*, 1994). The segment lacks the helix, and instead the segment 72-80 is disordered, segment 81-83 is extended with contacts to the β sheet, and segment 87-90 shows a well defined β turn. Besides this dramatic difference, the first hairpin loop at the binding site of cystatin (53-57) adopts a somewhat different conformation in the solution NMR structures. In the rest of the molecule however, the hydrogen bonding patterns and secondary structure designations of the X-ray and NMR structures are virtually identical (Engh *et al.*, 1994). Other minor differences, such as a decreased curvature of the β sheet in the NMR structure might be attributable to limitations of the method of simulated annealing with limited long range NOE data.

Viewed as a monomer, the X-ray structure at the segment 70-90 adopts a seemingly implausible structure, with a two turn α-helical segment only loosely associated with the rest of the inhibitor, connected through the disulphide bridge 71-81 and through weakly defined or undefined segments. At least three hydrophobic residues are exposed to solvent, including two residues in the helix, Phe83 and Leu78, and also Pro72. Looking at the structure in its crystalline environment, however, shows several interactions which are likely to stabilise such an arrangement (see Figure 2). Principal among these is the arrangement of the binding cleft (including the amino-terminal Gly9 and hairpin turns 53-57, 103-106), sandwiched between helix 1 and helix 2, and simultaneously, the embedding of the helix between the binding edge of one symmetry related molecule and the β-sheet region of another, including residues 45-48, 63-65, and 92-94, and the second hairpin loop of the binding edge of a third symmetry related molecule. These crystal images offer the side chain of Phe83 in the helix several hydrophobic partners, including Val12 $C^{\gamma 1}$, the hydrophobic atoms of the side chain of Arg52 (e.g. C^{γ}), and the side chains of Leu54 and Ile58.

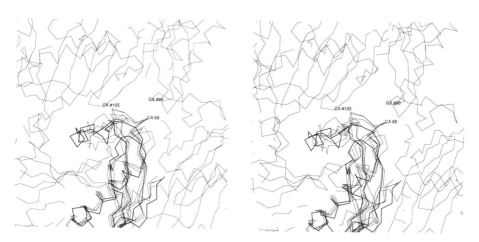

Figure 2. Superimposed on the crystal structure, the NMR structures nearly pack into the crystal without impossible contacts. However, the side chain of residue Glu88 would require adjustment, as would the second hairpin loop of the binding edge. Thus, the formation of a helix may be driven primarily by the availability of an external hydrophobic environment for Phe83 compared to the solution structures.

Although the NMR structures cannot be fitted to the crystal structure without producing impossible crystal contacts, these might be alleviated without refolding the main chain (Figure 2). Thus, several interproton distances clearly defined among the NMR-NOE data suggest an alternate model for the crystal structure, which must be considered, especially since such flexible regions as the segment around helix 2 are susceptible to errors from model bias. These NOE restraints include especially distance restraints between the side chain of Phe83 to the side chain of Tyr92 and several other residues, NOE restraints linking the main chain nitrogen of His84 to the main chain of Thr69 and C^β of Ala90, a strong NOE between C^α of Phe83 and Thr69 and the hydrogen bond suggested by NMR between Thr70 NH and Glu82 CO.

The electron density shows evidence both for and against such an alternate model. A single strong electron density peak (in both 2Fo-Fc and Fo-Fc maps) at the site for a potential hydrogen partner to Thr70 NH exists, but is solvent-like, without any indication of further covalent bonds. Larger volumes of electron density are seen at the putative NMR alternative site for Phe83, but these also show little evidence of further connectivity, even in further cycles of refinement when phenylalanine has been fitted to the density. Furthermore, the precise location of this density was largely incompatible with the restraint of the hydrogen bond between Glu82 CO and Thr70 NH and the further constraint of the disulphide bond to Cys81. Joint refinement including both the set of NOE restraints and the X-ray diffraction restraint energy function also fails to produce a structure satisfying all restraints. However, the presence of some electron density compatible with the NMR structure and the corresponding indications of disorder throughout the flexible regions might suggest that conformations similar to the NMR structures are present in a significant fraction of the molecules in the crystal. Experiments with improved data resolution of new crystals are underway to examine the extent of the disorder.

The large scale conformational changes seen in the segment 71-91, demonstrate a structural variability which may have functional significance. Proteins are known with similarly mutable helices for which a function is known or proposed. The complement proteins C3a/C5a show variations in content and position of terminal helices (Chazin *et al.*, 1988; Huber *et al.*, 1980). The *trp* repressor (Schevitz *et al.*, 1985; Lawson *et al.*, 1988; Arrowsmith *et al.*, 1991) possesses flexible helices apparently necessary for DNA binding. Glucagon is flexible in dilute solution but the monomers self-associate at increasing concentrations into helical trimers and higher oligomers, suggested to be the mechanism for concentration dependent receptor recognition (Blundell & Humbel, 1980). The serpins also show a propensity for helix formation suggested to have functional significance (Schulze *et al.*, 1992; Stein *et al.*, 1990; see below). For cystatin, the conservation of this segment among the family 2 inhibitors suggests some function, although none has been identified. One group of researchers has pointed out an intriguing, if weak, similarity of the 71-81 loop to small serine proteinase inhibitors, the larger serpins, and in particular the Bowman-Birk inhibitors (Saitoh *et al.*, 1991; see below). Tests for serine protease inhibition using appropriate cyclic peptides of the 71-91 loop showed very weak binding (Saitoh *et al.*, 1991), although tests of serine protease inhibition by chicken cystatin (Family 2) have shown no measurable inhibition (Anastasi *et al.*, 1983; Machleidt, personal communication).

Conformation at the Binding Edge

The binding edge of cystatin comprises residues near the N-terminus, and the first and second hairpin loops which link strands of the β-sheet. The structure of the first hairpin loop differs between the crystal and solution structures, although the difference is small and restricted to residues 55 and 56 (Figure 3). The NMR solution structure most closely resembles the conformation in the inhibitor-protease complex of stefin B--papain (Stubbs *et al.*, 1990). The structural difference probably arises from the crystal contacts at the binding edge. It is noteworthy that this structural alteration did not interfere with the prediction of the structure of the complex by computer docking (Bode *et al.*, 1988), although any attempts to predict hydrogen bonding in the complex from the original crystal structure would have required successful prediction both of the solvent molecules involved and the change in the backbone conformation at residues 55 and 56.

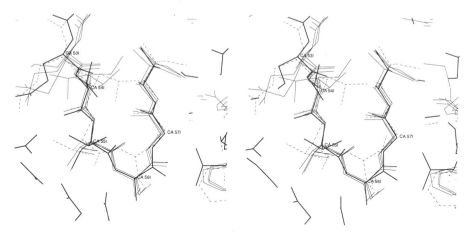

Figure 3. Superposition of the X-ray structure of cystatin (dashed lines), the X-ray structure of stefin B as complexed with papain (thick solid line) and NMR structures (thin solid lines) at the first hairpin loop of the binding edge. The backbone of the NMR structures very closely resemble that of stefin B and probably represents the binding conformation of cystatin as well.

THE BOWMAN-BIRK INHIBITOR FROM MUNG BEAN

The Bowman-Birk type inhibitors (see Ikenaka and Norioka, 1986; Birk, 1985) are quite small proteins (60-90 residues) and characteristically inhibit, independently and simultaneously, two serine proteinases, typically trypsin and chymotrypsin. Their covalent structure (Odani and Ikenaka, 1973) includes seven disulphide bridges and two tandem repeat sequences, each with an active inhibitory binding loop. The inhibitory specificity of the Bowman-Birk inhibitors indicated that they inhibit via the standard mechanism (Laskowski and Kato, 1980), by which inhibitors with lysine or arginine at the P1 residue typically bind trypsin, and those with an aromatic or large hydrophobic residue bind chymotrypsin. Structures of serine proteinase inhibitor complexes (Bode and Huber, 1992; Read and James, 1986) show that these inhibitors mimic substrates, whereby the binding loop adopts the 'canonical conformation' at the reactive site and the adjoining residues. This conformation is common to all the standard inhibitors despite quite different folding motifs and differing stabilising interactions between the reactive site and the rest of the inhibitor.

Information on the spatial structure of the Bowman-Birk inhibitors has come from low resolution X-ray crystal structures of a 1:1 complex of azuki bean inhibitor with trypsin (Tsunogae, *et al.*, 1986), the peanut Bowman-Birk inhibitor (Suzuki *et al.*, 1987), 2.5 Å resolution structures of the PI-II inhibitor from soybean (Chen *et al.*, 1992) and of the mung bean inhibitor in an inhibitory ternary (1:2) complex with trypsin (Lin *et al.*, 1993), and a solution NMR structure of BBI-I from soybean (Werner and Wemmer, 1992). The two tandem repeat segments fold identically, and also are related by an approximate two-fold symmetry axis. (Other Bowman-Birk inhibitors will be somewhat less symmetric due to a two-residue insertion at the binding site).

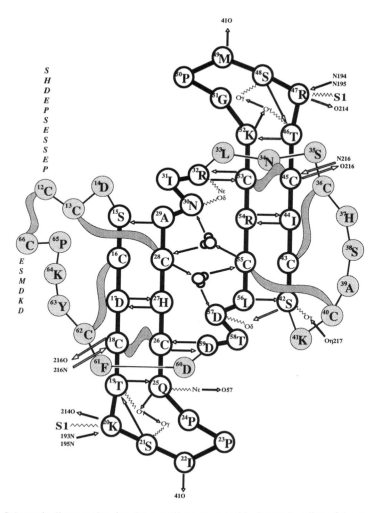

Figure 4. Schematic diagram showing the overall structure and hydrogen bonding of the mung bean inhibitor. The S-like curves denote disulphide bridges, open arrows denote hydrogen bonds (donor→acceptor), the solid line shows the polypeptide fold with the amino acids identified with the one-letter code, and side chains are shown as a sawtooth curve. Hydrogen bonds to trypsin are also shown; the trypsin hydrogen bond partners are identified by residue number and atom name. The unresolved carboxy- and amino-terminal sequences are identified in italics. The heavy solid line along with two central water molecules shows the moderately twisted β-sheet like structure of the inhibitor; the thin lines show the remaining segments which, in the orientation of the figure, lie in front of the planar region.

The inhibitor isolated from mung bean is among the most symmetric of the Bowman-Birk inhibitors, binding trypsin at each binding site (see Figure 4), rather than the more common trypsin-chymotrypsin pair. Each of these two proteinase binding regions is contained in the turn region between two antiparallel β strands. A *cis*-proline residue at the P3' position in each binding region (Pro23, Pro50) creates a turn classified as type VIb (Richardson, 1981). The pairs of β strands are in turn related by a pseudo symmetry rotation of 180°, corresponding to the crystallographic two-fold symmetry axis which relates the two trypsin molecules of the ternary complex. The two pairs of β strands form a structure similar to an extended, five stranded β sheet (Figure 5) with a moderate right-handed twist (viewed along the strands), where however the central strand is missing and its hydrogen bonds are provided by two solvent molecules and four other residues (Asn30, Arg32, Asp57 and Asp59). Residues Asn30 and its symmetry related Asp57 make hydrogen bonds with the two central water molecules. These residues are very strongly conserved as aspartic acid residues throughout the Bowman-Birk inhibitors. The pseudo β-sheet forms an exposed face of the inhibitor on one side, and on the other is covered by the N-terminal and C-terminal segments, and by the segment connecting the two domains.

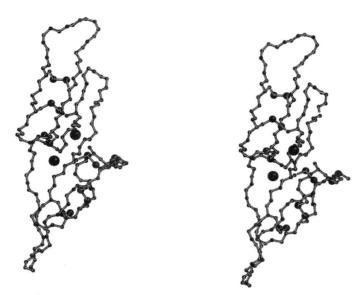

Figure 5. Stereo plot of the main chain fold of the mung bean inhibitor, including disulphide bonds and integral water molecules. The binding loops are located at the upper and lower extrema of the plot. The two water molecules participate in six protein-water hydrogen bonds, stabilizing the two β-ribbons in an overall β-sheet like structure.

The mung bean inhibitor structure was determined in ternary complex with two trypsin proteinases, and the symmetry of the inhibitor is reflected in the symmetry of the ternary complex as well. It is arranged approximately as a prolate spheroid, with the two trypsin molecules along the principle axis, bridged at the middle by the double-headed inhibitor beside the principle axis, and in contact with each other on the other side. A crystallographic two-fold rotation axis (in the space group $P3_121$) exists perpendicular to the principle axis, relating the two trypsin molecules and the corresponding binding regions of the inhibitor to each other. The crystallographic symmetry requires that each trypsin is equivalent, and that the inhibitor is disordered, present in the two possible orientations. One face of the mung bean inhibitor is packed against the trypsin-trypsin contact region; the other is accessible to solvent. This arrangement leads to crystal packing contacts only between trypsin molecules (see Figure 6); the highly symmetrical inhibitor looks free to arrange in either of the two possible orientations in the ternary complex.

This symmetric arrangement and the high symmetry of the inhibitor complicates the refinement of the X-ray structure. The position of the single trypsin molecule in the

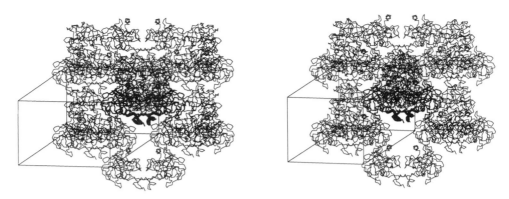

Figure 6. Stereo plot of the crystal packing of the trypsin-mung bean inhibitor-trypsin ternary complex. In each complex, the inhibitor is recognizable as the connecting strands (thick in the central complex) between the trypsin molecules. Crystal contacts occur only between trypsin molecules, so any orienting forces between complexes would arise directly only through long range electrostatic interactions, and indirectly only propagated through the trypsin structure. No evidence of such ordering is found however, and the structure is refined assuming statistically disordered orientations.

asymmetric unit is unambiguous, and the position of the ternary complex is thereby also unambiguous. The orientation of the inhibitor in the complex is however not so clear, and there are three most likely possibilities. First, one orientation of the inhibitor may be preferred over another. In this case, the two-fold rotation axis through the ternary complex cannot be a crystallographic axis, and the complex should be refined in the lower symmetry space group $P3_1$ without this axis. However, this would require that during crystal formation, the ternary complex leaving the solution phase must be oriented by its new crystal neighbours such that the inhibitor should adopt the preferred orientation. Since only trypsin molecules are involved in crystal contacts, and since there is no detectable asymmetry between the trypsin molecules of the trimer, including at the well-defined contact regions, such an orienting force must come from purely long-range electrostatic interactions between the inhibitors themselves. Comparisons of X-ray structure refinement using different symmetries ($P3_1$ and $P3_121$) indicate however that the two-fold rotation axis through the trimer is indeed a crystallographic symmetry axis.

A second possibility is that any asymmetry in the forces between trimers, either long-range electrostatic interactions or asymmetry propagated through the trypsin structures, is much smaller than thermal disorder and that the trimers adopt random orientations with respect to the aforementioned two-fold symmetry axis. In this case, X-ray refinement must proceed with the assumption of equal populations of each orientation of the trimer, and indeed such refinement shows evidence of its correctness (Lin *et al.*, 1993).

The third possibility is somewhere in between. Asymmetric forces might induce further ordering between ternary complexes, but not merely to reduce the symmetry to space group $P3_1$. Some kind of superordering of the trimers might be present, either throughout the crystal or in smaller domains. Such an ordering might in principle be detectable in the diffraction data, but in practice would be difficult to detect.

The two binding loops of the mung bean inhibitor are very similar in structure and in their interactions with trypsin. The structure and interactions conform to the usual characteristics of the standard mechanism inhibitors with two prominent exceptions. First, the occurrence of a proline at position P3', while conserved among Bowman-Birk inhibitors (and common among serpins, see serpin section below), is unique among standard mechanism inhibitors with known structures. Its *cis* conformation creates a turn such that the following residues form the canonical secondary binding segment in a new way (Bode & Huber, 1992).

The second distinct feature of this serine proteinase-inhibitor complex is the close approach of the catalytic O^γ of Ser195 to the P1 carbonyl carbon. At 2.10 Å, this is the closest seen among the X-ray structures of serine proteinase--protein inhibitor complexes, which usually show a 'sub van der Waals' contact distance of 2.70±0.10 Å (Bode and Huber, 1992). Continuous electron density at a high contouring level also appears between these atoms.

Marquardt *et al.* (1983) tabulated deviations from planarity at the reactive site carbonyl carbon according to the angle χC, defined by $\chi C = \omega_1 - \omega_3 + \pi$, where ω_1 is the C^α--C--N--C^α torsion angle and ω_3 is the O--C--N--C^α torsion angle. In the mung bean inhibitor, χC is approximately 40°, nearer the ideal tetrahedral value of 60° than the planar value of 0°. Plotting Δ, the distance of C to the plane defined by $C\alpha$, N, and O vs. the bond length of C--O provides an indication of the character of intermediates in the nucleophilic addition process (Bürgi *et al.*, 1973); Δ values of 0.3 to 0.4 Å for MBI's bond length of 2.10 Å are appropriate for a fully tetrahedral intermediate with shorter bond lengths (Marquardt *et al.*, 1983). Indeed, further refinement with weak symmetry restraints lead to shorter bond distances in one binding loop. However, in spite of the evidence for the distortion at the reactive site, uncertainties introduced by the inhibitor disorder and the relatively low data resolution preclude further analysis, and the first serine proteinase--inhibitor structures at resolutions of 2.80 Å (pancreatic trypsin inhibitor--bovine trypsin, Rühlmann *et al.*, 1973) and 2.60 Å (soybean trypsin inhibitor---porcine trypsin, Sweet *et al.*, 1974) were inadequately resolved to determine the precise structure at the binding site, subsequently shown by higher resolution structures (Huber *et al.*, 1974).

SERPIN 'SLEIGHT OF STRAND'

The subtle and complex behaviour of proteins is stunningly demonstrated by the proteinase inhibitors of the serpin superfamily. There has been strong interest in the superfamily since its identification in the early 1980's (Hunt and Dayhoff, 1980, Carrell and Travis, 1985), generated by their ubiquity and vital physiological roles, and this interest is growing as their unique folding properties are revealed. Several dozen members have been identified among plants, viruses, insects, and higher animals; they are mostly serine proteinase inhibitors, although several have been identified with other functions. In vertebrates, the serpins are the major plasma proteins, and make up about 10% of the total protein content in plasma. They play critical regulatory roles in a diverse range of physiological processes, including zymogen activation, complement activation, blood coagulation, fibrinolysis, digestive processes, tumour migration, and invasion during metastasis. A number of reviews have appeared in the past decade, dealing with their general mechanism of inhibition (Travis and Salvesen, 1983), physiological importance (Shapira and Patston, 1991), their role in genetic disorders (Crystal, 1991), and others.

Most discussion of spatial structure (Huber and Carrell, 1989) has relied on the original X-ray structure of proteolytically modified α1PI (Löbermann *et al.*, 1984). More recently, however, several new X-ray structures have been determined, and a variety of experiments have been performed to deduce aspects of serpin structure not yet observed in crystal studies. In addition to proteolytically modified a1PI, the structures of its most common deficiency mutant, the S-variant (Engh *et al.*, 1989), proteolytically modified inhibitors α-1-antichymotrypsin (a1AC: Baumann *et al.*, 1991), and the equine elastase inhibitor (HLEI: Baumann *et al.*, 1992) have been solved. The structure of the non-inhibitory serpin ovalbumin has been solved (PLAK: Wright *et al.*, 1990, OVAL: Stein *et al.*, 1990), while

Mottonen *et al.* (1992) have determined the structure of the uncleaved but latent form of the plasminogen activator inhibitor PAI-1. Since no crystal structure of an inhibitory form of a serpin has yet been determined, non-crystallographic studies have also yielded indirect evidence about important aspects of the inhibitory conformations (Schulze *et al.*, 1990, 1991, 1992, Mast *et al.*,1992, Evans *et al.*,1992, Björk *et al.*, 1992).

Many serpins act in conjunction with cofactors. Heparin is the most studied of these, interacting with antithrombin III (ATIII), heparin cofactor II, protease nexin I, protein C inhibitor, and PAI-1. Each of these serpins inhibit relatively weakly alone, but become efficient inhibitors upon heparin binding. Heparin binding to other serpins is distinct from binding to ATIII, reinforcing the view that its interactions are subtle and complex. Binding to heparin cofactor II (HCII) also occurs at or near helix D (Pratt *et al.*, 1992), but the interaction of HCII with thrombin in the presence of glycosaminoglycans differs from ATIII. The binding site for heparin on the protein C inhibitor has been assigned to helix H, completely different from ATIII, and the properties of complex formation are correspondingly different (Pratt and Church, 1992). Vitronectin, reviewed in detail by Preissner (1991), also forms various complexes with serpins, heparin, and target proteases. It acts as a non-competitive inhibitor of the heparin-accelerated reaction of ATIII with thrombin and factor Xa (Preissner *et al.*, 1985; Podack *et al.*, 1986). Vitronectin binds directly to PAI-1, however, mainly through ionic interactions (Mimuro and Loskutoff, 1989a), which stabilises PAI-1 in its inhibitory form (Wiman *et al.*, 1988; Declerck, 1988; Mimuro and Loskutoff, 1989b; Salonen *et al.*, 1989), although other sites have also been suggested (Seiffert and Loskutoff, 1991). Vitronectin itself binds heparin at a site adjacent to its PAI-1 binding site.

A major function of serpins is to provide a tag for the removal of proteolytic enzymes from circulation via cellular receptors (Pizzo, 1989), supposed to occur when complexation with the serpin reveals a receptor recognition site (Imber and Pizzo, 1981; Mast *et al.*, 1991). Two types of receptor interactions have been proposed for specific recognition of serpins: The α-2-macroglobulin/LDL receptor and a proposed serpin-enzyme complex (SEC) receptor. The α-2-macroglobulin/low density lipoprotein receptor is implicated in hepatic uptake of proteinase inhibitor complexes from circulation and extracellular space (Herz *et al.*, 1992, Nykjaer *et al.*, 1992). The α2M/LDL-receptor binds, internalises and degrades uPA-PAI-1-complexes but does not bind uPA alone. A specific serpin enzyme complex (SEC) receptor has been proposed to recognise a specific pentapeptide sequence highly conserved among serpins (Joslin *et al.*, 1991, 1992). This sequence, buried and completely inaccessible in crystal structures of serpins, is proposed to become accessible after a serpin conformational change upon complexation with the protease. This putative change must be similar to that seen upon cleavage, since cleaved α1PI competes with the α1PI-elastase complex for binding (Joslin *et al.*, 1993), which is not easily reconcilable with the known crystal structure of cleaved α1PI.

X-ray Structures

The first X-ray structure determination of a serpin was a reactive-site modified form of the normal (M variant) human α-1-proteinase inhibitor (Löbermann *et al.*, 1984). The structure showed the residues adjacent to the cleavage site to be separated by 69 Å in well ordered β-sheet structures, obviating a major post-cleavage structural rearrangement.

The cleaved inhibitor is folded into a highly ordered structure, with almost all of its residues found in three β sheets (A-C), nine alpha helices (A-I) and six helical turns (Figure 7). β-sheets A and B are the most extensive structures of the molecule, each with six β–strands. Three strands of β-sheet B and helix B form a highly conserved hydrophobic core. β-sheet A, almost devoid of twist in the major portion, is packed on β-sheet B against

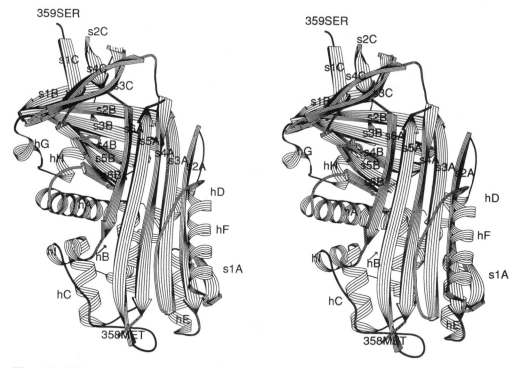

Figure 7. Ribbon type stereo plot (Priestle, 1988) of the cleaved form of α1PI. The strand s4A of sheet A is mobile, and is at least partially removed from sheet A in the inhibitory conformation.

this core. Helix F and the subsequent residues form a kind of flap which is found in front of the lower part of sheet A. β-sheet C is found on top of β-sheet B, exposed to the solvent on one side and packed on the other against β sheet B. Strands 3 and 4 of this sheet form a flexible β ribbon. Much of the rest of the molecule is helical.

The residues at the cleavage site, Met358 and Ser359, are fixed at opposite poles of the molecule by sheets A and C; the segment which forms the new intermediate carboxy terminus (s4A) after cleavage at Met358--Ser359 is incorporated into the middle of sheet A, and the segment forming the new intermediate amino terminus is incorporated onto the end of sheet C (sheet s1C). Thus, at least one of the β strands (s4A or s1C) would have to adopt a different conformation in the intact structure. As suggested by Loebermann *et al.* (1984), models of intact inhibitory forms were constructed assuming a reorganisation of strand s4A (Engh *et al.*, 1990). Higher resolution structures of other cleaved serpins (Baumann *et al.*, 1991, Baumann *et al.*, 1992) were very similar.

Ovalbumin was considered a possible model for the uncleaved structure, since it showed no evidence of the structural transition (Stein *et al.*, 1989; Gettins, 1989). The structure of the modified form (plakalbumin; Wright *et al.*, 1990) showed a five strand β sheet A consistent with a transition involving β sheet s4A, whereby the translation of one half of β sheet A accompanies the insertion of sheet s4A. Residue 345, at the N-terminal end of strand s4A, is an arginine in ovalbumin and angiotensinogen, but is otherwise conserved as Thr, Ser, Val, or Ala among all inhibitory serpins, explaining the different behaviour of ovalbumin. The conservation of the residue among inhibitors suggests the importance of possible insertion of strand s4A for inhibition. The X-ray structure of intact ovalbumin (Stein *et al.*, 1990), the first for an intact serpin, showed an α-helical conformation for the strand excised in the plakalbumin structure (P9 to P1'). This conformation is not consistent with the canonical conformation supposed to occur in the inhibitory conformation of serpins as well as other serine proteinase inhibitors (Engh *et al.*, 1990; Bode and Huber, 1992). However, it is consistent with the propensity for α-helix formation of the corresponding strand (s4A) in other serpins.

The X-ray structure of the latent form of human plasminogen activator inhibitor-1 (Mottonen, *et al.*, 1992), known to spontaneously and reversibly adopt a 'latent' conformation (Hekman and Loskutoff, 1985), showed a rearrangement of both strands s4A and s1C compared to cleaved serpins. Strand s4A was completely inserted into sheet A, and residues corresponding to strand s1C in a1PI were extended from s4A to strand s5B of sheet B. If the inhibitory conformation of PAI-1 is to resemble that suggested by the plakalbumin and ovalbumin structures, strands s3C and s4C of the β ribbon must move to accommodate the transition to the latent form; these were found to be disordered in latent PAI.

The X-ray structures of other intact serpins have been announced, including antithrombin III (Schreuder *et al.*, 1993; Wardell *et al.*, 1993) and antichymotrypsin (Wei *et al.*, 1992). Although these structures are likely to clarify aspects of serpin structures, the influence of crystallisation conditions on the flexible segments will encumber their interpretation.

What is the Inhibitory Structure?

Until three-dimensional structures of serpin-enzyme complexes are determined directly, modelling based on indirect experimental evidence is required. The most informative experiments have include thermal stability measurements, characterisations by fluorescence emission, CD, NMR, and FT-IR spectroscopy, proteolytic cleavage experiments, and aggregation studies. One particularly significant result is that α-2-antiplasmin inhibits trypsin and chymotrypsin at adjacent reactive sites (Potempa *et al.*, 1988; Enghild *et al.*, 1993). Perhaps the most revealing and unique experiments involve the complexation of small polypeptide with intact serpins. These peptides mimic strand s4A and can induce or prevent the conformational transitions, depending on their composition.

A series of observations has led to the conclusion that A1PI acts as an inhibitor only when partial insertion of strand s4A is possible. The first of these was the peptide annealing experiment (analogous to the peptide activation of trypsinogen, Huber and Bode, 1978), which involves the complexation of the serpin with a synthetic peptide mimicking β strand s4A (Schulze *et al.*, 1990). The complex shows CD spectra and denaturation stability similar to that of the cleaved form, consistent with peptide binding as the central strand of β sheet A. Another experiment was the replacement of Thr345 with Arg in α1PI by site directed mutagenesis (Schulze *et al.*, 1991). An arginine at this position corresponds to ovalbumin, and like the peptide annealing experiment, prevents the insertion of s4A into sheet A. Each alteration eliminated inhibitory activity, allowing rapid cleavage at the binding site. This suggested that the inhibitory conformation requires at least a partial insertion of the β-strand s4A into sheet A.

The extent of this insertion in the protease inhibitor complex remains uncertain, although an important result from further peptide annealing experiments significantly narrows the possibilities (Schulze *et al.*, 1992; Schulze, 1993), at least for a1PI. Beginning with the full length tetradecamer peptide, peptides progressively shortened at the N-terminal end were synthesised, complexed with intact α1PI, and tested for inhibitory activity, watching for the onset of normal inhibition. Complexation with a unadecamer peptide blocked with an acetyl protecting group showed normal inhibitory behaviour, supporting a model of partial insertion of Thr345 and maximally all or part of residue 346 (Figure 8).

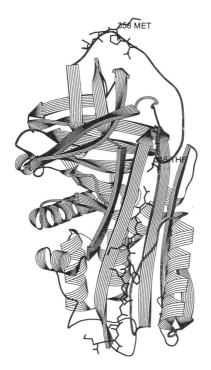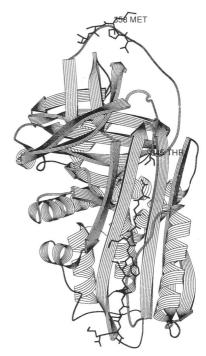

Figure 8. Ribbon-type stereo representation of a model of intact inhibitory α1PI. The reactive site has been modelled by assuming an insertion of strand s4A to Thr345, a canonical conformation at the reactive site (the P1 residue Met358 is labelled) in an orientation which may be docked to serine proteases, and an identical composition of sheet C as in the cleaved inhibitor. Sheet A is in the extended (six-stranded) conformation, identical to the cleaved form, as presumably induced by complexation with a peptide identical in sequence to strand s4A.

This is consistent with several other recent experiments. Mast *et al.* (1992) have used several enzymes to cleave the reactive site of α1PI at different sites. Aggregation of the cleaved species, assumed to be by insertion of the reactive site loop of one inhibitor into sheet A of the other, indicated a partial insertion to no more than P13. Based on studies of naturally occurring mutants of the C1 inhibitor, in particular mutations at the P12 and P10 sites which render the C1 inhibitor a substrate, Skriver *et al.* (1991) proposed that the insertion includes residues from P14 to approximately the P10 position. In this case, it is unclear whether the C1 inhibitor behaves differently from α1PI and has a longer insertion, or whether other effects cause the variation.

Such a model for an inhibitory form is subject to further constraints, besides that suggested by the peptide annealing experiments. First, the conformation of the binding loop in the serpin-inhibitor complex most likely adopts the 'canonical conformation' (Bode and Huber, 1992) of standard mechanism inhibitors (Travis and Salvesen, 1983). A weaker assumption, supported by the observation that the X-ray structures of cleaved serpins have very similar structures and conserved hydrogen bonding patterns in the sheet C region, is that this structure is preserved in the serpin-inhibitor complex. Finally, the model must fit to the target protease, in particular for α1PI it must fit to its natural target leukocyte elastase and to thrombin, targeted by the Pittsburgh deficiency variant (Met358→Arg, Owen *et al.*, 1983). A detailed model including atomic interactions between strand s4A and the body of the protein is currently not possible. Some deviations are possible in the precise location above sheet C. Progressively larger deviations are allowed in the event that the structure of sheet C is altered.

A more detailed aspect of the model involves the sequence after the cleavage site, namely Ser-Ile-Pro-Pro-Glu---. The second proline is the first residue of β strand s1C which is hydrogen bonded to the rest of sheet C. The proline-proline sequence has a small number of possible conformations, linking the 'canonical conformation' region to sheet C. Again with the assumption of a stable structure at sheet C, this further constrains the number of possible conformations at the binding site (Engh *et al.*, 1990). The first proline (s-i-PRO-p-e--) is relatively conserved among serpins, and is part of a weak homology to the binding loop of the Bowman-Birk inhibitors. In the Bowman-Birk inhibitors, it adopts a *cis* conformation (Tsunogae *et al.*, 1986; Suzuki *et al.*, 1987; Chen *et al.*, 1992; Lin *et al.*, 1993), as it is also found in the structure of cleaved α1PI. It also occurs in PAI-1, and the X-ray structure of the latent form shows this proline to adopt a *trans* conformation (E. Goldsmith, personal communication). At present, the ramifications of these observations are unknown.

The structure of β sheet A in intact serpins in an inhibitory conformation is unknown. Löbermann *et al.* (1984) speculated that this sheet is destabilised, allowing insertion, but did not suggest a specific structure. As a model for an intact form, the structure of modified ovalbumin (Wright *et al.*, 1990) suggests an intact five-stranded β-sheet structure, as does the recently solved structure of intact antichymotrypsin (Wei and Christianson, personal communication). Molecular dynamics simulations (Engh *et al.*, 1990) suggest the possibility of this structure in α1PI as well, while details of this transition has been analysed by Stein and Chothia (1991). The inhibitory complex with a serpin-peptide binary complex (Figure 8) however most likely has a six-stranded β sheet structure, where the peptide adopts the conformation of the strand s4A as seen in cleaved a1PI.

Comparison of aligned sequences shows that the serpins, although highly similar, have much greater variability in the binding loop. This variation encompasses not only a hypervariability as seen in the specificity sequences of small serine proteinase inhibitors (Laskowski *et al.*, 1987), but also in the length of the variable segment. Since serpins vary widely in function, including inhibitors as well as transport molecules, and involving different cofactors at different binding sites, and possibly different receptors, the length of the flexible sequence may have been a more productive evolutionary variable than typical for more rigid structures.

ACKNOWLEDGEMENTS

We are grateful especially to Andreas Schulze, Ennes Auerswald, Hartmut Oschkinat, Thorsten Dieckmann, Ulli Baumann, and Damir Zucic for collaboration, discussions, and other assistance in producing this manuscript.

REFERENCES

Abrahamson, M., Ritonja, A., Brown, M.A., Grubb, A., Machleidt, W. and Barrett, A.J., 1987, Identification of the probable inhibitory reactive sites of the cysteine proteinase inhibitors human cystatin C and chicken cystatin, *J. Biol. Chem.* 262:9688-9694.

Anastasi, A., Brown, M.A., Kembhavi, A.A., Nicklin, M.J.H., Sayers, C.A., Sunter, D.C. and Barrett, A.J., 1983, Cystatin, a protein inhibitor of cysteine proteinases: Improved purification from egg-white, characterization, and detection in chicken serum, *Biochem. J.* 211:129-138.

Arrowsmith, C., Pachter, R., Altman, R. and Jardetzky, O., 1991, The solution structures of Escherichia coli trp repressor and trp aporepressor at an intermediate resolution, *Eur. J. Biochem.* 202:53-66.

Barrett, A.J., Rawlings, N.D., Davies, M.E., Machleidt, W., Salvesen, G. and Turk, V., 1986a, Cysteine proteinase inhibitors of the cystatin superfamily, in: "Proteinase Inhibitors", A.J. Barrett and G. Salvesen, eds., Elsevier, Amsterdam, pp. 301-336.

Barrett, A.J., Fritz, H., Müller-Esterl, W., Grubb, A., Isemura, S., Järvinen, M., Katunuma, N., Machleidt, W., Sasaki, M. and Turk, V., 1986b, Nomenclature and classification of the proteins homologous with the cysteine proteinase inhibitor chicken cystatin, in: "Cysteine Proteinases and Their Inhibitors", V. Turk, ed., Walter de Gruyter, Berlin, pp. 1-2.

Baumann, U., Huber, R., Bode, W., Grosse, D., Lesjak, M. and Laurell, C.-B., 1991, Crystal structure of cleaved α-1-antichymotrypsin at 2.7Å resolution and its comparison with other serpins, J. Mol. Biol. 218:595-606.

Baumann, U., Bode, W., Huber, R., Travis, J. and Potempa, J., 1992, Crystal structure of cleaved equine leukocyte elastase inhibitor determined at 1.95Å resolution, J. Mol. Biol. 226:1207-1218.

Birk, Y., 1985, The Bowman-Birk inhibitor: Trypsin- and chymotrypsin-inhibitor from soybeans, Int. J. Peptide Protein Res. 25:113-131.

Blundell, T.L. and Humbel, R.E., 1980, Hormone families: pancreatic hormones and homologous growth factors, Nature 287:781-787.

Bode, W., Engh, R., Musil, D., Laber, B., Stubbs, M., Huber, R. and Turk, V., 1990, Mechanism of interaction of cysteine proteinases and their protein inhibitors as compared to the serine proteinase-inhibitor interaction, Biol. Chem. Hoppe-Seyler 371S: 111-118.

Bode, W., Engh, R., Musil, D., Thiele, U., Huber, R., Karshikov, A., Brzin, J., Kos, J. and Turk, V., 1988, The 2.0 Å X-ray crystal structure of chicken egg white cystatin and its possible mode of interaction with cysteine proteinases, EMBO J. 7: 2593-2599.

Bode, W., Musil, D., Engh, R., Huber, R., Brzin, J., Kos, J. and Turk, V., 1989, The 2.0 Å X-ray crystal structure of chicken egg white cystatin and its probable interaction with papain, in: "Intracellular Proteolysis: Mechanisms and Regulations," N. Katanuma and E. Kominami, eds., Japan Scientific Societies Press, Tokyo, pp. 297-304.

Bode, W. and Huber, R., 1992, Natural protein proteinase inhibitors and their interaction with proteinases, Eur. J. Biochem 204:433-451.

Björk, I., Nordling, K., Larsson, I. and Olson, S.T., 1992, Kinetic characterization of the substrate reaction between a complex of antithrombin with a synthetic reactive-bond loop tetradecapeptide and four target proteinases of the inhibitor, J. Biol. Chem. 267:19047-19050.

Bürgi, H.B., Dunitz, J.D. and Shefter, E., 1973, Geometrical reaction coordinates: II. Nucleophilic addition to a carbonyl group, J. Am. Chem. Soc. 95:5065-5067.

Carrell, R.W. and Travis, J., 1985, α-1-antitrypsin and the serpins: variation and countervariation, Trends Biochem. Sci. 10:20-24.

Chazin, W.J., Hugli, T.E. and Wright, P.E., 1988, Biochemistry 27:9139-9148.

Chen, P., Rose, J., Love, R., Wei, C.H. and Wang, B., 1992, Reactive sites of an anticarcinogenic Bowman-Birk proteinase inhibitor are similar to other trypsin inhibitors, J. Biol. Chem. 267:1990-1994.

Crystal, R. G., 1991, α-1-antitrypsin deficiency, emphysema, and liver disease, J. Clin. Invest. 85:1343-1352.

Declerck, P.J., DeMol, M., Alessi, M.-C., Baudner, S., Paques, E.-P., Preissner, K.T., Müller-Berghaus, G. and Collen, D., 1988, Purification and characterization of a plasminogen activator inhibitor 1 binding protein from human plasma. Identification as a multimeric form of S protein (vitronectin), J. Biol. Chem. 263:15454-15461.

Dieckmann, T., Mitschang, L., Hofmann, M., Kos, J., Turk, V., Auerswald, E.A., Jaenicke, R. and Oschkinat, H., 1994, The structures of native phosphorylated chicken cystatin and of a recombinant unphosphorylated variant in solution, J. Mol. Biol., in press.

Engh, R.A., Wright, H.T. and Huber, R., 1990, Modelling the intact form of the α-1-proteinase inhibitor, Protein Eng. 3:469-477.

Engh, R.A., Dieckmann, T., Bode, W., Auerswald, E., Turk, V., Huber, R. and Oschkinat, H., 1994, Conformational variability of chicken cystatin: comparison of structures determined by X-ray diffraction and NMR spectroscopy, J. Mol. Biol., in press.

Engh, R.A., Löbermann, H., Schneider, M., Wiegand, G., Huber, R. and Laurell, C.-B., 1989, The S-variant of human α-1-antitrypsin, structure and implication for function and mutations, Prot. Eng. 2:407-415.

Enghild, J.J., Valnickova, Z., Thogersen, I.B., Pizzo, S.V., and Salvesen, G., 1993, An examination of the inhibitory mechanism of serpins by analysing the interaction of trypsin and chymotrypsin with α2-antiplasmin, Biochem. J. 291:933-938.

Evans, D.Ll., Marshall, C.J., Christey, P.B. and Carrell, R.W., 1992, Heparin binding site, conformational change, and activation of antithrombin, Biochemistry 31:12629-12642.

Gettins, P., 1989, Absence of large scale conformational change upon limited proteolysis of ovalbumin, the prototypic serpin, J. Biol. Chem. 264:3781-3785.

Hekman, C.M. and Loskutoff, D.J., 1985, Endothelial cells produce a latent inhibitor of plasminogen activators that can be activated by denaturation, J. Biol. Chem. 260:11581-11587.

Herz, J., Clouthier, D.E. and Hammer, R.E., 1992, LDL-receptor related protein internalizes and degrades uPA-PAI-1 complexes and is essential for embryo implantation, *Cell* 71:411-421.

Huber, R., Kukla, D., Bode, W., Schwager, P., Bartels, K., Deisenhofer, J. and Steigemann, W., 1974, Structure of the complex formed by bovine trypsin and bovine pancreatic trypsin inhibitor. II. Crystallographic refinement at 1.9Å resolution, *J. Mol. Biol.* 89:73-101.

Huber, R. and Bode, W., 1978, Structural basis of the activation and action of trypsin, *Acc. Chem. Res.* 11:114-122.

Huber, R. and Carrell. R.W., 1989, Implications of the three-dimensional structure of α-1-antitrypsin for structure and function of serpins, *Biochemistry* 28:8951-8966.

Huber, R., Scholze, H., Pâques, E.P., and Deisenhofer, J., 1980, Crystal structure analysis and molecular model of human C3a anaphylatoxin, *Hoppe-Seyler's Z. Physiol. Chem.* 361:1389-1399.

Hunt L.T. and Dayhoff, M.O., 1980, A surprising new protein superfamily containing ovalbumin, antithrombin III and alpha-1-proteinase inhibitor, *Biochem. Biophys. Res. Commun.* 95:864-871.

Ikenaka, T. and Norioka, S., 1986, Bowman-Birk family serine proteinase inhibitors, *in:* "Proteinase Inhibitors", A. J. Barrett and G. Salvesen, eds., Elsevier, Amsterdam, pp. 361-374.

Imber, M.J. and Pizzo, S.V., 1981, Clearance and binding of two electrophoretic "fast" forms of human α-2-macroglobulin, *J. Biol. Chem.* 256:8134-8139.

Joslin, G., Fallon, R.J., Bullock, J., Adams, S. and Perlmutter, D.H., 1991, The SEC-receptor recognizes a pentapeptide neodomain of α-1-antitrypsin-proteinase complex, *J. Biol. Chem.* 266:11282-11288.

Joslin, G., Griffin, G.L., August, A.M., Adams, S., Fallon, R.J., Senior, R.M. and Perlmutter, D.H., 1992, The serpin-enzyme complex (SEC) receptor mediates the neutrophil chemotactic effect of α-1-antitrypsin-elastase complex and amyloid b peptide, *J. Clin. Invest.* 90:1150-1154.

Joslin, G., Wittwer, A., Adams, S., Tollefsen, D.M., August, A. Perlmutter, D.H., 1993, Cross competition for binding of α-1-antitrypsin (a1AT)-elastase complexes to serpin-enzyme complex receptor by other serpin enzyme complexes and by proteolytically modified α-1-AT*, *J. Biol. Chem.* 268:1886-1893.

Laskowski, J.M. and Kato, I., 1980, Protein inhibitors of proteinases, *Ann. Rev. Biochem.* 49:593-626.

Laskowski, M., Jr., Kato, I., Ardelt, W., Cook, J., Denton, A. Empie, M.W., Kohr, W.J., Park, S.J., Parks, K., Schatzley, B.L. Schoenberger, O.L., Tashiro, M., Vichot, G., Whatley, H.E. Wieczorek, A. and Wieczorek, M., 1987, Ovomucoid Third Domains from 100 Avian Species: Isolation, Sequences, and Hypervariability of Enzyme-Inhibitor Contact Residues, *Biochemistry* 26:202-221.

Lawson, C.L., Zhang, R., Schevitz, R.W., Otwinowski, Z., Joachimiak, A. and Sigler, P.B., 1988, Flexibility of the DNA-binding domains of trp repressor, *Proteins: Str., Fun., and Gen.* 3:18-31.

Lin. G., Bode, W., Huber, R. Chi, C. and Engh, R.A., 1993, The 0.25-nm X-ray structure of the Bowman-Birk type inhibitor from mung bean in ternary complex with porcine trypsin, *Eur. J. Biochem.* 212:549-555.

Löbermann, H., Tokuoka, R., Deisenhofer, J. and Huber, R., 1984, Human α-1 Proteinase inhibitor. Crystal structure analysis of two modifications. Molecular model and preliminary analysis of the implications for function, *J. Mol. Biol.* 177:531-556.

Machleidt, M., Thiele, U., Laber, B., Assfalg-Machleidt, I., Esterl, A., Wiegand, G., Kos, J., Turk, V. and Bode, W., 1989, Mechanism of inhibition of papain by chicken egg white cystatin. Inhibition constants of N-terminally truncated forms and cyanogen bromide fragments of the inhibitor *FEBS Lett.* 243:234-238

Marquardt, M., Walter, J., Deisenhofer, J., Bode, W. and Huber, R., 1983, The geometry of the reactive site and of the peptide groups in trypsin, trypsinogen, and its complexes with inhibitors, *Acta Crystallogr.* B39:480-490.

Mast, A.E., Enghild, J.J. and Pizzo, S.V., 1991, Analysis of the plasma elimination kinetics and conformational stabilities of native, proteinase complexed and reactive site cleaved serpins: comparison of α-1-proteinase inhibitor, α-1-antichymotrypsin, antithrombin III, α-2-antiplasmin, angiotensinogen and ovalbumin. *Biochemistry* 30:1225-1230.

Mast, A.E., Enghild, J.J. and Salvesen, G., 1992, Conformation of the reactive site loop of α-1- Proteinase Inhibitor probed by limited proteolysis, *Biochemistry* 31:2720-2728.

Mimuro, J. and Loskutoff, D.J., 1989a, Binding of type 1 plasminogen activator inhibitor to the extracellular matrix of cultured bovine endothelial cells, *J. Biol. Chem.* 264:5058-5063.

Mimuro, J. and Loskutoff, D.J., 1989b, Purification of a protein from bovine plasma that binds to type 1 plasminogen-activator inhibitor and prevents its interaction with extracellular-matrix - evidence that the protein is vitronectin, *J. Biol. Chem.* 264:936-939.

Mottonen, J., Strand, A., Symersky, J., Sweet, R.M., Danley, D.E., Geoghegan, K.F., Gerard, R.D. and Goldsmith, E.J., 1992, Structural basis of latency in plasminogen activator inhibitor-1, Nature 355, 270-273.

Murzin, A.G., 1993, Sweet-tasting protein monellin is related to the cystatin family of thiol proteinase inhibitors, *J. Mol. Biol.* 230: 689-694.

Nykjaer, A., Petersen, C.M., Moller, J., Jensen, P.H., Moestrup, S.K., Holtet, T.L., Etzrodt, M., Thogersen, H.C., Munch, M., Andreasen, P.A. and Gliemann, J., 1992, Purified α-2Macroglobulin receptor/LDL receptor-related protein binds urokinase-plasminogen activator inhibitor type-1 complexes, *J. Biol. Chem.* 267:14543-14546.

Odani, S. and Ikenaka, T., 1973, Studies on soybean trypsin inhibitors: VIII. Disulfide bridges in soybean Bowman-Birk inhibitor, *J. Biochem.* 74:697-715.

Owen, M.C., Brennan, S.O., Lewis, J.H. and Carrell, R.W., 1983, Mutation of antitrypsin to antithrombin. Antitrypsin Pittsburgh (358Met-Arg), a fatal bleeding disorder, *N. Engl. J. Med.* 309:694-698.

Pizzo, S.V., 1989, Serpin receptor 1. a hepatic receptor that mediates the clearance of antithrombin III-proteinase complexes, *Amer. J. Med.* 87 3B:10S-13S.

Podack, E-R., Dahlböck, B. and Griffin, J.H., 1986, Interaction of S-protein of complement with thrombin and antithrombin III during coagulation. Protection of thrombin by S-protein from antithrombin III inactivation, *J. Biol. Chem.* 261:7387-7392.

Potempa, J., Shieh, B.-H. and Travis, J., 1988, Alpha-2-antiplasmin: A serpin with two separate but overlapping reactive sites, *Science* 241:699-700.

Pratt, C.W. and Church, C., 1992, Heparin binding to protein C inhibitor, *J. Biol. Chem.* 267:8783- 8794.

Pratt, C.W. and Whinna, H.C. and Church, C., 1992, A comparison of the heparin binding serine proteinase inhibitors, *J. Biol. Chem.* 267:8795-8801.

Preissner, K.T., Wassmuth, R. and Müller-Berghaus, G., 1985, Physicochemical characterisation of human S-protein and its function in the blood coagulation system, *Biochem. J.* 231:349-355.

Preissner, K.T., 1991, Structure and biological role of vitronectin, *Annu. Rev. Cell Biol.* 7:275-310.

Priestle, J.P., 1988, Ribbon: A stereo cartoon drawing program for proteins, *J. Appl. Cryst.* 21:572-576.

Rawlings, N.D. and Barrett, A.J., 1990, Evolution of proteins of the cystatin superfamily, *J. Mol. Evol.* 30:60-71.

Read, R. and James, M.N.G., 1986, Introduction to the protein inhibitors: X-ray crystallography, *in:* "Proteinase Inhibitors", A. J. Barrett and G. Salvesen, eds., Elsevier, Amsterdam, pp. 301-336.

Richardson, J.S., 1981, The Anatomy and Taxonomy of Protein Structure, *Adv. Protein Chem.* 34:167-339.

Rühlmann, A., Kukla, D., Schwager, P., Bartels, K. and Huber, R., 1973, Structure of the complex formed by bovine trypsin and bovine pancreatic trypsin inhibitor: Crystal structure determination and stereochemistry of the contact region, *J. Mol. Biol.* 77:417-436.

Saitoh, E., Isemura, S., Sanada, K. and Ohnishi, K., 1991, Cystatins of family II are harboring two domains which retain inhibitory activities against the proteinases, *Biochem. Biophys. Res. Comm.* 175:1070-1075.

Salonen, E.-M., Vaheri, A, Pöllänen, J., Stephens, R. and Andreasen, P., 1989, Interaction of plasminogen activator inhibitor (PAI-1) with vitronectin, *J. Biol. Chem.* 264, 6339-6343.

Schapira, M. and Patston, P.A., 1991, Serine protease inhibitors (Serpins), *Trends Card. Med.* 1:146-151.

Schevitz, R.W., Otwinowski, Z., Joachimiak, A., Lawson, C.L. and Sigler, P.B., 1985, The three-dimensional structure of trp repressor, *Nature* 317: 782-786.

Schreuder, H., deBoer, H., Pronk, S., Hol., W., Dijkema, R., Moulders, J. and Theunissen, H., 1993, Crystallization and preliminary X-ray analysis of human antithrombin III, *J. Mol. Biol.* 229:249-250.

Schulze, A.J., Baumann, U., Knof, S., Jäger, E., Huber, R. and Laurell, C.-B., 1990, Structural transition of α-1-antitrypsin by a peptide sequentially similar to β-strand s4A, *Eur. J. Biochem.* 194:51-56.

Schulze A. J., Huber, R., Degryse, E., Speck, D. and Bischoff, R., 1991, Inhibitory activity and conformational changes in α-1-proteinase inhibitor variants, *Eur. J. Biochem.* 202:1147-1155.

Schulze, A.J., Frohnert, P.W., Engh, R.A. and Huber, R., 1992, Evidence for the extent of insertion of the active site loop of intact α-1-Proteinase Inhibitor in β-sheet A, *Biochemistry* 31:7560-7565

Schulze, A. J., 1993, Untersuchungen zur Struktur und Funktion von Serinproteinase Inhibitoren und Serinproteinasen, PhD Thesis, Technische Universität München.

Seiffert, D. and Loskutoff, D.J., 1991, Kinetic analysis of the interaction between type-1 plasminogen activator inhibitor and vitronectin and evidence that the bovine inhibitor binds to a thrombin-derived aminoterminal fragment of bovine vitronectin, *Biochem. Biophys. Acta* 1078:2223-2230.

Skriver, K., Wikoff, W.R., Patson, P.A., Tausk, F., Schapira, M., Kaplan, A.P., and Bock, C.S., 1991, Substrate properties of C1 inhibitor Ma (alanine 434→glutamic acid), *J. Biol. Chem.* 266, 9216-9221.

Stein, P.E. and Chothia, C., 1991, Serpin tertiary structure transformation, *J. Mol. Biol.* 221:615-621.

Stein, P.E., Tewkesbury, C. and Carrell., R.W., 1989, Ovalbumin and angiotensinogen lack serpin S-R conformational change, *Biochem. J.* 262:103-107.

Stein, P.E., Leslie, A.J., Finch, J.T., Turnell, W.G., McLaughlin, P.J. and Carrell, R.W., 1990, Crystal structure of ovalbumin as a model for the reactive centre of serpins, *Nature* 347:99-102.

Stubbs, M. T., Laber, B., Bode, W., Huber, R., Jerala, R., Lenarcic, B., Turk, V., 1990, The refined 2.4 Å X-ray crystal structure of recombinant human stefin B in complex with the cysteine proteinase papain: a novel type of proteinase inhibitor interaction, *EMBO J.* 9, 1939-1947

Sweet, R.M., Wright, H.T., Janin, J., Chothia, C.H. and Blow, D.M., 1974, Crystal structure of the complex of porcine trypsin with soybean trypsin inhibitor (Kunitz) at 2.6 Å resolution, *Biochemistry* 13:4212-4228.

Suzuki, A., Tsunogae, Y., Tanaka, I., Yamane, T., Ashida, T., Norioka, S., Hara, S. and Ikenaka, T., 1987, The structure of Bowman-Birk type protease inhibitor A-II from peanut (Arachis hypogaea) at 3.3 Å resolution, *J. Biochem.* 101:267-274.

Travis, A.J. and Salvesen, G., 1983, Plasma proteinase inhibitors, *Annu. Rev. Biochem.* 52:655-709.

Tsunogae, Y., Tanaka, I., Yamane, T., Kikkawa, J., Ashida, T., Ishikawa, C. and Takahashi, K., 1986, Structure of the trypsin-binding domain of Bowman-Birk type protease inhibitor and its interaction with trypsin, *J. Biochem.* 100:1637-1646.

Turk, V. and Bode, W., 1991, The cystatins: protein inhibitors of cysteine proteinases, *FEBS Lett.* 285:213-219.

Turk, V., Brzin, J., Lenarcic, B., Locnikar, P., Popovic, T., Ritonja, A., Babnik, J., Bode, W. and Machleidt, W., 1985, *in:* "Intracellular Protein Catabolism V," E. Khairallah and J. Bond, eds., Alan R. Liss, New York, pp. 91-103.

Wardell, M.R., Abrahams, J.-P., Bruce, D., Skinner, R. and Leslie, A.G.W., 1993, Crystallization and Preliminary X-ray Diffraction Analysis of Two Conformations of Intact Human Antithrombin, *J. Mol. Biol.*, in press.

Werner, M.H. and Wemmer, D.E., 1992, Three-dimensional structure of soybean trypsin/chymotrypsin Bowman-Birk inhibitor in solution, *Biochemistry* 31:999-1010.

Wei, A., Rubin, H., Cooperman, B.S., Schechter, N. and Christianson, D.W., 1992, Crystallization, activity assay and preliminary X-ray diffraction analysis of the uncleaved serpin antichymotrypsin, *J. Mol. Biol.* 226:273-276.

Wiman, B., Almquist, A., Sigurdardottir, O. and Lindahl, T., 1988, Plasminogen activator inhibitor 1 (PAI-1) is bound to vitronectin in plasma, *FEBS Lett.* 242:125-128.

Wright, T., Qian, H.-X. and Huber, R., 1990, Crystal structure of plakalbumin, a proteolytically nicked form of ovalbumin. Its relationship to the structure of cleaved alpha-1-antitrypsin, *J. Mol. Biol.* 213:513-528.

STRUCTURE AND tRNA[Phe]-BINDING PROPERTIES OF THE ZINC FINGER MOTIFS OF HIV-1 NUCLEOCAPSID PROTEIN

Yves Mély[1], Etienne Piémont[1], Monica Sorinas-Jimeno[1], Hugues de Rocquigny[2], Nathalie Jullian[2], Nelly Morellet[2], Bernard P. Roques[2] and Dominique Gérard[1]

[1] Laboratoire de Biophysique, URA 491 du CNRS, Université Louis Pasteur, 67401 Illkirch Cedex, France
[2] Département de Chimie Organique, INSERM U266 - CNRS UA 448, UFR des Sciences Pharmaceutiques et Biologiques, 75270 Paris Cedex 06, France

INTRODUCTION

The nucleocapsid protein NCp7 of the human immunodefiency virus type 1 (HIV-1) is a 72 amino acid peptide containing two zinc fingers of the type $CX_2CX_4HX_4C$. NCp7 is thought to be a key component of the retrovirus life cycle since it activates both viral RNA dimerization and replication primer tRNA[Lys,3] annealing to the initiation site of reverse transcription[1,2]. NCp7 constitutes thus a potential target for antiviral therapy.

Within the amino acid sequence of both zinc finger motifs, the conserved aromatic amino acids : Phe 16 and Trp 37 are probably of critical importance since their replacement even by a hydrophobic moiety led to a large decrease in NCp7 activities[3].

In order to clarify the structural and dynamic properties of these aromatic amino acids, the steady-state and time-resolved fluorescence properties of two zinc-saturated 18-residue synthetic peptides with the amino acid sequence of the N-terminal (NCp7 13-30 F16W where the natural occurring Phe was replaced by a Trp residue) and the C-terminal (NCp7 34-51) zinc finger motifs (Fig. 1) were investigated. Moreover, using Trp fluorescence we investigated the involvement of these aromatic amino acids in the interaction of both fingers with the heterologous tRNA[Phe].

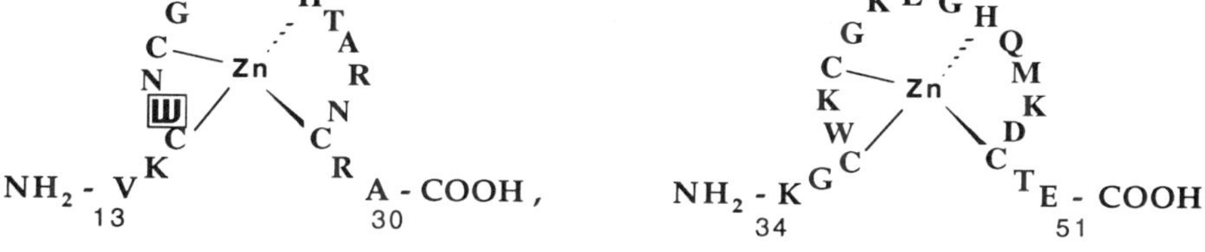

Figure 1. Amino acid sequences of NCp7 13-30 F16W and NCp7 34-51.

Statistical Mechanics, Protein Structure, and Protein Substrate Interactions
Edited by S. Doniach, Plenum Press, New York, 1994

MATERIALS AND METHODS

NCp7 13-30 F16W and NCp7 34-51 were synthesized by a solid phase method. Experiments were performed in 10 mM Hepes pH 7.5 in the presence of saturating zinc concentrations. Fluorescence steady-state measurements were performed with a SLM 48000 spectrofluorometer. Fluorescence lifetimes were measured using a frequency-doubled rhodamine 6G laser, dumped and synchronously pumped by a mode-locked Ar laser.

RESULTS AND DISCUSSION

Conformational and dynamic properties

The fluorescence intensity decays of both NCp7 13-30 F16W and NCp7 34-51 were biexponential but the fluorescence lifetimes of the latter were significantly higher than those of the former (Table I). Analysis of the fluorescence decays according to the wavelength and to the temperature is entirely consistent with the existence in both peptides of a ground-state equilibrium as previously proposed for Trp itself[4] and related nucleocapsid proteins[5]. By resolving the steady-state fluorescence spectrum into two components (DAS) associated with each lifetime, we found that in both peptides the fluorescence of the long-lived component contributes to about 90 % of the total fluorescence intensity and was 5 to 10 nm red-shifted as compared to the short-lived one. As subnanosecond protein motions have been largely reported, the two decay times and DAS were assigned to the existence of two classes of conformations rather than to two unique precise conformations. The equilibrium constant between the two classes, obtained from the integrated fluorescence intensities of the DAS is about 2.2 at 20°C for both peptides. Moreover, since no excited-state reaction could be observed by time-resolved fluorescence and since the indole moiety protons gave sharp resonances by ^{1}H-NMR[6], we infer that the interconversion rates between the two classes are between 10^3 and 5.10^7 s^{-1}.

TABLE 1. Fluorescence decay parameters of NCp7 13-30 F16W and NCp7 34-51 [1]

	τ_i (ns)	α_i (%)	f_i (%)	k_f x 10^{-7} (s^{-1})	k_{nri} x 10^{-8} (s^{-1})	Ea (kJ.mol^{-1})	f x 10^{-11} (s^{-1})
NCp7 13-30 F16W	1.1	32	8	4.3	8.7	-	-
	5.52	68	92	4.3	1.4	19.9	3.8
NCp7 34-51	1.9	32	12	4.2	4.8	16	4
	6.32	68	88	4.2	1.2	25.1	26

[1] Peptides were about 20 μM. Excitation and emission wavelengths were 295 and 350 nm, respectively. Lifetime components τ_i, normalized preexponential terms α_i, and fractional intensities f_i were obtained using a time-correlated single-photon-counting technique. The non radiative rate constant k_{nri} was calculated from : $k_{nri} = 1/\tau_i - k_f$ with k_f being the radiative rate constant. The activation energy Ea and frequency factor f were obtained from an Arrhenius plot by plotting $\ln(k_{nri})$ versus 1/T.

In both peptides, the two classes essentially differed in their non-radiative rate constants. As ^{1}H-NMR data indicated that the closest quenchers to Phe16 and Trp37 are peptide carbonyl groups[7] we infer that the main non-radiative pathway is a charge transfer between Trp and the carbonyl groups. Arrhenius plots of the non-radiative rate constants over the 10-60°C range showed that the ground-state classes differed by their activation energies (Table I) suggesting that the fluorescence quenchers involved in the two classes differed to some extent. In contrast, the frequency factors were low and similar in agreement with the highly constrained peptide chain conformation.

Finally the combination of fluorescence and ^{1}H-NMR data suggested that the ground-state heterogeneity is probably correlated to rotamers about the χ^1 angle of the $C\alpha$-$C\beta$ bond of the aromatic residues.

tRNA^{Phe}-binding properties

Addition of tRNA^{Phe} to both NCp7 13-30 F16W and NCp7 34-51 induced : (i) a very large decrease of the fluorescence intensity (more than 90% at the highest concentration of tRNA^{Phe}), (ii) no changes in the fluorescence intensity decay parameters, and (iii) a large positive peak at 295 nm in the difference absorption spectra[7]. These data clearly suggested a stacking of the indole moiety of Trp with the bases of tRNA^{Phe}.

TABLE 2. tRNA^{Phe}-binding parameters of NCp7 13-30 F16W and NCp7 34-51 [1]

	K (M^{-1}) [2]	n	m'	K$_{NE}$ (M^{-1})
NCp7 13-30 F16W	1.3 ($\pm$0.2).10^5	3.0 ($\pm$0.5)	0.85 ($\pm$0.07)	3.4 ($\pm$0.2).10^3
NCp7 34-51	8.4 ($\pm$0.7).10^3	2.9 ($\pm$0.7)	0.23 ($\pm$0.04)	3.2 ($\pm$0.6).10^3

[1] Peptide solutions (5 μM) were in 5 mM Hepes pH 7.5. m' is the number of ion pairs formed between the peptide and the nucleic acid. K$_{NE}$ is the extrapolated value of K at 1 M NaCl and is assigned to the nonelectrostatic component of the binding affinity.
[2] Each parameter was expressed as mean $\pm$ standard error of the mean using at least 5 independent experiments.

The binding of both fingers to tRNA^{Phe} was non cooperative with similar site size (3 nucleotide residues/peptide) but the affinity of the N-terminal finger domain in low-ionic strength buffer was one order of magnitude larger than the C-terminal one (Table II). Salt-reversal experiments showed that the nonelectrostatic component of the binding affinity was similar in both peptides and primarily drove the formation of the tRNA^{Phe}-peptide complex. From sequence comparison (Fig. 1) we infer that the aromatic residues are probably of crucial importance in these interactions since the other residues are either not hydrophobic or not conserved. The higher affinity of the N-terminal finger is correlated to additional electrostatic interactions probably involving Lys14 or/and Arg29 since these two residues are replaced by non charged residues in the C-terminal finger.

Nevertheless the affinity of the two peptides for tRNA^{Phe} is much lower than the NCp7 one (K=2.10^6 M^{-1} in the presence of 0.1 M NaCl) in keeping with a critical involvement of basic amino acids located outside of the finger domains[8].

REFERENCES

1. J.L. Darlix, C. Gabus, M.T. Nugeyre, F. Clavel, and F. Barré-Sinoussi *J. Mol. Biol.* 216:689-699 (1990).
2. C. Barat, V. Lullien, O. Schatz, G. Keith, M.T. Nugeyre, F. Grüninger-Leitch, F. Barré-Sinoussi, S.F.J. Le Grice, and J.L. Darlix, *Embo J.* 8:3279-3285 (1989).
3. C. Méric, and S.P. Goff, *J. Virol.* 63:1558-1568 (1989).
4. A.G. Szabo, and D.M. Rayner, *J. Am. Chem. Soc.* 102:554-563 (1980).
5. Y. Mély, H. De Rocquigny, E. Piémont, H. Déméné, N. Jullian, M.C. Fournié-Zaluski, B.P. Roques and D. Gérard, *Biochim. Biophys. Acta* 1161:6-18 (1993).
6. N. Morellet, N. Jullian, H. de Rocquigny, B. Maigret, J.L. Darlix, and B.P. Roques, *Embo J.* 11:3059-3065 (1992).
7. Y. Mély, E. Piémont, M. Sorinas-Jimeno, D. Gérard, H. De Rocquigny, N. Jullian, N. Morellet and B.P. Roques, *Biophys. J.* in press (1993).
8. H. de Rocquigny, C. Gabus, A. Vincent, M.C. Fournié-Zaluski, B.P. Roques and J.L. Darlix, *Proc. Natl. Acad. Sci. USA* 89:6472-6476 (1992).

ELECTROSTATIC COMPLEMENTARITY
IN PROTEIN-SUBSTRATE INTERACTIONS AND LIGAND DESIGN

J.M. Goodman[1] and P.-L. Chau[2]

[1]Department of Chemistry, University of Cambridge
Lensfield Road, Cambridge CB2 1EW United Kingdom
[2]to whom correspondence should be addressed at:
Department of Pharmacology, University of Cambridge
Tennis Court Road, Cambridge CB2 1QJ United Kingdom

INTRODUCTION

The analysis of protein-substrate interactions has been used as a model for ligand-receptor interactions. It is an important research problem, and an understanding of it will enable us to design new molecules to bind to receptors. In our previous work (Chan *et al.* 1992), we presented a method, called YING, for the assignment of atom-centred point charges for a ligand, based on the electrostatic potential of the receptor. These point charges are chosen to give the best possible complementarity to the receptor electrostatic potential over the van der Waals surface of the ligand.

In the present study we have used this method with two minor modifications. We have used all of the ligand surface to find the best atom-centred point charges, instead of deleting those points which are not adjacent to any protein atoms. Inclusion of these points does not appreciably change the complementarity between the ligand and the receptor (Chau and Dean, manuscript in preparation). Previously, a least-squares method was used to assess the complementarity between the ligand and receptor electrostatic potentials, where the ligand potential was treated as a dependent variable. In this study, an orthogonal least-squares method is used to assess the complementarity between the ligand and receptor potentials (Riggs *et al.* 1978).

STRUCTURES INVESTIGATED

The entire Brookhaven Protein Databank was surveyed for ligand-protein co-crystals, but structures containing metals and disordered structures were excluded. Only data with a resolution of 2.5Å or better were considered. The partial charges of the protein molecules were assigned from the ECEPP charges (Némethy *et al.* 1983), while those of the ligands were calculated using CNDO/2 (Pople *et al.* 1965; Pople and Segal 1965). Some of the ligands failed to optimize on CNDO/2, and they were excluded from this study. Table 1 shows the eleven ligand-protein pairs that were used in this study.

Statistical Mechanics, Protein Structure, and Protein Substrate Interactions
Edited by S. Doniach, Plenum Press, New York, 1994

373

Table 1. Brookhaven Protein Databank structures investigated.

PDB code	description	reference
1AK3	bovine heart mitochondria adenylate kinase and AMP[1]	Diederichs & Schulz (1991)
1BBP	bilin-binding protein and biliverdin IX-γ[2]	Huber *et al.* (1987)
1CSC	citrate synthetase holoenzyme and L-malate	Karpusas *et al.* (1990)
2CSC	citrate synthetase holoenzyme and D-malate	Karpusas *et al.* (1990)
3CSC	citrate synthetase holoenzyme and L-malate	Karpusas *et al.* (1990)
4CSC	citrate synthetase holoenzyme and D-malate	Karpusas *et al.* (1990)
1FNR	ferredoxin-NADP$^+$ reductase and FAD	Karplus *et al.* (1991)
1DHF	dihydrofolate reductase and folate[3]	Davies *et al.* (1990)
1LDM	apo-lactate dehydrogenase and oxamate[4]	Abad-Zapatero *et al.* (1987)
2YPI	triosephosphate isomerase and 2-phosphoglyco-late[5]	Lolis & Petsko (1990)
1RBP	human serum retinol binding protein and retinol	Cowan *et al.* (1990)

[1]1AK3 contains two subunits, A and B. Each subunit binds one AMP molecule.
[2]1BBP contains four subunits, A to D. Each subunit binds one biliverdin molecule.
[3]1DHF contains two subunits, A and B. Each subunit binds one folate molecule.
[4]1LDM contains one protein that binds two oxamate molecules.
[5]2YPI contains two subunits, A and B. Each subunit binds one 2-phosphoglycolate molecule.

METHODS AND RESULTS

For each structure, the partial atomic charges on the protein atoms were assigned, and the charges on the ligand atoms calculated. The electrostatic potentials on the van der Waals surface of the ligand generated by the protein and the ligand, respectively, were calculated using a Coulombic summation. We then calculated the Pearson product moment correlation coefficient and Spearman's rank correlation coefficient between the protein and ligand potentials, and performed a regression analysis using an orthogonal least-squares method. In cases of perfect complementarity, both correlation coefficients and the slope of regression would be −1, while the *y*-intercept would be 0 (Dean *et al.* 1992).

We then performed a YING analysis (Chan *et al.* 1992) on the ligands to assign the partial charges for optimal electrostatic complementarity. The electrostatic potential generated by the YING partial charges on the ligand van der Waals surface is compared with the protein potential. A similar statistical analysis was performed.

Comparisons of the statistical analyses are shown in table 2 below.

Table 2. Regression analysis of electrostatic potential generated by natural ligands and YING ligands against the protein potential. The first row of each two show the natural ligand data, the second row the YING ligand data. Legend to abbreviations: r = Pearson's product moment correlation coefficient, r_{rank} = Spearman's rank correlation coefficient, m_r = slope of regression, c_r = y-intercept of regression, e_c = c_r as percentage of total YING potential.

ligand-protein pair	r_{rank}	r	m_r	c_r	e_c
AMP(a)-1AK3(a)	-0.754	-0.707	-2.979	762	75
YING	-0.910	-0.877	-1.310	236	33
AMP(b)-1AK3(b)	-0.700	-0.670	-2.789	729	69
YING	-0.812	-0.786	-0.929	-141	24
BLV(a)-1BBP(a)	-0.102	-0.264	-0.644	-184	26
YING	-0.944	-0.952	-1.035	-51	4
BLV(b)-1BBP(b)	-0.112	-0.257	-0.468	-239	33
YING	-0.953	-0.958	-1.046	-52	4
BLV(c)-1BBP(c)	-0.260	-0.136	-0.432	-264	37
YING	-0.963	-0.963	-0.972	50	4
BLV(d)-1BBP(d)	-0.121	-0.246	-0.472	-247	19
YING	-0.948	-0.955	-0.977	59	5
L-malate-1CSC	-0.677	-0.639	-0.291	-858	162
YING	-0.928	-0.888	-0.827	-128	12
D-malate-2CSC	-0.671	-0.638	-0.283	-835	157
YING	-0.964	-0.952	-1.136	-158	11
L-malate-3CSC	-0.413	-0.358	-0.172	-723	187
YING	-0.921	-0.911	-0.932	29	4
D-malate-4CSC	-0.455	-0.409	-0.298	-602	103
YING	-0.945	-0.928	-0.956	57	6

Table 2. (Continued).

ligand-protein pair	r_{rank}	r	m_r	c_r	e_c
FAD-1FNR	−0.870	−0.812	−1.322	−607	63
YING	−0.991	−0.988	−1.009	22	3
folate(a)-1DHF(a)	−0.845	−0.794	−1.800	−519	62
YING	−0.991	−0.987	−0.984	−9	1
folate(b)-1DHF(b)	−0.874	−0.822	−1.563	−470	55
YING	−0.995	−0.992	−0.994	7	1
oxamate(a)-1LDM	−0.948	−0.922	−0.717	277	61
YING	−0.915	−0.873	−0.917	−51	8
oxamate(b)-1LDM	−0.653	−0.575	−1.567	595	131
YING	−0.680	−0.605	−0.764	1	0.2
2PGA(a)-2YPI(a)	−0.467	−0.384	−1.286	−862	163
YING	−0.919	−0.911	−0.949	−46	11
2PGA(b)-2YPI(b)	−0.148	−0.029	0.055	−1386	259
YING	−0.891	−0.876	−0.883	−113	22
retinol-1RBP	−0.289	−0.214	−0.147	−15	11
YING	−0.320	−0.410	−3.955	−467	108

From the table, it can be seen that YING is useful in assigning a set of atom-centred partial charges that produce a better electrostatic complementarity between the ligand and the receptor. In all but one case, the YING ligand exhibits a better correlation coefficient and rank correlation coefficient. This shows that it has a higher electrostatic complementarity than the natural ligand. In the case of oxamate(a) and apo-lactate dehydrogenase 1LDM, the complementarity is very slightly decreased, but the slope of the regression is nearer to −1, and the y-intercept is much nearer to 0. This is because YING optimizes the sum of squares of the difference between the ligand and protein potentials (Chan *et al.* 1992), and so would decrease the correlation to obtain a better regression.

We have considered biliverdin and its interaction with the 1BBP C-subunit in detail. This molecule is present in the tegument of the cabbage white butterfly *Pieris brassicae* (Rüdiger *et al.* 1968). The ligand-protein complex is most probably responsible for the change in ATP production in the larval stage period when different lighting schedules are imposed on the insect (Vuillaume and Dattée 1980). Figure 1 shows the scattergram of the natural ligand electrostatic potential against the protein potential, while figure 2 shows the scattergram YING-ligand electrostatic potential against the same protein potential.

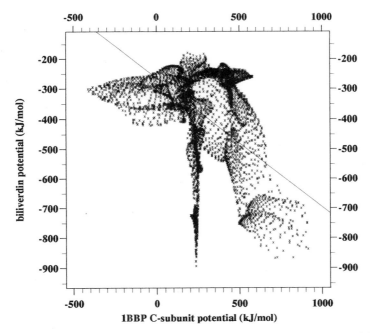

Figure 1. Scattergram showing the biliverdin IX-γ potential against the 1BBP protein potential (subunit C). $r = -0.260$, $r_{rank} = -0.136$, $m_r = -0.432$, $c_r = -264$.

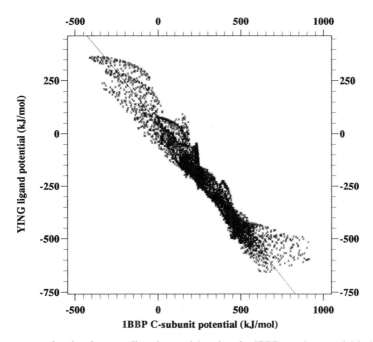

Figure 2. Scattergram showing the YING ligand potential against the 1BBP protein potential (subunit C). $r = -0.963$, $r_{rank} = -0.963$, $m_r = -0.972$, $c_r = 50$.

By inspecting the atom-centred partial charges assigned by YING to see how to inprove complementarity, we modified biliverdin IX-γ by changing three hydrogen atoms to fluorine atoms. The electrostatic potential on the van der Waals surface of this fluorinated biliverdin was compared with that from the protein, subunit C of 1BBP. The scattergram is shown in figure 3 below.

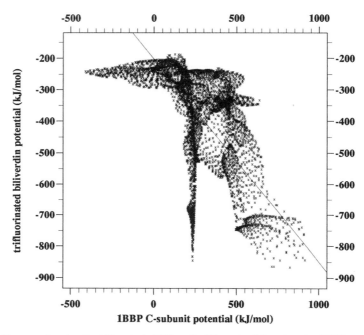

Figure 3. Scattergram showing the trifluorinated biliverdin IX-γ potential against the 1BBP protein potential (subunit C). $r = -0.611$, $r_{rank} = -0.517$, $m_r = -0.647$, $c_r = -204$.

It can be seen that the correlation is better for the trifluorinated biliverdin IX-γ than for biliverdin IX-γ, the natural ligand. More work is being done to improve the electrostatic complementarity even further.

CONCLUSION

Nature matches the complementarity of active sites rather well, but it does not do it perfectly. This may be due to other restrictions on the ligands, such as stability, solvation-desolvation effects and the limits of biosynthesis. In this work, we have demonstrated that it is possible to design a new ligand which have better electrostatic complementarity than that achieved by Nature, by using molecules that cannot be readily biosynthesized, such as the fluorinated derivatives of natural ligands. The optimal charge distribution for different ligands in the same receptor is compared, and has been used to suggest a new ligand that may have a higher affinity for the receptors. Regression analysis and YING can be used to explore the electrostatic complementarity between proteins and their substrates, and also to design new ligands that possess a higher degree of this complementarity.

ACKNOWLEDGMENTS

JMG thanks Clare College, Cambridge, for a research fellowship. PLC is a Rhône-Poulenc Rorer research fellow. We also thank Dr. S.L. Chan for useful discussions, and Miss M.T. Barakat for making her software available. Part of this work was supported by the Science and Engineering Research Council Cambridge Centre for Molecular Recognition.

REFERENCES

Abad-Zapatero, C., Griffith, J.P., Sussman, J.L. and Rossmann, M.G. (1987) "Refined crystal structure of dogfish M_4 apo-lactate dehydrogenase", *J. Mol. Biol.*, **198**, 445-467.

Chan, S.L., Chau, P.-L. and Goodman, J.M. (1992) "Ligand atom partial charges assignment for complementary electrostatic potentials", *J. Comput.-Aided. Mol. Design*, **6**, 461–474.

Cowan, S.W., Newcomer, M.E. and Jones, T.A. (1990) "Crystallographic refinement of human serum retinol binding protein at 2Å resolution", *Proteins: structure, function and genetics*, **8**, 44-61.

Davies, J.F., Delcamp, T.J., Prendergast, N.J., Ashford, V.A., Freisheim, J.H. and Kraut, J. (1990) "Crystal structures of recombinant human dihydrofolate reductase complexed with folate and 5-deazafolate", *Biochemistry*, **29**, 9467–9479.

Dean, P.M., Chau, P.-L. and Barakat, M.T. (1992) "Development of quantitative methods for studying electrostatic complementarity in molecular recognition and drug design", *J. Mol. Struct. (Theochem)*, **256**, 75-89.

Diederichs, K. and Schulz, G.E. (1991) "The refined structure of the complex between adenylate kinase from beef heart mitochondrial matrix and its substrate AMP at 1.85Å resolution", *J. Mol. Biol.*, **217**, 541–549.

Huber, R., Schneider, M., Mayr, I., Müller, R., Deutzmann, R., Suter, F., Zuber, H., Falk, H. and Kayser, H. (1987) "Molecular structure of the bilin binding protein (BBP) from *Pieris brassicae* after refinement at 2.0Å resolution", *J. Mol. Biol.*, **198**, 499–513.

Karpusas, M., Branchaud, B. and Remington, S.J. (1990) "Proposed mechanism for the condensation reaction of citrate synthase: 1.9Å structure of the ternary complex with oxaloacetate and carboxymethyl coenzyme A", *Biochemistry*, **29**, 2213–2219.

Karplus, P.A., Daniels, M.J. and Herriott, J.R. (1991) "Atomic structure of ferredoxin-$NADP^+$ reductase: prototype for a structurally novel flavoenzyme family", *Science*, **251**, 60-66.

Lolis, E. and Petsko, G.A. (1990) "Crystallographic analysis of the complex between triose-phosphate isomerase and 2-phosphoglycolate at 2.5Å resolution: implications for catalysis", *Biochemistry*, **29**, 6619–6625.

Némethy, G., Pottie, M.S. and Scheraga, H.A. (1983) "Energy parameters in polypeptides. 9. Updating of geometrical parameters, nonbonded interactions, and hydrogen bond interactions for the naturally-occurring amino acids", *J. Phys. Chem.*, **87**, 1883–1887.

Pople, J.A., Santry, D.P. and Segal, G.A. (1965) "Approximate self-consistent molecular orbital theory. I. Invariant procedures", *J. Chem. Phys.*, **43**, 5129–5135.

Pople, J.A. and Segal, G.A. (1965) "Approximate self-consistent molecular orbital theory. II. Calculation with complete neglect of differential overlap", *J. Chem. Phys.*, **43**, 5136–5151.

Riggs, D.S., Guarnieri, J.A. and Addelman, S. (1978) "Fitting straight lines when both variables are subject to error", *Life Sciences*, **22**, 1305–1360.

Rüdiger, W., Klose, W., Vuillaume, M. and Barbier, M. (1968) "On the structure of pterobilin, the blue pigment of *Pieris brassicae*", *Experientia*, **24**, 1000.

Vuillaume, M. and Dattée, Y. (1980) "Photostimulation de la production d'ATP dans le tégument d'un insecte: interprétation physiologique des analyses *in vivo* et *in vitro*", *Archives de zoologie expérimentale et générale*, **121**, 159–172.

MOLECULAR ENGINEERING IN THE PREPARATION OF BIOACTIVE PEPTIDES

Ettore Benedetti

Biocrystallography Centre, C.N.R. and C.I.R.PE.B.
Department of Chemistry
University of Napoli "Federico II", Napoli, Italy

INTRODUCTION

The main goal of structure-activity relationship studies in bioactive peptides, relevant to medicinal chemistry, is the understanding of the biological phenomena at molecular level in order to produce and possibly develop new materials which might mimic biological processes by enhancing or somehow modulating their effects. The peptide pharmaceutical targets are usually hormones, enzymes, transport systems, neurotransmitters, ion channel ionophores antibiotics, antigens. The advantages in the use of peptides as pharmaceuticals are based on the fact that: (i) they should be considered "natural" products; (ii) opportunely modified analogs could possibly show increased potency and enhanced specificity; (iii) they are easy to synthesize and (iv) they could present extraordinary ranges of biological properties coupled with minimal non-mechanism-based toxicity. In this area, constrained non-coded α-amino acid residues are of great interest as "building blocks" for the preparation of analogs, which should not only retain the pharmacological properties of the native peptide, but they should also exhibit enhanced resistance to biodegradation with improved bioavailability and pharmacokinetics.

Structural changes can be induced in a peptide by selectively substituting along the sequence specific residues with other residues or by substituting certain residues of the sequence with non-coded α-amino acid residues. The aim of these efforts is the preparation of analogues with different chemical structure and possibly modified conformational preferences, responsible of differences in the biological activity.

Conformational restrictions can be achieved by various methods. Analogs are usually peptides modified in the side chain, whereas the amide nature of the peptide bond is not changed. Other analogs, called pseudopeptides or peptidomimetics, present modifications in the peptide backbone chain.

Several solid state studies have been carried[1-6] out to define the conformational preferences in solution and in the solid-state of a specific class of non-coded α-amino acids: the symmetrical and unsymmetrical α,α-disubstituted glycines (α,α-dialkylated amino acid residues) . The incorporation of such constrained residues in the sequence introduces specific steric modifications

Statistical Mechanics, Protein Structure, and Protein Substrate Interactions
Edited by S. Doniach, Plenum Press, New York, 1994

381

giving rise to particular conformational effects. A growing "arsenal" of non-coded α-amino acids of this class has been built by mean of which specific structural changes can now be induced in bioactive molecules.

With a sort of "molecular engineering" procedure one can synthesize by selective insertion in a peptide sequence of the appropriate "building blocks" with definite conformational preferences, molecules, which hopefully can modulate or appropriately mimic specific biological effects. In addition, with protein structures increasingly becoming available, a stage is set to systematically exploit the knowledge of the three-dimensional structures of protein in a rational approach to design specific inhibitors, hormones, substrates and other effector molecules to achieve accurate interactions and induce or modulate biological functions. Research studies are currently being proposed in peptide chemistry, concerning with the synthesis and conformational characterization of linear, rigid, helical chains (3_{10}- or α-helices) or folded and cyclic molecules, essential architecture units of natural products, useful as molecular spacers (or templates) to build up molecules with predetermined biological properties (neurotransmitters, esterasis, ion carriers) or spectroscopic properties (ecciton properties, electronic transfers, charge transfers) due to the spatial proximity of particular side chains appropriately located.

In this review article we summarize the solid state structural results and the conformational preferences of some of these "building blocks" with particular enphasis on: i) the linear and cyclic symmetrically α,α-disubstituted glycines in which either two identical n-alkyl groups replace the hydrogen atoms of the glycine residue or a cyclic aliphatic side-chain system is formed including the α-carbon, respectively; ii) the unsymmetrically α,α-disubstituted glycines, derived by α-methylation of naturally occurring residues. Other modifications introduced by the use of unusual α-amino acids or peptidomimetics will be briefly discussed. Examples of the use of some of these non coded residues or peptidomimetics in the study of the bioactive conformation of cyclolinopeptide A and in the determination of a three-dimensional model for sweet taste are reported.

SYMMETRICAL C$^{\alpha,\alpha}$-DIALKYLATED GLYCINES WITH LINEAR SIDE-CHAINS

The prototype of this class of non-coded residues is certainly α-aminoisobutyric acid (Aib), (also called α-methyl-alanine or α,α-dimethylglycine), which is commonly found in natural antibiotics produced by microbial sources, such as chlamydocin, alamethicin, tricorzianine, antiamoebin, emerimicin, zervamicin.[7-10] Other C$^{\alpha,\alpha}$-dialkylated glycines with linear side chains, which have been the object of conformational studies [4,11-14] are those given in Figure 1.

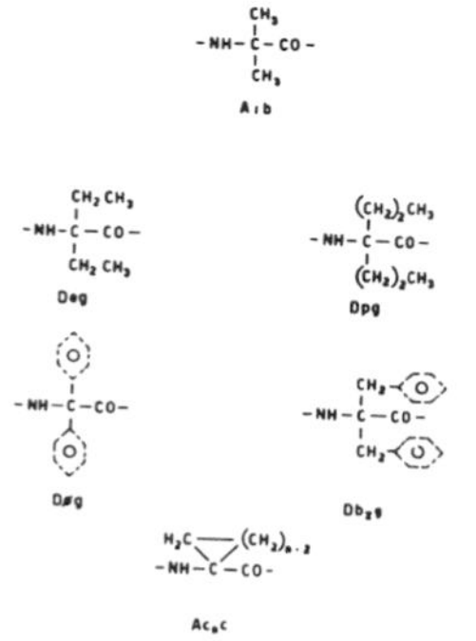

Figure 1. Symmetrical C$^{\alpha,\alpha}$-disubstituted glycine residues.

The replacement of the hydrogen atoms of the C^α carbon of glycine with methyl groups, as in the Aib residue, produces severe restrictions on the conformational freedom with values for the conformational angles φ and ψ occurring only in very restricted regions located in the neighborhood (within 20° for both φ and ψ) of (+55°, +45°) and (-55°, -45°), corresponding to left- and right-handed helical screw sense, respectively. At least two well defined helical types have been characterized: the 3_{10}-helix (φ = 60°, ψ = 30° or φ = -60°, ψ = -30° for right- or left-handed screw sense, respectively) and the α-helix (φ = 55°, ψ = 45° or φ = -55°, ψ = -45° for right- and left-handed screw sense, respectively). Because of the absence of chirality the Aib residue can be accommodated in helices with both handedness. The 3_{10^-} and the α-helical types, given in Figure 2, differ mainly for the stabilizing intramolecular

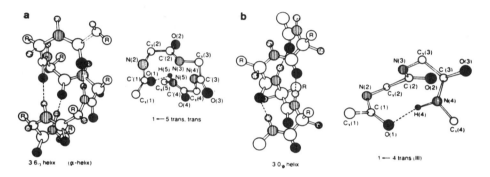

Figure 2. (a) The 3.6_{13}-helix (α-helix) and its building block, one of the 1----5 *trans,trans* intramolecularly hydrogen bonded peptide conformations (also termed α-bend or C_{13}-conformation). (b) The 3.0_{10}-helix and its building block, one of the 1----4 *trans* intramolecularly hydrogen bonded peptide conformation (also termed type III β-bend or C_{10}-conformation.

hydrogen bonding scheme, which involves the donor N-H's and the acceptor C=O's groups in successive type III β-turns (1---- 4, C_{10} intramolecularly hydrogen bonded structures) in the 3_{10}-helix or successive α-turns (1---- 5, C_{13} intramolecularly hydrogen bonded structures) in the α-helix, respectively. The solid state molecular structure of the $(Aib)_n$ series of homopeptides with n=1-12 has been studied. To date almost 100 structures containing this residue have been reported in the literature, most of them from our laboratory[15-22]. With only three exceptions[23,24] all homopeptides of Aib, assumes a 3_{10}-helical structure stabilized by 1---- 4 intramolecular N-H····O=C H-bonds characteristic of the type III (or the enantiomeric type III') β-bends (or C_{10} conformation).[25] The incipient (at the trimer level) 3_{10}-helix is fully developed at the pentamer level and is mantained up to very long Aib sequences and, under certain conditions, is the energetically more stable conformation for the homopolymer.[26] All these results point to the conclusion that the Aib residue itself or stretches of a polypeptide chain with very high content of Aib strongly and preferentially assume the 3_{10^-} helical conformation in the solid state[27] and that main chain length in the homopeptide series is not an overriding factor in directing the type of helical folding.[22] However, for a long peptide chain formed by C^α-monosubstituted α-

aminoacids with the same chirality, the 3_{10}-helix is energetically less favourable than the α-helix.[28,29] Therefore, although the occurrence of this type of helix has been recently well documented in protein crystal structure analyses,[30] it is not surprising that only relatively short peptide segments, particularly near the N- and C-termini of a α-helix, have been authenticated in the solid state. The factor governing the transition between 3_{10}- and α-helices appear to be quite subtle.

The structural preference of the Aib residues for helical conformations has also been systematically investigated in relation to the Aib content of the peptide and to the position of the Aib residues along the polypeptide chain. To delineate the effect induced by peptide main-chain lengthening on the 3_{10}---α-helix conformational interconversion, we have analyzed[31,32] the terminally blocked sequential oligopeptides pBrBz-(Aib-L-Ala)$_n$-OMe (n=1-6) (where pBrBz- is p-bromobenzoyl and -OMe is methoxy). These peptide all contain the same fraction of the highly helicogenic Aib and Ala residues. By diffraction analyses we have shown that the hexapeptide structure (n=3) is fully characterized as a 3_{10}-helix, the octapeptide (n=4) has a six residues long α-helical region, spanning the central and C-terminal parts while the deca- and dodeca-peptides (n=5 and n=6, respectively) confirm the preferential folding of the peptide chain in an α-helical structure. In all peptides the right-handed screw sense of the helix is dictated by the presence of the L-Ala residues. This study represents the first experimental unambiguous proof for a 3_{10}----α-helix conversion induced by peptide main-chain lengthening.

These studies have great relevance for the understanding of the forces responsible of the 3_{10}----α-helical transition with relation to the mechanism of biological action shown by the Aib-rich, membrane-active, channel-forming, ion-transporting peptaibol antibiotics. In the solid state alamethicin[33] and two long segments and model compounds[34,35] form mixed $3_{10}/\alpha$-helices with the concomitant occurrence of 1---4 and 1---5 intramolecular H-bonds. Two analogs of zervamicin[36,37] also adopt a similar mixed helical structure, while the Aib-rich, synthetic 2-9 octapeptide segment of emerimicin[38] revealed the presence of a slightly distorted 3_{10}-helix and the corresponding 1-9 nonapeptide[39] showed instead the onset of more than two and one-half turns of a α-helix. Tricorzianine A III C, a linear antifungic peptide 19 residues long, is essentially α-helical with a slight bend of the chain at the proline level.[40]

We have recently carried out a general survey of crystal structure analyses of Aib-rich peptides, in which helical molecules have been found.[41] Out of 54 independent molecules, 32 of them were found to be 3_{10}-helices and 22 α-helices. The minimal main-chain length required for a peptide to form an α-helix corresponds to seven residues. By contrast there is no critical main-chain length dependence for 3_{10}-helix formation, i.e. incipient 3_{10}-helices are formed at the lowest possible level (the N-blocked tripeptide). In peptides of eight or more residues the α-helix is preferred over the 3_{10}-helix if the percentage of Aib residues does not exceed 50%. However one or more 3_{10}-helical residues may be observed at either end of the α-helical stretch: these short bits of 3_{10}-helix tighten up the ends of the α-helix by moving the related peptide groups nearer the helical axis. The average number of α-helical residues in undeca and longer peptides is seven, corresponding to about two turns of the helix.

Theoretical studies by conformational energy computations and extensive experimental investigations, by diffraction analyses in the solid state or spectroscopic techniques in the solution state, have been carried out on compounds containing symmetrical α,α-disubstituted glycine residues other than

Aib, such as Deg (di-ethyl glycine), Dpg (di-n-propyl glycine), Dϕg (di-phenyl glycine), Db$_z$g (di-benzyl glycine), and Ac$_n$c (1-aminocycloalkane-1-carboxylic acid).[42-49] The allowed more stable conformations for all symmetrical $\alpha,\alpha-$ disubstituted glycine residues (including Aib) are found either in the fully extended region or in the helical region, corresponding to 3$_{10}$- or α-helical structures. The repeating motif of the fully extended conformation, characterized by φ and ψ angles close to 180°, is the 2---- 2 intraresidue, intramolecularly H-bonded structure, in which the N-H and the C=O groups of the same residue are interacting with each other, giving rise to a pentacyclic structure called C$_5$. The extended and helical conformations were computed for Aib, Deg, and Dpg residues as a function of the N-C$^\alpha$-C' bond angle τ:[50] while for Aib the helical structures remain the most stable, by contrast, for Deg and Dpg the stability order of the helical and extended conformations depends markedly on the angle τ, with extended conformation being favoured when $\tau < 107°$.

Among the vast literature of Aib containing compounds only one example has been found in which this residue adopts the C$_5$ fully extended conformation,[51] otherwise the helical structures are overwhelmingly represented (Ac$_n$c residues behave similarly). Instead, as shown in Figure 3, for the Cα,α-dialkylated Deg and Dpg residues the intramolecularly hydrogen bonded C$_5$ structure is the more commonly observed conformation:[52] it has been found in 19 out of 21 residues in

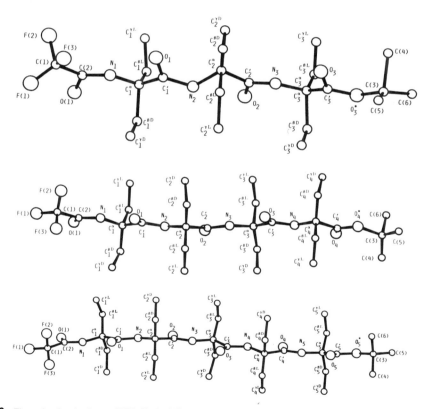

Figure 3. The molecular structures of TFA-(Deg)$_3$-OtBu (upper), TFA-(Deg)$_4$-OtBu (center), and TFA-(Deg)$_5$-OtBu (bottom) examples of fully extended peptide conformations stabilized by 2----2 intramolecular hydrogen bonds (also termed C$_5$ conformation).

the solid state. In homopeptides of Deg and Dpg consecutive C_5 forms stabilize the fully extended conformation as shown in Figure 3. In this structures the N-H and C=O groups, characterizing the consecutive C_5 forms, are not involved in any intermolecular H-bonds, probably because steric hindrance prevents adjacent chains to approach sufficiently. It is worth noting that in these experimentally observed extended structures the critical τ bond angle is narrowed down to 103°. The ethyl and propyl side chains are also fully extended in order to relieve unfavourable intramolecular side-chain to main-chain and side-chain to side-chain non-bonded interactions.

Recently the structural analyses[48,49] of Dϕg and Db$_z$g containing compounds have shown that these aromatic residues also tend to parallel the conformational behaviour seen in the aliphatic analogues Deg and Dpg. In fact, in the solid state most of the derivatives and peptides of Dϕg and Db$_z$g are found to adopt the typical C_5 structure. These studies have clearly demonstrated that, when the two side chain C^α atoms are symmetrically substituted, but not interconnected in a cyclic system (as in Ac$_n$c residues), the fully extended conformation is by far preferred over the helical structures.

SYMMETRICAL $C^{\alpha,\alpha}$-DISUBSTITUTED GLYCINE WITH CYCLIC SIDE CHAINS

Peptides containing Ac$_n$c residues with n=3,5,and 6 have also been recently investigated both theoretically and experimentally in the solid as well as in the solution state.[52-62] This family of compounds has shown again the great conformational versatility of α,α-disubstituted glycines. In fact, the three-membered ring Ac$_3$c residue shows a marked structural preference for the "bridge" region of the conformational space ($\varphi = 90°$, $\psi = 0°$), that is for position i+2 of type-I (or I') and type-II (or II') β-bends, but it can be accomodated in distorted type-III (or III') β-bends and then in 3$_{10}$-helices. In the Ac$_3$c φ,ψ map the most striking feature is the absence of fully extended structure, due to steric repulsions between the carbonyl moiety and the β-methylene hydrogens.

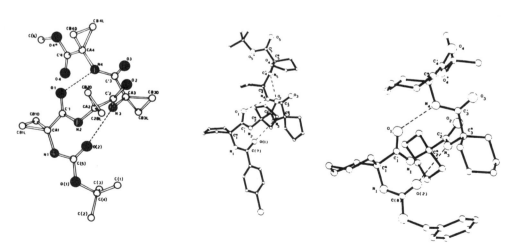

Figure 4. The structures found for t-Boc-(Ac$_3$c)4-OMe (left), pBrBz(Ac$_5$c)5OtBu (center) and Z-(Ac$_5$c)4OtBu, triclinic modification (right). All structures show helical conformations stabilized by 1----4 intramolecular hydrogen bonds (C$_{10}$-conformation).

The less strained penta- and hexa-cyclic residues Ac_5c and Ac_6c show theoretically and experimentally[53-55,61,62] a conformational behaviour very similar to that of the prototype Aib residue. Conformations of the helical type due to the strong tendency to induce type-II and type-III (or II' and III') β-bends are an usual observation in derivatives and homopeptides of these residues. In Figure 4 examples of the conformations observed in the solid state for an Ac_3c, an Ac_5c and an Ac_6c residue are represented.

As for the side chain conformation, the cyclopentyl rings adopt either the envelope or half-chair conformation, while the cyclohexyl rings show almost perfect chair conformations with predominantly axial arrangement of the amino substituent.

Solution studies carried out by IR absorption, CD and NMR spectroscopies on peptides containing, along their sequence, α,α-disubstituted glycines strongly support the idea that in solvents of low polarity Aib, and Ac_nc residues assume and possibly induce folded, helical conformations, being easily accomodated in C_{10} or C_{13} intramolecularly hydrogen bonded cyclic structures. By contrast, for Deg, Dpg, Dϕg and Db$_z$g residues the largely prevailing conformation is fully extended with the formation of intraresidue, intramolecularly hydrogen bonded C_5 structures.

UNSYMMETRICAL $C^{\alpha,\alpha}$-DISUBSTITUTED GLYCINES: α-AMINO ACIDS METHYLATED AT THE α-CARBON

The structural preferences of few chiral α-methylated residues as Iva (isovaline or α-methyl,α-ethyl-glycine), (αMe)Val, (αMe)Leu and (αMe)Phe, given in Figure 5, have been determined by conformational energy computations, x-

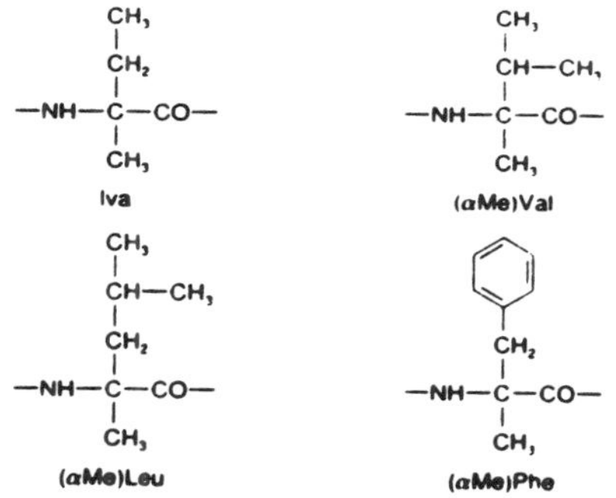

Figure 5. The a-methylated amino acids for which the conformational properties have been investigated in the solid state.

ray diffraction analyses, and NMR spectroscopy.[63] The results indicate that helical structures are preferentially adopted by peptides rich in these a-amino acids. In general, tripeptides and longer peptides containing these residues are folded in type I (I') and III (III') β-bends conformations or in a 3_{10}-helix, depending upon main chain length. However, conditions known to be required for a peptide to fold in a α-helical conformation (peptide chain longer than seven residues) have not been examined so far.

Fully extended structures are rarely seen in Iva and (αMe)Phe containing peptides, while more commonly observed in (αMe)Leu compounds. These structures have never been observed in (αMe)Val compounds. Also, β-pleated sheet structures have not been found in any of the peptides examined so far. These characteristics closely match those described for Aib rich peptides. Representative examples of solid state structure for a-methylated residues are given in Figure 6.

As for the relationships between a-carbon chirality and screw sense of the b-bend or 3_{10}-helical structure that is formed, Iva residues, with linear side chains may be classified as a screw sense indifferent residue. (αMe)Val, having a β-

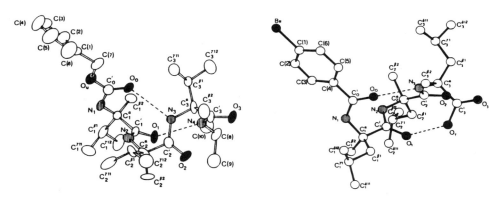

Figure 6. Molecular models for Z-[D-(αMe)Val]$_3$-NHiPr (left, iPr=isopropyl) and pBrBz-[D-(αMe)Leu]$_3$-OH (rigth, pBrBz= para-bromobenzoyl).

branched side chain exhibits the normal behaviour shown by the unmethylated residue, while the behaviour of (αMe)Leu and (αMe)Phe, having γ-branched side chains, tend to be opposite to that of protein amino acids. Conformational versatility seems to be a typical properties of peptides rich in the γ-branched.

Symmetrical and unsymmetrical α,α-dialkylated glycine residues have been used to synthesize analogues of biologically active peptides. Some of these compounds reported in the following have shown once again the great potential of these residue in elucidating and possibly defining at molecular level the bioactive conformation of naturally occurring peptides of medicinal and pharmaceutical interest. The understanding of the molecular mechanism of biological action is the necessary step for the rational design of "better" materials with highly specific activity.

EXAMPLES OF MOLECULAR ENGINEERING

1. The search for the bioactive conformation: cyclolinopeptide A

Cyclolinopeptide A (CLA), a homodetic cyclic nonapeptide from linseed, of structure c(Pro-Pro-Phe-Phe-Leu-Ile-Ile-Leu-Val) defends liver cells[63] against several poisonous substances, such as phalloidin, ethanol, cysteamine and DMSO. The strong cytoprotective activity, shared with antamanide and somatostatin, relies on the high affinity of these peptides for the same liver membrane proteins responsible for the uptake of toxic compounds and bile salts. There are evidences that the CLA ability to inhibit the uptake of cholate by hepatocytes is linked to well defined sequential features included in the Pro1-Pro2-Phe3-Phe4 moiety.[64] Recently we have carried out the conformational analysis of CLA,[65,66] both in the solid state and in solution. The molecular structure of CLA is characterized by a Pro-Pro cis peptide bond and five intramolecular N-H···O=C hydrogen bonds across the annular skeleton of the peptide. These H-bonds stabilize the structure with the formation of one C7, two type-II C10 and one C13 cyclic structures. In solution of different solvents CLA shows the presence of a large number of conformers; however we were able to detect a single conformational state in CDCl$_3$ solution by lowering the temperature to 214 K. The solid state and solution structures are almost identical, with the Leu-Ile-Ile-Leu sequence being part of an incipient 3$_{10}$-helix and the Leu residue at position 8 showing conformational angles characteristic of a left-handed α-helix, typically allowed to D residues. Other crystalline modifications[67] of CLA have shown that, regardless of different cocrystallized solvent molecules, the same overall structure, including the side chains, is mantained. On this basis we have designed and prepared among others, a CLA analog (CLAIB), which presents modifications at three residues of the sequence in the portion of the molecules not supposedly involved directly in the interaction with the receptor. The substitutions of Leu^5 and Ile^6 by two Aib residues and of Leu^8 by a D-Ala residue at position 8 were introduced in order to rigidify the molecular conformation observed both in the solid state and in the conformer detected in the NMR experiments. The changes should also mantain the 3$_{10}$-helical structure in the 5-8 portion of the molecule. As shown in Figure 7,[68] the overall solid state structure of CLAIB is very similar to that found for the three CLA crystalline modifications. Furthermore, the conformational analysis of CLAIB in CDCl$_3$ solution at room temperature indicates that solution and solid state structures are essentially the same and that the insertions of constrained residues such as Aib and D-Ala effectively quenches the very high flexibility of CLA.

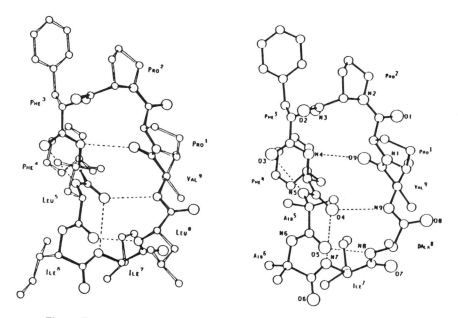

Figure 7. The molecular models of CLA and CLAIB as found in the solid state by x-ray diffraction.

The extent of the rigidity of CLAIB can be inferred by the similarity of ROESY spectra in CDCl$_3$ at three different temperatures 270, 295 and 315 K. Moreover, cations as Ba^{2+} which induce dramatic changes in the CD and NMR spectra[69] and hence in the structure of CLA, are practically ineffective in inducing significant modifications of the CLAIB conformation . The biological activity of CLA and CLAIB, measured as peptide concentration required to inhibit the cholate uptake by 50% (CD$_{50}$), increases from 0.84 µM for CLA to 30 µM for CLAIB. The conformational rigidity, imposed to this analog by the presence of two consecutive Aib residues and a D-alanine residue, which freeze the solution conformation, making it identical to that observed in the solid state, is clearly one of the reason for the lowering of the biological activity, demonstrating that i) the structure observed in the solid state for the three crystalline modification of CLA perhaps does not correspond to the bioactive one and ii) the flexibility, demonstrated by the NMR experiments in this peptide system plays an essential role in the explication of the bioactive function, which perhaps involves an induced fit mechanism for the peptide-receptor interaction. In the distance geometry calculations and molecular dynamics simulations performed on CLA at 214 and 300 K it has been possible[70] to observe the occurrence of hydrogen bonding patterns different from those found in the crystalline state. The dynamic behaviour in the H-bond formation could be the cause of the occurrence of exchanging multiple conformations of CLA in solution at room temperature (a pronounced line broadening was observed in the NMR spectrum).

2. A three dimensional model for sweet taste: Dipeptide sweeteners of the aspartame family

The transduction of taste is believed to be initiated by receptor proteins located on the surface of the taste cell. In terms of structural biology, the taste ligand and the receptor protein form a guest-host complex which produces a sense of taste. Since the taste receptors have their own structures, which have not been identified yet, the taste ligand must assume a specific three-dimensional structure.

A wide variety of unrelated compounds are known to elicit a common sweet taste. There have been numerour attempts to generalize structural features among sweet molecules. The results of these efforts have led to the general agreement that the sweet molecules contain a hydrogen bond donor (AH) and a hydrogen bond acceptor (B)[71] and that highly potent sweet molecules have a hydrophobic site (X)[72] along with the AH/B functions. In the case of aspartame, L-Asp-L-Phe-OMe, which is a typical example of peptide sweetener, the α-amino and β-carboxyl groups of the N-terminal aspartyl moiety are assigned as the AH and B elements, respectively. The hydrophobic X group is represented by the benzyl side chain of the Phe residue.

The C-terminal residue of sweet dipeptides consists of two hydrophobic groups with different sizes, the amino acid side chain and the ester or amide group. In the L-configuration of the C-terminal residue, the amino side chain is required to be larger than the ester or amide group. This relationship is reversed in the D-configuration, where the side chain is small and the ester or amide group is large.

We have developed a three-dimensional model for sweet and bitter taste by investigating conformations of a series of peptide-based taste ligands by x-ray crystallography, NMR spectroscopy and computer simulations.[73] The compounds involved in our study are dipeptides taste ligands of the aspartame family which incorporate stereoisomeric residues, peptidomimetics and constrained residues. Each of these compounds display unique conformational preferences. The chemical structures of the dipeptides used in the investigation is given in Figure 9, while in Table 1 the topochemical arrays available to each dipeptide are listed.

The results[73] of the x-ray crystal structure analyses of the dipeptides of Table 1 are to be considered rather ambiguous in determining the relationship between taste and structure of the ligand. In fact, while throughout the series of dipeptides investigated the structure and conformation of the N-terminal Asp moiety, which provides the AH/B functions, remains unchanged, the orientations of the hydrophobic X groups with respect to the AH/B elements are substantially different from one to another even in the sweet analogs. The conformation (or conformations) assumed by each molecule in the crystal is (or are) mainly the

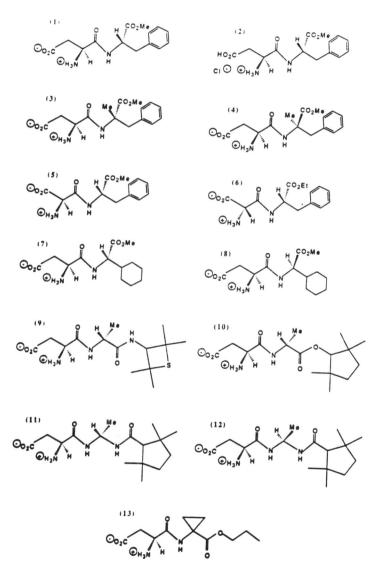

Figure 8. Chemical structures of the dipeptide taste ligands investigated.

Table 1. Dipeptide taste ligands with the topochemical arrays of the glucophores elements AH, B, and X available to them as determined by NMR spectroscopy and computer simulations.

	Analog	Taste	Accessible Topochemical Arrays				
			I	II	III	IV	V
1	L-Asp-L-Phe-OMe	Sweet	+	+	+	+	+
2	HCl.L-Asp-L-Phe-OMe*	Sweet	+	+	+	+	+
3	L-Asp-L(αMe)Phe-OMe	Sweet	+	+	+	+	-
4	L-Asp-D-(αMe)Phe-OMe	Bitter	-	+	+	+	+
5	(R)-Ama-L-Phe-OMe	Sweet	+	+	+	+	-
6	(S)-Ama--L-Phe-OEt	Tasteless	-	+	+	+	-
7	L-Asp-(S)-Chg-OMe	Sweet	+	-	-	+	-
8	L-Asp-(R)-Chg-OMe	Bitter	-	+	-	-	+
9	L-Asp-D-Ala-TAN	Sweet	+	+	+	+	-
10	L-Asp-D-Ala-OTMCP	Sweet	+	+	+	+	-
11	L-Asp-(R)-gAla-TMCP**	Sweet	+	+	+	+	-
12	L-Asp-(S)-gAla-TMCP**	Sweet	+	+	+	+	-
13	L-Asp-Ac₃c-OnPr	Sweet	+	+	+	+	-

*The hydrochloride of aspartame in solution obviously behave identically with aspartame itself while on the contrary in the solid state it shows a conformation for the N-terminal L-Asp moiety different from all other analogs, since its zwitteionic nature has been changed.

** Retro-inverso analog: gAla = NH_2-CH(CH_3)-NH_2 ; TMCP = tetramethylcyclopentanyl

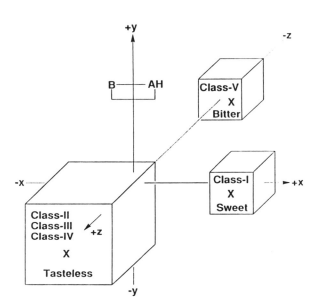

Figure 9. Schematic representation of the relationship between topochemical arrays of the AH (hydrogen bond donor), B (hydrogen bond acceptor), and X (hydrophobic element) glucophores and tastes. The α-carbon of the second residue defines the origin of the cartesian coordinate system.

result of packing forces rather than being inherent to the molecule, especially for charged compounds such as the molecules investigated, which are all present as zwitterions in the solid state. These results clearly indicate that the solid conformations by themselves cannot explain the taste properties of these molecules.

The NMR and molecular modeling studies provided the accessible conformations of the molecules.[73] For each analog of Table 1 the results of these studies show that one of the preferred conformations is very similar or identical to the x-ray structure. In general, however, this structure is not the lowest energy conformation. The comparison of accessible conformations possessed by sweet, bitter and tasteless analogs enables us to identify five classes of topochemical arrays for the three elements AH, B, and X., as shown in Figure 10. For a sweet tasting ligand the "bioactive conformation" is described with an "L" shape structure, where the AH and B containing zwitterionic ring of the N-terminal residue forms the stem of the "L" in the y axis and the hydrophobic X group projects out along the base of the "L" in the x axis. The plane of the zwitterionic ring is almost identical to the plane of the "L". A bitter molecule assumes characteristic conformation in which the hydrophobic X group project into the -z dimension, i.e. behind the stem of the "L". All other topochemical arrays lead to tasteless molecules. The "L" shaped structures of aspartame and alitame are given in Figure 11.

Among the analogs, those containing cyclohexylglycine (Chg) are particularly informative,[74] Figure 12. The sweet L-Asp-(S)-Chg-OMe can assume only two topologically different structures, the "L" shape structure and the other in which

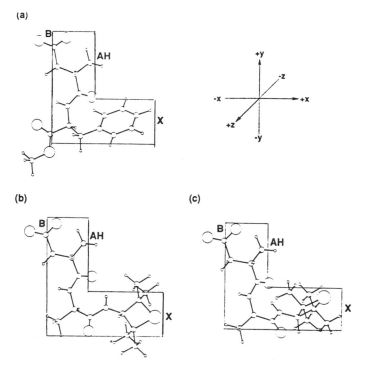

Figure 10. The "L" shaped structures of aspartame (a), and alitame (b) and (c), which produce a sweet taste. In the former the Phe side chain functions as the hydrophobic X glucophore, while the C-terminal TAN group (2,2,4,4-tetramethylthietanylamide) serves as the X glucophore in the latter.

the hydrophobic cyclohexyl side chain (X) projects into the +z dimension. No extended structures are accessible to this analog because a conformationally rigid cyclohexyl ring is directly attached to the α-carbon of the second residue. The sweet analog L-Asp-Ac$_3$c-OnPr is also interesting.[75] According to the constrained nature of the Ac$_3$c residue, this analog adopts the "L" shape, the reversed "L" shape, the extended structures, but the structures with hydrophobic n-propyl ester (X) projecting into the +z dimension, as shown in Figure 13.

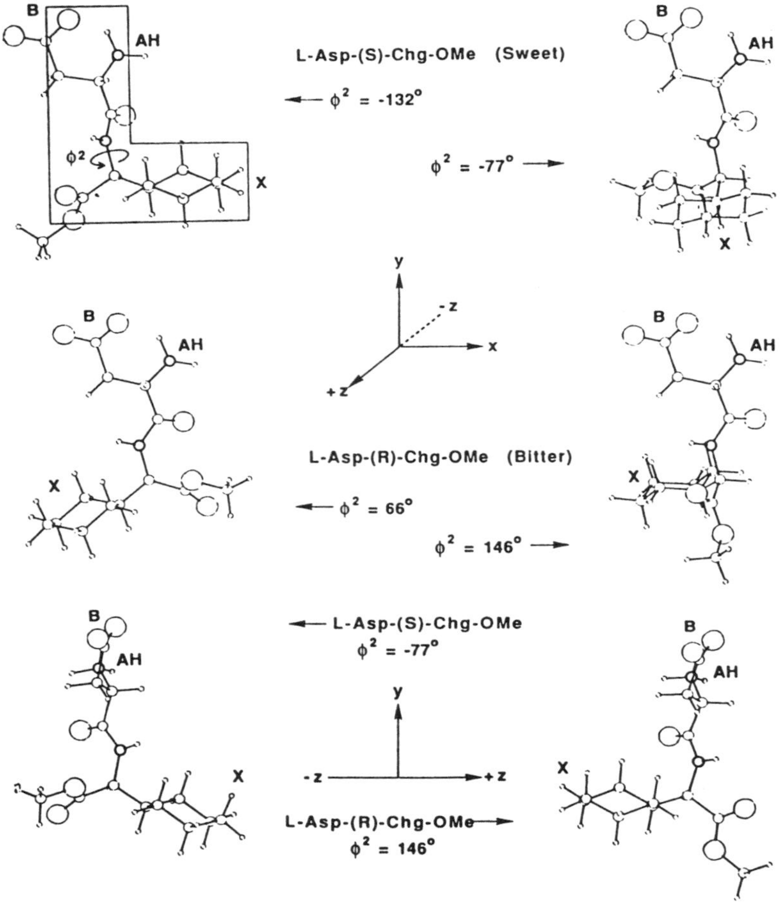

L-Asp-(S)-Chg-OMe (Sweet)

$\leftarrow \phi^2 = -132°$

$\phi^2 = -77°\ \rightarrow$

L-Asp-(R)-Chg-OMe (Bitter)

$\leftarrow \phi^2 = 66°$

$\phi^2 = 146°\ \rightarrow$

$\leftarrow$ L-Asp-(S)-Chg-OMe
$\phi^2 = -77°$

L-Asp-(R)-Chg-OMe $\rightarrow$
$\phi^2 = 146°$

Figure 11. Preferred topochemical arrays of glucophores (AH, B, and X) for L-Asp-(S)-Chg-OMe and L-Asp-(R)-Chg-OMe by NMR and computer simulations.

Therefore, only the "L" shape structure is common to both sweet molecules. Similarly all sweet analogs can assume the "L" shape structure among others. On the contrary, no "L" shape structures are accessible to the tasteless (S)-Ama-L-Phe-OEt[73] and the bitter L-Asp-(R)-Chg-OMe. The tastes of the dipeptide derivatives of Table 1 as well as that of many other similar compound containing α,α-dialkylated residues are correctly explained by the "L" shape model for sweet taste ligands and by the reversed "L" shape model for bitter taste ligands. Molecules to which neither of the two type of topochemical arrays are accessible are tasteless.

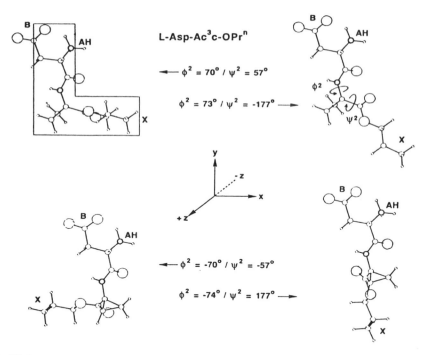

Figure 12. Preferred topochemical array of glucophores (AH, B, and X) for L-Asp-Ac3c-OnPr by NMR spectroscopy and computer simulations.

ACKNOWLEDGEMENTS

The authors gratefully acknowledge the financial support of the Ministry of Education of Italy (40% and 60%) and the Progetto Finalizzato "Chimica Fine II" of the C.N.R. (grant PF 91.1657)

REFERENCES

1. C.Toniolo, Structure of conformationally constrained peptides from model to bioactive peptides, *Biopolymers* 28:247(1989).
2. C.Toniolo, and E.Benedetti, Structures of polypeptides from α-amino acids disubstituted at the α-carbon, *Macromolecules* 24:4004 (1991).
3. C.Toniolo and E.Benedetti,Old and new structures from studies of synthetic peptides rich in Cα,α–disubstituted glycines, *ISIAtlasSci.:Biochem.* 1:225(1988).
4. C.Toniolo, G.M.Bonora, A.Bavoso, E. Benedetti, B.DiBlasio, V.Pavone, and C.Pedone, Preferred conformations of peptides containing α,α-disubstituted α-aminoacids,*Biopolymers* 22:205(1983).
5. B.V.V.Prasad and P.Balaram, The stereochemistry of peptides containing α–aminoisobutyric acid , *C.R.C.Crit.Rev.Biochem.* 16:307(1984).
6. B. Di Blasio, V. Pavone, A.Lombardi, C.Pedone and E.Benedetti, Noncoded residues as building blocks in the design of specific secondary structures:symmetrically disubstituted glycines and β-alanine, *Biopolymers* 33:1037(1993); and references therein.

7. R.C.Pandey, J.C.Cook,Jr., and K.L.Rinehart, Structures of the peptides antibiotics Emerimicins III and IV, *J.Am.Chem.Soc.* 99:5205(1977).

8. B.Bachet, C.Brassy, I.Morize, E.Surcouf, J.P.Mornon, B.Bodo and S.Rebuffat, Crystallization and preliminary x-ray diffraction results of trichorzianine A 1, a peptide with nineteen residues from trichoderma harzianum, *J.Mol.Biol.* 170:795 (1983).

9. E.K.S.Vijayakumar and P.Balaram, Solution conformations of penta and heptapeptides containing repetitive α-aminoisobutyryl-L-alanyl and α-aminoisobutyryl-L-valyl sequences *Tetrahedron* 39:2725(1983).

10. E.K.S.Vijayakumar, and P.Balaram, Stereochemistry of a-aminoisobutyric acid peptidesin solution: helical conformations of protected decapeptides with repeating Aib-L-Ala and Aib-L-Val sequences, *Biopolymers* 22:2133(1983).

11. R.Nagaraj and P.Balaram, Alamethicin, a transmembrane channel, *Acc.Chem. Res.*14:356(1981).

12. E.Benedetti, C.Pedone, and C.Toniolo, First crystal structure analysis of a complete homo-oligopeptide series, in: "Peptides 1980", K.Brunfeldt, ed., Scriptor, Copenhagen (1981).

13. R.Bosch, G.Jung, H.Schmitt, and W.Winter, Crystal structure of the α-helical undecapeptide Boc-Ala-Aib-Ala-Aib-Ala-Glu(OBzl)-Ala-Aib-Ala-Aib-Ala-OMe, *Biopolymers* 24:961(1985).

14. A.W.Burgess and S.J.Leach, An obligatory α-helical amino acid residue, *Biopolymers* 12:2599(1973).

15. N.Shamala, R.Nagaraj, and P.Balaram, The 3_{10}helical conformation of a pentapeptide containing α-aminoisobutyric acid (Aib): x-ray crystal structure of Tos-(Aib)$_5$-OMe, *J.Chem.Soc.Chem.Commun.* 996(1978).

16. C.Pulla Rao, N.Shamala, R.Nagaraj, C.N.R.Rao, and P.Balaram, Hydrophobic channel in crystals of an α-aminoisobutyric acid pentapeptide, *Bioch.Biophys.Res. Commun.* 103:898(1981).

17. E.Benedetti, A.Bavoso, B.Di Blasio, V.Pavone, C.Pedone, M.Crisma, G.M.Bonora, and C.Toniolo, Solid-state and solution conformation of homo oligo(α-aminoisobutyric acids) from tripeptide to pentapeptide: evidence for a 3_{10} helix, *J.Am.ChemSoc.* 104:2437(1982).

18. A.Bavoso, E.Benedetti, B.Di Blasio, V.Pavone, C.Pedone, C.Toniolo, and G.M.Bonora, Long polypeptides 3_{10}-helices at atomic resolution, *Proc.Natl.Acad. Sci. USA* 83, 1988(1986).

19. C.Toniolo, G.M.Bonora, A.Bavoso, E.Benedetti, B.Di Blasio, V.Pavone, and C.Pedone, A long regular polypeptide 3_{10} helix, *Macromolecules* 19:472(1986).

20. B.Di blasio, A.Santini, V.Pavone, C.Pedone, E.Benedetti, V.Moretto, M.Crisma, and C.Toniolo, Crystal-state conformation of homo-oligomers of α-aminoisobutyric acid: molecular and crystal structure of pBrBz-(Aib)$_6$-OMe, *Struct.Chem.* 2:523 (1990).

21. V.Pavone, B.Di Blasio, C.Pedone, A.Santini, E.Benedetti, F.Formaggio, M.Crisma, and C.Toniolo, Preferred conformation of homo-oligomers of α-aminoisobutyric acid: molecular and crystal structure of Z-(Aib)$_7$-OMe, *Gazz.Chim. Ital.* 121:21(1990).

22. C.Toniolo, M.Crisma, G.M.Bonora, E.Benedetti, B.Di Blasio, V.Pavone, C.Pedone, and A.Santini, Preferred conformation of the terminally blocked (Aib) homo-oligopeptide: a long 3_{10}-helix, *Biopolymers* 31:129(1991).

23. G.Valle, C.Toniolo, and G.Jung, New peptide conformation in crystals of H-Aib-Aib-OTBu and TFA-Aib-Aib-OTBu *Gazz.Chim.Ital.* 117:549(1987).

24. P.Van Roey, G.D.Smith, T.M.Balasubramanian, and G.R.Marshall, Tert-butyloxycarbonyl-α-aminoisobutyrate benzyl ester, $C_{20}H_{30}N_2O_5$, *Acta Cryst.* C39: 894(1983).

25. C.M.Venkatachalam, Stereochemical criteria for polypeptides and proteins.V. Conformation of a system of three linked peptide units, *Biopolymers* 6:1425(1968).

26. C.Aleman, J.A.Subirana, and J.J.Perez, A molecular mechanical study of the structure of poly(α-aminoisobutyric acid), *Biopolymers* 32:621(1992).

27. C.Toniolo, and E.Benedetti, The polypeptide 3_{10}-helix, *TIBS* 16:350(1991).

28. G.N.Ramachandran, C.M.Venkatachalam, and S.Krimm, Stereochemical criteria for polypeptide and protein chain conformations.III. Helical and hydrogen-bonded polypeptide chains, *Biophys.J.* 6:849(1966).

29. E.N.Baker, and R.E.Hubbard, Hydrogen bonding in globular proteins, *Prog. Biophys.Mol.Biol.* 44:97(1984).

30. D.J.Barlow, and J.M.Thornton, Helix geometry in proteins, *J.Mol.Biol.* 201:601 (1988).

31. V.Pavone, E.Benedetti, B.Di Blasio, C.Pedone, A.Santini, A.Bavoso, C.Toniolo, M.Crisma,, and L.Sartore, Critical main-chain length for conformational conversion from 3_{10}-helix to α-helix in polypeptides, *J.Biomol.Struct.Dyn.* 7:1321 (1990).

32. E.Benedetti, B.Di Blasio, V.Pavone, C.Pedone, A.Santini, A.Bavoso, C.Toniolo, M.Crisma, and L.Sartore, Linear oligopeptides. Part 227. X-ray crystal and molecular structures of two α-helix forming (Aib-Ala) sequential oligopeptides, pBrBz-(Aib-L-Ala)$_5$-OMe and pBrBz-(Aib-L-Ala)$_6$-OMe, *J.Chem.Soc. Perkin Trans.* 2:1829(1990).

33. R.O.Fox, and F.M.Richards, A voltage-gated ion channel model inferred from the crystal structure of alamethicin at 1.5 A resolution, *Nature* 300:325(1982).

34. T.Butters, P.Hutter, G.Jung, P.Pauls, H.Schmitt, G.M.Sheldrick, and W.Winter, On the structure of the Helical N-terminus in Alamethicin - α-Helix or 310 Helix?, *Angew.Chem.* Int. Ed. Engl. 20:889(1981).

35. A.K.Francis, M.Iqbal, P.Balaram, and M.Vijayan, The crystal structure of a 3_{10} helical decapeptide containig α-aminoisobutyric acid, *FEBS Lett.* 155:230(1983).

36. I.L.Karle, M.Sukumar, and P.Balaram, Parallel packing of α-helices in crystals of the zervamicin IIA analog Boc-Trp-Ile-Ala-Aib-Ile-Val-Aib-Leu-Aib-Pro-OMe. 2H2O, *Proc.Natl.Acad.Sci.USA* 83:9284(1986).

37. I.L.Karle, J.L. Flippen-Anderson, M.Sukumar, and P.Balaram, Conformation of a 16-residue zervamicin IIA analog peptide containing three different structural features: 3_{10}-helix, α-helix, and β-bend ribbon, *Proc.Natl.Acad.Sci.USA* 84:5087 (1987).

38. C.Toniolo, G.M.Bonora, A.Bavoso, E.Benedetti, B.Di Blasio, V.Pavone, and C.Pedone, Molecular structure of peptaibol antibiotics: solution conformation and crystal structure of the octapeptide corresponding to the 2-9 sequence of emerimicin III and IV, *J.Biomol.Struct.Dyn.* 3:585(1985).

39. G.R. Marshall, E.E.Hodgkin, D.A.Lang, G.D.Smith, J.Zabrocki, and M.T.Leplawy, Factors governing helical preference of peptides containing multiple α,α-dialkyl amino acids, *Proc.Natl.Acad.Sci.USA* 87:487(1988).

40. M.LeBars, B.Bachet, and J.P.Mornon, Structure of a helical 19 peptide (trichorzianine A IIIC). Modelling of trans membrane channels, *Zeit.Kristall.* 185: 588(1988).

41. E.Benedetti, B.Di Blasio, V.Pavone, C.Pedone, A.Santini, M.Crisma, and C.Toniolo, The 3_{10}- and α-helical conformations in peptides, in: "Molecular conformation and Biological Interactions", P.Balaram and S.Ramaseshan, eds., Indian Academy of Science, Bangalore (1991).

42. E.Benedetti, C.Toniolo, P.Hardy, V.Barone, A.Bavoso, B.Di Blasio, P.Grimaldi, F.Lelj, V.Pavone, C.Pedone, G.M.Bonora, and I.Lingham, Folded and extended structures of homooligopeptides from α,α-dialkylated glycines. Conformational energy computation and x-ray diffraction study, *J.Am.Chem.Soc.* 106:8146(1984).

43. G.M.Bonora, C.Toniolo, B.Di Blasio, V.Pavone, C.Pedone, E.Benedetti, I.Lingham, and P.Hardy, Folded and extended structures of homooligopeptides from α,α-dialkylated α-amino acids. An infrared absorption and ^{1}H nuclear magnetic resonance study, *J.Am.Che.Soc.* 106:8152(1984).

44. C.Toniolo, G.M.Bonora, V.Barone, A.Bavoso, E.Benedetti, B.Di Blasio, P.Grimaldi, F.Lelj, V.Pavone, and C.Pedone, Conformation of pleionomers of α-aminoisobutyric acid, *Macromolecules* 18:895(1985).

45. G.Valle, G.M.Bonora, C.Toniolo, P.M.Hardy, M.T. Leplawy, and A.Redlinski, Intramolecularly hydrogen-bonded peptide conformations. Preferred crystal-state and solution conformation of N-monochloroacetylated glycines dialkylated at the α-carbon atom, *J.Chem.Soc. Perkin Trans.* 2:885(1986).

46. E.Benedetti, V.Barone, A.Bavoso, B.Di Blasio, F.Lelj, V.Pavone, C.Pedone, G.M.Bonora, C.Toniolo, M.T. Leplawy, K.Kaczmarek, and A.Redlinski, Structural versatility of peptides from Cα,α-dialkylated glycines. I . A conformational energy computation and x-ray diffraction study of homo-peptides from Cα,α-diethylglycine, *Biopolymers* 27:357(1988).

47. C.Toniolo, G.M.Bonora, A.Bavoso, E.Benedetti, B.Di Blasio, V.Pavone, C.Pedone, V.Barone, F.Lelj, M.T.Leplawy, K.Kaczmarek, and A.Redlinski, Structural versatility of peptides from Cα,α-dialkylated glycines. II. An IR absorption and ^{1}H-NMR study of homo-oligopeptides from Cα,α-diethylglycine, *Biopolymers* 27:373(1988).

48. M.Crisma, G.Valle, G.M.Bonora, E.De Menego, C.Toniolo, F.Lelj, V.Barone, F.Fraternali, Structural versatility of peptides from Cα,α-disubstituted glycine: preferred conformation of Cα,α-diphenylglycine residue, *Biopolymers* 30:1(1990).

49. G.Valle, M.Crisma, G.M.Bonora, C.Toniolo, F.Lelj, V.Barone, F.Fraternali, P.M.Hardy, A.Langram-Goldsmith, and H.L.S.Maia, Structural versatility of peptides from Cα,α-disubstituted glycines. Preferred conformation of the Cα,α-dibenzylglycine residue, *J.Chem.Soc. Perkin Trans II* 1481(1990).

50. V.Barone, F.Lelj, A.Bavoso, B.Di Blasio, P.Grimaldi, V.Pavone, and C.Pedone, Conformational behaviour of α,α-dialkylated peptides, *Biopolymers* 24:1759(1985).

51. C.Toniolo, and E.Benedetti, The fully extended polypeptide conformation, in: "Molecular Conformation and Biological Interactions", P.Balaram and S.Ramaseshan, eds., Indian Academy of Science, Bangalore (1991).

52. C.Toniolo, M.Crisma, G.Valle, G.M.Bonora, V.Barone, E.Benedetti, B.Di Blasio, V.Pavone, C.Pedone, A.Santini, and F.Lelj, Structural versatility of peptides from Cα,α-dialkylated glycines: Ac$_3$c-rich peptides, in "Peptides Chemistry", T.Shiba and S.Sakakibara, eds., Protein Res.Foundation, Osaka(1988).

53. R.Bardi, A.M.Piazzesi, C.Toniolo, M.Sukumar, and P.Balaram, Stereochemistry of peptides containing 1-aminocyclopentanecarboxylic acid (Ac5c): solution and solid-state conformation of Boc-Ac5c-Ac5c-NHMe, *Biopolymers*, 25:1635(1986).

54. A.Santini, V.Barone, A.Bavoso, E.Benedetti, B.Di Blasio, F.Fraternali, F.Lelj, V.Pavone, C.Pedone, M.Crisma, G.M.Bonora, and C.Toniolo, Structural versatility of peptides from Cα,α-dialkylated glycines: a conformational energy calculation and x-ray diffraction study of homopeptides from 1-aminocyclopentane-1-carboxylic acid, *Int.J.Biol.Macromol.* 10:292(1988).

55. E.Benedetti, V.Barone, A.Bavoso, B.Di Blasio, F.Lelj, V.Pavone, C.Pedone, C.Toniolo, M.Crisma, and G.M.Bonora, Structural versatility of peptides from Cα,α-dialkylated glycines: Ac5c- and Ac6c-containing peptides, in: "Peptides 1986", D. Theodoropoulos, ed., De Gruyter, Berlin(1986).

56. V.Barone, F.Fraternali, P.L.Cristinziano, F.Lelj, and A.Rosa, Conformational behaviour of α,α-dialkylated peptides: *ab initio* and empirical results for cyclopropylglycine, *Biopolymers* 27:1673(1988).

57. M.Crisma, G.M.Bonora, C.Toniolo, V.Barone, E.Benedetti, B.Di Blasio, V.Pavone, C.Pedone, A.Santini, F.Fraternali, A.Bavoso, and F.Lelj, Structural versatility of peptides containing Cα,α-dialkylated glycines. Conformational energy computations, i.r. absorption and 1H-Nuclear magnetic resonance analysis of 1-aminocyclopropane-1-carboxylic acid homopeptides, *Int.J.Biol.Macromol.* 11:345 (1989).

58. G.Valle, M.Crisma, C.Toniolo, E.M.Holt, M.Tamura, J.Bland, and C.H.Stammer, Crystallographic characterization of conformation of 1-aminocyclopropane-1-carboxylic acid residue (Ac3c) in simple derivatives and peptides, *Int.J.Peptide Proteins Res.* 34:56(1989).

59. E.Benedetti, B.Di Blasio, V.Pavone, C.Pedone, A.Santini, M.Crisma, G.Valle, and C.Toniolo, Structural versatility of peptides from Cα,α-dialkylated glycines. Linear Ac3c homo-oligopeptides, *Biopolymers* 28:175(1989).

60. E.Benedetti, B.Di Blasio, V.Pavone, C.Pedone, A.Santini, V.Barone, F.Fraternali, F.Lelj, A.Bavoso, M.Crisma, and C.Toniolo, Structural versatility of peptides containing Cα,α-dialkylated glycines. An x-ray diffraction study of six 1-aminocyclopropane-1-carboxylic acid rich peptides, *Int.J.Biol.Macrom.* 11:353 (1989).

61. V.Pavone, E.Benedetti, V.Barone, B.Di Blasio, F.Lelj, C.Pedone, A.Santini, M.Crisma, G.M.Bonora, and C.Toniolo, Structural versatility of peptides from Cα,α-dialkylated glycines. A conformational energy computation and x-ray diffraction study of homopeptides from 1-aminocyclohexane-1-carboxylic acid, *Macromolecules*, 21:2064(1988).

62. E.Benedetti, B.Di Blasio, V.Pavone, C.Pedone, A.Santini, M.Crisma, and C.Toniolo, Molecular and crystal structure of benzyloxycarbonyl-1-aminocyclohexane-1-carboxylyl-1-aminocyclohexane-1-carboxylic acid tert-butyl ester, *Acta Cryst.* C45:634(1989).

63. H.Kessler, M.Kelin, A.Muller, K.Wagner, J.W.Bats, K.Ziegler, and M.Frimmer, Conformational prerequisites for the in vitro inhibition of cholate uptake in hepatocytes by cyclic analogues of antamanide and somatostatin, *Angew.Chem. Int.Ed.Engl.* 25:997(1986).

64. H.Kessler, J.W.Bats, C.Griesinger, S.Koll, M.Will, and K.Wagner, Peptide conformation. 46. Conformational analysis of a superpotent cytoprotective cyclic somatostatin analogue, *J.Am.Chem.Soc.* 110:1033(1986).

65. B.Di Blasio, E.Benedetti, V.Pavone, C.Pedone, and M.Goodman, Conformations of bioactive peptides: cyclolinopeptide A, *Biopolymers* 26:2099 (1986).

66. B.Di Blasio, F.Rossi, E.Benedetti, V.Pavone, C.Pedone, P.A.Temussi, G.Zanotti, and T.Tancredi, Bioactive peptides: solid-state and solution conformation of cyclolinopeptide A, *J.Am.Chem.Soc.* 111:9089(1989).

67. B.S.Neela, M.V. Manijula, S.Ramakumar, D.Balasubramanian, and M.A.Viswamitra, Conformation of cyclolinopeptide dihydrate: an antamanide analogue, *Biopolymers* 29:1499(1990).

68. B.Di Blasio, F.Rossi, E.Benedetti, V.Pavone, M.Saviano, C.Pedone, G.Zanotti, and T.Tancredi, Bioactive peptides: x-ray and NMR conformational study of [Aib^{5,6}-D-Ala^8]cyclolinopeptide A.*J.Am.Chem.Soc.* 114:8277(1992).

69. T.Tancredi, E.Benedetti, M.Grimaldi, C.Pedone, F.Rossi, M.Saviano, P.A.Temussi, and G.Zanotti, Ion binding of Cyclolinopeptide A: an NMR and CD conformational study, *Biopolymers* 31:761(1991).

70. M.A.Castiglione-Morelli, A.Pastore, C.Pedone, P.A.Temussi, G.Zanotti and T.Tancredi, Conformational study of cyclolinopeptide A. A distance-geometry and molecular dynamics approach, *Int.J.Peptide Protein Res.* 37:81(1991).

71. R.S.Shallenberger, and T.Acree, Molecular theory of sweet taste, *Nature* 216:480(1967).

72. E.W.Deutsch, and C.Hansch, Dependence of relative sweetness on hydrophobic bonding, *Nature* 211:75(1966).

73. T.Yamazaki, E.Benedetti, D.Kent, and M.Goodman, A three-dimensional model for Taste utilizing stereoisomeric peptides and peptidomimetics, *Angew.Chem.Int. Ed.Engl.*, in press; and references therein.

74. R.D.Feinstein, A.Polinski, A.J.Douglas, C.M.C.F.Beijer, R.K.Chadha, E.Benedetti, and M.Goodman, Conformational analysis of the dipeptide sweetener alitame and two stereoisomers by proton NMR, computer simulations, and x-ray crystallography, *J.Am.Chem.Soc.* 113:3467(1991).

75. C.Mapelli, M.G.Newton, C.E.Ringold, and C.H.Stammer, Cyclopropane amino acid ester dipeptide sweeteners, *Int.J.Peptide Protein Res.* 30:498(1987).

AUTHOR INDEX

SUBJECT INDEX